Theory and Experiment in Gravitational Physics

The 2015 centenary of the publication of Einstein's general theory of relativity and the first detection of gravitational waves have focused renewed attention on the question of whether Einstein was right.

This review of experimental gravity provides a detailed survey of the intensive testing of Einstein's theory of gravity, including tests in the emerging strong-field dynamical regime. It discusses the theoretical frameworks needed to analyze gravitational theories and interpret experiments. Completely revised and updated, this new edition features coverage of new alternative theories of gravity, a unified treatment of gravitational radiation, and the implications of the latest binary pulsar observations. It spans the earliest tests involving the solar system to the latest tests using gravitational waves detected from merging black holes and neutron stars. It is a comprehensive reference for researchers and graduate students working in general relativity, cosmology, particle physics, and astrophysics.

Clifford M. Will is Distinguished Professor of Physics at the University of Florida and Chercheur Associé at the Institut d'Astrophysique de Paris. He is a member of the US National Academy of Sciences and a Fellow of the American Physical Society, the American Academy of Arts and Sciences, and the International Society on General Relativity and Gravitation.

Theory and Experiment in Gravitational Physics

CLIFFORD M. WILL

University of Florida

Shaftesbury Road, Cambridge CB2 8EA, United Kingdom

One Liberty Plaza, 20th Floor, New York, NY 10006, USA

477 Williamstown Road, Port Melbourne, VIC 3207, Australia

314–321, 3rd Floor, Plot 3, Splendor Forum, Jasola District Centre, New Delhi – 110025, India

103 Penang Road, #05–06/07, Visioncrest Commercial, Singapore 238467

Cambridge University Press is part of Cambridge University Press & Assessment, a department of the University of Cambridge.

We share the University's mission to contribute to society through the pursuit of education, learning and research at the highest international levels of excellence.

www.cambridge.org
Information on this title: www.cambridge.org/9781107117440

DOI: 10.1017/9781316338612

First published 2018

A catalogue record for this publication is available from the British Library

ISBN 978-1-107-11744-0 Hardback

For Leslie

Contents

Preface

The year 2015 marked the 100th anniversary of the publication of Einstein's general theory of relativity, and relativists worldwide celebrated this historic occasion. As if this were not enough, on September 14, 2015, scientists at the LIGO gravitational-wave observatories in the United States detected, for the first time, gravitational waves passing the Earth, emitted by a pair of merging black holes over a billion light years away. This event provided a kind of fairy-tale capstone to a remarkable century.

Indeed, some popular accounts of the history of general relativity read like a fairy tale, going something like this: in 1905, Einstein discovered special relativity. He then turned his attention to general relativity and after ten years of hard work, he got general relativity in November 1915. In 1919, Eddington verified the theory by measuring the bending of starlight during a solar eclipse. Einstein became famous. And everybody lived happily ever after.

The real history of general relativity is rather more complex. At the time of Eddington's measurements of light bending, there was considerable skepticism about the results. There were major conceptual difficulties with the theory; it was very hard to understand what this new theory was and what it really predicted. And finally, there was an abiding sense that the theory mainly predicted some very tiny corrections to Newtonian gravity, and that it really wasn't all that important for physics.

As a result, within about ten years of its development, general relativity entered a period of decline, dubbed the "low-water mark" by Jean Eisenstaedt (2006), so that by the end of the 1950s, general relativity was considered to be in the backwaters of physics and astronomy, not a fit subject for a serious scientist to pursue.

But during the 1960s there began a remarkable renaissance for the theory. This was driven in part by the discovery of quasars, pulsars, and the cosmic background radiation, systems where it became clear that general relativity would play a central role. It was also fueled by the beginnings of a worldwide effort to put the theory to the test using new precision tools such as atomic clocks and radio telescopes, together with the emerging space program. And gravitational theorists developed a variety of tools that allowed them to clarify what the theory and its competitors predicted, analyze the new experimental results, and devise new tests.

Today, general relativity is fully integrated into the mainstream of physics, and in fact is central to some of the key scientific questions of today, such as: How did the universe begin and what is its future? What governs physics at the shortest distances and the longest distances? Do black holes really exist and how do they affect their surroundings? How can we reconcile gravity and quantum mechanics?

Yet, is it the correct theory of gravity? Was Einstein really right?

By the time of the centenary of general relativity, Einstein's theory had been tested in many ways and to high precision, and had passed every test. So far, no experiment has been able to put an unambiguous dent in the armor of general relativity. And yet "experimental gravity" is as active and exciting a field as it was in 1981, when the first edition of this book came out. This is motivated in part by the ongoing mysteries and conundra associated with the acceleration of the universe, the apparent existence of dark matter, and the difficulty of marrying general relativity with quantum mechanics.

But it is also motivated by our newfound ability to explore regimes for testing Einstein's theory far beyond the relatively weak and benign realms of the laboratory and the solar system. This exploration began, of course, with the discovery of the binary pulsar in 1974, leading to the first tests involving neutron stars, but in recent years it has accelerated dramatically.

As we look toward the second century of general relativity, two important themes are going to be (i) testing general relativity in the strong-gravity regime near black holes and neutron stars, going beyond the weak-gravity conditions of the solar system and (ii) testing general relativity in the highly dynamical regime, where gravitational radiation is both a phenomenon to be scrutinized and a tool for studying dynamical, strong-gravity sources.

There is no better illustration of this new era for testing general relativity than the outpouring of papers following the first detections of gravitational waves by the LIGO-Virgo collaboration, showing how the data place new and compelling constraints on a wide range of alternative theories, in ways that would not have been possible using solar-system measurements. It is my hope that this book will serve not only as an update of the 1981 edition but also as a foundation for students and researchers who wish to join in this new effort to check whether Einstein was right.

Acknowledgments

Over my fifty-year-long adventure in the world of testing general relativity, it has been my great fortune to have had three inspiring mentors: Kip Thorne, Subrahmanyan Chandrasekhar, and Bob Wagoner. Kip assigned me the PhD project of figuring out what had to be done to test general relativity better. He also taught by example the ways of research, teaching, scientific writing, and presenting science to broad audiences. From Chandra I learned a style of meticulous attention to detail in calculations that has been a hallmark of my own research. He also gave me a deep appreciation for science history. Bob's infectious enthusiasm for physics and astronomy taught me that, if you aren't having fun doing this work, you're in the wrong business. I have always had fun!

I also want to acknowledge some of the colleagues, collaborators and students who had an impact directly or indirectly on my career in experimental gravity, and who thus, in one way or another, left a mark on this book: Eric Adelberger, John Anderson, K. G. Arun, Emanuele Berti, Bruno Bertotti, Luc Blanchet, Alessandra Buonanno, Thibault Damour, Robert Dicke, Douglas Eardley, Francis Everitt, William Fairbank, Mark Haugan, Timothy Krisher, Bala Iyer, Larry Kidder, Ryan Lang, David Lee, Alan Lightman, Craig Lincoln, Saeed Mirshekari, Wei-Tou Ni, Anna Nobili, Ken Nordtvedt, Ho-Jung Paik, Michael Pati, Eric Poisson, Bob Reasenberg, Peter Saulson, Bernard Schutz, Irwin Shapiro, Adam Stavridis, Joe Taylor, Robert Vessot, Martin Walker, Joseph Weber, Alan Wiseman, Nicolás Yunes, and Helmut Zaglauer.

I am grateful to the Institut d'Astrophysique de Paris for its hospitality during annual stays in 2015–2017, where large parts of this book were written. This work was supported in part by the US National Science Foundation under Grants No. PHY 13-06069 and PHY 16-00188.

1 Introduction

On September 14, 1959, twelve days after passing through her point of closest approach to the Earth, the planet Venus was bombarded by pulses of radio waves sent from Earth. A handful of anxious scientists at Lincoln Laboratories in Massachusetts waited to detect the echo of the reflected waves. To their initial disappointment, neither the data from this day, nor from any of the days during that month-long observation, showed any detectable echo near inferior conjunction of Venus. However, a later, improved reanalysis of the data showed a bona fide echo in the data from one day: September 14. Thus occurred the first recorded radar echo from a planet.

Exactly fifty-six years later, on September 14, 2015, a rather different signal was received by scientists, this time in Hanford, Washington, and Livingston, Louisiana. The signal was not electromagnetic but instead was a wave in the fabric of spacetime itself. It was the final burst of gravitational waves from two black holes that merged to form a single black hole somewhere in the southern sky some 1.3 billion years ago. The signal was recognized within minutes by automated data processing software. This time, the scientists, numbering over 1,000, were anxious lest the signal be an unlucky artifact of instrumental noise. But on February 11, 2016, after an intensive and secretive five months of detailed analysis, checks and cross-checks, they announced at a Washington, DC, press conference that the Laser Interferometer Gravitational-Wave Observatory (LIGO) had made the first direct detection of gravitational waves.

As if the 100th anniversary of the general theory of relativity during 2015 was not already something to celebrate, the detection of gravitational waves was icing on the cake.[1] It was also the capstone of a half-century period during which general relativity experienced a remarkable renaissance, from a subject relegated to the backwaters of physics and astronomy, to one that is regarded as central to the major scientific questions of the day, from the nature of the fundamental particles to the fate of the universe. It was a perfect illustration of how the field was transformed from one that was once called "a theorist's paradise and an experimentalist's purgatory," to one in which experimentalists and theorists work hand in hand. It highlighted the field's evolution from the world of "small science," where individuals or small groups scratched out mathematical formulas in their tiny offices, to that of "big science," where worldwide collaborations of scientists conduct their affairs via telecon and Skype, spend budgets measured in units of megabucks or megaeuros, and require project managers to keep matters on track.

[1] And a scoop of ice cream on top of the icing was the award of the 2017 Nobel Prize in Physics to three of LIGOs founders, Rainer Weiss, Barry Barish, and Kip Thorne. The other pioneer of the project, Ron Drever, had already died in March of that year.

The origins of this remarkable transformation of general relativity from an obscure niche of mathematics and physics to a major subfield of physics and astronomy, today called "Gravitational Physics," can be found in a set of events of the academic year 1959–1960, beginning with that first radar echo from Venus. Four key events followed.

On March 9, 1960, the editorial office of *Physical Review Letters* received a paper by Robert Pound and Glen Rebka Jr., entitled "Apparent Weight of Photons." The paper reported the first successful laboratory measurement of the gravitational redshift of light. The paper was accepted and published in the April 1 issue.

In June 1960, there appeared in volume 10 of the *Annals of Physics* a paper on "A Spinor Approach to General Relativity" by Roger Penrose. It outlined a streamlined calculus for general relativity based upon "spinors" rather than upon tensors.

Later that summer, Carl Brans, a young Princeton graduate student working with Robert Dicke, began putting the finishing touches on his PhD thesis, entitled "Mach's Principle and a Varying Gravitational Constant." Part of that thesis was devoted to the development of a "scalar-tensor" alternative to the general theory of relativity. Although its authors never referred to it this way, it came to be known as the Brans-Dicke theory.

On September 26, 1960, just over a year after the recorded Venus radar echo, astronomers Thomas Matthews and Allan Sandage and coworkers at Mount Palomar used the 200-inch telescope to make a photographic plate of the star field around the location of the radio source 3C48. Although they expected to find a cluster of galaxies, what they saw at the precise location of the radio source was an object that had a decidedly stellar appearance, an unusual spectrum, and a luminosity that varied on a timescale as short as fifteen minutes. The name quasistellar radio source or "quasar" was soon applied to this object and to others like it.

These disparate and seemingly unrelated events of the academic year 1959–1960, in fields ranging from experimental physics to abstract theory to astronomy, signaled the beginning of a new era for general relativity. This era was to be one in which general relativity not only would become an important theoretical tool of the astrophysicist but also would have its validity challenged as never before. Yet it was also to be a time in which experimental tools would become available to test the theory in unheard-of ways and to unheard-of levels of precision.

The optical identification of 3C48 (Matthews and Sandage, 1963) and the subsequent discovery of the large redshifts in its spectral lines and in those of 3C273 (Greenstein and Matthews, 1963; Schmidt, 1963) presented theorists with the problem of understanding the enormous output of energy from a region of space compact enough to permit the luminosity to vary systematically over timescales as short as days or hours. Many theorists turned to general relativity and to the strong relativistic gravitational fields it predicts, to provide the mechanism underlying such violent events. This was the first use of the theory's strong-field aspect, in an attempt to interpret and understand observations. The subsequent discovery of the cosmic microwave background (CMB) radiation in 1964, of pulsars in 1967, and of the first black hole candidate in 1971 showed that it would not be the last. However, the use of relativistic gravitation in astrophysical model building forced theorists and experimentalists to address the question: Is general relativity the correct relativistic theory of gravitation? It would be difficult to place much confidence in models for such phenomena as quasars and pulsars if there were serious doubt about one of the basic underlying physical theories. Thus, the growth of "relativistic astrophysics" intensified the need to strengthen the empirical evidence for or against general relativity.

The publication of Penrose's spinor approach to general relativity (Penrose, 1960) was one of the products of a new school of relativity theorists that came to the fore in the late 1950s. These relativists applied the elegant, abstract techniques of pure mathematics to physical problems in general relativity, and demonstrated that these techniques could also aid in the work of their more astrophysically oriented colleagues. The bridging of the gaps between mathematics and physics and mathematics and astrophysics by such workers as Bondi, Dicke, Sciama, Pirani, Penrose, Sachs, Ehlers, Misner, and others changed the way that research (and teaching) in relativity was carried out, and helped make it an active and exciting field of physics. Yet again the question had to be addressed: Is general relativity the correct basis for this research?

The other three events of 1959–1960 contributed to the rebirth of a program to answer that question, a program of experimental gravitation that had been semi-dormant for forty years.

The Pound-Rebka (1960) experiment, in addition to verifying the principle of equivalence and the gravitational redshift, demonstrated the powerful use of quantum technology in gravitational experiments of high precision. The next two decades would see further uses of quantum technology in such tools as atomic clocks, laser ranging, superconducting gravimeters, and gravitational-wave detectors, to name only a few. Recording radar echos from Venus (Smith, 1963) opened up the solar system as a laboratory for testing relativistic gravity. The rapid development of the interplanetary space program during the early 1960s made radar ranging to both planets and artificial satellites a vital new tool for probing relativistic gravitational effects. Coupled with the theoretical discovery in 1964 of the relativistic time-delay effect (Shapiro, 1964), it provided new and accurate tests of general relativity. For the next decade and a half, until the summer of 1974, the solar system would be the primary arena for high-precision tests of general relativity. Finally, the development of the Brans-Dicke (1961) theory provided a viable alternative to general relativity. Its very existence and agreement with the experimental results of the day demonstrated that general relativity was not a unique theory of gravity. Some even preferred it over general relativity on aesthetic and theoretical grounds. At the very least, it showed that discussions of experimental tests of relativistic gravitational effects should be carried on using a broader theoretical framework than that provided by general relativity alone. It also heightened the need for high-precision experiments because it showed that the mere *detection* of a small general relativistic effect was not enough. What was now required was measurements of these effects to accuracies of 10 percent, 1 percent, or fractions of a percent and better, to distinguish among competing theories of gravitation.

To appreciate more fully the regenerative effect that these events had on gravitational theory and its experimental tests, it is useful to review briefly the history of general relativity in the forty-five years following Einstein's publication of the theory.

In deriving general relativity, Einstein was not particularly motivated by a desire to account for unexplained experimental or observational results. Instead, he was driven by theoretical criteria of elegance and simplicity. His primary goal was to produce a gravitation theory that incorporated the principle of equivalence and special relativity in a natural way. In the end, however, he had to confront the theory with experiment. This confrontation was based on what came to be known as the "three classical tests."

One of these tests was an immediate success – the ability of the theory to account for the anomalous perihelion shift of Mercury. This had been an unsolved problem in celestial mechanics for over half a century, since the announcement by Urbain Jean Joseph Le Verrier in 1859 that, after the perturbing effects of the planets on Mercury's orbit had been accounted for, there remained in the data an unexplained advance in the perihelion of Mercury. The modern value for this discrepancy is about forty-three arcseconds per century. A number of ad hoc proposals were made in an attempt to account for this excess, including the existence of a new planet, dubbed "Vulcan," near the Sun, and a deviation from the inverse-square law of gravitation. A half century of astronomical searches for Vulcan yielded numerous claimed sightings, but in the end, no solid evidence for the planet was found. And while a change in the Newtonian inverse-square law proposed by Simon Newcombe, from the power 2 to the power 2.000000157, could account for the perihelion advance of Mercury, it ultimately conflicted with data on the motion of the Moon.

Einstein was well aware of the problem of Mercury, and, in fact, he used it as a way to test his early attempts at a theory of gravity; for example, he finally rejected the 1912 "Entwurf" or "draft" theory that he had developed with Marcel Grossmann in part because it gave the wrong perihelion advance. But when he thought he had obtained the final theory in November 1915, the fact that it gave the correct advance convinced him that he had succeeded.

The next classical test, the deflection of light by the Sun, was not only a success, it was a sensation. Shortly after the end of World War I, two expeditions organized by Arthur Stanley Eddington set out from England: one for Sobral, in Brazil; and one for the island of Principe off the coast of Africa to observe the solar eclipse of May 29, 1919. Their goal was to measure the deflection of light as predicted by general relativity: 1.75 arcseconds for a ray that grazes the Sun. The observations had to be made in the path of totality of a solar eclipse, during which the Moon would block the light from the Sun and reveal the field of stars around it. Photographic plates taken of the star field during the eclipse were compared with plates of the same field taken when the Sun was not present, and the angular displacement of each star was determined. The results were 1.13 ± 0.07 times the Einstein prediction for the Sobral expedition, and 0.92 ± 0.17 for the Principe expedition (Dyson et al., 1920). The November 1919 announcement of these results confirming the theory caught the attention of a war-weary public and helped make Einstein a celebrity. Nevertheless, the experiments were plagued by systematic errors, and subsequent eclipse expeditions did little to improve the situation.

The third classical test was actually the first proposed by Einstein (1908): the gravitational redshift of light. But by contrast with the other two tests, there was no reliable confirmation of it until the 1960 Pound-Rebka experiment. One possible test involved the red shift of spectral lines from the sun. A 1917 measurement by astronomer Charles St. John (1917) failed to detect the effect, sowing considerable doubt about the validity of the theory. Thirty years of such measurements revealed mainly that the observed shifts in solar spectral lines are dominated by Doppler shifts due to radial mass motions in the solar photosphere, and by line shifts due to the high pressures in the solar atmosphere, making detection of the Einstein shift very difficult. It would be 1962 before a reliable solar redshift measurement would be made. Similarly inconclusive were attempts to measure the gravitational redshift of spectral lines from white dwarfs, primarily from Sirius B and 40 Eridani B, both members of binary systems. Because of uncertainties in the determination of the masses and radii of these stars, and because of possible complications in their spectra due to scattered light from their companions, reliable, precise measurements were not possible. Furthermore, by the late 1950s, it was being suggested that the gravitational red shift was not a true test of general relativity after all. According to Leonard Schiff and Robert Dicke, the gravitational red shift was a consequence purely of the principle of equivalence, and did not test the specific field equations of gravitational theory.

Cosmology was one area where general relativity could conceivably be confronted with observation. Initially the theory met with success in its ability to account for the observed expansion of the universe, yet by the 1940s there was considerable doubt about its applicability. According to pure general relativity, the expansion of the universe originated in a dense primordial explosion called the "big bang." However, at that time, the measured value of the expansion rate was so high that working backward in time using the cosmological solutions of general relativity led to the conclusion that the age of the universe was less than that of the Earth! One result of this doubt was the rise in popularity during the 1950s of the steady-state cosmology of Herman Bondi, Thomas Gold, and Fred Hoyle. This model avoided the big bang altogether, and allowed for the expansion of the universe by the continuous creation of matter. But by the late 1950s, revisions in the cosmic distance scale had reduced the expansion rate by a factor of five, and had thereby increased the age of the universe in the big bang model to a more acceptable level. Nevertheless, cosmology was still in its infancy, hardly suitable as an arena for testing theories of gravity. The era of "precision cosmology" would not begin until the launch of the Cosmic Background Explorer (COBE) satellite in 1989 followed by its precise measurements of the spectrum and fluctuations of the cosmic background radiation.

Meanwhile, a small "cottage industry" had sprung up, devoted to the construction of alternative theories of gravitation. Some of these theories were produced by such luminaries as Henri Poincaré, Alfred North Whitehead, Edward Arthur Milne, George Birkhoff, Nathan Rosen, and Frederick Belinfante. Many of these authors expressed an uneasiness with the notions of general covariance and curved spacetime, which were built into general relativity, and responded by producing "special relativistic" theories of gravitation. Many of these theories considered spacetime itself to be governed by special relativity, and treated gravitation as a field on that background. As of 1960, it was possible to enumerate at least twenty-five such alternative theories, as found in the primary research literature between 1905 and 1960; for a partial list, see Whitrow and Morduch (1965).

Thus, by 1960, it could be argued that the validity of general relativity rested on the following empirical foundation: one test of moderate precision (the perihelion shift, approximately 10 percent), one test of low precision (the deflection of light, approximately 25 percent), one inconclusive test that was not a real test anyway (the gravitational redshift), and cosmological observations that could not distinguish between general relativity and the steady-state theory. Furthermore, a variety of alternative theories laid claim to viability.

In addition, the attitude toward the theory seemed to be that, whereas it was undoubtedly important as a fundamental theory of nature, its observational contacts were limited. This view was present for example in the standard textbooks on general relativity of this period, such as those by Møller (1952), Synge (1960), and Landau and Lifshitz (1962). As a consequence, general relativity was cut off from the mainstream of physics. It was during this period that one newly minted graduate of the California Institute of Technology was advised not to pursue this subject for his graduate work, because general relativity "had so little connection with the rest of physics and astronomy" (his name: Kip Thorne).

However, the events of 1959–1960 changed all that. The pace of research in general relativity and relativistic astrophysics began to quicken and, associated with this renewed effort, the systematic high-precision testing of gravitational theory became an active and challenging field, with many new experimental and theoretical possibilities. These included new versions of old tests, such as the gravitational red shift and deflection of light, with accuracies that were unthinkable before 1960. They also included brand new tests of gravitational theory, such as the gyroscope precession, the time delay of light, and the "Nordtvedt effect" in lunar motion, all discovered theoretically after 1959.

Because many of the experiments involved the resources of programs for interplanetary space exploration and observational astronomy, their cost in terms of money and manpower was high and their dependence upon increasingly constrained government funding agencies was strong. Thus, it became crucial to have as good a theoretical framework as possible for comparing the relative merits of various experiments, and for proposing new ones that might have been overlooked. Another reason that such a theoretical framework was necessary was to make some sense of the large (and still growing) number of alternative theories of gravitation. Such a framework could be used to classify theories, elucidate their similarities and differences, and compare their predictions with the results of experiments in a systematic way. It would have to be powerful enough to be used to design and assess experimental tests in detail, yet general enough not to be biased in favor of general relativity.

A leading exponent of this viewpoint was Dicke (1964). It led him and others to perform several high-precision null experiments that greatly strengthened the empirical support for the foundations of gravitation theory. Within this viewpoint one asks general questions about the nature of gravity and devises experiments to test them. The most important dividend of the Dicke framework is the understanding that gravitational experiments can be divided into two classes. The first consists of experiments that test the foundations of gravitation theory, one of these foundations being the principle of equivalence. These experiments (Eötvös experiment, Hughes-Drever experiment, gravitational redshift experiment, and others) accurately verify that gravitation is a phenomenon of curved spacetime, that is, it must be described by a "metric theory" of gravity, at least to a high level of precision. General relativity and Brans-Dicke theory are examples of metric theories of gravity.

The second class of experiments consists of those that test metric theories of gravity. Here another theoretical framework was developed that takes up where the Dicke framework leaves off. Known as the "Parametrized post-Newtonian" or PPN formalism, it was pioneered by Kenneth Nordtvedt Jr. (1968b), and later extended and improved by Will (1971c), Will and Nordtvedt (1972), and Will (1973). The PPN framework takes the slow motion, weak field, or post-Newtonian limit of metric theories of gravity, and characterizes that limit by a set of ten real-valued parameters. Each metric theory of gravity has particular values for the PPN parameters. The PPN framework was ideally suited to the analysis of solar system gravitational experiments, whose task then became one of measuring the values of the PPN parameters and thereby delineating which theory of gravity is correct. A second powerful use of the PPN framework was in the discovery and analysis of new tests of gravitation theory, examples being the Nordtvedt effect (Nordtvedt, 1968a), preferred-frame effects (Will, 1971b), and preferred-location effects (Will, 1971b, 1973). The Nordtvedt effect, for instance, is a violation of the equality of acceleration of massive bodies, such as the Earth and Moon, in an external field; the effect is absent in general relativity but present in many alternative theories, including the Brans-Dicke theory. The third use of the PPN formalism was in the analysis and classification of alternative metric theories of gravitation. After 1960, the invention of alternative gravitation theories did not abate but changed character. The crude attempts to derive Lorentz-invariant field theories described previously were mostly abandoned in favor of metric theories of gravity, whose development and motivation were often patterned after that of the Brans-Dicke theory. A "theory of gravitation theories" was developed around the PPN formalism to aid in their systematic study. The PPN formalism thus became the standard theoretical tool for analyzing solar system experiments, looking for new tests, and studying alternative metric theories of gravity.

But by the middle 1970s it became apparent that the solar system could no longer be the sole testing ground for gravitation theories. The reason was that many alternative theories of gravity agreed with general relativity in their weak-field, slow-motion limits closely enough to pass all solar system tests. But they did not necessarily agree in other predictions, such as neutron stars, black holes, gravitational radiation, or cosmology, phenomena that involved strong or dynamical gravity.

This was confirmed in the summer of 1974 with the discovery by Joseph Taylor and Russell Hulse of the binary pulsar (Hulse and Taylor, 1975). Here was a system that featured, in addition to significant post-Newtonian gravitational effects, highly relativistic gravitational fields associated with the pulsar (and possibly its companion) and the possibility of the emission of gravitational radiation by the binary system. The role of the binary pulsar as a new arena for testing relativistic gravity was confirmed four years later with the announcement (Taylor et al., 1979) that the rate of change of the orbital period of the system had been measured. The result agreed with the prediction of general relativity for the rate of orbital energy loss due to the emission of gravitational radiation. But it disagreed strongly with the predictions of many alternative theories, even some with post-Newtonian limits identical to that of general relativity.

By 1981, when the first edition of this book was published, it was not uncommon to describe the period 1960–1980 as a "golden era" for experimental gravity. Many of the events of that period were described for a lay audience in my 1986 book *Was Einstein Right?* (Will, 1986). But the phrase "golden era" suggests that it was downhill from that time forward. Quite the opposite was true.

Solar-system tests of relativistic gravity continued, with highlights including dramatically improved measurements of light deflection and the Shapiro time delay, measurements of "frame-dragging" by the Gravity Probe B and the Laser Geodynamics Satellite (LAGEOS) experiments, and steadily improving lunar laser ranging. Binary pulsar tests continued, aided by remarkable discoveries, including the famous "double pulsar" and a pulsar in a triple system.

At the same time, the central thrust of testing gravity began to shift away from the weak-field limit. Two themes began to emerge as the key themes for the future.

The first theme is Dynamical Gravity. This involves phenomena in which the variation with time of the spacetime geometry plays an important role. In the solar system, velocities are small compared to the speed of light and the masses of the planets are small compared to the mass of the sun, so the underlying spacetime geometry can be viewed either as being stationary or as evolving in a quasistationary manner. But in the binary pulsar, for example, the two bodies have almost the same mass and are orbiting each other ten times faster than planets in the solar system, and consequently the varying spacetime geometry that they both generate devolves into gravitational waves propagating away from the system, causing it to lose energy. A more dramatic example is the final inspiral of the two black holes whose gravitational signal was detected by LIGO in 2015. The black holes are made of pure curved spacetime, and the manner in which that geometry evolved during the final fractions of a second of the inspiral and merger left its imprint on the gravitational waves that were detected. The final black hole that was left over even oscillated a few times, emitting a specific kind of gravitational radiation called ringdown waves. This is the regime of dynamical gravity. Dynamical gravity often goes hand in hand with gravitational radiation.

The second theme is Strong Gravity. Much like modern art, the term "strong" means different things to different people. To someone steeped in general relativity, the principal figure of merit that distinguishes strong from weak gravity is the quantity $\epsilon \sim Gm/c^2r$, where m is the characteristic mass scale of the phenomenon, r is the characteristic distance scale, and G and c are the Newtonian gravitational constant and the speed of light, respectively. Near the event horizon of a nonrotating black hole, or for the expanding observable universe, $\epsilon \sim 1$; for neutron stars, $\epsilon \sim 0.2$. These are the regimes of strong gravity. For the solar system, $\epsilon < 10^{-5}$; this is the regime of weak gravity.

An alternative view of "strong" gravity comes from the world of particle physics. Here the figure of merit is $Gm/c^2r^3 \sim \ell^{-2}$, where the curvature of spacetime associated with the phenomenon, represented by the left-hand side, is comparable to the inverse square of a favorite length scale ℓ. If ℓ is the Planck length $(\hbar G/c^3)^{1/2} \sim 10^{-35}$ m, this would correspond to the regime where one expects conventional quantum gravity effects to come into play. Another possible scale for ℓ is the TeV scale associated with many models for unification of the forces, or models with extra spacetime dimensions. From this viewpoint, strong gravity is where the radius of curvature of spacetime is comparable to the fundamental length. Weak gravity is where the radius of curvature is much larger than this. The universe at the Planck time is strong gravity. Just outside the event horizon of an astrophysical black hole is weak gravity.

We will adopt the relativist's view of strong gravity.

The boundary between dynamical gravity and strong gravity is somewhat fuzzy. One can explore strong gravity alone by studying the motion of a star around a static supermassive black hole or of gas around a neutron star. Gravitational waves can be emitted by a binary system of white dwarfs, well characterized by weak gravity. However, the strongest waves tend to come from systems with compact, strongly gravitating bodies, because only such bodies can get close enough together to reach the relativistic speeds required to generate strong gravitational waves. And the universe as a whole can be thought of as both "strong gravity" and dynamical, yet because of the high degree of symmetry, gravitational waves do not play a major role in its evolution. By contrast, primordial gravitational waves could be detectable, in fluctuations of the cosmic background radiation, for example. Regardless of the specific context, testing general relativity in the strong-field and dynamical regimes will dominate this field for some time to come.

As a young student of seventeen at the Polytechnical Institute of Zürich, Einstein studied the work of Helmholtz, Maxwell, and Hertz, and ultimately used his deep understanding of electromagnetic theory as a foundation for special and general relativity. He appears to have been especially impressed by Hertz's confirmation that light and electromagnetic waves are one and the same (Schilpp, 1949). The electromagnetic waves that Hertz studied were in the radio part of the spectrum, at 30 MHz. It is amusing to note that, sixty years later, the "golden age" for testing relativistic gravity began with radio waves, the 440 MHz waves reflected from Venus, and ended with radio waves, the pulsed signals from the binary pulsar, observed at 430 MHz. We are now in a new era for testing general relativity, an era in which we can exploit and study an entirely new kind of wave, a wave in the fabric of spacetime itself.

During the half-century that closed on the centenary of Einstein's formulation of general relativity, the empirical foundations of his great theory were strengthened as never before. The question then arises, why bother to continue to test it? One reason is that gravity is a fundamental interaction of nature, and as such requires the most solid empirical underpinning we can provide. Another is that all attempts to quantize gravity and to unify it with the other forces suggest that the standard general relativity of Einstein may not be the last word. Furthermore, the predictions of general relativity are fixed; the pure theory contains no adjustable constants, so nothing can be changed. Thus every test of the theory is either a potentially deadly test or a possible probe for new physics. Although it is remarkable that this theory, born 100 years ago out of almost pure thought, has managed to survive every test, the possibility of finding a discrepancy will continue to drive experiments for years to come. These experiments will search for new physics beyond Einstein in many different directions: the large distance scales of the cosmological realm; scales of very short distances or high energy; and the realms of strong and dynamical gravity.

Box 1.1 Units and conventions

Throughout this book, we will adopt the units and conventions of standard textbooks such as Misner, Thorne, and Wheeler (1973) (hereafter referred to as MTW) or Schutz (2009). For a pedagogical development of many of the topics presented here, such as Newtonian gravity, post-Newtonian theory, and gravitational radiation, we will refer readers to Poisson and Will (2014) (hereafter referred to as PW). Although we have attempted to produce a reasonably self-contained account of gravitation theory and gravitational experiments, the reader's path will be greatly smoothed by a familiarity with general relativity at the level of one of these texts.

We will use "geometrized units," in which $G = c = 1$ (except in Chapter 2) and in which mass and time have the same units as distance. Greek indices on vectors and tensors will run over the four spacetime dimensions, while Latin indices will run only over spatial dimensions. We will use the Einstein summation convention, in which one sums repeated indices over their range. Multi-index objects, such as products $x^j x^k x^l \ldots$ will be denoted using capital superscripts, e.g., x^N, where N is the number of indices. Partial derivatives and covariant derivatives will be denoted by commas and semicolons preceding indices, respectively. Parentheses enclosing indices will denote symmetrization, while square brackets will denote antisymmetrization.

2 The Einstein Equivalence Principle

The Principle of Equivalence has played a central role in the development of gravitation theory. Newton regarded this principle as such a cornerstone of mechanics that he devoted the opening paragraphs of his masterwork *Philosophiae Naturalis Principia Mathematica* (often called the "Principia") to a detailed discussion of it (Newton, 1686). On page 1, Definition 1, he defined the "quantity" of matter to be its mass, and also defined the "weight" of a body. He asserted that the mass "is proportional the weight, as I have found by experiments on pendulums, very accurately made, which shall be shown hereafter." To Newton, the Principle of Equivalence demanded that the "mass" of any body, namely that property of a body (inertia) that regulates its response to an applied force, be equal to its "weight," that property that regulates its response to gravitation. Bondi (1957) coined the terms "inertial mass" $m_{\rm I}$, and "passive gravitational mass" $m_{\rm P}$, to refer to these quantities, so that Newton's second law and the law of gravitation take the forms

$$\boldsymbol{F} = m_{\rm I}\boldsymbol{a}\,, \qquad \boldsymbol{F} = m_{\rm P}\boldsymbol{g}\,, \tag{2.1}$$

where $\boldsymbol{g}$ is the gravitational field. The Principle of Equivalence can then be stated succinctly: for any body,[1]

$$m_{\rm P} = m_{\rm I}\,. \tag{2.2}$$

An alternative statement of this principle is that all bodies fall in a gravitational field with the same acceleration regardless of their mass or internal structure. Newton's equivalence principle is now generally referred to as the "Weak Equivalence Principle" (WEP).

It was Einstein who added the key element to WEP that revealed the path to general relativity. If all bodies fall with the same acceleration in an external gravitational field, then to an observer in a freely falling elevator in the same gravitational field, the bodies should be unaccelerated, except for possible tidal effects due to inhomogeneities in the gravitational field. Tidal effects can be made as small as one pleases by confining everything a sufficiently small elevator. Thus, insofar as their mechanical motions are concerned, the bodies will behave as if gravity were absent. Einstein went one step further. He proposed that not only should mechanical laws behave in such an elevator as if gravity were absent but also so should all the laws of physics, including, for example, the laws of electrodynamics. This new principle led Einstein to general relativity. It is now called the "Einstein Equivalence Principle" (EEP).

Yet, it was only in the 1960s that we gained a deeper understanding of the significance of these principles of equivalence for gravitation and experiment. Largely through the work

[1] Although Newton asserted only that $m_{\rm P}$ and $m_{\rm I}$ be proportional to each other, they can be made equal by suitable choice of units for $\boldsymbol{a}$ and $\boldsymbol{g}$

of Robert Dicke, we have come to view principles of equivalence, along with experiments such as the Eötvös experiment and the gravitational redshift experiment, as probes more of the foundations of gravitation theory than of general relativity itself. This viewpoint is part of what has come to be known as the Dicke Framework, to be described in Section 2.1, allowing one to discuss at a very fundamental level the nature of spacetime and gravity. Within this framework one asks questions such as: Do all bodies respond to gravity with the same acceleration? Does energy conservation imply anything about gravitational effects? What types of fields, if any, are associated with gravitation – scalar fields, vector fields, tensor fields ...? In Section 2.2, we argue that the Einstein Equivalence Principle is the foundation for all gravitation theories that describe gravity as a manifestation of curved spacetime, the so-called metric theories of gravity. In Section 2.3 we describe the empirical support for EEP from a variety of experiments.

Einstein's generalization of the Weak Equivalence Principle may not have been a generalization at all, according to a conjecture based on the work of Leonard Schiff. In Section 2.4 we discuss Schiff's conjecture, which states that any complete and self-consistent theory of gravity that satisfies WEP necessarily satisfies EEP. Schiff's conjecture and the Dicke Framework have spawned a number of concrete theoretical formalisms for comparing and contrasting metric theories of gravity with nonmetric theories, for analyzing experiments that test EEP, and for proving Schiff's conjecture. These include the $TH\epsilon\mu$ and c^2 formalisms, presented in Section 2.5, and the Standard Model Extension (SME), discussed in Section 2.6.

What would happen if a violation of one of these principles were observed? One possibility is that the entire edifice of metric theories of gravity, including general relativity, would come tumbling down. Another possibility is that the apparent violation would actually signal the presence of some field or interaction that lies outside the standard model of strong, electromagnetic and weak interactions, plus gravity. This latter viewpoint has proven to be fruitful, exploiting the ultrahigh precision apparatus developed to test equivalence principles in order to search for and ultimately place limits on new physics. We will describe a few examples of this approach in Section 2.7.

2.1 The Dicke Framework

The Dicke Framework for analyzing experimental tests of gravitation was spelled out in appendix 4 of Dicke's Les Houches lectures (Dicke, 1964). It makes two main assumptions about the type of mathematical formalism to be used in discussing gravity:

1. Spacetime is a four-dimensional differentiable manifold, with each point in the manifold corresponding to a physical event. The manifold need not *a priori* have either a metric or an affine connection. The hope is that experiment will force us to conclude that it has both.
2. The equations of gravity and the mathematical entities in them are to be expressed in a form that is independent of the particular coordinates used, i.e., in covariant form.

Notice that even if there is some physically preferred coordinate system or reference frame in spacetime, the theory can still be put into covariant form. For example, if a theory has a preferred cosmic time coordinate, one can introduce a scalar field $T(\mathcal{P})$, whose numerical values are equal to the values of the preferred time t according to $T(\mathcal{P}) = t(\mathcal{P})$, where $\mathcal{P}$ is a point in spacetime. If spacetime is endowed with a metric, one might also demand that $\boldsymbol{\nabla} T$ be a timelike vector field and be consistently oriented toward the future (or the past) throughout spacetime by imposing the covariant constraints $\boldsymbol{\nabla} T \cdot \boldsymbol{\nabla} T < 0$ and $\boldsymbol{\nabla} \otimes \boldsymbol{\nabla} T = 0$. where $\boldsymbol{\nabla}$ is a covariant derivative with respect to the metric. Other types of theories have "flat background metrics" $\boldsymbol{\eta}$; these can also be written covariantly by defining $\boldsymbol{\eta}$ to be a second-rank tensor field whose Riemann tensor vanishes everywhere, that is, **Riem** $(\boldsymbol{\eta}) = 0$ and by defining covariant derivatives and contractions with respect to $\boldsymbol{\eta}$. In most cases, this covariance is achieved at the price of the introduction into the theory of "absolute" or "prior geometric" elements $(T, \boldsymbol{\eta})$, that are not determined by the dynamical equations of the theory. Some authors regard the introduction of absolute elements as a failure of general covariance (Einstein would be one example), however we will adopt the weaker assumption of coordinate invariance alone. (For further discussion of prior geometry, see Section 3.3.)

Having laid down this mathematical viewpoint, Dicke then imposes two constraints on all acceptable theories of gravity. They are:

1. Gravity must be associated with one or more fields of tensorial character (scalars, vectors, and tensors of various ranks).
2. The dynamical equations that govern gravity must be derivable from an invariant action principle.

These constraints strongly confine acceptable theories. For this reason we should accept them only if they are fundamental to our subsequent arguments. For most applications of the Dicke Framework only the first constraint is needed. It is a fact, however, that the most successful gravitation theories, and *all* theories of current interest, are those that satisfy both constraints.

The Dicke Framework is particularly useful for asking questions such as what types of fields are associated with gravity, and how do they interact with the fundamental fields of the standard model of electromagnetic, weak and strong interactions. For example, there is strong evidence from elementary particle physics for at least one symmetric second-rank tensor field that is approximated by the Minkowski metric $\boldsymbol{\eta}$ when gravitational effects can be ignored. The Hughes-Drever experiment rules out or strongly constrains the existence of more than one second-rank tensor field, each coupling directly to matter, and various laboratory tests of Lorentz invariance rule out a long-range vector field coupling directly to matter. However, this is not the only powerful use of the Dicke Framework.

The general unbiased viewpoint embodied in the Dicke Framework has allowed theorists to formulate a set of fundamental criteria that any gravitation theory should satisfy if it is to be viable [here we do *not* impose constraints (1) and (2) above]. Two of these criteria are purely theoretical, whereas two are based on experimental evidence.

(i) It must be complete, that is, it must be capable of analyzing from "first principles" the outcome of any experiment of interest. It is not enough for the theory to postulate that bodies made of different material fall with the same acceleration. The theory must incorporate a complete set of electrodynamic and quantum mechanical laws, which can be used to calculate the detailed behavior of real bodies composed of nucleons and electrons in gravitational fields. This demand should not be extended too far, however. In areas such as quantum gravity, unification with the standard model of particle physics, spacetime singularities, and cosmic initial conditions, even special and general relativity are not regarded as being complete or fully developed. We also do not regard the presence of "absolute elements" and arbitrary parameters in gravitational theories as a sign of incompleteness, even though they are generally not derivable from "first principles," rather we view them as part of the class of cosmic boundary conditions. The most common and successful way of formulating a "complete" theory is to use an action principle that combines the standard model action (as it is currently known) for the nongravitational sector with an action for the gravitational "fields" (including the spacetime metric), together with some coupling between them.

(ii) It must be self-consistent, that is, its prediction for the outcome of every experiment must be unique. When one calculates predictions by two different, though equivalent methods, one must get the same results. An example is the bending of light computed either in the geometrical optics limit of Maxwell's equations or in the zero-rest-mass limit of the motion of test particles.

(iii) It must be relativistic, that is, in the limit as gravity is "turned off" compared to other physical interactions, the nongravitational laws of physics must reduce to the laws of special relativity, either perfectly or to a high degree of precision. The evidence for this comes from more than a century of successes of special relativity in areas ranging from high-energy physics to atomic physics (see Box 2.1). This does not necessarily imply a blind or perfect acceptance of Lorentz invariance and special relativity, and in fact vigorous experimental searches for potential violations of Lorentz invariance are continuing, in part to search for relic signatures of quantum gravity or of weak cosmic fields that couple to matter (see Section 2.6).

The fundamental theoretical object that enters these laws is the Minkowski metric $\boldsymbol{\eta}$, which has orthonormal tetrads related by Lorentz transformations, and which determines the ticking rates of atomic clocks and the lengths of laboratory rods. If we view $\boldsymbol{\eta}$ as a field, then we conclude that there must exist at least one second-rank tensor field in the Universe, a symmetric tensor $\boldsymbol{\psi}$, that is well approximated by $\boldsymbol{\eta}$ when gravitational effects can be ignored.

Let us examine what the evidence for special relativity does and does not tell us about the tensor field $\boldsymbol{\psi}$. First, it does *not* guarantee the existence of *global* Lorentz frames, that is, coordinate systems extending throughout spacetime in which $\boldsymbol{\psi} = \boldsymbol{\eta} = \mathrm{diag}(-1, 1, 1, 1)$. Nor does it demand that at each event $\mathcal{P}$, there exist local frames related by Lorentz transformations, in which the laws of nongravitational physics take on their special relativistic forms. Special relativity only demands that, in the limit as gravity is "turned off" the nongravitational laws of physics reduce to the laws of special relativity.

We will henceforth assume the existence of the tensor field $\boldsymbol{\psi}$.

(iv) It must have the correct Newtonian limit, that is, in the limit of weak gravitational fields and slow motions, it must reproduce Newton's laws. The overwhelming majority of phenomena in the universe can be very adequately described by the laws of Newtonian

Box 2.1 Tests of special relativity

Special relativity has been so thoroughly integrated into the fabric of modern physics that its validity is rarely challenged, except by cranks and crackpots. But we should remember that it does rest on a strong empirical foundation, including a number of classic tests.

The Michelson-Morley (1887) experiment and its many descendents (Shankland et al., 1955; Champeney et al., 1963; Jaseja et al., 1964; Brillet and Hall, 1979; Riis et al., 1988; Krisher et al., 1990b) failed to find evidence of a variation of the speed of light with the Earth's velocity through a hypothetical "aether."

Several classic experiments were performed to verify that the speed of light is independent of the speed of the emitter. If the speed of light were given by $\boldsymbol{c} + k\boldsymbol{v}$, where $\boldsymbol{v}$ is the velocity of the emitter, and k is a parameter to be measured, then orbits of binary-star systems would appear to have an anomalous eccentricity unexplainable by normal Newtonian gravity. This test is not unambiguous at optical wavelengths, however, because light is absorbed and reemitted by the intervening interstellar medium, thereby losing the memory of the speed of the source, a phenomenon known to astronomers as extinction. But at X-ray wavelengths, the path length of extinction is tens of kiloparsecs, so Brecher (1977) used three nearby X-ray binary systems in our galaxy to obtain a bound $|k| < 2 \times 10^{-9}$, for typical orbital velocities $v/c \sim 10^{-3}$.

At the other extreme, a 1964 experiment at CERN used neutral pions moving at $v/c \geq 0.99975$ as the source of light. Photons produced by the decay $\pi^0 \to \gamma + \gamma$ were collimated and timed over a flight path of 30 meters. The agreement of the photons' speed with the laboratory value set a bound $|k| < 10^{-4}$ for $v \approx c$ (Alväger et al., 1964).

The observational evidence for time dilation is overwhelming. Ives and Stilwell (1938) measured the frequency shifts of radiation emitted in the forward and backward direction by moving ions of H_2 and H_3 molecules. The first-order Doppler shift cancels out from the sum of the forward and backward shifts, revealing the second-order time-dilation effect, which was found to agree with theory. (Ironically, Ives was a die-hard opponent of special relativity.)

The classic Rossi-Hall (1941) experiment showed that the lifetime of μ-mesons was prolonged by the Lorentz factor $\gamma = (1 - v^2/c^2)^{-1/2}$. Muons are created in the upper atmosphere when cosmic-ray protons collide with nuclei of air, producing pions, which decay quickly to muons. With a rest half-life of 2.2×10^{-6} s, and with no time dilation, a muon travelling near the speed of light should travel only 2/3 of a kilometer on average before decaying to a harmless electron or positron and two neutrinos. Yet muons are the primary component of cosmic rays detected at sea level. Rossi and Hall measured the distribution of muons as a function of altitude and also measured their energies, and confirmed the time-dilation formula.

In an experiment performed in 1966 at CERN, muons in a storage ring moving at $v/c = 0.997$ were observed to have lifetimes 12 times larger than muons at rest, in agreement with the prediction to 2 percent (Farley et al., 1966). Also, since the storage ring was 5 meters in diameter, the muons' accelerations were greater than the gravitational acceleration on the Earth's surface by a factor of 10^{19}; these accelerations had no apparent effect on their decay rates.

The incorporation of Lorentz invariance into quantum mechanics provided further support for special relativity. The achievements include the prediction of anti-particles and elementary particle spin, and the many successes of relativistic quantum field theory.

For a pedagogical review written on the occasion of the 2005 centenary of special relativity, see Will (2006). We will describe contemporary tests of what is today called "Local Lorentz Invariance" in Section 2.3.2.

gravity. To a high degree of accuracy, Newton rules the Sun, the Earth, the solar system, all normal stars, galaxies and clusters of galaxies. On laboratory scales, the inverse-square law of Newton has been validated down to distances as small as 100 microns.

Thus, to have a shot at being viable, a gravitation theory must be complete, self-consistent, relativistic, and compatible with Newtonian gravity.

2.2 The Einstein Equivalence Principle

The Einstein Equivalence Principle is the foundation of all curved spacetime or "metric" theories of gravity, including general relativity. It is a powerful tool for dividing gravitational theories into two distinct classes: metric theories, those that embody EEP, and nonmetric theories, those that do not embody EEP. For this reason, we will discuss it in some detail and devote the next section to the experimental evidence that supports it.

We begin by stating the Weak Equivalence Principle in more precise terms than those used before. WEP states that *if an uncharged test body is placed at an initial event in spacetime and given an initial velocity there, then its subsequent trajectory will be independent of its internal structure and composition.* By "uncharged test body" we mean an electrically neutral body that has negligible self-gravitational energy (as estimated using Newtonian theory) and that is small enough in size so that its coupling to inhomogeneities in external fields can be ignored. In the same spirit, it is also useful to define a "local nongravitational test experiment" to be any experiment: (i) performed in a freely falling laboratory that is shielded and is sufficiently small that inhomogeneities in the external fields can be ignored throughout its volume and (ii) in which self-gravitational effects are negligible. For example, a measurement of the fine structure constant is a local nongravitational test experiment; a Cavendish experiment to measure Newton's constant G is not.

The Einstein Equivalence Principle then states:

1. WEP is valid.
2. The outcome of any local nongravitational test experiment is independent of the velocity of the (freely falling) apparatus.
3. The outcome of any local nongravitational test experiment is independent of where and when in the universe it is performed.

This principle is at the heart of gravitation theory, for it is possible to argue convincingly that if EEP is valid, then gravitation must be a curved-spacetime phenomenon, that is, must satisfy the postulates of Metric Theories of Gravity. These postulates state:

1. Spacetime is endowed with a metric.
2. The world lines of test bodies are geodesics of that metric.
3. In local freely falling frames, the nongravitational laws of physics are those of special relativity.

The argument proceeds as follows. The validity of WEP endows spacetime with a family of preferred trajectories, the world lines of freely falling test bodies. In a local frame that follows one of these trajectories, test bodies have unaccelerated motions. Furthermore, the results of local nongravitational test experiments are independent of the velocity of the frame. In two such frames located at the same event $\mathcal{P}$ in spacetime, but moving relative to each other, all the nongravitational laws of physics must make the same predictions for identical experiments, that is, they must be Lorentz invariant. We call this aspect of EEP Local Lorentz Invariance (LLI). Therefore, there must exist in the universe one or more second-rank tensor fields $\psi^{(1)}$, $\psi^{(2)}$, ... that reduce in a local freely falling frame to fields that are proportional to the Minkowski metric, $\phi^{(1)}(\mathcal{P})\boldsymbol{\eta}$, $\phi^{(2)}(\mathcal{P})\boldsymbol{\eta}$, ..., where $\phi^{(A)}(\mathcal{P})$ are scalar fields that can vary from event to event. Different members of this set of fields may couple to different nongravitational fields, such as boson fields, fermion fields, electromagnetic fields, and so on. However, the results of local nongravitational test experiments must also be independent of the spacetime location of the frame. We call this Local Position Invariance (LPI). There are then two possibilities. (i) The local versions of $\psi^{(A)}$ have constant coefficients, that is, the scalar fields $\phi^{(A)}(\mathcal{P})$ are actually constants, independent of $\mathcal{P}$. It is therefore possible by a simple universal rescaling of coordinates and coupling constants (such as the unit of electric charge) to set each scalar field equal to unity in every local frame. (ii) The scalar fields $\phi^{(A)}(\mathcal{P})$ must be constant multiples of a single scalar field $\phi(\mathcal{P})$, that is, $\phi^{(A)}(\mathcal{P}) = c^A\phi(\mathcal{P})$. If this is true, then physically measurable quantities, being dimensionless ratios, will be location independent (essentially, the scalar field will cancel out). One example is a measurement of the fine structure constant; another is a measurement of the length of a rigid rod in centimeters, since such a measurement is a ratio between the length of the rod and that of a standard rod whose length is defined to be one centimeter. Thus, a combination of a rescaling of coupling constants to set the c^A's equal to unity (redefinition of units), together with a "conformal" transformation to a new field $\bar{\psi} = \phi^{-1}\psi$ guarantees that the local version of $\bar{\psi}$ will be $\boldsymbol{\eta}$.

In either case, we conclude that there exist fields that reduce to $\boldsymbol{\eta}$ in every local freely falling frame. Elementary differential geometry then shows that these fields are one and the same: a unique, symmetric second-rank tensor field that we now denote by "the metric" $\boldsymbol{g}$. This $\boldsymbol{g}$ has the property that it possesses a family of preferred worldlines called geodesics, and that at each event $\mathcal{P}$ there exist local frames, called local Lorentz frames, that follow these geodesics, in which

$$g_{\mu\nu}(\mathcal{P}) = \eta_{\mu\nu} , \qquad \frac{\partial g_{\mu\nu}}{\partial x^\alpha}(\mathcal{P}) = 0 . \tag{2.3}$$

However, geodesics are straight lines in local Lorentz frames, as are the trajectories of test bodies in local freely falling frames, hence the test bodies move on geodesics of $\boldsymbol{g}$ and the Local Lorentz frames coincide with the freely falling frames. The idea that a single field $\boldsymbol{g}$ interacts with all the nongravitational fields in a unique manner is often called "universal coupling." We will discuss the implications of the postulates of metric theories of gravity and of universal coupling in more detail in Chapter 3. Because EEP is so crucial to this conclusion about the nature of gravity, we turn now to the supporting experimental evidence.

Box 2.2 Choice of units

Throughout this chapter we will *avoid* units in which $c = 1$. The reason for this is that if EEP is not valid then the speed of light may depend on the nature of the devices used to measure it. Thus, to be precise we should denote c as the speed of light as measured by some *standard* experiment. However, once we accept the validity of EEP in Chapter 3 and beyond, then c has the same value in every local Lorentz frame, independently of the method used to measure it, and thus c can be set equal to unity by an appropriate choice of units.

2.3 Experimental Tests of the Einstein Equivalence Principle

2.3.1 Tests of the Weak Equivalence Principle

A direct test of WEP is the comparison of the acceleration of two laboratory-sized bodies of different composition in an external gravitational field, today often called the Eötvös experiment. If WEP were invalid, then the accelerations of different bodies would differ. The simplest way to quantify such possible violations of WEP in a phenomenological form suitable for comparison with experiment is to suppose that for a body of inertial mass $m_{\rm I}$, the passive mass $m_{\rm P}$ is no longer equal to $m_{\rm I}$. Now the inertial mass of a typical laboratory body is made up of several types of mass energy: rest energy of the constituent particles, electromagnetic energy, weak-interaction energy, strong-interaction energy, and so on. If one of these forms of energy contributes to $m_{\rm P}$ differently than it does to $m_{\rm I}$, a violation of WEP would result. One could then write

$$m_{\rm P} = m_{\rm I} + \sum_{\rm A} \eta^{\rm A} \frac{E^{\rm A}}{c^2} , \tag{2.4}$$

where $E^{\rm A}$ is the internal energy of the body generated by interaction A, $\eta^{\rm A}$ is a dimensionless parameter that measures the strength of the violation of WEP induced by that interaction, and c is the speed of light. For two bodies, the acceleration is then given by

$$a_1 = \left(1 + \sum_{\rm A} \eta^{\rm A} \frac{E_1^{\rm A}}{m_1 c^2} \right) g , \qquad a_2 = \left(1 + \sum_{\rm A} \eta^{\rm A} \frac{E_2^{\rm A}}{m_2 c^2} \right) g , \tag{2.5}$$

where we have dropped the subscript "I" on m_1 and m_2.

A measurement of the relative difference in acceleration then yields a quantity called the "Eötvös ratio" given by

$$\eta \equiv \frac{2|a_1 - a_2|}{|a_1 + a_2|} = \sum_{\rm A} \eta^{\rm A} \left(\frac{E_1^{\rm A}}{m_1 c^2} - \frac{E_2^{\rm A}}{m_2 c^2} \right) . \tag{2.6}$$

Thus, experimental limits on η place limits on the WEP-violation parameters $\eta^{\rm A}$.

One way to place a bound on η is to drop two different objects in the Earth's gravitational field and compare how long they take to fall. Legend has it that Galileo Galilei originated and verified WEP by dropping objects off the Leaning Tower of Pisa around 1590. In fact, the principle of equal acceleration had already been asserted as early as 400 CE by Ioannes

Philiponus, and more recently by Benedetti in 1553, and experimental tests were performed by Stevin in 1586. If Galileo did indeed drop things off the Tower (and the only information about this comes from his last student and biographer Viviani, who wasn't even born until 1622), he may simply have been performing a kind of classroom demonstration of an established fact for his students. Unfortunately, this approach is plagued by experimental errors, such as the difficulty of releasing the objects at exactly the same time, by the effects of air drag, and by the short time available for timing the drop.

A better approach is to balance the gravitational force (which depends on m_P) by a support force (which depends on m_I); the classic model is the pendulum experiments performed by Newton and reported in his *Principia*; improved versions were carried out by Bessel (1832) and Potter (1923). The period of the pendulum depends on m_P/m_I, g, and the length of the pendulum. These experiments are also troubled by air drag, by errors in measuring or controlling the length of the pendulum, and by errors in timing the swing.

The best approach for laboratory tests was pioneered by Baron Roland von Eötvös, a Hungarian geophysicist working around the turn of the 20th century. He developed the *torsion balance*, schematically consisting of a rod suspended by a wire near its mid-point, with objects comprised of different materials attached at each end. The point where the wire is attached to achieve a horizontal balance depends only on the gravitational masses of the two objects, so this configuration does not tell us anything. But if an additional gravitational force can be applied in a direction not parallel to the supporting wire, then there will be a torque induced on the wire, given by

$$N \propto \eta r(\boldsymbol{g} \times \boldsymbol{e}_w) \cdot \boldsymbol{e}_r \,, \tag{2.7}$$

where r is the length of the rod, $\boldsymbol{g}$ is the gravitational acceleration, and $\boldsymbol{e}_w$ and $\boldsymbol{e}_r$ are unit vectors along the wire and the rod, respectively (see Figure 2.1). If the entire apparatus is rotated about some direction with angular velocity ω, the torque will be modulated with period $2\pi/\omega$.

In the experiments of Eötvös and his collaborators (Eötvös et al., 1922), the wire and $\boldsymbol{g}$ were not quite parallel by about $0.1°$ because of the centripetal acceleration on the apparatus due to the Earth's rotation; the apparatus was rotated about the direction of the wire. In the classic experiments by the groups headed by Dicke and Braginsky (Roll et al., 1964; Braginsky and Panov, 1972), $\boldsymbol{g}$ was that of the Sun, and the rotation of the Earth provided the modulation of the torque at a period of 24 hr. During the late 1980s, the search for an intermediate-range "fifth force" (Section 2.7) exploited nearby masses such as cliffs, mountains, and water reservoirs (for a bibliography of experiments up to 1991, see Fischbach et al. (1992)). In the quest for higher precision, the standard torsion balance rod was replaced by a horizontal tray on which masses of different composition were carefully fabricated and arranged so as to suppress torques due to normal quadrupole and higher moment coupling between the masses and nearby laboratory apparatus (including the experimenters). The "Eöt-Wash" experiments carried out at the University of Washington used such sophisticated arrangements to compare the accelerations of various materials toward local topography on Earth, movable laboratory masses, the Sun and the galaxy, and established a bound on the Eötvös parameter at the level of 2×10^{-13} (Su et al., 1994; Baeßler et al., 1999; Adelberger, 2001; Schlamminger et al., 2008; Wagner et al., 2012).

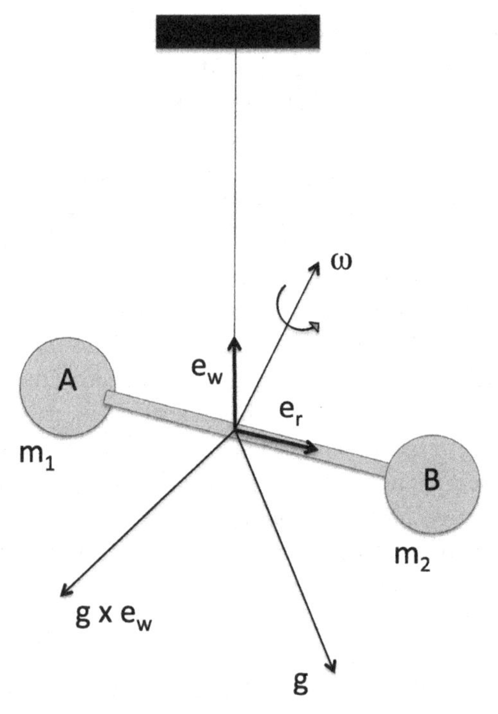

Fig. 2.1 Schematic arrangement of a torsion-balance Eötvös experiment.

Galileo has not been completely forgotten however: in a "free-fall" experiment performed at the University of Colorado, the relative acceleration in a drop tower of two bodies made of uranium and copper was measured using a laser interferometric technique (Niebauer et al., 1987).

Going into space makes possible a "perpetual" Galileo experiment in which the two bodies never hit the ground, permitting long integration times as a way to reduce errors. An example is MICROSCOPE, designed to test WEP to parts in 10^{15}. Developed by the French space agency CNES, it was launched on April 25, 2016, for a two-year mission. The two masses are nested coaxial cylinders made of different materials, one a Platinum-Rhodium alloy and the other a Titanium-Aluminum-Vanadium alloy. Another pair of nested cylinders are made of the same Platinum-Rhodium alloy, in order to account for differential accelerations induced by other phenomena such as gravity gradients. The drag-free satellite is in a circular, Sun-synchronous orbit at 720 km altitude. In late 2017 the MICROSCOPE team reported the result of an analysis of the first year of data, with a bound on the order of a part in 10^{14} (Touboul et al., 2017).

Other concepts include tests on orbiting satellites, such as Galileo-Galilei (Nobili et al., 2012) and STEP (Overduin et al., 2012), and on sounding rockets, such as SR-POEM (Reasenberg et al., 2012).

The recent development of atom interferometry has yielded tests of WEP, albeit to modest accuracy, comparable to that of the original Eötvös experiment. In these experiments, one measures the local acceleration of the two separated wavefunctions of

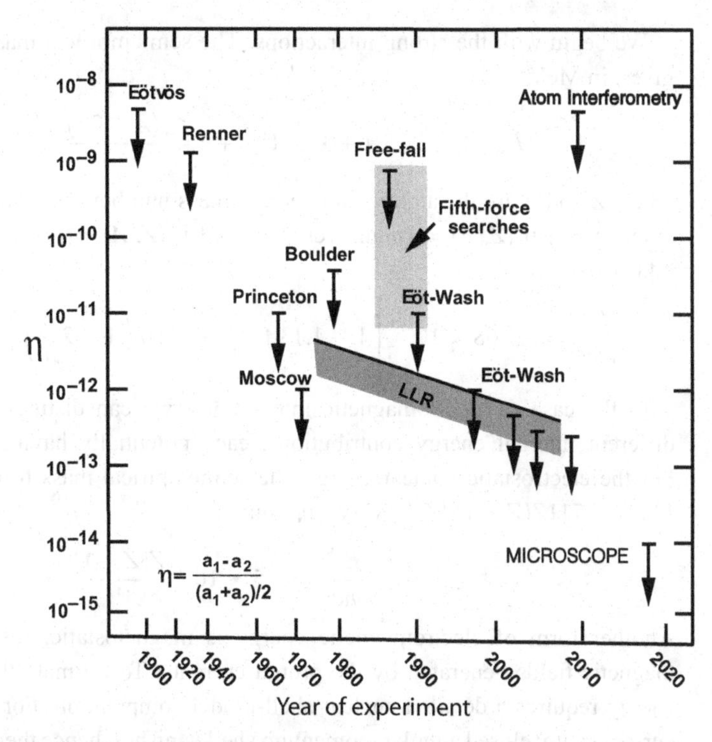

Fig. 2.2 Selected tests of the Weak Equivalence Principle, showing bounds on the Eötvös ratio η. The light grey region represents many experiments originally performed to search for a fifth force, including the "free-fall" and the initial Eöt-Wash experiments. The dark grey band shows evolving bounds on η for gravitating bodies from lunar laser ranging (LLR) (see Section 8.1 for discussion).

an atom such as Cesium by studying the interference pattern when the wavefunctions are recombined, and compares that with the acceleration of a nearby macroscopic object of different composition (Merlet et al., 2010; Müller et al., 2010).[2] Other ongoing experiments include dropping cold atom interferometers at the 141 meter drop tower in Bremen, Germany, with the ultimate idea of putting such devices in space (Herrmann et al., 2012), and testing the free-fall acceleration of anti-Hydrogen (Doser et al., 2012; Perez and Sacquin, 2012). The recent discovery of a pulsar in orbit with two white-dwarf companions (Ransom et al., 2014) may provide interesting new tests of WEP, because of the strong difference in composition between the neutron star and the white dwarfs, as well as precise tests of the Nordtvedt effect (see Section 8.1).

The current upper limits on η are summarized in Figure 2.2.

To determine the limits placed on the individual parameters η^{A} by the best of the current experiments, we must estimate the internal energies E^{A}/mc^2 for the different interactions and for different materials. For laboratory-sized bodies, the dominant contribution to E^{A} comes from the atomic nucleus.

[2] The claim by Müller et al. (2010) that these experiments test the gravitational redshift was subsequently shown to be incorrect (Wolf et al., 2011).

We begin with the strong interactions. The semiempirical mass formula (Rohlf, 1994) gives, in MeV,

$$E^{\mathrm{S}} = -15.75A + 17.8A^{2/3} + 23.7\frac{(A-2Z)^2}{A} + 11.18\frac{\delta}{A^{1/2}}, \tag{2.8}$$

where Z and A are the atomic number and mass number, respectively, of the nucleus, and where $\delta = 1$ if $(Z, A) = $ (odd, even), $\delta = -1$ if $(Z, A) = $ (even, even), and $\delta = 0$ if A is odd. Then

$$\frac{E^{\mathrm{S}}}{mc^2} = -1.68 \times 10^{-2}\left[1 - 1.13A^{-1/3} - 1.50\left(1 - 2\frac{Z}{A}\right)^2 - 0.71\frac{\delta}{A^{3/2}}\right]. \tag{2.9}$$

In the case of electromagnetic interactions, we can distinguish among a number of different internal energy contributions, each potentially having its own η^{A} parameter. For the electrostatic nuclear energy, the semiempirical mass formula yields the estimate $E^{\mathrm{ES}} = 0.711Z(Z-1)/A^{1/3}$ MeV, and thus

$$\frac{E^{\mathrm{ES}}}{mc^2} = 7.5 \times 10^{-4}\frac{Z(Z-1)}{A^{4/3}}. \tag{2.10}$$

Another form of electromagnetic energy is magnetostatic, resulting from the nuclear magnetic fields generated by the proton currents. To estimate the nuclear magnetostatic energy requires a detailed nuclear shell-model computation. For example, the net proton current in any closed angular momentum shell vanishes, hence there is no energy associated with the magnetostatic interaction between such a closed shell and any particle outside the shell. For aluminum and platinum, the materials used in the Dicke and Braginsky experiments, Haugan and Will (1977) showed that $(E^{\mathrm{MS}}/mc^2)_{\mathrm{Al}} = 4.1 \times 10^{-7}$ and $(E^{\mathrm{MS}}/mc^2)_{\mathrm{Pt}} = 2.4 \times 10^{-7}$; however no general formula in terms of Z and A is known.

There is also the hyperfine energy of interaction between the spins of the nucleons and the magnetic fields generated by the proton and neutron magnetic moments. Computations by Haugan (1978) yielded the estimate $E^{\mathrm{HF}}/mc^2 = (2\pi/V)\mu_N^2[g_p^2Z^2 + g_n^2(A-Z)^2]$, where V is the nuclear volume, μ_N is the nuclear magneton, and $g_p = 2.79$ and $g_n = -1.91$ are the gyromagnetic ratios of the proton and neutron, respectively. Then,

$$\frac{E^{\mathrm{HF}}}{mc^2} = 2.1 \times 10^{-5}\left(\frac{g_p^2Z^2 + g_n^2(A-Z)^2}{A^2}\right). \tag{2.11}$$

For weak interactions, while the parity nonconserving part makes no contribution to the nuclear ground-state energy to first order in the weak coupling constant G_W, the parity-*conserving* part does (Haugan and Will, 1976). In the Weinberg-Salam model for weak and electromagnetic interactions, the result is

$$\begin{aligned} \frac{E^{\mathrm{W}}}{mc^2} &= 2.2 \times 10^{-8}\frac{NZ}{A^2}[1 + g(N,Z)], \\ g(N,Z) &= 0.295\left[\frac{(N-Z)^2}{2NZ} + 4\sin^2\theta_W + \frac{Z}{N}\sin^2\theta_W(2\sin^2\theta_W - 1)\right], \end{aligned} \tag{2.12}$$

where $N = A - Z$ is the neutron number, and θ_W is the Weinberg angle, with $\sin^2\theta_W = 0.222$.

Table 2.1 Bounds on η^A parameters from the Eöt-Wash experiments.

Energy type	Be-Ti	Be-Al
Strong	4.9×10^{-11}	6.5×10^{-11}
Electrostatic	1.3×10^{-10}	2.1×10^{-11}
Hyperfine	2.3×10^{-7}	0.8×10^{-7}
Weak	2.0×10^{-2}	1.5×10^{-2}

Gravitational interactions are specifically excluded from WEP and EEP. In Chapter 3, we will extend these two principles to incorporate local gravitational effects, thereby defining the Gravitational Weak Equivalence Principle (GWEP) and the Strong Equivalence Principle (SEP). These two principles will be useful in classifying alternative metric theories of gravity. In any case, for laboratory Eötvös experiments, gravitational interactions are totally irrelevant, since for an atomic nucleus

$$\frac{E^{\rm G}}{mc^2} \sim \frac{Gm_{\rm nucleus}}{c^2 R_{\rm nucleus}} \sim 10^{-39} A^{2/3} \, . \tag{2.13}$$

To test for gravitational effects in GWEP, it will be necessary to employ objects containing substantial self-gravity, such as planets or neutron stars (Section 8.1).

Assuming the 1σ bound of $|\eta| < 2 \times 10^{-13}$ from the latest summary of Eöt-Wash experiments (Wagner et al., 2012), which compared Beryllium with Titanium and Aluminum, we show the corresponding bounds on the various η^A parameters in Table 2.1.

2.3.2 Tests of Local Lorentz Invariance

In Box 2.1 we reviewed the many experiments that support special relativity. One might therefore ask "what is there left to test?" Plenty, as it turns out. In recent years, an extensive theoretical and experimental effort has taken place to find or constrain violations of special relativity, or in more contemporary language, violations of Local Lorentz Invariance. The motivation for this effort is not a desire to repudiate special relativity, but to look for evidence of new physics "beyond" special relativity, such as apparent, or "effective" violations of Lorentz invariance that might result from certain models of quantum gravity, or from the existence of feeble long-range fields in the universe that might couple to matter. Some formulations of quantum gravity assert that there is a fundamental length scale given by the Planck length, $\ell_{\rm Pl} = (\hbar G/c^3)^{1/2} = 1.6 \times 10^{-35}$ m, but since length is not an invariant quantity, then there could be a violation of Lorentz invariance at some level in quantum gravity. In brane-world scenarios, while physics may be locally Lorentz invariant in the higher dimensional world, the confinement of the interactions of normal physics to our four-dimensional "brane" could induce apparent Lorentz-violating effects. And in models such as string theory, the presence of additional scalar, vector, and tensor long-range fields that couple weakly to matter of the standard model could induce effective violations of Lorentz symmetry. These and other ideas have motivated a serious reconsideration of how to test Local Lorentz Invariance with better precision and in new ways.

In many ways, the model for this new approach to testing Local Lorentz Invariance is the classic Hughes-Drever experiment. It was actually two experiments, performed in 1959–1960 independently by Hughes and collaborators at Yale University and by Drever at Glasgow University (Hughes et al., 1960; Drever, 1961). The experiments examined the $J = 3/2$ ground state of the ^{7}Li nucleus in an external magnetic field. The state is split into four levels by the magnetic field, with equal spacing in the absence of external perturbations, so the transition line is a singlet. Any external perturbation associated with a preferred direction in space that has a quadrupole ($\ell = 2$) component will destroy the equality of the energy spacing and will split the transition lines. Using NMR techniques, the experiments set a limit of 0.04 Hz (1.7×10^{-16} eV) on the separation in frequency (energy) of the lines. The original interpretation of this result, based on an idea by Cocconi and Salpeter (1958) is that it sets a limit on a possible anisotropy in inertial mass induced by some tensor field associated with the galaxy.

Phenomenologically, we can express that anisotropy in the energy of the system in the form

$$\delta E^{ij} \sim \sum_{\mathrm{A}} \delta^{\mathrm{A}} E^{\mathrm{A}} , \tag{2.14}$$

where E^{A} is the internal energy affected by the interaction, and δ^{A} is a dimensionless parameter that measures the strength of the anisotropy induced by the interaction. Using the formulae from Section 2.3.1, we can then make estimates of E^{A} for ^{7}Li and obtain the following limits

$$|\delta^{\mathrm{S}}| < 4 \times 10^{-24} , \qquad |\delta^{\mathrm{HF}}| < 5 \times 10^{-22} ,$$
$$|\delta^{\mathrm{ES}}| < 8 \times 10^{-23} , \qquad |\delta^{\mathrm{W}}| < 4 \times 10^{-18} . \tag{2.15}$$

A different interpretation treats these experiments as tests of Local Lorentz Invariance. If there are vector or tensor fields in the universe that are linked to the cosmological distribution of matter, then in the mean rest frame of that matter distribution (as represented by the frame in which the cosmic microwave background is isotropic), those fields will take on privileged forms, for example $(K, 0, 0, 0)$, for a timelike vector field K^μ. Because the Earth moves at 368 km s^{-1} relative to that frame, then there will be a preferred spatial direction associated with those fields. If any of those fields then interacts with local matter, then there could be shifts in the energy levels of atoms and nuclei that depend on the orientation of the quantization axis of the state relative to our universal velocity vector, and on the quantum numbers of the state, leading to an apparent "anisotropy" of the energy or mass of the state. If the induced anisotropies are a quadrupole effect, one would expect the net effect to be proportional to the square of the velocity w of the laboratory, if they are a dipole effect, one would expect the net effect to be linear in the velocity. If δ_0^{A} is a parameter that measures the "bare" strength of the LLI violation, then one would expect

$$\delta^{\mathrm{A}} \sim w^\ell \delta_0^{\mathrm{A}} . \tag{2.16}$$

where $\ell = 1$ or 2, depending on the nature of the anisotropy.

In order to make this discussion more concrete, we will use a specific theoretical framework, called the "c^2 formalism," to be discussed in more detail in Section 2.5.5. In this formalism, which is restricted to electromagnetic interactions, one assumes that the net effect of a nonmetric coupling of electrodynamics to some external fields is that the effective speed of electromagnetic radiation c is no longer equal to the limiting speed c_0 of test particles or of the unbroken Lorentz invariance. In the preferred universal rest frame, light propagates at speed c in all directions. Such a Lorentz-non-invariant electromagnetic interaction would cause the speed of light to depend on the observer's motion relative to the preferred frame (a modern version of "ether drift"). The altered interatomic electromagnetic fields would cause moving rods to shrink in a manner that is given by the Lorentz-Fitzgerald factor $(1 - v^2/c_0^2)^{1/2}$, using c_0, not c. The alterations could lead to anisotropies in energy levels of atoms and nuclei. In the c^2 formalism the size of the effects induced turns out to be proportional to the velocity w relative to the preferred frame to an appropriate power, and to the coefficient

$$(\delta_0)_{c^2\ \mathrm{formalism}} = 1 - \frac{c_0^2}{c^2} \,. \tag{2.17}$$

Using this framework, one can analyze a variety of tests of Local Lorentz Invariance and place bounds on δ_0; the bounds are summarized in Figure 2.3. The Hughes-Drever

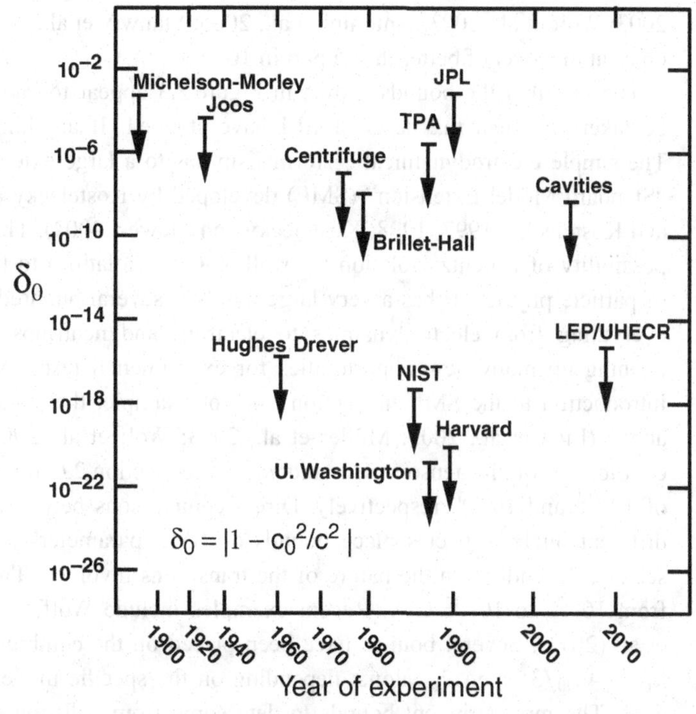

Fig. 2.3 Selected tests of Local Lorentz Invariance, showing bounds on the parameter δ_0. which measures the degree of violation of Lorentz invariance in electromagnetism. The Michelson-Morley, Joos, Brillet-Hall, and cavity experiments test the isotropy of the round-trip speed of light. The centrifuge, two-photon absorption (TPA) and JPL experiments test the isotropy of light speed using one-way propagation. The most precise experiments test isotropy of atomic energy levels, including the Hughes-Drever experiments, and the University of Washington, NIST, and Harvard experiments, which exploited cold trapped atoms. The limits assume a speed of Earth of 368 km s^{-1} relative to the mean rest frame of the universe.

experiments yielded extremely accurate bounds on δ_0. Dramatic improvements were made in the 1980s using laser-cooled trapped atoms and ions (Prestage et al., 1985; Lamoreaux et al., 1986; Chupp et al., 1989). This technique made it possible to reduce the broadening of resonance lines caused by collisions, leading to improved bounds on δ_0 shown in Figure 2.3 (experiments labelled U. Washington, NIST, and Harvard).

Also included for comparison are the corresponding limits obtained from Michelson-Morley type experiments [for a review, see Haugan and Will (1987)]. In those experiments, when viewed from the preferred frame, the speed of light down the two arms of the moving interferometer is c, while it can be shown using the electrodynamics of the c^2 formalism, that the compensating Lorentz-FitzGerald contraction of the parallel arm is governed by the speed c_0. Thus the Michelson-Morley experiment and its descendants also measure the coefficient δ_0. One of these is the Brillet-Hall experiment (Brillet and Hall, 1979), which used a Fabry-Pérot laser interferometer. A number of experiments addressed the isotropy of the one-way speed of light, labelled Centrifuge, TPA and JPL (for a review and references, see Will (1992a)). In a recent series of experiments, the frequencies of electromagnetic cavity oscillators in various orientations were compared with each other or with atomic clocks as a function of the orientation of the laboratory (Lipa et al., 2003; Müller et al., 2003; Wolf et al., 2003; Antonini et al., 2005; Stanwix et al., 2005). These placed bounds on δ_0 at the level of better than a part in 10^9.

The fact that the bounds shown in Figure 2.3 appear to end around 2010 should not be taken to mean that tests of LLI have stopped. If anything, they have intensified. The simple electrodynamical c^2 formalism has to a large extent been supplanted by the "Standard Model Extension" (SME) developed by Kostelecký and colleagues (Colladay and Kostelecký, 1997, 1998; Kostelecký and Mewes, 2002). This framework extends the possibility of Lorentz violation (as well as CPT violation) to the entire standard model of particle physics. It has a very large number (several hundred) of parameters covering everything from electrodynamics to electrons and neutrinos, to hadrons and gluons, opening up many new opportunities for experimental tests. We will give a very brief introduction to the SME in Section 2.6. For example, the cavity experiments described above (Lipa et al., 2003; Müller et al., 2003; Wolf et al., 2003) placed bounds on the coefficients of the tensors $\tilde{\kappa}_{e-}$ and $\tilde{\kappa}_{o+}$ (see Section 2.6 for definitions) at the levels of 10^{-14} and 10^{-10}, respectively. Direct comparisons between atomic clocks based on different nuclear species place bounds on SME parameters in the neutron and proton sectors, depending on the nature of the transitions involved. The bounds achieved range from 10^{-27} to 10^{-32} GeV. Recent examples include Wolf et al. (2006) and Smiciklas et al. (2011). Several bounds have been placed on the combination of SME parameters $\tilde{\kappa}_{\rm tr} - 4c_{00}/3$ or on $\tilde{\kappa}_{\rm tr}$ alone; depending on the specific model, these are equivalent to $\delta_0/2$. The most stringent bounds to date come from collision experiments at the Large Electron-Positron Collider (LEP) (Altschul, 2009), and from the absence of certain decay channels in the observation of a clean ultrahigh-energy cosmic ray (UHECR) event at the Pierre Auger Observatory (Klinkhamer and Schreck, 2008; Klinkhamer and Risse, 2008). For other bounds on these parameters and on all the parameters of the SME, from the photon sector to the Higgs sector, see Kostelecký and Russell (2011).

Astrophysical observations have been used to bound Lorentz violations. For example, if photons satisfy the Lorentz violating dispersion relation

$$E^2 = p^2c^2 + E_{\rm Pl} f^{(1)}|p|c + f^{(2)}p^2c^2 + \frac{f^{(3)}}{E_{\rm Pl}}|p|^3c^3 \dots , \tag{2.18}$$

where $E_{\rm Pl} = (\hbar c^5/G)^{1/2}$ is the Planck energy, and $f^{(n)}$ are dimensionless coefficients, then the speed of light $v_\gamma = \partial E/\partial p$ would be given, to linear order in the $f^{(n)}$ by

$$\frac{v_\gamma}{c} \approx 1 + \frac{1}{2}\sum_{n\geq 1}(n-1)f^{(n)}\left(\frac{E}{E_{\rm Pl}}\right)^{n-2} . \tag{2.19}$$

Such a Lorentz-violating dispersion relation could be a relic of quantum gravity, for instance. By bounding the difference in arrival time of high-energy photons from a burst source at large distances, one could bound contributions to the dispersion for $n > 2$. One limit, $|f^{(3)}| < 128$ comes from observations of 1 and 2 TeV gamma rays from a flare associated with the active galaxy Markarian 421 (Biller et al., 1999). Another limit comes from birefringence in photon propagation: in many Lorentz violating models, different photon polarizations may propagate with different speeds, causing the plane of polarization of a wave to rotate. If the frequency dependence of this rotation has a dispersion relation similar to Eq. (2.18), then by studying "polarization diffusion" of light from a polarized source in a given bandwidth, one can effectively place a bound $|f^{(3)}| < 10^{-4}$ (Gleiser and Kozameh, 2001). Measurements of the spectrum of ultra-high-energy cosmic rays using data from the HiRes and Pierre Auger observatories show no evidence for violations of Lorentz invariance (Stecker and Scully, 2009; Bi et al., 2009).

For thorough surveys of both the theoretical frameworks and the experimental results for tests of LLI see the reviews by Mattingly (2005), Liberati (2013) and Kostelecký and Russell (2011). The last article gives "data tables" showing experimental bounds on all the various parameters of the SME.

Local Lorentz Invariance can also be violated in gravitational interactions; these will be discussed under the rubric of "preferred-frame effects" in Chapter 8.

2.3.3 Tests of Local Position Invariance

The two main tests of Local Position Invariance are gravitational redshift experiments, which test the existence of spatial dependence on the outcomes of local experiments, and measurements of the constancy of the fundamental nongravitational constants, which test for temporal dependence.

Gravitational redshift experiments

The quintessential gravitational redshift experiment measures the frequency or wavelength shift $Z \equiv \Delta\nu/\nu = -\Delta\lambda/\lambda$ between two identical frequency standards (clocks) placed at rest at different heights in a static gravitational field. To illustrate how such an experiment tests Local position Invariance, we will assume that the remaining parts of EEP, namely, WEP and Local Lorentz Invariance, are valid. (In Sections 2.4 and 2.5, we will discuss this question under somewhat different assumptions.) WEP guarantees that there exist local freely falling frames whose acceleration $\boldsymbol{g}$ relative to the static gravitational field is the same as that of test bodies. Local Lorentz Invariance guarantees that in these frames, the proper time measured by an atomic clock is related to the Minkowski metric by

$$c^2 d\tau^2 \propto -\eta_{\mu\nu} dx_{\rm F}^{\mu} dx_{\rm F}^{\nu} \propto c^2 dt_{\rm F}^2 - dx_{\rm F}^2 - dy_{\rm F}^2 - dz_{\rm F}^2 , \tag{2.20}$$

where $x_{\rm F}$ are coordinates attached to the freely falling frame. However, in a local freely falling frame that is momentarily at rest with respect to the atomic clock, we permit its rate to depend on its location (violation of Local Position Invariance), that is, relative to a chosen atomic time standard based on a clock whose fundamental structure is different than the one being analyzed, the proper time between ticks is given by

$$\tau = \tau(U) , \tag{2.21}$$

where U is a gravitational potential whose gradient is related to the test-body acceleration by $\boldsymbol{\nabla} U = \boldsymbol{g} = -g\boldsymbol{e}_z$.

Now the emitter, receiver, and gravitational field are assumed to be static. We define a static coordinate system in which the emitter and receiver are at rest in the spatial coordinates $\boldsymbol{x}_{\rm S}$, and in which the time coordinate $t_{\rm S}$ is defined such that the interval of time $\Delta t_{\rm S}$ between ticks (passage of wave crests) of the emitter and of the receiver are equal. In other words, the time coordinate is defined by counting crests from the emitter as they pass a given point and ascribing a chosen "unit" of time to that interval. The choice of coordinates is arbitrary, but in this case it is a natural choice because of the time translation invariance associated with the static nature of the problem. The static coordinates are not the freely falling coordinates of Eq. (2.20), but are accelerated upward (in the $+z$ direction) relative to the freely falling frame, with acceleration g. Thus, for $|gt_{\rm S}/c| \sim |gz_{\rm S}/c^2| \ll 1$ (i.e., for g uniform over the distance between the clocks), a sequence of Lorentz transformations yields (MTW, section 6.6)

$$\begin{aligned} ct_{\rm F} &= (z_{\rm S} + c^2/g) \sinh(gt_{\rm S}/c) , \\ z_{\rm F} &= (z_{\rm S} + c^2/g) \cosh(gt_{\rm S}/c) , \\ x_{\rm F} &= x_{\rm S} , \\ y_{\rm F} &= y_{\rm S} . \end{aligned} \tag{2.22}$$

Thus the time measured by the atomic clocks (relative to the standard clock) is given by

$$\begin{aligned} c^2 d\tau^2 &= \tau^2(U) \left[c^2 dt_{\rm F}^2 - dx_{\rm F}^2 - dy_{\rm F}^2 - dz_{\rm F}^2 \right] , \\ &= \tau^2(U) \left[\left(1 + \frac{gz_{\rm S}}{c^2} \right)^2 c^2 dt_{\rm S}^2 - dx_{\rm S}^2 - dy_{\rm S}^2 - dz_{\rm S}^2 \right] . \end{aligned} \tag{2.23}$$

Since the emission and reception rates are the same $(1/\Delta t_{\rm S})$ when measured in static coordinate time, and since $dx_{\rm S} = dy_{\rm S} = dz_{\rm S} = 0$ for both clocks, the measured rates $(\nu = \Delta\tau^{-1})$ are related by

$$Z = \frac{\nu_{\rm rec} - \nu_{\rm em}}{\nu_{\rm rec}} = 1 - \left[\frac{\tau(U_{\rm rec}) \left(1 + gz_{\rm rec}/c^2 \right)}{\tau(U_{\rm em}) \left(1 + gz_{\rm em}/c^2 \right)} \right] . \tag{2.24}$$

For small separations $\Delta z = z_{\rm rec} - z_{\rm em}$, we can expand $\tau(U)$ in the form

$$\tau(U_{\rm rec}) \simeq \tau_0 - \tau_0' g \Delta z + O(\Delta z^2) , \tag{2.25}$$

where $\tau_0 \equiv \tau(U_{\rm em})$ and $\tau'_0 \equiv d\tau/dU|_{\rm em}$. Then

$$Z = (1+\alpha)\Delta U/c^2 , \tag{2.26}$$

where $\alpha \equiv -c^2\tau'_0/\tau_0$ and $\Delta U = \mathbf{g}\cdot\Delta\mathbf{x} = -g(z_{\rm rec}-z_{\rm em})$. If there is no location dependence in the clock rate, then $\alpha = 0$, and the redshift is the standard prediction, that is,

$$\Delta Z = \Delta U/c^2 . \tag{2.27}$$

Although there were several attempts following the publication of the general theory of relativity to measure the gravitational redshift of spectral lines from the Sun and from white dwarf stars, the results were inconclusive (see Bertotti et al. (1962) for review). During the early years of general relativity, the failure to measure this effect in solar lines was seized upon by some as reason to doubt the theory (see Crelinsten (2006) for an engaging history of this period). Unfortunately, the measurement is not simple. Solar spectral lines are subject to the "limb effect," a variation of spectral line wavelengths between the center of the solar disk and its edge or "limb"; this effect is actually a Doppler shift caused by complex convective and turbulent motions in the photosphere and lower chromosphere, and is expected to be minimized by observing at the solar limb, where the motions are predominantly transverse to the line of sight. Some atoms are also subject to line shifts caused by the enormous pressure of the solar atmosphere. The secret is to use strong, symmetrical lines, leading to unambiguous wavelength measurements. Successful measurements were finally made in the 1960s and 1970s (Brault, 1962; Snider, 1972); LoPresto et al. (1991) measured the solar shift in agreement with LPI to about 2 percent by observing the oxygen triplet lines both in absorption in the limb and in emission just off the limb. Roca Cortés and Pallé (2014) used helioseismology data from 1976 to 2013 to measure the solar shift to about 5 percent; the error was dominated by effects due to the asymmetry of the spectral line used.

The first successful, high-precision redshift measurement was the series of Pound-Rebka-Snider experiments of 1960–1965 (Pound and Rebka, 1960; Pound and Snider, 1965), which measured the frequency shift of γ-ray photons from ^{57}Fe as they ascended or descended the Jefferson Physical Laboratory tower at Harvard University. The high accuracy of 1 percent was obtained by making use of the recently discovered Mössbauer effect, whereby the recoil momentum of the nucleus from the emitted γ-ray was absorbed by the entire lattice, in order to produce a narrow resonance line whose shift could be accurately determined.

But the 1970s saw the beginning of a new era in redshift experiments, with the development of atomic frequency standards of ultrahigh stability, reaching parts in 10^{15}–10^{16} over averaging times of 10–100 seconds and longer. Examples included atomic clocks based on Cesium, Rubidium and Hydrogen, and clocks based on stabilized microwave and optical cavities. The development of laser cooling of atoms and molecules in the 1980s led to even more stable frequency standards, based on cold atoms, Bose-Einstein condensates, atom fountains, and atom interferometry. These technological developments led to new and more precise tests of LPI.

The first of these was the famous Hafele-Keating (1972a; 1972b) experiment, which took cesium-beam atomic clocks around the world on commercial airline flights, one eastward and one westward, and compared the readings on the traveling clocks upon their return with the readings on identical stay-at-home clocks. The differences in elapsed times, which included both the gravitational redshift and special relativistic time dilation agreed with LPI to about 10 percent.

The most precise standard redshift test to date was the Vessot-Levine rocket experiment known as Gravity Probe A (GPA) that took place in June 1976 (Vessot et al., 1980). A hydrogen-maser atomic clock was flown on a rocket to an altitude of about 10,000 km and its frequency compared to a hydrogen-maser clock on the ground. The experiment took advantage of the masers' frequency stability (parts in 10^{15} over 100 s averaging times) by monitoring the frequency shift as a function of altitude. A sophisticated data acquisition scheme accurately eliminated all effects of the first-order Doppler shift due to the rocket's motion, while tracking data were used to determine the payload's location and the velocity, in order to evaluate the potential difference ΔU and the special relativistic time dilation throughout the flight. Analysis of the data yielded a limit $|\alpha| < 2 \times 10^{-4}$. This could be improved by an order of magnitude by measuring the redshift between the ground and hydrogen-maser clocks on two European global navigation satellites (Galileo 5 and 6). Because these satellites were mistakenly launched into eccentric orbits instead of the intended circular orbits, integrating the modulations of the redshift over a long time will improve the statistics (Delva et al., 2015).

The varying gravitational redshift of Earth-bound clocks relative to the highly stable millisecond pulsar B1937+21 was measured to about 10 percent (Taylor, 1987). This effect is caused by the Earth's monthly oscillation with an amplitude of about 4000 km in the solar gravitational field resulting from its motion around the Earth–Moon center of mass. Two measurements of the redshift using stable oscillator clocks on spacecraft were made at the 1 percent level: one used the Voyager spacecraft in Saturn's gravitational field (Krisher et al., 1990a), while another used the Galileo spacecraft in the Sun's field (Krisher et al., 1993).

Advances in stable clocks also made possible a new type of redshift experiment that is a direct test of Local Position Invariance (LPI): a "null" gravitational redshift experiment that compares two different types of clocks, side by side, in the same laboratory. If LPI is violated, then not only can the proper ticking rate of an atomic clock vary with position, but the variation must depend on the structure and composition of the clock, otherwise all clocks would vary with position in a universal way and there would be no operational way to detect the effect (since one clock must be selected as a standard and ratios taken relative to that clock). Thus, we must write, for a given clock type A,

$$\tau = \tau^A(U) = \tau_0^A \left(1 - \alpha^A U/c^2\right) . \tag{2.28}$$

Then a comparison of two different clocks at the same location would measure

$$\frac{\tau^A}{\tau^B} = \left(\frac{\tau^A}{\tau^B}\right)_0 \left[1 - \left(\alpha^A - \alpha^B\right)\frac{U}{c^2}\right] , \tag{2.29}$$

where $(\tau^A/\tau^B)_0$ is the constant ratio between the two clock times observed at a chosen initial location.

The first null redshift experiment of this type was performed in April 1978 at Stanford University. The rates of two hydrogen maser clocks and of an ensemble of three superconducting-cavity stabilized oscillator (SCSO) clocks were compared over a 10-day period (Turneaure et al., 1983). During this time, the solar potential U in the laboratory changed sinusoidally with a 24-hour period by 3×10^{-13} because of the Earth's rotation, and changed linearly at 3×10^{-12} per day because the Earth is 90° from its perihelion in April. Analysis of the data set an upper limit on both effects, leading to a limit on the LPI violation parameter

$$|\alpha^{\rm H} - \alpha^{\rm SCSO}| < 10^{-2} . \tag{2.30}$$

Over the following years, this bound was improved using more stable frequency standards, such as atomic fountain clocks (Godone et al., 1995; Prestage et al., 1995; Bauch and Weyers, 2002; Blatt et al., 2008). The best current bounds, from comparing a ^{87}Rubidium atomic fountain with either a ^{133}Cesium fountain or a hydrogen maser (Guéna et al., 2012; Peil et al., 2013), and from comparing transitions of two different isotopes of Dysprosium (Leefer et al., 2013), hover around the one part per million mark.

Figure 2.4 summarizes the bounds on the parameter α from standard redshift experiments or on the difference $\alpha_A - \alpha_B$ for different types of clocks from null redshift experiments.

The Atomic Clock Ensemble in Space (ACES) project will place both a cold trapped atom clock based on Cesium called PHARAO (Projet d'Horloge Atomique par Refroidissement d'Atomes en Orbite), and an advanced hydrogen maser clock on the International Space Station to measure the gravitational redshift to parts in 10^6, as well as to carry out a number of fundamental physics experiments and to enable improvements in global timekeeping (Reynaud et al., 2009). Launch is currently scheduled for late 2018 or early 2019.

Modern advances in navigation using Earth-orbiting atomic clocks and accurate time-transfer must routinely take gravitational redshift and time-dilation effects into account. For example, the US Global Positioning System (GPS) provides absolute positional accuracies of around 15 m (even better in its military mode), and 50 nanoseconds in time transfer accuracy, anywhere on Earth. Yet the difference in rate between satellite and ground clocks as a result of relativistic effects is a whopping 39 microseconds per day (46 μs from the gravitational redshift, and -7 μs from time dilation). If these effects were not accurately accounted for, GPS would fail to function at its stated accuracy. This represents a welcome practical application of general relativity! For the role of general relativity in GPS, see Ashby (2002, 2003); for a popular essay, see Will (2000).

A final example of the almost "everyday" implications of the gravitational redshift is a remarkable measurement using optical clocks based on trapped aluminum ions of the frequency shift over a height of 1/3 of a meter (Chou et al., 2010).

Variations of the fundamental constants

Local Position Invariance also refers to position in time. If LPI is satisfied, the fundamental constants of nongravitational physics should be constants in time. If LPI is not satisfied,

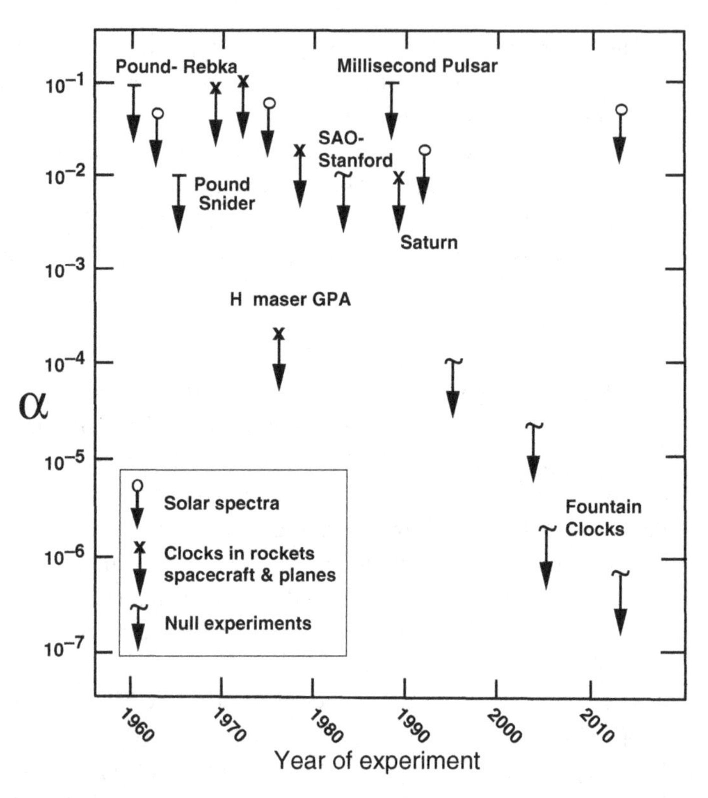

Fig. 2.4 Selected tests of Local Position Invariance via gravitational redshift experiments, showing bounds on α, which measures the degree of deviation of redshift from the formula $\Delta\nu/\nu = \Delta U/c^2$. In null redshift experiments, the bound is on the difference in α between different kinds of clocks.

then some constants could vary with time as one or more external fields that couple to matter in a nonmetric way evolve with time as the universe expands.

Early ideas about the possibility of time variations of the constants originated with Dirac, and were based on the so-called "large numbers coincidence," in which a certain dimensionless combination of fundamental constants produced the very large number 10^{40}, while another combination of constants together with the age of the universe produced the same large number. Was this merely a remarkable coincidence related to the specific conditions of the present epoch (such as the existence of physicists to make such speculations), or was it a hint of a fundamental law that would therefore require at least one constant to vary along with the age of the universe in order to maintain the equality between the large numbers? Dyson (1972) gives a detailed review of these early ideas. For a comprehensive recent reviews both of experiments and of theoretical ideas that underlie proposals for varying constants, see Uzan (2011) and Martins (2017). Here we will focus on recent observational bounds on cosmological variations of selected dimensionless combinations of the nongravitational constants (we will discuss variations of the Newtonian gravitational constant in Section 8.5).

The bounds on varying constants come in two types: bounds on the present rate of variation, and bounds on the difference between today's value and a value in the distant past. The main example of the former type is the clock comparison test, in which highly stable atomic clocks of different fundamental type are intercompared over periods ranging from months to years (variants of the null redshift experiment). If the frequencies of the clocks depend differently on the electromagnetic fine structure constant $\alpha_{\rm EM}$, the weak interaction constant $\alpha_{\rm W}$, the electron-proton mass ratio m_e/m_p or the proton gyromagnetic ration g_p, for example, then a limit on a drift of the ratio of the frequencies translates into a limit on a drift of the constant(s). The dependence of the frequencies on the constants may be quite complex, depending on the atomic species involved. Recent tests have exploited the techniques of laser cooling and trapping, and of atom fountains, in order to achieve extremely high long-term clock stability. Specific experiments compared the ground-state hyperfine transition in ^{133}Cesium to the ^{87}Rubidium hyperfine transition (Marion et al., 2003), the ^{199}Mercury ion electric quadrupole transition (Bize et al., 2003), the atomic Hydrogen 1S-2S transition (Fischer et al., 2004), and an optical transition in ^{171}Ytterbium- (Peik et al., 2004). More recent experiments have used ^{87}Strontium atoms trapped in optical lattices compared with Cesium (Blatt et al., 2008) to obtain $\dot\alpha_{\rm EM}/\alpha_{\rm EM} < 6 \times 10^{-16}\,{\rm yr}^{-1}$, compared ^{87}Rubidium and ^{133}Cesium fountains (Guéna et al., 2012) to obtain $\dot\alpha_{\rm EM}/\alpha_{\rm EM} < 2.3 \times 10^{-16}\,{\rm yr}^{-1}$, or compared two isotopes of Dysprosium (Leefer et al., 2013) to obtain $\dot\alpha_{\rm EM}/\alpha_{\rm EM} < 1.3 \times 10^{-16}\,{\rm yr}^{-1}$.

The second type of bound involves measuring the relics of or signal from a process that occurred in the distant past and comparing the inferred value of the constant with the value measured in the laboratory today. One subtype uses astronomical measurements of spectral lines at large redshift, while the other uses fossils of nuclear processes on Earth to infer values of constants early in geological history.

Earlier comparisons of spectral lines of different atoms or transitions in distant galaxies and quasars produced bounds on variations of $\alpha_{\rm EM}$ or $g_p(m_e/m_p)$ on the order of 1 part in 10 per Hubble time (Wolfe et al., 1976). Dramatic improvements in the precision of astronomical and laboratory spectroscopy, in the ability to model the complex astronomical environments where emission and absorption lines are produced, and in the ability to reach large redshift have made it possible to improve the bounds significantly. In fact, in 1999, Webb and collaborators (Webb et al., 1999; Murphy et al., 2001) announced that measurements of absorption lines in Magnesium, Aluminum, Silicon, Chromium, Iron, Nickel and Zinc in quasars in the redshift range $0.5 < Z < 3.5$ indicated a smaller value of $\alpha_{\rm EM}$ in earlier epochs, namely $\Delta\alpha_{\rm EM}/\alpha_{\rm EM} = (-0.72 \pm 0.18) \times 10^{-5}$, corresponding to $\dot\alpha_{\rm EM}/\alpha_{\rm EM} = (6.4 \pm 1.4) \times 10^{-16}\,{\rm yr}^{-1}$ (assuming a linear drift with time). The Webb group continues to report data supporting changes in $\alpha_{\rm EM}$ over large redshifts (King et al., 2012; Bainbridge and Webb, 2017). Measurements by other groups have so far failed to confirm this nonzero effect (Quast et al., 2004; Chand et al., 2005); an analysis of Magnesium absorption systems in quasars at $0.4 < Z < 2.3$ gave $\dot\alpha_{\rm EM}/\alpha_{\rm EM} = (-0.6 \pm 0.6) \times 10^{-16}\,{\rm yr}^{-1}$ (Srianand et al., 2004). More recent studies have also yielded no evidence for a variation in $\alpha_{\rm EM}$ (Kanekar et al., 2012; Lentati et al., 2013).

Table 2.2 Bounds on cosmological variations of nongravitational constants.

Constant k	Limit on $\dot{k}/k$ (yr^{-1})	Redshift	Method
Fine structure	$<1.3 \times 10^{-16}$	0	Clock comparisons
constant	$<0.5 \times 10^{-16}$	0.15	Oklo natural reactor
($\alpha_{EM} = e^2/\hbar c$)	$<3.4 \times 10^{-16}$	0.45	^{187}Re decay
	$<1.2 \times 10^{-16}$	0.4–2.3	Spectra in distant quasars
Weak interaction	$<1 \times 10^{-11}$	0.15	Oklo natural reactor
constant	$<5 \times 10^{-12}$	10^9	Big Bang nucleosynthesis
($\alpha_W = G_f m_p^2 c/\hbar^3$)			
e-p mass ratio	$<3.3 \times 10^{-15}$	0	Clock comparisons
	$<3 \times 10^{-15}$	2.6–3.0	Spectra in distant quasars

Another important set of bounds arises from studies of the "Oklo" phenomenon, a group of natural, sustained ^{235}Uranium fission reactors that occurred in the Oklo region of Gabon, Africa, around 1.8 billion years ago. Measurements of ore samples yielded an abnormally low value for the ratio of two isotopes of Samarium, ^{149}Sm/^{147}Sm. Neither of these isotopes is a fission product, but ^{149}Sm can be depleted by a flux of neutrons. Estimates of the neutron fluence (integrated dose) during the reactors' "on" phase, combined with the measured abundance anomaly, yield a value for the neutron cross-section for ^{149}Sm 1.8 billion years ago that agrees with the modern value. However, the capture cross-section is extremely sensitive to the energy of a low-lying level ($E \sim 0.1$ eV), so that a variation in the energy of this level of only 20 meV over a billion years would change the capture cross-section from its present value by more than the observed amount. This was first analyzed by Shlyakter (1976). Recent reanalyses of the Oklo data (Damour and Dyson, 1996; Fujii, 2004; Petrov et al., 2006) lead to a bound on $\dot{\alpha}_{EM}$ at the level of around 5×10^{-17} yr^{-1}.

In a similar manner, reanalyses of decay rates of ^{187}Rhenium in ancient meteorites (4.5 billion years old) gave the bound $|\dot{\alpha}_{EM}/\alpha_{EM}| < 3.4 \times 10^{-16}$ yr^{-1} (Olive et al., 2004). Finally, bounds on any variation of the weak interaction coupling constant α_W have been set by comparing predicted abundances of the light elements produced during Big-Bang nucleosynthesis and the observationally inferred primordial abundances (Malaney and Mathews, 1993; Reeves, 1994). The current best bounds are summarized in Table 2.2.

2.4 Schiff's Conjecture

Because the three parts of the Einstein Equivalence Principle discussed above are so very different in their empirical consequences, it is tempting to regard them as independent theoretical principles. However, any complete and self-consistent gravitation theory must possess sufficient mathematical machinery to make predictions for the outcomes of experiments that test each principle, and because there are limits to the number of ways that gravitation can be meshed with the standard model of particles and their interactions, one might not be surprised if there were theoretical connections between the three subprinciples. For instance, the same mathematical formalism that produces equations describing the free fall of a hydrogen atom in a gravitational field must also produce equations that determine the energy levels of hydrogen in that gravitational field, and thereby determine the ticking rate of a hydrogen maser clock. Hence a violation of EEP in the fundamental machinery of a theory that manifests itself as a violation of WEP might also be expected to lead to a violation of Local Position Invariance. Around 1960, Leonard Schiff conjectured that this kind of connection was a necessary feature of any self-consistent theory of gravity.

More precisely, Schiff's conjecture stated that any complete, self-consistent theory of gravity that embodies WEP necessarily embodies EEP. In other words, the validity of WEP alone guarantees the validity of Local Lorentz and Position Invariance, and thereby of EEP. This form of Schiff's conjecture is an embellished classical version of his original 1960 quantum mechanical conjecture (Schiff, 1960). His interest in this conjecture was rekindled in November 1970 by a vigorous argument with Kip Thorne at a conference on experimental gravitation held at the California Institute of Technology. Unfortunately, his untimely death in January, 1971 cut short his renewed effort.

Schiff's conjecture, taken literally, implies that the Eötvös experiments are the direct empirical foundation for EEP, and for the interpretation of gravity as a curved-spacetime phenomenon, and that gravitational redshift experiments, for example, are weak tests of gravitation theory and thus unnecessary. For these reasons, much effort went into "proving" Schiff's conjecture. Of course, a rigorous proof of such a conjecture is impossible, and in fact some counterexamples are known. And even if Schiff was right, this has not stopped people from carrying out gravitational redshift experiments or tests of Local Lorentz Invariance, motivated at least in part by the notion that any confirmed violation of these two principles could be the signal of entirely new physics.

Nevertheless, Schiff's conjecture provides a very useful platform for viewing the deep connections among phenomena described by the parts of EEP, and therefore we will devote some time to analyzing a range of arguments that support the conjecture.

We first restrict attention to theories of gravity based on a Lagrangian. We consider an idealized composite body made up of structureless test particles that interact by some nongravitational force to form a bound system and focus on the Hamiltonian governing the system. For a body that moves sufficiently slowly in a weak, static gravitational field, the laws governing its motion can be put into a quasi-Newtonian form (we assume the theory has a Newtonian limit). In particular, the conserved energy of the system, as obtained from the Hamiltonian will be assumed to be of the form

$$E_C = M_R c^2 + \frac{1}{2} M_R V^2 - M_R U(\boldsymbol{X}) + c^{-2} O(M_R U^2, M_R V^4, M_R V^2 U), \tag{2.31}$$

where $\boldsymbol{X}$ and $\boldsymbol{V}$ are the quasi-Newtonian location and velocity of the center of mass of the body, M_R is the "rest" mass of the body, U is the external gravitational potential, and c is a fundamental speed used to convert units of mass into units of energy. If EEP is violated, we must allow for the possibility that the speed of light and the limiting speed of material particles may differ in the presence of gravity; to maintain this possibility we do not set $c = 1$ automatically in Eq. (2.31) (see also Box 2.2). Note that V is the velocity relative to some preferred frame. In problems involving external, static gravitational potentials, the preferred frame is generally the rest frame of the external potential, while in problems involving cosmological gravitational effects where localized potentials can be ignored, the preferred frame is that of the universe rest frame. [In problems involving both kinds of effects, the simple form of Eq. (2.31) no longer holds.]

The possible occurrence of EEP violations arises when we write the rest mass $M_{\rm R}$ in the form

$$M_{\rm R} = M_0 - c^{-2}E_{\rm B}(\boldsymbol{X}, V) , \tag{2.32}$$

where M_0 is the sum of the rest masses of the structureless constituent particles, and $E_{\rm B}$ is the binding energy of the body. It is the position and velocity dependence of $E_{\rm B}$, a dependence that in general is a function of the structure of the system, which signals the breakdown of EEP. Roughly speaking, an observer in a freely falling frame can monitor the binding energy of the system, thereby detecting the effects of his location and velocity in local nongravitational experiments. Haugan (1979) made this more precise by showing that in fact it is the possible functional difference in $E_{\rm B}(\boldsymbol{X}, V)$ between the system under study and a "standard" system arbitrarily chosen as the basis for the units of measurement that leads to measurable effects. Because the location and velocity dependence in $E_{\rm B}$ is a result of the external gravitational environment, it is useful to expand it in powers of U and V^2. To an order consistent with the quasi-Newtonian approximation in Eq. (2.31), we write

$$E_{\rm B}(\boldsymbol{X}, V) = E_{\rm B}^0 + \delta m_{\rm P}^{jk} U^{jk}(\boldsymbol{X}) - \frac{1}{2}\delta m_{\rm I}^{jk} V^j V^k + c^{-2}O(E_{\rm B}^0 U^2, \dots) , \tag{2.33}$$

where U^{jk} is the external gravitational potential tensor [see Eq. (4.33)]; it is of the same order as U and satisfies $U^{jj} = U$. Summation over repeated spatial indices j and k is assumed. The quantities $\delta m_{\rm P}^{jk}$ and $\delta m_{\rm I}^{jk}$ are called the anomalous passive gravitational and inertial mass tensors, respectively. They are expected to be of order $\eta E_{\rm B}^0$, where η is a dimensionless parameter that characterizes the strength of EEP-violating effects; they depend upon the detailed internal structure of the composite body. We now assume that, associated with the Hamiltonian that governs the composite system in the external environment is a conjugate momentum and a conserved energy of the form

$$\begin{aligned} P^j &= (M_0 - c^{-2}E_{\rm B}^0)V^j + \delta m_{\rm I}^{jk} V^k , \\ E_{\rm C} &= M_0 c^2 - E_{\rm B}^0 + \frac{1}{2}\left[(M_0 - c^{-2}E_{\rm B}^0)\delta^{jk} + \delta m_{\rm I}^{jk}\right] V^j V^k \\ &\quad - \left[(M_0 - c^{-2}E_{\rm B}^0)\delta^{jk} + \delta m_{\rm P}^{jk}\right] U^{jk} . \end{aligned} \tag{2.34}$$

We first show that this yields violations of Local Position and Lorentz Invariance. We imagine that our test system makes a transition from one internal state to another, emitting a photon of frequency ν. The change in the rest energy of the test system is

$$\Delta M_R c^2 = \left[(M_R)_i - (M_R)_f\right] c^2 = -\Delta E_{\rm B}^0 \equiv h\nu_0 , \tag{2.35}$$

where h is Planck's constant. Conservation of energy and momentum, including the emitted photon implies that

$$h\nu = \Delta E_C \,,$$
$$c^{-1} h\nu \boldsymbol{e}_\gamma = \Delta \boldsymbol{P} , \tag{2.36}$$

where $\boldsymbol{e}_\gamma$ is a unit vector in the direction of the emitted photon. Combining these conservation laws, and including the changes $\Delta \boldsymbol{V}$, $\Delta\delta m_{\rm P}^{jk}$, and $\Delta\delta m_{\rm I}^{jk}$ between the initial and final state, and working to first order in the small changes, we find that the emitted frequency of the photon is given by

$$\nu = \frac{\nu_0}{1 - \boldsymbol{V}\cdot\boldsymbol{e}_\gamma/c}\left[1 - \frac{1}{2c^2}\left(\delta^{jk} + \alpha_{\rm I}^{jk}\right)V^j V^k - \frac{1}{c^2}\left(\delta^{jk} + \alpha_{\rm P}^{jk}\right)U^{jk}\right], \tag{2.37}$$

where

$$\alpha_{\rm P}^{jk} \equiv \frac{\Delta(\delta m_{\rm P}^{jk})c^2}{h\nu_0},$$
$$\alpha_{\rm I}^{jk} \equiv \frac{\Delta(\delta m_{\rm I}^{jk})c^2}{h\nu_0}, \tag{2.38}$$

and $\boldsymbol{V}$ is the final velocity of the test system. The factor in the denominator in Eq. (2.37) is the usual kinematical first-order Doppler effect.

As it stands, however, the result of Eq. (2.37) is not very informative, because we have no clear way to measure the frequency ν_0; even far from the Earth, we cannot be certain that there is not some residual potential U^{jk} due to the galaxy or the universe, or any residual motion relative to some preferred frame. However, what can be measured unambiguously is the ratio of the frequency of the photon to that of a photon emitted by test system of a different internal composition, that is adjacent to the first system and at rest with respect to it. This system will be chosen to be the "standard" of frequency. The ratio between the two frequencies would then be given by

$$\frac{\nu_2}{\nu_1} = \frac{(\nu_0)_2}{(\nu_0)_1}\left\{1 - \frac{1}{2}\left[(\alpha_{\rm I}^{jk})_2 - (\alpha_{\rm I}^{jk})_1\right]V^j V^k - \left[(\alpha_{\rm P}^{jk})_2 - (\alpha_{\rm P}^{jk})_1\right]U^{jk}\right\}. \tag{2.39}$$

If EEP is valid, so that $\alpha_{\rm I}^{jk} = \alpha_{\rm P}^{jk} = 0$ for all systems, then the result is the universal value of the frequency of the transition of system 2 in units of the standard frequency. If the anomalous passive gravitational mass tensor $\delta m_{\rm P}^{jk}$ is nonzero, it can produce preferred-location effects in a null gravitational redshift experiment. If the anomalous inertial mass tensor $\delta m_{\rm I}^{jk}$ is nonzero, it can produce preferred-frame effects in a Hughes-Drever experiment.

A different argument, based only upon the assumption of energy conservation, can be used to demonstrate the link between WEP and LPI and LLI. This assumption allows one to perform very simple cyclic *gedanken* experiments in which the energy at the end of the cycle must equal that at the beginning of the cycle. This approach was pioneered by Dicke (1964), and subsequently generalized by Nordtvedt (1975) and Haugan (1979).

We begin with a set of n free particles of mass m_0 at rest at $\boldsymbol{X} = \boldsymbol{h}$. From Eq. (2.34), the conserved energy is simply $nm_0c^2[1 - U(\boldsymbol{h})/c^2]$. We then form a composite body out of these particles and convert the released energy to free particles of rest mass m_0, stored in a massless reservoir. The conserved energy of the composite body is

$$E_{\rm comp}(\boldsymbol{h}) = \bar{M}_R c^2 - \left[\bar{M}_R \delta^{jk} + \delta m_{\rm P}^{jk}\right] U^{jk}(\boldsymbol{h}) , \tag{2.40}$$

where $\bar{M}_R \equiv nm_0 - c^{-2}E_B^0$, and that of the reservoir is

$$E_{\rm res}(\boldsymbol{h}) = E_B^0\left(1 - \frac{U(\boldsymbol{h})}{c^2}\right) + \delta m_P^{jk} U^{jk}(\boldsymbol{h}) \,. \tag{2.41}$$

The composite body falls freely to $\boldsymbol{X} = 0$ with an acceleration assumed to be $\boldsymbol{A}$, while the stored test particles fall with acceleration $\boldsymbol{g} = \boldsymbol{\nabla} U$ (by definition), reaching $\boldsymbol{X} = 0$ with velocities $(-2\boldsymbol{A}\cdot\boldsymbol{h})^{1/2}$ and $(-2\boldsymbol{g}\cdot\boldsymbol{h})^{1/2}$, respectively (we assume that $\boldsymbol{g}$, $\boldsymbol{V}$ and $\boldsymbol{h}$ are parallel). At $\boldsymbol{X} = 0$, the energies of the composite particle and the test particles are

$$E_{\rm comp}(0) = \bar{M}_R c^2 - \left[\bar{M}_R\delta^{jk} + \delta m_P^{jk}\right] U^{jk}(\boldsymbol{0}) - \left[\bar{M}_R\delta^{jk} + \delta m_I^{jk}\right] A^j h^k \,, \tag{2.42a}$$

$$E_{\rm res}(0) = E_B^0\left[1 - \frac{U(\boldsymbol{0})}{c^2} - \frac{\boldsymbol{g}\cdot\boldsymbol{h}}{c^2}\right] + \delta m_P^{jk} U^{jk}(\boldsymbol{h})[1 + O(U/c^2)] \,, \tag{2.42b}$$

The particle is brought to rest and the acquired kinetic energy $-[\bar{M}_R\delta^{jk} + \delta m_I^{jk}]A^j h^k$ is stored in the reservoir. We now extract from the reservoir enough energy $E_B^0[1 - U(\boldsymbol{0})/c^2] + \delta m_P^{jk} U^{jk}(\boldsymbol{0})$ to disassemble the composite particle into its n constituent test masses, plus enough energy $-nm_0\boldsymbol{g}\cdot\boldsymbol{h}$ to return them to $\boldsymbol{X} = \boldsymbol{h}$, arriving at rest with the same conserved energy $nm_0c^2[1 - U(\boldsymbol{h})/c^2]$ as they had before. The reservoir now contains

$$\begin{aligned} E_{\rm res}(0) &= -[\bar{M}_R\delta^{jk} + \delta m_I^{jk}]A^j h^k + \left[nm_0 - \frac{E_B^0}{c^2}\right]\boldsymbol{g}\cdot\boldsymbol{h} + \delta m_P^{jk}\left[U^{jk}(\boldsymbol{h}) - U^{jk}(\boldsymbol{0})\right] \\ &= \left[\bar{M}_R(g^i - A^i) - \delta m_I^{ij} A^j + \delta m_P^{jk} U^{jk}(\boldsymbol{0})_{,i}\right] h^i \,. \end{aligned} \tag{2.43}$$

If energy is to be conserved, the reservoir must be empty, and thus, working to first order in the small difference $\boldsymbol{A} - \boldsymbol{g}$, we obtain

$$A^i = g^i - \beta_I^{ij} g^j + \beta_P^{jk} U^{jk}_{,i} \,, \tag{2.44}$$

where

$$\begin{aligned} \beta_P^{jk} &\equiv \frac{\delta m_P^{jk}}{\bar{M}_R} \,, \\ \beta_I^{jk} &\equiv \frac{\delta m_I^{jk}}{\bar{M}_R} \,, \end{aligned} \tag{2.45}$$

where $\bar{M}_R \equiv M_0 - c^{-2}E_B^0$. Again, we observe the deep link between violations of LPI and LLI and violations of WEP. If WEP is satisfied ($\delta m_P^{jk} = \delta m_I^{jk} = 0$ for all systems), then so are LPI and LLI.

A variant of this cyclic *gedanken* experiment was devised by Nordtvedt (1975) in order to derive the gravitational redshift between systems separated by a height $\boldsymbol{h}$. We begin as before with n free particles of mass m_0 at rest at $\boldsymbol{X} = \boldsymbol{h}$ (for simplicity we will assume that the overall velocity relative to a preferred frame vanishes). We again form a composite body with conserved energy given by Eq. (2.40). This time we convert the released energy into a photon, which propagates downward to $\boldsymbol{X} = 0$. Its energy is compared to that of a photon emitted from an identical composite system residing at $\boldsymbol{X} = 0$. We will assume that the energy of the arriving photon is given by $(1 - Z)[E_B^0(1 - U(\boldsymbol{0})/c^2) + \delta m_P^{jk} U^{jk}(\boldsymbol{0})]$, where Z is the "redshift," to be determined. This energy is added to the reservoir along with the kinetic energy $-[\bar{M}_R\delta^{jk} + \delta m_I^{jk}]A^j h^k$ acquired by the falling composite body, less an amount of energy required to disassemble the body and return it to $\boldsymbol{X} = \boldsymbol{h}$. Substituting Eq. (2.44) for A^i, we find that the reservoir contains

$$E_{\rm res}(0) = -Z\left\{E_{\rm B}^0\left[1 - \frac{U(\mathbf{0})}{c^2}\right] + \delta m_{\rm P}^{jk} U^{jk}(\mathbf{0})\right\} + E_{\rm B}^0 \boldsymbol{g}\cdot\boldsymbol{h} - \delta m_{\rm P}^{jk} U^{jk}(\mathbf{0})_{,i} h^i \,. \tag{2.46}$$

The reservoir must be empty, and so solving for Z to first order in $\boldsymbol{g}\cdot\boldsymbol{h} \sim U$ we obtain

$$Z = \frac{1}{c^2}\left[\Delta U - \frac{\delta m_{\rm P}^{jk}}{E_{\rm B}^0}\Delta U^{jk}\right], \tag{2.47}$$

where $\Delta U = \boldsymbol{g}\cdot\boldsymbol{h}$, and $\Delta U^{jk} = \boldsymbol{\nabla} U^{jk}\cdot\boldsymbol{h}$.

By similar analyses, one can derive the effect of LLI violations on frequency shifts between moving emitters and receivers undergoing general motions or orbits (Haugan, 1979; Wolf and Blanchet, 2016). In particular the work of Wolf and Blanchet (2016) yielded new insight into the so-called "noon-midnight" redshift. It has long been known that in any metric theory of gravity, there is no frequency shift due to the Sun's gravitational field between two clocks on opposite sides of the Earth. Since the Earth is in free fall around the Sun, the only effect of the Sun's field is a tidal effect, leading to a quadrupolar noon–6 pm shift, but with an amplitude far too small to be detectable with current technology. But if EEP is violated, then the anomalous mass tensors $\delta m_{\rm P}^{jk}$ and $\delta m_{\rm I}^{jk}$ *can* contribute to a noon-midnight effect when different clocks are compared, and that can be tested to interesting levels, such as in an orbiting clock experiment called STE-QUEST, that is currently under study (Altschul et al., 2015).

Thorne, Lee, and Lightman (1973) proposed a more qualitative "proof" of Schiff's conjecture for that class of gravitation theories that are based on an invariant action principle. They begin by defining the concept of "universal coupling": a generally covariant Lagrangian-based theory is universally coupled if it can be put into a mathematical form (representation) in which the action for matter and nongravitational fields $I_{\rm NG}$ contains precisely one gravitational field: a symmetric, second-rank tensor $\boldsymbol{\psi}$ with signature $+2$ that reduces to $\boldsymbol{\eta}$ when gravity is turned off; and when $\boldsymbol{\psi}$ is replaced by $\boldsymbol{\eta}$, $I_{\rm NG}$ becomes the action of special relativity. Clearly, among all Lagrangian-based theories, one is universally coupled if and only if it is a metric theory.

Let us illustrate this point with a simple example. Consider a Lagrangian-based theory of gravity that possesses a globally flat background metric $\boldsymbol{\eta}$ and a symmetric, second-rank tensor gravitational field $\boldsymbol{h}$. The nongravitational action for charged point particles of rest mass m_o and charge e, and for electromagnetic fields has the form

$$I_{\rm NG} \equiv I_0 + I_{\rm int} + I_{\rm em}\,, \tag{2.48}$$

where

$$\begin{aligned} I_0 &= -m_0 \int d\tau\,, \quad d\tau^2 = -(\eta_{\mu\nu} + h_{\mu\nu})dx^\mu dx^\nu\,, \\ I_{\rm int} &= e\int A_\mu dx^\mu\,, \\ I_{\rm em} &= -\frac{1}{16\pi}\int (\eta^{\mu\alpha} - h^{\mu\alpha})(\eta^{\beta\nu} - h^{\beta\nu})F_{\mu\nu}F_{\alpha\beta}\left(1 + \frac{1}{2}h\right) d^4x\,, \end{aligned} \tag{2.49}$$

where $F_{\mu\nu} = A_{\nu,\mu} - A_{\mu,\nu}$, and where

$$||\eta^{\mu\nu}|| \equiv ||\eta_{\mu\nu}||^{-1}, \quad h^{\mu\nu} \equiv \eta^{\mu\alpha}\eta^{\nu\beta}h_{\alpha\beta}, \quad h \equiv \eta^{\mu\nu}h_{\mu\nu}, \tag{2.50}$$

We work in a coordinate system in which $\boldsymbol{\eta} = \text{diag}(-1, 1, 1, 1)$. To see whether this theory is universally coupled, the obvious step is to assume that the single gravitational field $\psi_{\mu\nu}$ is given by

$$\psi_{\mu\nu} = \eta_{\mu\nu} + h_{\mu\nu}. \tag{2.51}$$

This would make I_0 and I_{int} appear to be universally coupled. However, in the electromagnetic Lagrangian, we obtain, for example,

$$\eta^{\mu\alpha} - h^{\mu\alpha} = \psi^{\mu\alpha} - h^{\mu\beta}h^{\alpha}_{\beta} + O(h^3), \tag{2.52}$$

where $||\psi^{\mu\alpha}|| = ||\psi_{\mu\alpha}||^{-1}$. Thus, there is no way to combine $\eta_{\mu\nu}$ and $h_{\mu\nu}$ into a single gravitational field in I_{NG}, hence the theory is not universally coupled. To see that the theory is also not a metric theory, we transform to a frame in which at an event $\mathcal{P}$,

$$\psi_{\hat{\mu}\hat{\nu}}(\mathcal{P}) = \eta_{\hat{\mu}\hat{\nu}} \quad \psi_{\hat{\mu}\hat{\nu},\hat{\alpha}}(\mathcal{P}) = 0. \tag{2.53}$$

Note that in this frame, $h_{\hat{\mu}\hat{\nu}}(\mathcal{P}) \neq 0$ in general, thus the action can be put into the form

$$I_{\text{NG}} = I_{\text{SRT}} + \Delta I, \tag{2.54}$$

where

$$\begin{aligned} I_{\text{SRT}} &= -m_0 \int d\tau + e \int A_{\hat{\mu}} dx^{\hat{\mu}} - \frac{1}{16\pi} \int \eta^{\hat{\mu}\hat{\alpha}}\eta^{\hat{\beta}\hat{\nu}} F_{\hat{\mu}\hat{\nu}} F_{\hat{\alpha}\hat{\beta}} (-\hat{\eta})^{1/2} d^4\hat{x}, \\ \Delta I &= -\frac{1}{16\pi} \int \left[2h^{\hat{\mu}\hat{\delta}} h^{\hat{\alpha}}_{\hat{\delta}} \eta^{\hat{\beta}\hat{\nu}} + \frac{1}{8} \left(\hat{h}^2 - 2h^{\hat{\gamma}\hat{\delta}} h_{\hat{\gamma}\hat{\delta}} \right) \eta^{\hat{\mu}\hat{\alpha}}\eta^{\hat{\beta}\hat{\nu}} \right] F_{\hat{\mu}\hat{\nu}} F_{\hat{\alpha}\hat{\beta}} (-\hat{\eta})^{1/2} d^4\hat{x}. \end{aligned} \tag{2.55}$$

So in a local Lorentz frame, the laws of physics are *not* those of special relativity, so the theory is not a metric theory. Notice that in this particular case, for weak gravitational fields ($|h_{\mu\nu}| \ll 1$), the theory *is* metric to first order in $h_{\mu\nu}$, while the deviations from metric form occur at second order in $h_{\mu\nu}$. In the next section, we will present a mathematical framework for examining a class of theories with nonuniversal coupling and for making quantitative computations of its empirical consequences.

Consider now all Lagrangian-based theories of gravity, and assume that WEP is valid. WEP forces I_{NG} to involve one and only one gravitational field (which must be a second-rank tensor $\boldsymbol{\psi}$ which reduces to $\boldsymbol{\eta}$ far from gravitating matter). If I_{NG} were to involve some other gravitational fields ϕ, K_μ, $h_{\mu\nu}, \ldots$ they would all have to conspire to produce exactly the same acceleration for a body made largely of electromagnetic energy as for one made largely of nuclear energy, and so on. This is unlikely unless $\psi_{\mu\nu}$, and the other fields appear everywhere in I_{NG} in precisely the same form, for example, $f(\phi)\psi_{\mu\nu}$, if a scalar field is present, $\psi_{\mu\nu} + ah_{\mu\nu}$ if a tensor field is present, and so on. In this case, one can absorb these fields into a new field $g_{\mu\nu}$ and end up with only one gravitational field in I_{NG}. This means that the theory must be universally coupled, and therefore a metric theory, and must satisfy EEP.

One possible counterexample to Schiff's conjecture was proposed by Ni (1977): a pseudoscalar field ϕ that couples to electromagnetism in a Lagrangian term of the form $\phi\epsilon^{\mu\nu\alpha\beta}F_{\mu\nu}F_{\alpha\beta}$, where $\epsilon^{\mu\nu\alpha\beta}$ is the completely antisymmetric Levi-Civita symbol. Ni argued that such a term, while violating EEP, does not violate WEP in the sense of composition dependent accelerations, although it does have the observable effect of producing anomalous torques on systems of electromagnetically bound charged particles.

2.5 The $TH\epsilon\mu$ Formalism

The discussion of Schiff's conjecture presented in the previous section was very general, and perhaps gives compelling evidence for the validity of the conjecture. However, because of the generality of those arguments, there was little quantitative information. For example, no means was presented to compute explicitly the anomalous mass tensors $\delta m_{\rm P}^{jk}$ and $\delta m_{\rm I}^{jk}$ for various systems. In order to make these ideas more concrete, we need a model theory of the nongravitational laws of physics in the presence of gravity that incorporates the possibility of both metric and nonmetric (nonuniversal) coupling. This theory should be simple, yet capable of making quantitative predictions for the outcomes of experiments. One such "model" theory is the $TH\epsilon\mu$ formalism, devised by Lightman and Lee (1973). It restricts attention to the motions and electromagnetic interactions of charged structureless test particles in an external, static, and spherically symmetric (SSS) gravitational field. It assumes that the nongravitational laws of physics can be derived from an action $I_{\rm NG}$ given by

$$
\begin{aligned}
I_{\rm NG} = &-\sum_a m_{0a}\int (T - Hv_a^2)^{1/2}dt + \sum_a e_a \int A_\mu(x_a^\nu)v_a^\mu dt \\
&+ \frac{1}{8\pi}\int \left(\epsilon E^2 - \mu^{-1}B^2\right) d^4x ,
\end{aligned}
\tag{2.56}
$$

(here we use units in which $\boldsymbol{x}$ and t both have units of length) where m_{0a}, e_a, and $x_a^\mu(t)$ are the rest mass, charge, and world line of particle a; $x^0 \equiv t$, $v_a^\mu \equiv dx_a^\mu/dt$, $\boldsymbol{E} \equiv \boldsymbol{\nabla} A_0 - \boldsymbol{A}_{,0}$, $\boldsymbol{B} \equiv \boldsymbol{\nabla} \times \boldsymbol{A}$, and where scalar products between 3-vectors are taken with respect to the Cartesian metric δ^{ij}. The functions T, H, ϵ and μ are assumed to be functions of a single external gravitational potential Φ, but are otherwise arbitrary. For an SSS field in a given theory, T, H, ϵ, and μ will be particular functions of Φ.

For SSS fields, Eq. (2.56) is general enough to encompass all metric theories of gravitation. In a metric theory, the action $I_{\rm NG}$ is given by

$$
\begin{aligned}
I_{\rm NG} = &-\sum_a m_{0a}\int (-g_{\mu\nu}v_a^\mu v_a^\nu)^{1/2}dt + \sum_a e_a \int A_\mu(x_a^\nu)v_a^\mu dt \\
&- \frac{1}{16\pi}\int g^{\mu\alpha}g^{\nu\beta}F_{\mu\nu}F_{\alpha\beta}(-g)^{1/2}d^4x ,
\end{aligned}
\tag{2.57}
$$

where g is the determinant of $g_{\mu\nu}$, and $F_{\mu\nu} = A_{\nu,\mu} - A_{\mu,\nu}$. For a static spherically symmetric spacetime in isotropic coordinates, $g_{00} = -T$, $g_{0j} = 0$, and $g_{jk} = H\delta_{jk}$. It is then straightforward to show that the action in a metric theory takes the form

$$
\begin{aligned}
I_{\mathrm{NG}} = & -\sum_a m_{0a} \int (T - Hv_a^2)^{1/2} dt + \sum_a e_a \int A_\mu(x_a^\nu) v_a^\mu dt \\
& + \frac{1}{8\pi} \int \left[\left(\frac{H}{T}\right)^{1/2} E^2 - \left(\frac{T}{H}\right)^{1/2} B^2 \right] d^4x .
\end{aligned} \tag{2.58}
$$

Thus any metric theory can be cast in the $TH\epsilon\mu$ form but with ϵ and μ given by

$$
\epsilon = \mu = \left(\frac{H}{T}\right)^{1/2} . \tag{2.59}
$$

It also encompasses a wide class of nonmetric theories, such as the Belinfante Swihart (1957a, 1957b, 1957c) theory and the nonmetric theory discussed in Section 2.4.

2.5.1 Einstein Equivalence Principle in the $TH\epsilon\mu$ formalism

We begin by exploring in some detail the properties of the formalism as presented in Eq. (2.56). Later, we will apply it to the interpretation of experiments.

In order to examine the Einstein Equivalence Principle in this formalism we must work in a local freely falling frame. But we do not yet know whether WEP is satisfied by the $TH\epsilon\mu$ theory (and we suspect that it is not, in general), so we do not know to which freely falling trajectories local frames should be attached. We must therefore arbitrarily choose a set of trajectories: the most natural choice is the set of trajectories of neutral test particles, that is, particles governed only by the first term in I_{NG}, since their trajectories are universal and independent of the mass m_{0a}. We make a transformation to a coordinate system $x^{\hat{\alpha}} = (\hat{t}, \hat{\boldsymbol{x}})$, chosen according to the following criteria: (i) the origins of both coordinate systems coincide, that is, for a selected event $\mathcal{P}$, $x^\alpha(\mathcal{P}) = x^{\hat{\alpha}}(\mathcal{P}) = 0$, (ii) at $\mathcal{P}$, a neutral test body has zero acceleration in the new coordinates, that is, $d^2x^{\hat{\jmath}}/d\hat{t}^2|_{\mathcal{P}} = 0$, and in the neighborhood of $\mathcal{P}$ the deviations from zero acceleration are quadratic in the quantities $\Delta x^{\hat{\alpha}} = x^{\hat{\alpha}} - x^{\hat{\alpha}}(\mathcal{P})$, and (iii) the motion of the neutral test body is derivable from an action I_0. The required transformation, correct to first order in the quantities $\boldsymbol{g}_0\hat{t}$ and $\boldsymbol{g}_0 \cdot \hat{\boldsymbol{x}}$, both assumed small, is

$$
\begin{aligned}
\hat{t} &= T_0^{1/2} t \left[1 + \frac{T_0'}{2T_0} \boldsymbol{g}_0 \cdot \boldsymbol{x} \right] , \\
\hat{\boldsymbol{x}} &= H_0^{1/2} \left[\boldsymbol{x} + \frac{T_0'}{4H_0} \boldsymbol{g}_0\, t^2 + \frac{H_0'}{4H_0} \left(2\boldsymbol{x}\boldsymbol{g}_0 \cdot \boldsymbol{x} - \boldsymbol{g}_0 x^2 \right) \right] ,
\end{aligned} \tag{2.60}
$$

where the subscript (0) and superscript (′) on functions such as T, H, ϵ, or μ denote

$$
T_0 \equiv T(x^\alpha = x^{\hat{\alpha}} = 0) , \quad T_0' \equiv \partial T / \partial \Phi |_{x^\alpha = x^{\hat{\alpha}} = 0} , \tag{2.61}
$$

and where $\boldsymbol{g}_0 \equiv \boldsymbol{\nabla}\Phi$. The action I_{NG} in the new coordinates then has the form

$$
\begin{aligned}
I_{\mathrm{NG}} = & -\sum_a m_{0a} \int (1 - \hat{v}_a^2)^{1/2} d\hat{t} + \sum_a e_a \int A_{\hat{\mu}} v_a^{\hat{\mu}} d\hat{t} \\
& + \frac{1}{8\pi} \epsilon_0 \left(\frac{T_0}{H_0}\right)^{1/2} \int \left\{ \hat{E}^2 \left[1 + \frac{1}{2} \left(\frac{T_0'}{T_0 H_0^{1/2}} \right) \Gamma_0 \boldsymbol{g}_0 \cdot \hat{\boldsymbol{x}} \right] \right.
\end{aligned}
$$

$$
-\frac{H_0}{T_0\epsilon_0\mu_0}\hat{B}^2\left[1-\frac{1}{2}\left(\frac{T_0'}{T_0H_0^{1/2}}\right)\Lambda_0\,\boldsymbol{g}_0\cdot\hat{\boldsymbol{x}}\right]
$$

$$
-\frac{T_0'H_0^2}{T_0^{3/2}\epsilon_0\mu_0}\Upsilon_0\,\hat{t}\boldsymbol{g}_0\cdot\left(\hat{\boldsymbol{E}}\times\hat{\boldsymbol{B}}\right)\Bigg\}\,d^4\hat{x}+\left[\text{corrections }O(x^{\hat{\alpha}})^2\right], \tag{2.62}
$$

where $\hat{\boldsymbol{v}}_a = d\hat{\boldsymbol{x}}_a/d\hat{t}$, and

$$
A_{\hat{\mu}} \equiv \frac{\partial x^\alpha}{\partial x^{\hat{\mu}}}A_\alpha\,,\quad \hat{\boldsymbol{E}} \equiv \hat{\boldsymbol{\nabla}}A_{\hat{0}} - \hat{\boldsymbol{A}}_{,\hat{0}}\,,\quad \hat{\boldsymbol{B}} \equiv \hat{\boldsymbol{\nabla}}\times\hat{\boldsymbol{A}}\,. \tag{2.63}
$$

The quantities Γ_0, Λ_0, and Υ_0 are defined by

$$
\Gamma_0 \equiv \frac{2T_0}{T_0'}\left(\frac{\epsilon_0'}{\epsilon_0}+\frac{T_0'}{2T_0}-\frac{H_0'}{2H_0}\right), \tag{2.64a}
$$

$$
\Lambda_0 \equiv \frac{2T_0}{T_0'}\left(\frac{\mu_0'}{\mu_0}+\frac{T_0'}{2T_0}-\frac{H_0'}{2H_0}\right), \tag{2.64b}
$$

$$
\Upsilon_0 \equiv 1-\frac{T_0\epsilon_0\mu_0}{H_0}\,. \tag{2.64c}
$$

Note that our choice of the multiplicative factors $T_0^{1/2}$ and $H_0^{1/2}$ in the transformation equations (2.60) resulted in unit coefficients in the point-particle part of the action, making it look exactly like that of special relativity.

Let us now examine the consequences for EEP of the physics governed by $I_{\rm NG}$. We first focus on the form of $I_{\rm NG}$ at the event $\mathcal{P}(x^{\hat{\jmath}} = \hat{t} = 0)$, since local test experiments are assumed to take place in vanishingly small regions surrounding $\mathcal{P}$. Because such experiments are designed to be electrically neutral overall, we can assume that the $\boldsymbol{E}$ and $\boldsymbol{B}$ fields do not extend outside this region. Then at $\mathcal{P}$,

$$
I_{\rm NG} = -\sum_a m_{0a}\int(1-\hat{v}_a^2)^{1/2}d\hat{t}+\sum_a e_a\int A_{\hat{\mu}}v_a^{\hat{\mu}}d\hat{t}
$$

$$
+\frac{1}{8\pi}\epsilon_0\left(\frac{T_0}{H_0}\right)^{1/2}\int\left\{\hat{E}^2-\left(\frac{H_0}{T_0\epsilon_0\mu_0}\right)\hat{B}^2\right\}d^4\hat{x}\,. \tag{2.65}
$$

We first see that, in general, $I_{\rm NG}$ violates Local Lorentz Invariance. A simple Lorentz transformation of particle coordinates and fields in $I_{\rm NG}$ shows that $I_{\rm NG}$ is a Lorentz invariant if and only if

$$
\frac{H_0}{T_0\epsilon_0\mu_0}=1 \quad\text{or}\quad \epsilon_0\mu_0=\frac{H_0}{T_0} \quad\text{or}\quad \Upsilon_0=0\,. \tag{2.66}
$$

Since we have not specified the event $\mathcal{P}$, this condition must hold throughout the SSS spacetime. Notice that the quantity $(T_0^{-1}H_0\epsilon_0^{-1}\mu_0^{-1})^{1/2}$ plays the role of the speed of light in the local frame, or more precisely, of the ratio of the speed of light $c_{\rm light}$ to the limiting speed c_0 of neutral test particles, that is,

$$
\frac{H_0}{T_0\epsilon_0\mu_0}=\left(\frac{c_{\rm light}}{c_0}\right)^2\,. \tag{2.67}
$$

Our coordinate transformation was chosen in such away that, in the local freely falling frame, $c_0 = 1$; equivalently, in the original $TH\epsilon\mu$ coordinate system [see Eq. (2.56)]

$$c_0 = (T_0/H_0)^{1/2}, \quad c_{\text{light}} = (\epsilon_0\mu_0)^{-1/2}. \tag{2.68}$$

These speeds will be the same only if Eq. (2.66) is satisfied. If not, then the rest frame of the SSS field is a preferred frame in which I_{NG} takes its $TH\epsilon\mu$ form, and one can expect observable effects in experiments that move relative to this frame. Thus, the quantity Υ_0 plays the role of a preferred-frame parameter: if it is zero everywhere, the formalism is locally Lorentz invariant; if it is nonzero anywhere, there will be preferred-frame effects there. As we will see, the Hughes-Drever experiments and their descendents provide the most stringent limits on this preferred-frame parameter.

Next, we observe that the action in Eq. (2.62) is locally position invariant if and only if

$$\begin{aligned} \epsilon_0\left(\frac{T_0}{H_0}\right)^{1/2} &= [\text{constant, independent of}\,\mathcal{P}], \\ \mu_0\left(\frac{T_0}{H_0}\right)^{1/2} &= [\text{constant, independent of}\,\mathcal{P}]. \end{aligned} \tag{2.69}$$

Even if the theory is locally Lorentz invariant ($\Upsilon_0 = 0$, independent of $\mathcal{P}$), there may still be location-dependent effects if the quantities in Eq. (2.69) are not constant. This would correspond, for example, to the situation discussed in Section 2.2, in which different parts of the local physical laws in a freely falling frame couple to different multiples $\phi^{(1)}\boldsymbol{\eta}$, $\phi^{(2)}\boldsymbol{\eta}$, ... of the Minkowski metric; in this case, free particle motion coupling to $\boldsymbol{\eta}$ itself, electrodynamics coupling to the position-dependent tensor $\boldsymbol{\eta}^* = \epsilon T^{1/2}H^{-1/2}\boldsymbol{\eta}$ in the manner given by the field Lagrangian $(-\eta^*)^{1/2}\eta^{*\mu\alpha}\eta^{*\nu\beta}F_{\mu\nu}F_{\alpha\beta}$. The nonuniversality of this coupling violates EEP and leads to position-dependent effects, for example, in gravitational redshift experiments. An alternative way to characterize these effects in the case where Local Lorentz Invariance is satisfied is to renormalize the unit of charge and the vector potential at each event $\mathcal{P}$ according to

$$e_a^* \equiv e_a\left(\frac{H_0}{\epsilon_0^2 T_0}\right)^{1/4}, \quad A_\mu^* \equiv A_\mu\left(\frac{H_0}{\epsilon_0^2 T_0}\right)^{-1/4}. \tag{2.70}$$

Then the action of Eq. (2.65) takes the form

$$\begin{aligned} I_{\text{NG}} = &-\sum_a m_{0a}\int(1-\hat{v}_a^2)^{1/2}d\hat{t} + \sum_a e_a^*\int A_{\hat{\mu}}^* v_a^{\hat{\mu}} d\hat{t} \\ &+ \frac{1}{8\pi}\int\left[\hat{E}^{*2} - \hat{B}^{*2}\right]d^4\hat{x}. \end{aligned} \tag{2.71}$$

This action has the special relativistic form, except that the physically measured charge e_a^* now depends on location via Eq. (2.70), unless $\epsilon_0(T_0/H_0)^{1/2}$ is independent of $\mathcal{P}$. In the latter case, the units of charge can be effectively chosen so that everywhere in spacetime, $\epsilon_0(T_0/H_0)^{1/2} = 1$.

Note, however, that if LPI alone is satisfied [Eq. (2.69)], one can renormalize the charge and vector potential to make *either* $\epsilon_0(T_0/H_0)^{1/2} = 1$ *or* $\mu_0(T_0/H_0)^{1/2} = 1$, *but not both*, thus in general LLI need not be satisfied. Thus we see that a necessary and sufficient condition for both Local Lorentz Invariance and Local Position Invariance to be valid is

$$\epsilon_0 = \mu_0 = (H_0/T_0)^{1/2}, \text{ for all events } \mathcal{P}. \tag{2.72}$$

These constraints are equivalent to those of metric theories of gravity, Eq. (2.59).

Return to the terms in $I_{\rm NG}$, in Eq. (2.62) that depend on the first-order displacements $\hat{x}$ and $\hat{t}$ from the event $\mathcal{P}$. These occur only in the electromagnetic field part of the action, and presumably produce polarizations of the electromagnetic fields of charged bodies proportional to the external "acceleration" $\boldsymbol{g}_0 = \boldsymbol{\nabla}\Phi$. One would expect these polarizations to result in accelerations of composite bodies made up of charged test particles relative to the local freely falling frame (i.e., relative to neutral test particles), in other words, to result in violations of WEP. These terms are absent if $\Gamma_0 = \Lambda_0 = \Upsilon_0 = 0$, that is, if Eq. (2.69) holds. This discussion suggests that WEP alone may guarantee EEP. We now demonstrate this directly by carrying out an explicit calculation of the acceleration of a composite test body within the $TH\epsilon\mu$ framework. The resulting restricted proof of Schiff's conjecture was first formulated by Lightman and Lee (1973).

2.5.2 Proof of Schiff's conjecture

We work in the global $TH\epsilon\mu$ coordinate system in which $I_{\rm NG}$ has the form of Eq. (2.56). Variation of $I_{\rm NG}$ with respect to the particle and field variables gives a complete set of particle equations of motion given by

$$\frac{d}{dt}\left(\frac{H}{W}\boldsymbol{v}_a\right) + \boldsymbol{\nabla}W = \frac{e_a}{m_{0a}}\left[\boldsymbol{\nabla}\left(A_0 + \boldsymbol{v}_a\cdot\boldsymbol{A}\right) - \frac{d\boldsymbol{A}}{dt}\right], \tag{2.73}$$

where $W \equiv (T - Hv_a^2)^{1/2}$, $A_0 = A_0(\boldsymbol{x}_a)$, $\boldsymbol{A} = \boldsymbol{A}(\boldsymbol{x}_a)$, and "gravitationally modified" Maxwell (GMM) equations, given by

$$\begin{aligned} \boldsymbol{\nabla}\cdot(\epsilon\boldsymbol{E}) &= 4\pi\rho, \\ \boldsymbol{\nabla}\times\left(\mu^{-1}\boldsymbol{B}\right) &= 4\pi\boldsymbol{J} + \frac{\partial(\epsilon\boldsymbol{E})}{\partial t}, \end{aligned} \tag{2.74}$$

where $\rho \equiv \sum_a e_a\delta^3(\boldsymbol{x} - \boldsymbol{x}_a)$, $\boldsymbol{J} \equiv \sum_a e_a\boldsymbol{v}_a\delta^3(\boldsymbol{x} - \boldsymbol{x}_a)$. These equations are used to calculate the acceleration from rest of an electromagnetically bound composite body consisting of charged point particles. A number of approximations are necessary to make the computation tractable. First, the functions T, H, ϵ and μ, considered to be functions of Φ are expanded about the instantaneous center-of-mass location $\boldsymbol{X} = 0$ of the composite body, in the form

$$T(\Phi) = T_0 + T_0'\boldsymbol{g}_0\cdot\boldsymbol{x} + O[(\boldsymbol{g}_0\cdot\boldsymbol{x})^2, \nabla g_0 x^2, \ldots], \tag{2.75}$$

where $T_0 = T(\boldsymbol{x}=0)$, $T_0' = dT/d\Phi|_{\boldsymbol{x}=0}$. As long as the body is small compared to the scale over which Φ varies, we can work to first order in $\boldsymbol{g}_0\cdot\boldsymbol{x}$. Second, we assume that the internal particle velocities and electromagnetic fields are sufficiently small that we can expand the equations of motion and GMM equations in terms of the small quantities $v^2 \sim e^2/m_0 r \ll 1$ where r is a typical interparticle distance. By analogy with the post-Newtonian expansion to be described in Chapter 4, we call this a post-Coulombian expansion; for the purpose of the present discussion we will work to first post-Coulombian order. We expect the single-particle acceleration to contain terms that are $O(g_0)$ (bare gravitational acceleration), $O(v^2)$ (Coulomb interparticle acceleration), $O(g_0 v^2)$ (post-Coulombian gravitational acceleration), $O(v^4)$ (post-Coulombian interparticle acceleration), $O(g_0 v^4)$ (post-post-Coulombian gravitational acceleration), and so on. To $O(g_0 v^2)$, we obtain

$$\begin{aligned}\frac{d\boldsymbol{v}_a}{dt} &= -\frac{T_0'}{2H_0}\boldsymbol{g}_0 + \frac{H_0'}{2H_0}\boldsymbol{g}_0 v_a^2 + \left(\frac{T_0'}{T_0} - \frac{H_0'}{H_0}\right)\boldsymbol{v}_a(\boldsymbol{g}_0 \cdot \boldsymbol{v}_a) \\ &+ \frac{e_a}{m_{0a}}\frac{T_0^{1/2}}{H_0}\left[\boldsymbol{\nabla}(A_0 + \boldsymbol{v}_a \cdot \boldsymbol{A}) - \frac{d\boldsymbol{A}}{dt}\right]. \end{aligned} \tag{2.76}$$

To write the equation of motion directly in terms of particle coordinates, we must obtain the vector potential A_μ, in terms of coordinates and velocities to an appropriate order. In a gauge in which

$$\epsilon\mu A_{0,0} - \boldsymbol{\nabla}\cdot\boldsymbol{A} = 0\,, \tag{2.77}$$

the GMM equations take the form

$$\begin{aligned}\nabla^2 A_0 - \epsilon\mu A_{0,00} &= 4\pi\epsilon^{-1}\rho - \epsilon^{-1}\boldsymbol{\nabla}\epsilon\cdot(\boldsymbol{\nabla}A_0 - \boldsymbol{A}_{,0})\,, \\ \nabla^2\boldsymbol{A} - \epsilon\mu\boldsymbol{A}_{,00} &= -4\pi\mu\boldsymbol{J} + (\epsilon\mu)^{-1}\boldsymbol{\nabla}(\epsilon\mu)(\boldsymbol{\nabla}\cdot\boldsymbol{A}) - \mu^{-1}\boldsymbol{\nabla}\mu\times(\boldsymbol{\nabla}\times\boldsymbol{A})\,. \end{aligned} \tag{2.78}$$

These equations can be solved iteratively by writing $A_\mu = A_\mu^{(0)} + A_\mu^{(1)}$, where $A_\mu^{(1)}/A_\mu^{(0)} \sim O(g_0)$ and solving for each term to an appropriate order in v^2. The result is

$$\begin{aligned}A_0 &= -\phi - \frac{1}{2}\epsilon_0\mu_0 X_{,00} + \frac{1}{2}\epsilon_0'\epsilon_0^{-1}\left(2\phi\boldsymbol{g}_0\cdot\boldsymbol{x} - \boldsymbol{g}_0\cdot\boldsymbol{\nabla}X\right)\,, \\ \boldsymbol{A} &= \boldsymbol{V} + O(g_0)\,, \end{aligned} \tag{2.79}$$

where

$$\begin{aligned}\phi &\equiv \epsilon_0^{-1}\sum_a \frac{e_a}{|\boldsymbol{x}-\boldsymbol{x}_a|}\,, \\ X &\equiv \epsilon_0^{-1}\sum_a e_a|\boldsymbol{x}-\boldsymbol{x}_a|\,, \\ \boldsymbol{V} &\equiv \mu_0\sum_a \frac{e_a\boldsymbol{v}_a}{|\boldsymbol{x}-\boldsymbol{x}_a|}\,. \end{aligned} \tag{2.80}$$

We insert the resulting single-particle acceleration into a definition of the acceleration of the body's center of mass. It turns out that to post-Coulombian order, it suffices to use the simple center-of-mass definition

$$\boldsymbol{X} \equiv m^{-1}\sum_a m_{0a}\boldsymbol{x}_a\,, \quad m \equiv \sum_a m_{0a}\,. \tag{2.81}$$

We then compute, $d^2\boldsymbol{X}/dt^2$, substituting the single-particle equation of motion to the necessary order, and using the fact that at $t=0$, $\boldsymbol{X} = d\boldsymbol{X}/dt = 0$. The resulting expression is simplified by the use of virial identities that relate internal, structure-dependent quantities to each other via total time derivatives of other internal quantities. As long as we restrict attention to bodies in equilibrium, these time derivatives can be assumed to vanish when averaged over intervals of time that are long compared to internal timescales of the body. Errors generated by our definition of the center-of-mass similarly vanish. To post-Coulombian order, the required virial identity is

$$\langle T^{jk} \rangle + \frac{1}{2} T_0^{1/2} H_0^{-1} \epsilon_0^{-1} \langle \Omega^{jk} \rangle = 0 \,, \tag{2.82}$$

where angular brackets denote a time average, and where

$$T^{jk} \equiv \frac{1}{2} \sum_a m_{0a} v_a^j v_a^k \,, \quad \Omega^{jk} \equiv \frac{1}{2} \sum_{a,b} e_a e_b \frac{x_{ab}^j x_{ab}^k}{r_{ab}^3} \,, \tag{2.83}$$

where $\boldsymbol{x}_{ab} \equiv \boldsymbol{x}_a - \boldsymbol{x}_b$, $r_{ab} = |\boldsymbol{x}_{ab}|$, and the double sum over a and b excludes the case $a = b$. The final result is

$$\frac{d^2 X^j}{dt^2} = g^j - \Gamma_0 \left[\frac{E_{\rm B}^{\rm ES}}{mc_0^2} \right] g^j + \Upsilon_0 \left(\frac{E_{{\rm B}jk}^{\rm ES} + \delta_{jk} E_{\rm B}^{\rm ES}}{mc_0^2} \right) g^k \,, \tag{2.84}$$

where $c_0^2 = T_0/H_0$,

$$g^j = -\frac{T_0'}{2H_0} g_0^j \,, \tag{2.85}$$

and

$$E_{{\rm B}jk}^{\rm ES} \equiv -\frac{T_0^{1/2}}{H_0 \epsilon_0} \langle \Omega^{jk} \rangle \,, \quad E_{\rm B}^{\rm ES} \equiv -\frac{T_0^{1/2}}{H_0 \epsilon_0} \langle \Omega^{kk} \rangle \,, \tag{2.86}$$

are the "binding energy" tensor and scalar of the body. The first term in Eq. (2.84) is g^j, the universal acceleration of a neutral test body governed by the point particle action alone; the other two terms depend on the body's electromagnetic self-energy and self-energy tensor. These terms vanish for all bodies (i.e., WEP is satisfied) if and only if, at any event $\mathcal{P}$,

$$\Gamma_0 = \Upsilon_0 = 0 \,. \tag{2.87}$$

These conditions imply that $\epsilon_0 (T_0/H_0)^{1/2} = C$ and $\mu_0 (T_0/H_0)^{1/2} = C^{-1}$, where C is a constant, but as before, a simple constant rescaling of charges and the vector potential using Eq. (2.70) sets $C = 1$ which is equivalent to Eq. (2.72), hence

$$\text{WEP} \Rightarrow \text{LPI} \quad \text{and} \quad \text{LLI} \,, \tag{2.88}$$

and Schiff's conjecture is verified, at least within the confines of the $TH\epsilon\mu$ formalism.

It is useful to define the gravitational potential U whose gradient yields the test-body acceleration $\boldsymbol{g}$; modulo a constant,

$$U(\boldsymbol{x}) \equiv -(T_0'/2H_0) \boldsymbol{g}_0 \cdot \boldsymbol{x} \,. \tag{2.89}$$

If the functions T, H, ϵ and μ are now considered as functions of U instead of Φ, then Γ_0 and Λ_0 can be expressed in the useful forms

$$\begin{aligned} \Gamma_0 &= -c_0^2 \frac{d}{dU} \ln \left[\epsilon \, (T/H)^{1/2} \right]_{x=0} \,, \\ \Lambda_0 &= -c_0^2 \frac{d}{dU} \ln \left[\mu \, (T/H)^{1/2} \right]_{x=0} \,. \end{aligned} \tag{2.90}$$

2.5.3 Energy conservation and anomalous mass tensors

Because the $TH\epsilon\mu$ formalism is based on an action principle, it possesses conservation laws, and so is amenable to analysis using the conserved-energy framework described in Section 2.4. The main products of that framework are the anomalous inertial mass tensor $\delta m_{\rm I}^{jk}$ and passive gravitational mass tensor $\delta m_{\rm P}^{jk}$, obtained from the conserved energy. These two quantities then yield expressions for violations of WEP, Local Lorentz Invariance, and Local Position Invariance.

As a concrete example (Haugan, 1979), we consider a classical bound system of two charged particles. As in the "proof" of Schiff's conjecture we work to post-Coulombian order and to first order in $\boldsymbol{g}_0\cdot\boldsymbol{x}$. We first formulate the equations of motion in terms of a truncated action $\tilde{I}_{\rm NG}$ in which the field piece in Eq. (2.56) is dropped and the interaction piece is rewritten entirely in terms of particle coordinates by substituting the post-Coulombian solutions Eqs. (2.79) and (2.80) for A_μ. Variation of $\tilde{I}_{\rm NG}$ with respect to the particle coordinates then yields the complete particle equations of motion. We identify a Lagrangian L using the definition $\tilde{I}_{\rm NG} = \int L dt$. It is convenient to rescale L by the factor $T_0^{1/2}H_0^{-1}$; this has no effect on the dynamics but will simplify the subsequent analysis. We next make a change of variables in L from $\boldsymbol{x}_1$, $\boldsymbol{x}_2$, $\boldsymbol{v}_1$ and $\boldsymbol{v}_2$ to the center-of-mass and relative variables

$$\boldsymbol{X} \equiv M_0^{-1}(m_1\boldsymbol{x}_1 + m_2\boldsymbol{x}_2)\,, \quad \boldsymbol{V} \equiv d\boldsymbol{X}/dt\,,$$
$$\boldsymbol{x} \equiv \boldsymbol{x}_1 - \boldsymbol{x}_2\,, \quad \boldsymbol{v} \equiv d\boldsymbol{x}/dt\,, \tag{2.91}$$

where we define $M_0 \equiv m_1 + m_2$, $\mu = m_1m_2/M_0$. We construct a Hamiltonian H from the rescaled L using the standard method:

$$P^j \equiv \partial L/\partial V^j\,, \quad p^j \equiv \partial L/\partial v^j\,, \quad H \equiv P^jV^j + p^jv^j - L\,. \tag{2.92}$$

The result, to post-Coulombian order, is

$$\begin{aligned}
H &= M_0c_0^2\left(1 + \frac{T_0'}{2T_0}\boldsymbol{g}_0\cdot\boldsymbol{X}\right) + \frac{P^2}{2M_0} + \frac{p^2}{2\mu} + \frac{T_0^{1/2}}{H_0\epsilon_0}\frac{e_1e_2}{r} \\
&\quad + \left[\frac{T_0'}{2T_0} - \frac{H_0'}{H_0}\right]\frac{p^2}{2\mu}\boldsymbol{g}_0\cdot\boldsymbol{X} - \frac{T_0^{1/2}\epsilon_0'}{H_0\epsilon_0^2}\frac{e_1e_2}{r}\boldsymbol{g}_0\cdot\boldsymbol{X} \\
&\quad - \frac{[2(\boldsymbol{p}\cdot\boldsymbol{P})^2 + p^2P^2]}{4\mu M_0^2c_0^2} - \frac{T_0^{1/2}\mu_0}{H_0}\frac{e_1e_2}{r}\frac{[(\boldsymbol{n}\cdot\boldsymbol{P})^2 + P^2]}{2M_0^2} \\
&\quad + O(Pp^3) + O(p^4) + O(P^4)\,, \\
&= \bar{M}_Rc_0^2\left(1 + \frac{T_0'}{2T_0}\boldsymbol{g}_0\cdot\boldsymbol{X}\right) + \frac{P^2}{2\bar{M}_R} \\
&\quad - \boldsymbol{g}_0\cdot\boldsymbol{X}\left\{\left[\frac{H_0'}{H_0}\right]\frac{p^2}{2\mu} + \left[\frac{T_0'}{2T_0} + \frac{\epsilon_0'}{\epsilon_0}\right]\frac{T_0^{1/2}}{H_0\epsilon_0}\frac{e_1e_2}{r}\right\} \\
&\quad + \frac{P^jP^k}{2\bar{M}_R^2c_0^2}\Upsilon_0\frac{T_0^{1/2}}{\epsilon_0H_0}\frac{e_1e_2}{r}(\delta^{jk} + n^jn^k) \\
&\quad - \frac{P^jP^k}{2\bar{M}_R^2c_0^2}\left[\frac{p^jp^k}{\mu} + \frac{T_0^{1/2}}{\epsilon_0H_0}\frac{e_1e_2}{r}n^jn^k\right]\,,
\end{aligned} \tag{2.93}$$

where we define

$$\bar{M}_R \equiv M_0 + c_o^{-2}\left(\frac{p^2}{2\mu} + \frac{T_0^{1/2}}{\epsilon_0 H_0}\frac{e_1 e_2}{r}\right). \tag{2.94}$$

The post-Coulombian terms $O(p^4)$ and $O(P^4)$ neglected in Eq. (2.93) do not couple the internal motion and the center-of-mass motion and thus do not lead to violations of EEP. We now average H over several timescales for the internal motions of the bound two-body system, assumed short compared to the timescale for the center-of-mass motion. This allows us to drop terms that are odd in internal vectors, such as terms of $O(Pp^3)$, or $O(Ppe^2/r)$ since such terms vanish on averaging over a system with reflection symmetry.

The average is further simplified by using virial theorems analogous to Eq. (2.82); these are obtained from Hamilton's equations for the internal variables derived from H, namely

$$\left\langle\frac{p^j p^k}{\mu}\right\rangle + \frac{T_0^{1/2}}{H_0\epsilon_0}\left\langle\frac{e_1 e_2}{r}n^j n^k\right\rangle = 0\,. \tag{2.95}$$

modulo post-Coulombian corrections. Notice that although the post-Coulombian corrections in Eq. (2.95) may depend on the center of mass variables $\boldsymbol{P}$ or $\boldsymbol{X}$, this dependence does not affect the form of H: it is only the explicit dependence on $\boldsymbol{P}$ or $\boldsymbol{X}$ in Eq. (2.93) that generates the center-of-mass motion. Making use of Eqs. (2.64a) and (2.89) and defining

$$E^{\rm ES}_{{\rm B}jk} \equiv -\frac{1}{2}\frac{T_0^{1/2}}{H_0\epsilon_0}\left\langle\frac{e_1 e_2}{r}n^j n^k\right\rangle, \tag{2.96}$$

we obtain

$$\langle H\rangle = \bar{M}_R c_0^2 - \bar{M}_R U + \frac{P^2}{2\bar{M}_R} - 2\Gamma_0 U\frac{E^{\rm ES}_{\rm B}}{c_0^2} - \Upsilon_0\left(\frac{E^{\rm ES}_{\rm B}\delta_{jk} + E^{\rm ES}_{{\rm B}jk}}{\bar{M}_R c_0^2}\right)\frac{P^j P^k}{\bar{M}_R}, \tag{2.97}$$

where now $\bar{M}_R = M_0 - E^{\rm ES}_{\rm B}/c_0^2$.

Using $V^j = \partial\langle H\rangle/\partial P^j$, we can express the momentum P^j and the conserved energy $E_{\rm C} = \langle H\rangle$ in precisely the form of Eqs. (2.34), with

$$\delta m_{\rm I}^{jk} = 2\Upsilon_0\left(\frac{E^{\rm ES}_{\rm B}\delta_{jk} + E^{\rm ES}_{{\rm B}jk}}{c_0^2}\right), \tag{2.98a}$$

$$\delta m_{\rm P}^{jk} = 2\Gamma_0\frac{E^{\rm ES}_{\rm B}}{c_0^2}\delta^{jk}\,. \tag{2.98b}$$

Substitution of these formulae into Eq. (2.44) for the general center-of-mass acceleration of a composite system yields precisely Eq. (2.84).

2.5.4 Quantum systems and EEP

One advantage of the Hamiltonian approach is that it can also be applied to quantum systems (Will, 1974). This is especially useful in discussing gravitational redshift experiments since it is transitions between quantized energy levels that produce the photons whose redshifts are measured. For the idealized gravitational redshift experiments discussed in Section 2.4, only the anomalous passive mass tensor $\delta m_{\rm P}^{jk}$ is needed. The simplest quantum system of interest is that of a charged particle (electron) moving in a given external electromagnetic potential of a charged particle (proton) at rest in the SSS field, that is, a hydrogen atom. For such a system the truncated Lagrangian (2.56) has the form

$$L = -m_e(T_0 - H_0 v^2)^{1/2} - eA_\mu v^\mu \,, \tag{2.99}$$

where $m_0 = m_e$ and $e = |e|$ for the electron. We will ignore the spatial variation of T, H, ϵ, and μ across the atom, hence we evaluate each at $\boldsymbol{x} = 0$. The Hamiltonian obtained from L is given by

$$H = T_0^{1/2}\left[m_e^2 + H_0^{-1}|\boldsymbol{p} + e\boldsymbol{A}|^2\right]^{1/2} + eA_0 \,, \tag{2.100}$$

where $p_j \equiv \partial L/\partial v^j$. Introducing the Dirac matrices

$$\beta = \begin{pmatrix} I & 0 \\ 0 & -I \end{pmatrix}, \quad \boldsymbol{\alpha} = \begin{pmatrix} 0 & \boldsymbol{\sigma} \\ \boldsymbol{\sigma} & 0 \end{pmatrix}, \quad \tilde{I} = \begin{pmatrix} I & 0 \\ 0 & I \end{pmatrix}, \tag{2.101}$$

where I is the two-dimensional unit matrix and $\boldsymbol{\sigma}$ are the constant Pauli spin matrices, we perform the "square root" in H, and obtain

$$H = T_0^{1/2}\left[m_e\beta + H_0^{-1/2}\boldsymbol{\alpha}\cdot(\boldsymbol{p} + e\boldsymbol{A})\right] + eA_0\tilde{I} \,. \tag{2.102}$$

The gravitationally modified Dirac equation is then $H|\psi\rangle = i\hbar(\partial/\partial t)|\psi\rangle$. For most applications it is more convenient to use the semirelativistic approximation to H obtained by means of a Foldy-Wouthuysen transformation, yielding

$$\begin{aligned} H = T_0^{1/2}&\left[m_e + \frac{|\boldsymbol{p} + e\boldsymbol{A}|^2}{2H_0 m_e} - \frac{p^4}{8H_0^2 m_e^2} + \frac{e\hbar}{2H_0 m_e}\boldsymbol{\sigma}\cdot\boldsymbol{B}\right] + eA_0 \\ &- \frac{e\hbar}{4H_0 m_e^2}\boldsymbol{\sigma}\cdot\left(\boldsymbol{E}\times\boldsymbol{p} - \frac{1}{2}i\hbar\boldsymbol{\nabla}\times\boldsymbol{E}\right) + \frac{e\hbar^2}{8H_0 m_e^2}\boldsymbol{\nabla}\cdot\boldsymbol{E} + O(p^6) \,, \end{aligned} \tag{2.103}$$

where we have made the usual identification $\boldsymbol{p} \to -i\hbar\boldsymbol{\nabla}$ and have ignored the effects of spatial variations in T, H, ϵ, or μ on the atomic structure. For a charged particle with magnetic moment $\boldsymbol{M}_p$ at rest at the origin, the vector potential as obtained from the GMM equations is given (to the necessary accuracy) by

$$A_0 = -\frac{e}{\epsilon_0 r}, \quad \boldsymbol{A} = \frac{1}{2}\mu_0\frac{\boldsymbol{M}_p\times\boldsymbol{x}}{r^3} \,. \tag{2.104}$$

The Hamiltonian then takes the form

$$H \equiv H_{\rm r} + H_{\rm s} + H_{\rm f} + H_{\rm hf} + O(p^6) \,, \tag{2.105}$$

where

$$\begin{aligned} H_{\rm r} &= T_0^{1/2} m_e \,, \\ H_{\rm p} &= \frac{T_0^{1/2}}{H_0}\frac{p^2}{2m_e} - \frac{e^2}{\epsilon_0 r} \,, \\ H_{\rm f} &= -\frac{T_0^{1/2}}{H_0^2}\frac{p^4}{8m_e^2} - \frac{1}{H_0\epsilon_0}\left(\frac{\hbar e^2}{4m_e^2 r^3}\right)\boldsymbol{\sigma}\cdot\boldsymbol{L} \,, \\ H_{\rm hf} &= \frac{T_0^{1/2}}{H_0}\left(\frac{e\hbar}{2m_e}\right)\boldsymbol{\sigma}\cdot\boldsymbol{B} \,, \end{aligned} \tag{2.106}$$

where $\boldsymbol{L} = \boldsymbol{x}\times\boldsymbol{p}$ is the angular momentum of the electron. The four pieces of H are the usual rest mass ($H_{\rm r}$), principal ($H_{\rm p}$), fine-structure ($H_{\rm f}$) and hyperfine-structure ($H_{\rm hf}$)

contributions. We have ignored the Darwin term ($\propto \nabla \cdot \boldsymbol{E}$). The magnetic field produced by the proton is given by

$$\boldsymbol{B} = \nabla \times \boldsymbol{A} = -\frac{1}{2}\mu_0 \left[\frac{\boldsymbol{M}_p - 3\boldsymbol{n}(\boldsymbol{n} \cdot \boldsymbol{M}_p)}{r^3} - \frac{8\pi}{3}\boldsymbol{M}_p \delta^3(\boldsymbol{x}) \right] . \tag{2.107}$$

We must first identify the proton magnetic moment. From the hyperfine term $H_{\rm hf}$, it is clear that the magnetic moment of the electron is given by

$$\boldsymbol{M}_e = -\frac{T_0^{1/2}}{H_0} \frac{e\hbar}{2m_e} \boldsymbol{\sigma} . \tag{2.108}$$

It is then reasonable to assume that the magnetic moment of the proton has the same dependence on T_0 and H_0, so that

$$\boldsymbol{M}_p = \frac{T_0^{1/2}}{H_0} \frac{g_p e\hbar}{2m_p} \boldsymbol{\sigma}_p , \tag{2.109}$$

where g_p is the gyromagnetic ratio of the proton and m_p is its mass. Then

$$H_{\rm hf} = -\frac{1}{2} \frac{\mu_0 T_0}{H_0^2} \left(\frac{g_p e^2 \hbar^2}{4m_e m_p} \right) \boldsymbol{\sigma}_e \cdot \left[\frac{\boldsymbol{\sigma}_p - 3\boldsymbol{n}(\boldsymbol{n} \cdot \boldsymbol{\sigma}_p)}{r^3} - \frac{8\pi}{3} \boldsymbol{\sigma}_p \delta^3(\boldsymbol{x}) \right] . \tag{2.110}$$

Solving for the eigenstates of the Hamiltonian using perturbation theory yields

$$E = T_0^{1/2} \left[m_e + \left(\frac{H_0}{T_0 \epsilon_0^2} \right) \mathcal{E}_{\rm p} + \left(\frac{H_0}{T_0 \epsilon_0^2} \right)^2 \mathcal{E}_{\rm f} + \left(\frac{H_0}{T_0 \mu_0 \epsilon_0^3} \right) \mathcal{E}_{\rm hf} \right] , \tag{2.111}$$

where $\mathcal{E}_{\rm p}$, $\mathcal{E}_{\rm f}$ and $\mathcal{E}_{\rm hf}$ are the usual expressions for the principal, fine-structure, and hyperfine-structure energy levels in terms of atomic constants m_e, e, m_p, g_p, $\hbar$, and quantum numbers. In order to calculate the anomalous mass tensor $\delta m_{\rm P}^{jk}$, we must determine the manner in which E varies as the location of the atom is changed. Expanding E to first order in $\boldsymbol{g}_0 \cdot \boldsymbol{X}$, substituting Eq. (2.89), converting to the conserved energy function $E_{\rm C} = E(T_0^{1/2}/H_0)$, defining the binding energy $E_{\rm B}^0 = E_{\rm B}^{\rm ES} + E_{\rm B}^{\rm F} + E_{\rm B}^{\rm HF}$, where

$$E_{\rm B}^{\rm ES} \equiv -\epsilon_0^{-2} \mathcal{E}_{\rm p} , \quad E_{\rm B}^{\rm F} \equiv -H_0 T_0^{-1} \epsilon_0^{-4} \mathcal{E}_{\rm f} , \quad E_{\rm B}^{\rm HF} \equiv -\mu_0 \epsilon^{-3} \mathcal{E}_{\rm hf} , \tag{2.112}$$

we obtain Eq. (2.34) (with $V = 0$), with

$$\begin{aligned} \delta m_{\rm P}^{({\rm ES})jk} &= 2\Gamma_0 \frac{E_{\rm B}^{\rm ES}}{c_0^2} \delta^{jk} , \\ \delta m_{\rm P}^{({\rm F})jk} &= 4\Gamma_0 \frac{E_{\rm B}^{\rm F}}{c_0^2} \delta^{jk} , \\ \delta m_{\rm P}^{({\rm HF})jk} &= (3\Lambda_0 - \Gamma_0) \frac{E_{\rm B}^{\rm HF}}{c_0^2} \delta^{jk} . \end{aligned} \tag{2.113}$$

Compare $\delta m_{\rm P}^{({\rm ES})jk}$ here with Eq. (2.98b).

A useful fact that emerges from the solution for the energy eigenstates is that the Bohr radius is given by

$$a = \frac{T_0^{1/2} \epsilon_0}{H_0} \left(\frac{\hbar^2}{m_e e^2} \right) . \tag{2.114}$$

This is important in analyzing the gravitational redshift of microwave or optical cavities. From the quantization of the electromagnetic field in the $TH\epsilon\mu$ formalism, it is straightforward to see that the energy of the electromagnetic field in a cavity is proportional to $\hbar\omega$, where ω is the angular frequency of the wave mode in the cavity, and where the dispersion relation implied by the GMM equations implies that $|\boldsymbol{k}| = (\epsilon_0\mu_0)^{1/2}\omega$. The wavenumber $\boldsymbol{k}$ must satisfy $\boldsymbol{k}\cdot\boldsymbol{L} = n\pi$ for a stationary mode, where $|\boldsymbol{L}|$ is the length of the cavity. But L is proportional to an integer number of atoms times the interatomic spacing, whose fundamental scale is the Bohr radius. But Eq. (2.114) shows that this scales as $T_0^{1/2}\epsilon_0/H_0$. Thus $|\boldsymbol{k}| \propto H_0 T_0^{-1/2}\epsilon_0^{-1}$, and the energy of the mode can be written as

$$E = \mathcal{E}_{\text{cavity}}\frac{H_0}{T_0^{1/2}\epsilon_0^{3/2}\mu_0^{1/2}}, \tag{2.115}$$

where $\mathcal{E}_{\text{cavity}}$ depends only on atomic constants and integers. Expanding in terms of $\boldsymbol{g}_0\cdot\boldsymbol{x}$ and calculating the conserved energy function E_{C}, we obtain Eq. (2.34), with

$$\begin{aligned} E_B^{\text{cavity}} &= -\mathcal{E}_{\text{cavity}}\mu_0^{-1/2}\epsilon_0^{-3/2}, \\ \delta m_{\text{P}}^{(\text{cavity})jk} &= \frac{1}{2}(3\Gamma_0 + \Lambda_0)\frac{E_B^{\text{cavity}}}{c_0^2}\delta^{jk}. \end{aligned} \tag{2.116}$$

2.5.5 Local Lorentz Invariance and the c^2 formalism

If we wish to focus attention on tests of local Lorentz Invariance then the $TH\epsilon\mu$ formalism can be simplified (Haugan and Will, 1987; Gabriel and Haugan, 1990). Since most such tests do not concern themselves with the spatial variation of the functions T, H, ϵ, and μ, but rather with observations made in moving frames, we can treat them as spatial constants, so that $T_0' = H_0' = \epsilon_0' = \mu_0' = 0$. Then by rescaling the time and space coordinates using Eqs. (2.60), and the charges and the electromagnetic fields using Eqs. (2.70), we can put the action (2.56) into the form

$$\begin{aligned} I_{\text{NG}} = &-\sum_a m_{0a}\int(1 - v_a^2)^{1/2}dt + \sum_a e_a\int A_\mu v_a^\mu dt \\ &+ \frac{1}{8\pi}\int\left\{E^2 - c^2B^2\right\}d^4\hat{x}, \end{aligned} \tag{2.117}$$

where $c^2 = H_0/(T_0\epsilon_0\mu_0) = (1 - \Upsilon_0)^{-1}$. This amounts to using units in which the limiting speed c_0 of massive test particles is unity, and the speed of light is c. If $c \neq 1$, LLI is violated, and the form of the action above must be assumed to be valid only in some preferred universal rest frame. The natural candidate for such a frame is the rest frame of the microwave background.

The electrodynamic equations that follow from Eq. (2.117) yield the behavior of rods and clocks, just as in the full $TH\epsilon\mu$ formalism. For example, the length of a rod that moves with velocity $\boldsymbol{V}$ relative to the rest frame in a direction parallel to its length will be observed by a rest observer to be contracted relative to an identical rod perpendicular to the motion by a factor $1 - V^2/2 + O(V^4)$. Notice that c does not appear in this expression, because only electrostatic interactions are involved in determining the interatomic spacing in the rod, while c appears only in the magnetic sector of the action.

The energy and momentum of an electromagnetically bound body moving with velocity V relative to the rest frame are given by

$$P^j = M_R V^j + \delta m_{\rm I}^{jk} V^k ,$$
$$E_{\rm C} = M_R c^2 + \frac{1}{2} M_R V^2 + \frac{1}{2} \delta m_{\rm I}^{jk} V^j V^k , \tag{2.118}$$

where $M_R = M_0 - E_{\rm B}^{\rm ES}$,

$$\delta m_{\rm I}^{jk} = 2\Upsilon_0 \left(E_{\rm B}^{\rm ES} \delta_{jk} + E_{{\rm B}jk}^{\rm ES} \right) , \tag{2.119}$$

and where $E_{{\rm B}jk}^{\rm ES}$ is given by Eq. (2.96), with $E_{\rm B}^{\rm ES} = \delta^{jk} E_{{\rm B}jk}^{\rm ES}$.

Note that Υ_0 corresponds to the parameter δ_0 in Eq. (2.17) and plotted in Figure 2.3.

2.5.6 Application to tests of EEP

We now apply the $TH\epsilon\mu$ framework to some of the specific experimental tests of EEP described in Section 2.3. Most such experiments occur in Earth-bound or near-Earth orbiting laboratories, and they tend to single out the gravitational field of one body, the Earth or the Sun as the relevant field driving the phenomenon being studied. These fields are spherically symmetric to a high degree of accuracy, and because the timescales over which these field change are long compared to the internal timescales of the atoms that make up the apparatus being used, they can be treated as static.

Tests of WEP

Equation (2.84) gives the acceleration of a composite body through post-Coulombian order in an external SSS field. When specialized to composite bodies that are spherical on average (a good approximation for experimental situations), the resulting acceleration is given by

$$\frac{d^2 \boldsymbol{X}}{dt^2} = \boldsymbol{g} \left\{ 1 + \frac{E_{\rm B}^{\rm ES}}{M c_0^2} \left(2\Gamma_0 - \frac{8}{3} \Upsilon_0 \right) \right\} , \tag{2.120}$$

where $E_{\rm B}^{\rm ES}$ is given by Eq. (2.86). Then the Eötvös ratio $\eta^{\rm ES}$ defined in Eq. (2.6) is given by

$$\eta^{\rm ES} = 2\Gamma_0 - \frac{8}{3} \Upsilon_0 , \tag{2.121}$$

permitting us to translate experimental bounds summarized in Figure 2.2 into bounds on parameters of the $TH\epsilon\mu$ framework.

Haugan and Will (1977) extended the calculation of the acceleration to "post-post-Coulombian" order, finding violations of WEP dependent on "magnetostatic" energies proportional to $(e_a e_b / r_{ab}) v_a v_b$, with amplitudes dependent upon Λ_0 as well as on Γ_0 and Υ_0.

Tests of LLI

Here it is natural to use the c^2 formalism, and to let the velocity $\boldsymbol{V} = \boldsymbol{w}$, the velocity of the laboratory relative to the cosmic background radiation.

From the quantized electrodynamic sector of the action (2.117) the energy of a photon is $\hbar\omega$, while its momentum is $\hbar\omega/c$. Using this approach, and working in the presumed universal preferred frame, one finds that the difference in round trip travel times of light along the two arms of the interferometer in the Michelson-Morley experiment is given by $\Upsilon_0 L_0(w^2/c)$, where L_0 is the length of the arms Unlike the standard special relativistic case, the Lorentz-Fitzgerald contraction of the parallel arm ($\propto 1-w^2/2$) does *not* precisely cancel the time difference along the two paths because the latter depends on c, while the former does not.

The behavior of moving atomic clocks can be analyzed in detail, and bounds on Υ_0 can be placed using results from tests of time dilation and of the propagation of light. In some cases, it is advantageous to combine the c^2 framework with a "kinematical" viewpoint that treats a general class of boost transformations between moving frames. Such kinematical approaches have been discussed by Mansouri and Sexl (1977a,b,c), MacArthur (1986) and Will (1992a).

For example, in the "JPL" experiment (see Figure 2.3), in which the phases of two hydrogen masers connected by a fiberoptic link were compared as a function of the Earth's orientation, the predicted phase difference as a function of direction is, to first order in $\boldsymbol{w}$, the velocity of the Earth through the cosmic background,

$$\frac{\Delta\phi}{\phi} \approx \frac{4}{3}\Upsilon_0 \left(\boldsymbol{w}\cdot\boldsymbol{n} - \boldsymbol{w}\cdot\boldsymbol{n}_0\right), \tag{2.122}$$

where $\phi = 2\pi\nu L$, ν is the maser frequency, $L = 21$ km is the baseline, and $\boldsymbol{n}$ and $\boldsymbol{n}_0$ are unit vectors along the direction of propagation of the light at a given time and at the initial time of the experiment, respectively. The observed limit on a diurnal variation in the relative phase resulted in the bound $|\delta_0| = |\Upsilon_0| < 3\times10^{-4}$. Tighter bounds were obtained from a "two-photon absorption" (TPA) experiment, and a 1960s series of "Mössbauer-rotor" experiments, which tested the isotropy of time dilation between a gamma ray emitter on the rim of a rotating disk and an absorber placed at the center (Will, 1992a).

Similarly the anisotropy in energy levels is clearly illustrated by the tensorial terms in Eqs. (2.118). To illustrate this in a concrete setting, we will consider the Hughes-Drever experiment (Haugan, 1978). The anisotropic term in Eq. (2.118) could lead to energy shifts of states having different values of $\delta m_{\rm I}^{jk}$ and thus to observable effects in a quantum mechanical transition between these states. In the case of the Hughes-Drever experiment, the system, a ^{7}Li nucleus, can be approximated as a two-body system consisting of $J = 0$ core (two protons and four neutrons) of charge $+2$, and a valence proton in a ground state with angular momentum of 1. The spin of the proton couples to its angular momentum to yield a total angular momentum $J = 3/2$. In an applied magnetic field, the four magnetic substates $m_J = \pm1/2, \pm3/2$ are normally split equally in energy, giving a singlet emission line for transitions between the three pairs of states. How does the anistropic term alter the energies of these four states? The isotropic part of $\delta m_{\rm I}^{jk} \propto E_{\rm B}^{\rm ES}\delta^{jk}$ in Eq. (2.119) simply shifts all four levels equally, since $\langle JM_J|e_1e_2r^{-1}|JM_J\rangle$ is independent of M_J. However, the other contribution to $\delta m_{\rm I}^{jk} \propto E_{{\rm B}jk}^{\rm ES}$, does shift the levels unequally. We first decompose w^j into a component $w_\parallel$, parallel to the applied magnetic field and a component $w_\perp$, perpendicular to it. Then

$$\frac{x^j x^k}{r^3} w^j w^k = \frac{1}{r}\left(w_\parallel^2 \cos^2\theta + 2w_\perp w_\parallel \cos\theta \sin\theta \sin\phi + w_\perp^2 \sin^2\theta \sin^2\phi\right), \tag{2.123}$$

where θ and ϕ are polar coordinates appropriate to the orbital wave function $\phi_{1m} = f(r)Y_{1m}(\theta, \phi)$ in a basis where the magnetic field is in the z-direction. By combining the orbital wave function and spin states into eigenstates of total J, M_J, we then calculate the expectation value of $(1/2)\delta m_1^{jk} w^j w^k$ in states of different M_J. Taking the difference in the energy shifts between adjacent M_J, states, we find that the singlet line splits into a triplet with relative energies

$$\begin{aligned} h\nu(3/2 \to 1/2) &= \delta, \\ h\nu(1/2 \to -1/2) &= 0, \\ h\nu(-1/2 \to -3/2) &= -\delta, \end{aligned} \tag{2.124}$$

where

$$\delta = \frac{2}{15}\Upsilon_0 \frac{E_B^{ES}}{c^2}\left(w_\perp^2 - w_\parallel^2\right) \tag{2.125}$$

(recall that in the units of the c^2 formalism, c is a dimensionless number, presumably close to or equal to unity). In the notation of Eq. (2.14) we can write that

$$\delta^{ES} = \frac{2}{15}\Upsilon_0 \left(w_\perp^2 - w_\parallel^2\right). \tag{2.126}$$

By assuming that $w = 368\ \mathrm{km\ s^{-1}}$, we can infer the bound on Υ or δ_0 shown in Figure 2.3, for the Hughes-Drever experiments. For the later improved experiments using cold trapped atoms, analogous calculations must be done for each specific atomic species used in order infer the bounds shown.

Tests of LPI

Consider gravitational redshift experiments. Suppose, for example, that one measures the gravitational redshift of photons emitted from various transitions of hydrogen, such as principal transitions, fine-structure transitions within a principal level, a hyperfine transition such as the 21 cm line, the basis for hydrogen maser clocks, or of a photon from an electromagnetic cavity. Then, substituting Eqs. (2.113) and (2.116) into Eq. (2.47), we obtain (Will, 1974)

$$\begin{aligned} Z_p &= [1 - 2\Gamma_0]\,\Delta U/c_0^2, \\ Z_f &= [1 - 4\Gamma_0]\,\Delta U/c_0^2, \\ Z_{hf} &= [1 - (3\Gamma_0 - \Lambda_0)]\,\Delta U/c_0^2, \\ Z_{\text{cavity}} &= \left[1 - \frac{1}{2}(3\Gamma_0 + \Lambda_0)\right]\Delta U/c_0^2. \end{aligned} \tag{2.127}$$

Notice that the four redshifts are different in general. Thus the gravitational redshift depends on the nature of the clock whose frequency shift is being measured unless $\Gamma_0 = \Lambda_0 = 0$, that is, unless LPI is satisfied. From the hydrogen-maser Gravity Probe A experiment or the SAO-Stanford microwave cavity/H-maser comparison shown in Figure 2.4, it is straightforward to infer bounds on Γ_0 and Λ_0. In order to infer bounds on these parameters from the advanced null redshift experiments shown in Figure 2.4 that involve clocks based on Rubidium, Cesium, Dysprosium, or Aluminium, one must carry out more sophisticated atomic physics calculations in order to establish the analogues of Eqs. (2.127).

2.6 The Standard Model Extension

Despite their utility, the $TH\epsilon\mu$ and c^2 formalisms are restricted to electromagnetic interactions. What about the other nongravitational interactions—the strong and the weak interactions?

Kostelecký and collaborators developed an important framework for discussing violations of Local Lorentz Invariance in the context of the full Standard Model of particle physics (Colladay and Kostelecký, 1997, 1998; Kostelecký and Mewes, 2002). Called the Standard Model Extension (SME), it takes the standard $SU(3) \times SU(2) \times U(1)$ field theory of particle physics, and modifies the terms in the action by inserting a variety of tensorial quantities in the quark, lepton, Higgs, and gauge boson sectors that could explicitly violate LLI. SME extends the $TH\epsilon\mu$ and c^2 frameworks, and a related "$\chi - g$" framework of Ni (1977) to quantum field theory and particle physics. The modified terms split naturally into those that are odd under CPT (i.e., that violate CPT) and terms that are even under CPT. The result is a rich and complex framework, with hundreds of parameters to be analyzed and tested by experiment. Such details are beyond the scope of this chapter; for a review of SME and other frameworks, the reader is referred to the Living Review by Mattingly (2005) or the review by Liberati (2013). The review of the SME by Kostelecký and Russell (2011) provides "data tables" showing experimental bounds on all the many parameters of the SME.

Here we confine our attention to the QED sector, in order to illustrate some features of the SME and to link it with the c^2 framework discussed in Section 2.5.5. In a simplified version of the SME that preserves CPT, the action for an electron interacting with an electromagnetic field takes the form $I = \int \mathcal{L} d^4x$, with

$$\mathcal{L} = -\bar{\psi} m \psi + \bar{\psi} i \left[\eta_{\mu\nu} + c_{\mu\nu}\right] \gamma^\mu D^\nu \psi - \frac{1}{4}\left[\eta^{\mu\alpha}\eta^{\nu\beta} + k_{\rm F}^{\mu\nu\alpha\beta}\right] F_{\mu\nu} F_{\alpha\beta} , \tag{2.128}$$

where $D^\mu = \partial^\mu + ieA^\mu$, γ^μ are gamma matrices, $c_{\mu\nu}$ is a real tracefree tensor, and $k_{\rm F}^{\mu\nu\alpha\beta}$ is a tensor with the symmetries of the Riemann tensor, and with vanishing double trace. In this simple model there are $15 + 19 = 34$ free parameters in principle.

The tensor $k_{\rm F}^{\mu\nu\alpha\beta}$ can be decomposed into "electric," "magnetic," and "odd-parity" components, by defining

$$\begin{aligned}
\kappa_{\rm DE}^{jk} &\equiv -2k_{\rm F}^{0j0k} , \\
\kappa_{\rm HB}^{jk} &\equiv \frac{1}{2}\epsilon^{jpq}\epsilon^{krs}k_{\rm F}^{pqrs} , \\
\kappa_{\rm DB}^{kj} &\equiv -\kappa_{\rm HE}^{jk} \equiv \epsilon^{jpq}k_{\rm F}^{0kpq} .
\end{aligned} \tag{2.129}$$

In many applications it is useful to use the further decomposition

$$\begin{aligned}
\tilde\kappa_{\rm tr} &\equiv \frac{1}{3}\kappa_{\rm DE}^{jj}, \\
\tilde\kappa_{\rm e+}^{jk} &\equiv \frac{1}{2}(\kappa_{\rm DE}+\kappa_{\rm HB})^{jk}, \\
\tilde\kappa_{\rm e-}^{jk} &\equiv \frac{1}{2}(\kappa_{\rm DE}-\kappa_{\rm HB})^{jk} - \frac{1}{3}\delta^{jk}\kappa_{\rm DE}^{ii}, \\
\tilde\kappa_{\rm o+}^{jk} &\equiv \frac{1}{2}(\kappa_{\rm DB}+\kappa_{\rm HE})^{jk}, \\
\tilde\kappa_{\rm o-}^{jk} &\equiv \frac{1}{2}(\kappa_{\rm DB}-\kappa_{\rm HE})^{jk}.
\end{aligned} \tag{2.130}$$

The first expression is a single number, the next three are symmetric tracefree matrices, and the final is an antisymmetric matrix, accounting thereby for the 19 components of the original tensor $k_{\rm F}^{\mu\nu\alpha\beta}$.

In the rest frame of the universe, these tensors have some form that is established by the global nature of the solutions of the overarching theory being used, and could in principle be anisotropic. In a frame that is moving relative to the universe, the tensors will have components that depend on the velocity of the frame, and on the orientation of the frame relative to that velocity, thus leading to violations of LLI.

However, since the observable universe is isotropic to a high degree of accuracy (parts in 10^5), then it is not unreasonable to assume that, in whatever theory is responsible for the Lorentz violation, in the rest frame of the cosmic microwave background, the tensors $c_{\mu\nu}$, and $k_{\rm F}^{\mu\nu\alpha\beta}$ should be isotropic, so that $c_{jk} \propto \delta_{jk}$, $k_{\rm F}^{0j0k} \propto \delta^{jk}$, $k_{\rm F}^{0jkl} = 0$, and $k_{\rm F}^{ijkl} \propto \delta^{i[k}\delta^{l]j}$. Combining this with the tracefree constraint, it is simple to show that the tensors can be expressed in the general form, valid in any frame moving with four-velocity u^μ relative to the preferred frame:

$$\begin{aligned}
c_{\mu\nu} &= \tilde c\left(u_\mu u_\nu + \frac{1}{4}\eta_{\mu\nu}\right), \\
k_{\rm F}^{\mu\nu\alpha\beta} &= \tilde\kappa_{\rm tr}\left(4u^{[\mu}\eta^{\nu][\alpha}u^{\beta]} - \eta^{\mu[\alpha}\eta^{\beta]\nu}\right).
\end{aligned} \tag{2.131}$$

Working in the preferred frame, we write out the Lagrangian explicitly in terms of temporal and spatial derivatives and in terms of electric and magnetic fields, using $E_j \equiv F_{j0}$, $B_j \equiv \frac{1}{2}\epsilon_{jkl}F_{kl}$. We then rescale the coordinates by $t = \alpha\hat t$, $x^j = \beta x^{\hat\jmath}$, which induces the transformations $D^0 = \alpha D^{\hat 0}$, $D^j = \beta D^{\hat\jmath}$, $\boldsymbol{E} = (\alpha\beta)^{-1}\hat{\boldsymbol{E}}$ and $\boldsymbol{B} = \beta^{-2}\hat{\boldsymbol{B}}$. Finally, we rescale the electric charge and the vector potential by $e = e^*(1+\tilde\kappa_{\rm tr})^{1/2}/(\alpha\beta)$, and $A_{\hat\mu} = A^*_{\hat\mu}\alpha\beta/(1+\tilde\kappa_{\rm tr})^{1/2}$. Choosing $\alpha = (1-3\tilde c/4)^{-1}$ and $\beta = (1+\tilde c/4)^{-1}$, we put the Lagrangian (2.128) into the form (after dropping hats and asterisks),

$$\mathcal{L} = -\bar\psi m\psi + \bar\psi i\eta_{\mu\nu}\gamma^\mu D^\nu\psi - \frac{1}{2}\left(E^2 - c^2B^2\right), \tag{2.132}$$

which is the Lagrangian in the c^2 framework, with

$$c = \left(\frac{1-\tilde\kappa_{\rm tr}}{1+\tilde\kappa_{\rm tr}}\right)^{1/2}\left(\frac{1+\tilde c/4}{1-3\tilde c/4}\right). \tag{2.133}$$

To first order in the small parameters $\tilde{\kappa}_{tr}$ and $\tilde{c}$, we then have

$$\Upsilon_0 = -2(\tilde{\kappa}_{tr} - \tilde{c}) = -2\left(\tilde{\kappa}_{tr} - \frac{4}{3}c_{00}\right), \tag{2.134}$$

where c_{00} is the 00 component of $c_{\mu\nu}$ in the preferred frame. The latter expression corresponds to the parameter combination listed in table VIII of Kostelecký and Russell (2011).

2.7 EEP, Particle Physics, and the Search for New Interactions

Thus far, we have discussed EEP as a principle that strictly divides the world into metric and nonmetric theories, and have implied that a failure of EEP might invalidate metric theories (and thus general relativity). On the other hand, there is theoretical evidence to suggest that EEP *could* be violated at some level, whether by quantum gravity effects, by effects arising from string theory, or by hitherto undetected interactions. Roughly speaking, in addition to the pure Einsteinian gravitational interaction, which respects EEP, theories such as string theory predict other interactions which do not. In string theory, for example, such EEP-violating fields are known to exist. However, the theory is unable to provide a robust or unique prediction of their strength relative to gravity, or a determination of whether they are long range, like gravity, or short range, like the nuclear and weak interactions, and thus too short-range to be detectable, although numerous toy-model calculations exist. For examples of specific models, see Taylor and Veneziano (1988) and Damour and Polyakov (1994) Another class of nonmetric theories is the "varying speed of light (VSL)" set of theories; for a detailed review, see Magueijo (2003).

On the other hand, whether one views such effects as a violation of EEP or as effects arising from additional "matter" fields whose interactions, like those of the electromagnetic field, do not fully embody EEP, is to some degree a matter of semantics. Unlike the fields of the standard model of electromagnetic, weak and strong interactions, which couple to properties other than mass-energy and are either short range or are strongly screened, the fields inspired by string theory or other theories *could* be long range (if they remain massless by virtue of a symmetry, or at best, acquire a very small mass), and *can* couple to mass-energy, and thus can mimic gravitational fields. Still, there appears to be no way to make this precise.

As a result, EEP and related tests are now frequently viewed as ways to discover or place constraints on new physical interactions, or as a branch of "nonaccelerator particle physics," searching for the possible imprints of high-energy particle effects in the low-energy realm of gravity. Whether current or proposed experiments can actually probe these phenomena meaningfully is an open question at the moment, largely because of a dearth of firm theoretical predictions. For example, notwithstanding the impressive experimental bounds on the many parameters of the SME, no reasonable theory is known to have been killed by those constraints.

The "fifth" force

On the phenomenological side, the idea of using EEP tests in this way originated in the middle 1980s, with the search for a "fifth" force. In 1986, as a result of a detailed reanalysis of Eötvös' original data, Fischbach et al. (1986) suggested the existence of a fifth force of nature, with a strength of about a percent that of gravity, but with a range (as defined by the range λ of a Yukawa potential, $\propto r^{-1}e^{-r/\lambda}$) of a few hundred meters. This proposal dovetailed with earlier hints of a deviation from the inverse-square law of Newtonian gravitation derived from measurements of the gravity profile down deep mines in Australia, and with emerging ideas from particle physics, notably kaon decay, suggesting the possible presence of very low-mass particles with gravitational-strength couplings.

During the next four years numerous experiments looked for evidence of the fifth force by searching for composition-dependent differences in acceleration, with variants of the Eötvös experiment or with free-fall Galileo-type experiments. Although two early experiments reported positive evidence, the others all yielded null results. Over the range between one and 10^4 meters, the null experiments produced upper limits on the strength of a postulated fifth force between 10^{-3} and 10^{-6} of the strength of gravity. Interpreted as tests of WEP (corresponding to the limit of infinite-range forces), the results of two representative experiments from this period, the free-fall Galileo experiment and the early Eöt-Wash experiment, are shown in Figure 2.2. The associated gray region in that figure is meant to encapsulate the many experimental bounds reported during that period. At the same time, tests of the inverse-square law of gravity were carried out by comparing variations in gravity measurements up tall towers or down mines or boreholes with gravity variations predicted using the inverse square law together with Earth models and surface gravity data mathematically "continued" up the tower or down the hole. Despite early reports of anomalies, independent tower, borehole, and seawater measurements ultimately showed no evidence of a deviation. Analyses of orbital data from planetary range measurements, lunar laser ranging (LLR), and laser tracking of the LAGEOS satellite verified the inverse-square law to parts in 10^8 over scales of 10^3–10^5 km, and to parts in 10^9 over planetary scales of several astronomical units Talmadge et al. (1988). A consensus emerged that there was no credible experimental evidence for a fifth force of nature, of a type and range proposed by Fischbach *et al.* For reviews and bibliographies of this episode, see Fischbach et al. (1992), Fischbach and Talmadge (1992), Fischbach and Talmadge (1999), Adelberger et al. (1991), and Will (1990).

Short-range modifications of Newtonian gravity

Although the idea of an intermediate-range violation of Newton's gravitational law was dropped, new ideas emerged to suggest the possibility that the inverse-square law could be violated at very short ranges, below the centimeter range of existing laboratory verifications of the $1/r^2$ behavior. One set of ideas (Antoniadis et al., 1998; Arkani-Hamed et al., 1998; Randall and Sundrum, 1999a, 1999b) posited that some of the extra spatial dimensions that come with string theory could extend over macroscopic scales, rather than being rolled up at the Planck scale of 10^{-33} cm, which was then the conventional viewpoint. On laboratory distances large compared to the relevant scale of the extra dimension, gravity would fall off as the inverse square, whereas on short scales, gravity would fall off as $1/r^{2+n}$, where n is the number of large extra dimensions. Many models favored $n = 1$ or $n = 2$. Other possibilities for effective modifications of gravity at short range involved the exchange of light scalar particles.

Following these proposals, many of the high-precision, low-noise methods that were developed for tests of WEP were adapted to carry out laboratory tests of the inverse square law of Newtonian gravitation at millimeter scales and below. The challenge of these experiments was to distinguish gravitation-like interactions from electromagnetic and quantum mechanical (Casimir) effects. No deviations from the inverse square law have been found to date at distances between tens of nanometers and 10 mm (Long et al., 1999; Hoyle et al., 2001, 2004; Chiaverini et al., 2003; Long et al., 2003; Kapner et al., 2007; Adelberger et al., 2007; Tu et al., 2007; Geraci et al., 2008; Sushkov et al., 2011; Bezerra et al., 2011; Yang et al., 2012; Klimchitskaya et al., 2013). For a comprehensive review of both the theory and the experiments circa 2002, see Adelberger et al. (2003).

The Pioneer anomaly

In 1998, Anderson et al. reported the presence of an anomalous deceleration in the motion of the Pioneer 10 and 11 spacecraft at distances between 20 and 70 astronomical units from the Sun. Although the anomaly was the result of a rigorous analysis of Doppler data taken over many years, it might have been dismissed as having no real significance for new physics, where it not for the fact that the acceleration, of order 10^{-9} m/s^2, when divided by the speed of light, was strangely close to the inverse of the Hubble time. The Pioneer anomaly prompted an outpouring of hundreds of papers, most attempting to explain it via modifications of gravity or via physical interactions beyond the standard model of particle physics, with a small subset trying to explain it by conventional means.

Soon after the publication of the initial Pioneer anomaly paper, Katz (1999) pointed out that the anomaly could be accounted for as the result of the anisotropic emission of radiation from the radioactive thermal generators (RTG) that continued to power the spacecraft decades after their launch. At the time, there was insufficient data on the performance of the RTG over time or on the thermal characteristics of the spacecraft to justify more than an order-of-magnitude estimate. However, the recovery of an extended set of Doppler data covering a longer stretch of the orbits of both spacecraft, together with the fortuitous discovery of project documentation and of telemetry data giving on-board temperature information, made it possible both to improve the orbit analysis and to develop detailed thermal models of the spacecraft in order to quantify the effect of thermal emission anisotropies. Several independent analyses now confirm that the anomaly is almost entirely due to the recoil of the spacecraft from the anisotropic emission of residual thermal radiation (Rievers and Lämmerzahl, 2011; Turyshev et al., 2012; Modenini and Tortora, 2014) For a thorough review of the Pioneer anomaly published just as the new analyses were underway, see the *Living Review* by Turyshev and Toth (2010).

3 Gravitation as a Geometric Phenomenon

The overwhelming empirical evidence for the Einstein Equivalence Principle that we presented in Chapter 2 supports the conclusion that the only theories of gravity that have a hope of being viable are metric theories, or possibly theories that are metric apart from very weak or short-range non-metric couplings (as in string theory). Therefore, for the remainder of this book, we will turn our attention exclusively to metric theories of gravity. In Section 3.1, we review the concept of universal coupling, first defined in Section 2.2 and restate the postulates of metric theories of gravity. We then develop, in Section 3.2, the mathematical equations that describe the behavior of matter and nongravitational fields in curved spacetime. Every metric theory of gravity possesses these equations. Metric theories of gravity differ from each other in the number and type of additional gravitational fields they introduce and in the field equations that determine their structure and evolution; nevertheless, the only field that couples directly to matter is the metric itself. In Section 3.3, we discuss general features of metric theories of gravity, and present an additional principle, the Strong Equivalence Principle that is useful for classifying theories and for analyzing experiments.

3.1 Universal Coupling

The validity of the Einstein Equivalence Principle requires that every nongravitational field or particle should couple to the same symmetric, second rank tensor field of signature -2. In Section 2.2, we denoted this field $\boldsymbol{g}$, and saw that it was the central element in the postulates of metric theories of gravity: (i) there exists a metric $\boldsymbol{g}$, (ii) test bodies follow geodesics of $\boldsymbol{g}$, and (iii) in local Lorentz frames, the nongravitational laws of physics are those of special relativity.

The property that all nongravitational fields should couple in the same manner to a single gravitational field – universal coupling – allows one discuss the metric $\boldsymbol{g}$ as a property of spacetime itself rather than as a field over spacetime. This is because its properties may be measured and studied using a variety of different experimental devices, composed of different nongravitational fields and particles, and because of universal coupling, the results will be independent of the device. Consider, as a simple example, the proper time between two events as measured by two different clocks. To be specific, imagine a Hydrogen maser clock and an electromagnetic cavity clock at rest in a static spherically symmetric gravitational field. If each clock is governed by a Hamiltonian H, then the proper time (number of clock "ticks") between two events separated by coordinate time dt is given by

$$N = \nu dt = h^{-1}Edt\,, \tag{3.1}$$

where E is the eigenstate energy of the Hamiltonian (or energy difference, for a transition). The results of Section 2.5.4 show that if, for instance, the $TH\epsilon\mu$ formalism is applicable, and if EEP is satisfied, then $\epsilon_0 = \mu_0 = (H_0/T_0)^{1/2}$ everywhere, thus using Eqs. (2.111) and (2.115) we obtain for each clock

$$\begin{aligned} N_{\rm H} &\propto dt(H_0 T_0^{-1/2}\mu_0\epsilon_0^{-3}) = T_0^{1/2}dt\,, \\ N_{\rm cavity} &\propto dt(H_0 T_0^{-1/2}\mu_0^{-1/2}\epsilon_0^{-3/2}) = T_0^{1/2}dt\,, \end{aligned} \tag{3.2}$$

where the proportionality constants are fixed by calibrating each clock against a standard clock far from gravitating matter. Thus, each clock measures the same quantity T_0 (in metric theories of gravity, in SSS fields, $T_0 = -g_{00}$) and the proper time between two events is a characteristic of spacetime and of the location of the events, not of the clocks used to measure them.

Consequently, if EEP is valid, the nongravitational laws of physics may be formulated by taking their special relativistic forms in terms of the Minkowski metric $\boldsymbol{\eta}$ and simply "going over" to new forms in terms of the curved spacetime metric $\boldsymbol{g}$, using the mathematics of differential geometry. The details of this "going over" are the subject of the next section.

3.2 Nongravitational Physics in Curved Spacetime

3.2.1 Charged point-particle dynamics

In local Lorentz frames, the nongravitational laws of physics are those of special relativity. For charged point test particles coupled to electromagnetic fields, for example, these laws may be derived from the action

$$\begin{aligned} I_{\rm NG} = &-\sum_a m_{0a}\int\left(\eta_{\hat\mu\hat\nu}\frac{dx_a^{\hat\mu}}{d\hat t}\frac{dx_a^{\hat\nu}}{d\hat t}\right)^{1/2}d\hat t + \sum_a e_a\int A_{\hat\mu}(x_a^{\hat\nu})dx_a^{\hat\mu} \\ &- \frac{1}{16\pi}\int \eta^{\hat\mu\hat\alpha}\eta^{\hat\mu\hat\beta}F_{\hat\mu\hat\nu}F_{\hat\alpha\hat\beta}(-\hat\eta)^{1/2}d^4\hat x\,, \end{aligned} \tag{3.3}$$

where

$$\begin{aligned} F_{\hat\mu\hat\nu} &\equiv A_{\hat\nu,\hat\mu} - A_{\hat\mu,\hat\nu}\,, \\ \hat\eta &\equiv \det(\eta_{\hat\mu\hat\nu})\,. \end{aligned} \tag{3.4}$$

Here, $\eta_{\hat\mu\hat\nu}$ is the Minkowski metric, which in Cartesian coordinates has the form

$$\eta_{\hat\mu\hat\nu} = \mathrm{diag}(-1,\,1,\,1,\,1)\,. \tag{3.5}$$

In the local Lorentz frame, $\eta_{\hat\mu\hat\nu}$ is assumed to have this form only up to corrections of order $[x^{\hat\alpha} - x^{\hat\alpha}(\mathcal{P})]^2$, where $x^{\hat\alpha}(\mathcal{P})$ is the coordinate of a chosen fiducial event $\mathcal{P}$ in the local frame, in other words, $\eta_{\hat\mu\hat\nu}$ is described more precisely as

$$\eta_{\hat{\mu}\hat{\nu}}(\mathcal{P}) = \mathrm{diag}(-1,\, 1,\, 1,\, 1)\,,$$
$$\eta_{\hat{\mu}\hat{\nu},\hat{\alpha}}(\mathcal{P}) = 0\,. \tag{3.6}$$

According to the discussion in Section 3.1, the general form of these laws in any frame is obtained by a simple coordinate transformation from the freely falling frame to the chosen frame. This transformation is given by

$$x^{\hat{\mu}} = x^{\hat{\mu}}(x^{\alpha})\,. \tag{3.7}$$

Then the vectors and tensors that appear in I_{NG} transform according to

$$dx^{\hat{\mu}} = \frac{\partial x^{\hat{\mu}}}{\partial x^{\alpha}} dx^{\alpha}\,, \tag{3.8}$$
$$d^4\hat{x} = J(x^{\hat{\alpha}}, x^{\beta}) d^4x\,, \tag{3.9}$$
$$\eta_{\hat{\mu}\hat{\nu}} = \frac{\partial x^{\alpha}}{\partial x^{\hat{\mu}}} \frac{\partial x^{\beta}}{\partial x^{\hat{\nu}}} \eta_{\alpha\beta}\,, \tag{3.10}$$
$$A_{\hat{\mu}} = \frac{\partial x^{\alpha}}{\partial x^{\hat{\mu}}} A_{\alpha}\,, \tag{3.11}$$

where J is the Jacobian of the transformation. Partial derivatives of fields, as for example in the formula for $F_{\hat{\mu}\hat{\nu}}$, transform according to

$$A_{\hat{\mu},\hat{\nu}} = \frac{\partial x^{\beta}}{\partial x^{\hat{\nu}}} \frac{\partial}{\partial x^{\beta}} \left(\frac{\partial x^{\alpha}}{\partial x^{\hat{\mu}}} A_{\alpha} \right) = \frac{\partial x^{\alpha}}{\partial x^{\hat{\mu}}} \frac{\partial x^{\beta}}{\partial x^{\hat{\nu}}} A_{\alpha,\beta} + \frac{\partial^2 x^{\alpha}}{\partial x^{\hat{\mu}} x^{\hat{\nu}}} A_{\alpha}\,. \tag{3.12}$$

However, in the local frame, $\eta_{\hat{\mu}\hat{\nu},\hat{\omega}} = 0$. Thus,

$$0 = \eta_{\hat{\mu}\hat{\nu},\hat{\omega}} = \frac{\partial x^{\alpha}}{\partial x^{\hat{\mu}}} \frac{\partial x^{\beta}}{\partial x^{\hat{\nu}}} \frac{\partial x^{\gamma}}{\partial x^{\hat{\omega}}} \eta_{\alpha\beta,\gamma} + \frac{\partial^2 x^{\alpha}}{\partial x^{\hat{\mu}} x^{\hat{\omega}}} \frac{\partial x^{\beta}}{\partial x^{\hat{\nu}}} \eta_{\alpha\beta} + \frac{\partial^2 x^{\beta}}{\partial x^{\hat{\nu}} x^{\hat{\omega}}} \frac{\partial x^{\alpha}}{\partial x^{\hat{\mu}}} \eta_{\alpha\beta}\,. \tag{3.13}$$

Using the fact that

$$\frac{\partial x^{\alpha}}{\partial x^{\hat{\nu}}} \frac{\partial x^{\hat{\mu}}}{\partial x^{\alpha}} = \delta^{\mu}_{\nu}\,, \tag{3.14}$$

we obtain

$$\eta_{\alpha\beta,\gamma} = -\frac{\partial x^{\hat{\mu}}}{\partial x^{\alpha}} \frac{\partial x^{\hat{\omega}}}{\partial x^{\gamma}} \frac{\partial^2 x^{\delta}}{\partial x^{\hat{\mu}} x^{\hat{\omega}}} \eta_{\delta\beta} - \frac{\partial x^{\hat{\nu}}}{\partial x^{\beta}} \frac{\partial x^{\hat{\omega}}}{\partial x^{\gamma}} \frac{\partial^2 x^{\delta}}{\partial x^{\hat{\nu}} x^{\hat{\omega}}} \eta_{\alpha\delta}\,. \tag{3.15}$$

If we now define η in the new coordinates to be the metric for a nonlocal Lorentz frame,

$$g_{\alpha\beta} \equiv \eta_{\alpha\beta}\,, \tag{3.16}$$
$$g^{\alpha\beta} \equiv ||g_{\alpha\beta}||^{-1} = \frac{\partial x^{\alpha}}{\partial x^{\hat{\mu}}} \frac{\partial x^{\beta}}{\partial x^{\hat{\nu}}} \eta^{\hat{\mu}\hat{\nu}}\,, \tag{3.17}$$
$$\Gamma^{\alpha}_{\beta\gamma} \equiv -\frac{\partial x^{\hat{\mu}}}{\partial x^{\beta}} \frac{\partial x^{\hat{\nu}}}{\partial x^{\gamma}} \frac{\partial^2 x^{\alpha}}{\partial x^{\hat{\mu}} x^{\hat{\nu}}}\,, \tag{3.18}$$

then Eq. (3.15) can be written

$$g_{\alpha\beta,\gamma} = \Gamma^{\delta}_{\alpha\gamma} g_{\delta\beta} + \Gamma^{\delta}_{\beta\gamma} g_{\alpha\delta}\,. \tag{3.19}$$

Permuting the α, β and γ indices to generate two additional versions of this equation, and making use of the symmetry of $\Gamma^{\alpha}_{\beta\gamma}$ on the lower indices, which follows from

Eq. (3.18), and combining the three equations suitably, we obtain the standard formula for the Christoffel symbols $\Gamma^{\alpha}_{\beta\gamma}$ (also known as connection coefficients)

$$\Gamma^{\alpha}_{\beta\gamma} = \frac{1}{2} g^{\alpha\delta} \left(g_{\delta\beta,\gamma} + g_{\delta\gamma,\beta} - g_{\gamma\beta,\delta} \right) . \tag{3.20}$$

Then Eq. (3.12) becomes

$$A_{\hat{\mu},\hat{\nu}} = \frac{\partial x^{\alpha}}{\partial x^{\hat{\mu}}} \frac{\partial x^{\beta}}{\partial x^{\hat{\nu}}} \left(A_{\alpha,\beta} - \Gamma^{\gamma}_{\alpha\beta} A_{\gamma} \right) . \tag{3.21}$$

We define the covariant derivative ";" by

$$A_{\alpha;\beta} \equiv A_{\alpha,\beta} - \Gamma^{\gamma}_{\alpha\beta} A_{\gamma} , \tag{3.22}$$

and notice that it transforms as a tensor; it can be shown that

$$A^{\alpha}{}_{;\beta} = A^{\alpha}{}_{,\beta} + \Gamma^{\alpha}_{\gamma\beta} A^{\gamma} , \tag{3.23}$$

where $A^{\alpha} \equiv g^{\alpha\beta} A_{\beta}$. Taking the determinant of Eq. (3.10) yields

$$\eta = [\det(\partial x^{\alpha}/\partial x^{\hat{\mu}})]^2 g , \tag{3.24}$$

where $g \equiv \det(g_{\mu\nu})$. Recalling that $\det(\partial x^{\alpha}/\partial x^{\hat{\mu}}) = J(x^{\hat{\mu}}, x^{\alpha})^{-1}$, we obtain

$$J(x^{\hat{\mu}}, x^{\alpha}) = \frac{(-g)^{1/2}}{(-\eta)^{1/2}} . \tag{3.25}$$

Substituting these results into $I_{\rm NG}$ gives

$$\begin{aligned} I_{\rm NG} = &-\sum_a m_{0a} \int \left(g_{\mu\nu} \frac{dx^{\mu}_a}{dt} \frac{dx^{\nu}_a}{dt} \right)^{1/2} dt + \sum_a e_a \int A_{\mu}(x^{\nu}_a) dx^{\mu}_a \\ &- \frac{1}{16\pi} \int g^{\mu\alpha} g^{\mu\beta} F_{\mu\nu} F_{\alpha\beta} (-g)^{1/2} d^4x , \end{aligned} \tag{3.26}$$

where

$$F_{\mu\nu} \equiv A_{\nu;\mu} - A_{\mu;\nu} . \tag{3.27}$$

We notice that the transformation from a local inertial frame to an arbitrary frame has resulted simply in the replacements

$$\begin{aligned} \eta_{\mu\nu} \quad &\text{by} \quad g_{\mu\nu} \\ \text{“comma”} \quad &\text{by} \quad \text{“semicolon”} \\ (-\eta)^{1/2} d^4x \quad &\text{by} \quad (-g)^{1/2} d^4x . \end{aligned} \tag{3.28}$$

This is the mathematical manifestation of EEP. We must point out that the specific mathematical forms given for the Christoffel symbols, transformation laws, and so on are valid only in coordinate bases (see MTW, chapter 10, for further discussion). However, in this book we will work exclusively in coordinate bases.

Generally speaking, then, the procedure for implementing EEP is: put the local special relativistic laws into a frame-invariant form using Lorentz-invariant scalars, vectors, tensors, and so on, then make the above replacements. It is simple to show that the same rules apply to the field equations and equations of motion derived from the Lagrangian. In the local frame they are

$$m_{0a}\frac{d(u_{\hat\mu})_a}{d\tau} = eF_{\hat\mu\hat\nu}(u^{\hat\nu})_a\,,$$
$$F^{\hat\mu\hat\nu}{}_{,\hat\nu} = 4\pi J^{\hat\mu}\,, \tag{3.29}$$

where $d\tau = (-\eta_{\hat\mu\hat\nu}dx^{\hat\mu}dx^{\hat\nu})^{1/2}$, $u^{\hat\mu} = dx^{\hat\mu}/d\tau$, and the current is given by

$$J^{\hat\mu} = \sum_a e_a\delta^3(\hat{\boldsymbol{x}} - \hat{\boldsymbol{x}}_a)\frac{dx^{\hat\mu}}{d\hat t}\,. \tag{3.30}$$

However, these are not in a manifestly frame-invariant form. We must instead write them in the equivalent forms $du_{\hat\mu}/d\tau = u^{\hat\nu}u_{\hat\mu,\hat\nu}$ and

$$J^{\hat\mu} = \sum_a e_a\int(-\hat\eta)^{-1/2}\delta^4(x^{\hat\alpha} - x^{\hat\alpha}_a)\frac{dx^{\hat\mu}}{d\tau}d\tau\,. \tag{3.31}$$

where we have used the fact that, for the four-dimensional delta function, $(-\hat\eta)^{-1/2}\delta^4$ is invariant (since $\int\delta^4 d^4x = 1$ or 0 regardless of the frame). Then in the general frame the equations are

$$m_{0a}\frac{D(u_\mu)_a}{D\tau} = e_aF_{\mu\nu}(u^\nu)_a\,,$$
$$F^{\mu\nu}{}_{;\nu} = 4\pi J^\mu\,, \tag{3.32}$$

where $d\tau = (-g_{\mu\nu}dx^\mu dx^\nu)^{1/2}$, $u^\mu = dx^\mu/d\tau$,

$$\frac{D(u_\mu)_a}{D\tau} \equiv u^\nu u_{\mu;\nu}\,, \tag{3.33}$$

and the current is given by

$$J^\mu = \sum_a e_a(-g)^{-1/2}\delta^3(\boldsymbol{x} - \boldsymbol{x}_a)\frac{dx^\mu}{dt}\,. \tag{3.34}$$

However, there is a potential ambiguity in the application of EEP to electrodynamics if one writes Maxwell's equations (3.32), in terms of the vector potential A_μ. In the local Lorentz frame, Maxwell's equations have the special relativistic form

$$A^{\hat\mu,\hat\nu}{}_{,\hat\nu} - A^{\hat\nu,\hat\mu}{}_{,\hat\nu} = -4\pi J^{\hat\mu}\,. \tag{3.35}$$

It is always possible to choose a gauge (Lorenz gauge) in which $A^{\hat\nu}_{,\hat\nu} = 0$, thus, since $A^{\hat\nu,\hat\mu}{}_{,\hat\nu} = A^{\hat\nu}{}_{,\hat\nu}{}^{,\hat\mu}$, we have

$$\Box_{\hat\eta}A^{\hat\mu} \equiv A^{\hat\mu,\hat\nu}{}_{,\hat\nu} = -4\pi J^{\hat\mu}\,. \tag{3.36}$$

It is tempting then to apply the rules of EEP to this equation to obtain

$$\Box_g A^\mu \equiv A^{\mu;\nu}{}_{;\nu} \equiv g^{\nu\lambda}A^\mu{}_{;\nu\lambda} = -4\pi J^\mu\,,$$
$$A^\nu{}_{;\nu} = 0\,. \tag{3.37}$$

However there is another alternative. The curved-spacetime Maxwell equation, Eq. (3.32), yields

$$A^{\mu;\nu}{}_{;\nu} - A^{\nu;\mu}{}_{;\nu} = -4\pi J^{\mu} \,. \tag{3.38}$$

But covariant derivatives of vectors and tensors do not commute in curved spacetime, in fact in general

$$A^{\mu}{}_{;\beta\alpha} = A^{\mu}{}_{;\alpha\beta} + R^{\mu}_{\nu\alpha\beta}A^{\nu} \,, \tag{3.39}$$

where $R^{\mu}_{\nu\alpha\beta}$ is the Riemann curvature tensor, given by

$$R^{\mu}_{\nu\alpha\beta} = \Gamma^{\mu}_{\nu\beta,\alpha} - \Gamma^{\mu}_{\nu\alpha,\beta} + \Gamma^{\mu}_{\gamma\alpha}\Gamma^{\gamma}_{\nu\beta} - \Gamma^{\mu}_{\gamma\beta}\Gamma^{\gamma}_{\nu\alpha} \,. \tag{3.40}$$

Then

$$A^{\nu;\mu}{}_{;\nu} = A^{\nu}{}_{;\nu}{}^{;\mu} + R^{\mu}_{\beta}A^{\beta} \,, \tag{3.41}$$

where R^{μ}_{β} is the Ricci tensor, given by

$$R^{\mu}_{\beta} \equiv g^{\mu\nu}R_{\nu\beta} \,, \quad R_{\nu\beta} \equiv R^{\alpha}_{\nu\alpha\beta} \,. \tag{3.42}$$

This version of Maxwell's equations in Lorenz gauge becomes

$$\Box_g A^{\mu} - R^{\mu}_{\beta}A^{\beta} = -4\pi J^{\mu} \,,$$
$$A^{\nu}{}_{;\nu} = 0 \,. \tag{3.43}$$

It is generally agreed that this second version is correct (although there is no experimental evidence one way or the other). To resolve such ambiguities, the following rule of thumb should be applied: the simple replacements $\boldsymbol{\eta} \to \boldsymbol{g}$ and comma $\to$ semicolon should be used in equations formulated using physically measurable quantities in the local Lorentz frame ($F^{\mu\nu}$ is physically measurable while A^{μ} is not). For a fuller discussion, see MTW, box 16.1.

3.2.2 Geodesics

An uncharged test body follows a trajectory given by Eq. (3.32) with $e_a = 0$, namely, $Du^{\mu}_a/D\tau = 0$. This equation can easily be written in the form

$$\frac{d^2x^{\mu}}{d\tau^2} + \Gamma^{\mu}_{\alpha\beta}\frac{dx^{\alpha}}{d\tau}\frac{dx^{\beta}}{d\tau} = 0 \,. \tag{3.44}$$

This is the geodesic equation. It is also the curve that extremizes the invariant action

$$I_{\mathrm{NG}} = -\int \left(g_{\mu\nu}\frac{dx^{\mu}}{dt}\frac{dx^{\nu}}{dt}\right)^{1/2} dt = -\int d\tau \tag{3.45}$$

between two fixed events in spacetime.

The mathematics of measurements made by atomic clocks and rigid measuring rods follow the same rules since the structure of such measuring devices is governed by solutions of the nongravitational laws of physics. In special relativity, the proper time between two events separated by an infinitesimal coordinate displacement dx^α as measured by any atomic clock moving on a trajectory that connects the events, is given by

$$d\tau = (-\eta_{\mu\nu}dx^\mu dx^\nu)^{1/2}\,, \tag{3.46}$$

if the separation is timelike, that is if $\eta_{\mu\nu}dx^\mu dx^\nu < 0$. The proper distance between two events as measured by a rigid rod joining them is given by

$$d\ell = (\eta_{\mu\nu}dx^\mu dx^\nu)^{1/2}\,, \tag{3.47}$$

if the separation is spacelike, that is, if $\eta_{\mu\nu}dx^\mu dx^\nu > 0$. These results are independent of the coordinates used. Then in curved spacetime we have

$$\begin{aligned} d\tau &= (-g_{\mu\nu}dx^\mu dx^\nu)^{1/2} \quad [\text{timelike}] \Longleftrightarrow g_{\mu\nu}dx^\mu dx^\nu < 0\,, \\ d\ell &= (g_{\mu\nu}dx^\mu dx^\nu)^{1/2} \quad [\text{spacelike}] \Longleftrightarrow g_{\mu\nu}dx^\mu dx^\nu > 0\,. \end{aligned} \tag{3.48}$$

There is a third class of separation dx^α between events, those for which, in a local Lorentz frame,

$$\eta_{\mu\nu}dx^\mu dx^\nu = 0\,. \tag{3.49}$$

These are called null or lightlike separations, and pairs of events that satisfy this condition are connectible by light rays. It is a tenet of special relativity that light rays move along straight, null trajectories, that is, if $k^\mu = dx^\mu/d\sigma$ is a tangent vector to a light-ray trajectory, then

$$dk^\mu/d\sigma = 0\,, \quad \eta_{\mu\nu}k^\mu k^\nu = 0\,. \tag{3.50}$$

where σ is a parameter labeling points along the trajectory. It should not be forgotten, however, that this is not a postulate but rather is a consequence of Maxwell's equations, valid only in the "geometrical optics" limit, in which the characteristic wavelength $\lambda \sim (k^0)^{-1}$ is small compared to the scale $\mathcal{L}$ over which the amplitude of the wave changes. (For example, $\mathcal{L}$ might be the radius of curvature of a spherical wavefront.)

Since the first of Eqs. (3.50) can be written, in flat spacetime

$$dk^\mu/d\sigma = (dx^\nu/d\sigma)k^\mu{}_{,\nu} = k^\nu k^\mu{}_{,\nu} = 0\,, \tag{3.51}$$

then EEP yields the equations

$$k^\nu k^\mu{}_{;\nu} = 0\,, \quad g_{\mu\nu}k^\mu k^\nu = 0\,, \tag{3.52}$$

that is, the trajectories of light rays in the geometrical optics limit are null geodesics.

3.2.3 Actions and energy-momentum tensors

Another useful and important form of the equations of motion for matter and nongravitational fields can be derived in the case where the equations are obtained from a covariant action principle. This will essentially always be the case, for the following reason: in special relativity, all modern viable theories of nongravitational fields and their interactions take an action principle as their starting point, leading to an action $I_{\rm NG}$. The use of EEP does

Box 3.1 **Maxwell's equations and null geodesics**

It is useful to derive the equation for null geodesics directly from the curved-spacetime form of Maxwell's equations, in order to illustrate the role and the limits of validity of the geometrical-optics assumption. In curved spacetime, the geometrical-optics limit requires that λ be small compared both to $\mathcal{L}$ and to $\mathcal{R}$, the scale over which the background geometry changes ($\mathcal{R}$ is related to the Riemann curvature tensor), that is,

$$\frac{\lambda}{\min(\mathcal{L},\mathcal{R})} \equiv \frac{\lambda}{L} \ll 1. \tag{3.53}$$

In this limit, the electromagnetic vector potential can be written in terms of a rapidly varying real phase and a slowly varying complex amplitude in the form (see MTW, section 22.5, for details)

$$A_\mu = (a_\mu + \epsilon b_\mu + \dots)e^{i\theta/\epsilon}, \tag{3.54}$$

where θ is a real phase, a_μ, b_μ, ... are complex amplitudes, and ϵ is a formal expansion parameter that keeps track of the powers of λ/L. Ultimately, one takes only the real part of A_μ in any physical calculations. We define the wave vector

$$k_\mu \equiv \theta_{,\mu}, \quad k^\mu = g^{\mu\nu}\theta_{,\nu}. \tag{3.55}$$

Then Maxwell's equations in Lorenz gauge, Eqs. (3.43), yield

$$\begin{aligned} 0 = A^\mu{}_{;\mu} &= \left[\frac{1}{\epsilon}k_\mu(a^\mu + \epsilon b^\mu) + a^\mu{}_{;\mu} + O(\epsilon)\right]e^{i\theta/\epsilon}, \\ 0 = \Box_g A^\mu - R^\mu_\beta A^\beta & \\ &= \left[\frac{1}{\epsilon^2}k_\beta k^\beta(a^\mu + \epsilon b^\mu) + \frac{i}{\epsilon}\left(2k^\beta a^\mu{}_{;\beta} + k^\beta{}_{;\beta}a^\mu\right) + O(\epsilon^0)\right]e^{i\theta/\epsilon}. \end{aligned} \tag{3.56}$$

Setting the coefficients of each power of ϵ equal to zero, we obtain for the leading terms in each equation

$$k^\mu a_\mu = 0, \tag{3.57}$$

$$k^\mu k_\mu = 0, \tag{3.58}$$

in other words, the amplitude is orthogonal to the wave vector, and the wave vector is null. Taking the gradient of Eq. (3.58) and noting that $k_{\mu;\nu} = k_{\nu;\mu}$, since k_μ is itself a gradient, we get

$$k^\nu k^\mu{}_{;\nu} = 0 \tag{3.59}$$

which is the geodesic equation for k^μ. The trajectory $x^\mu(\sigma)$ of the ray can then be shown to be related to k^μ by the differential equation $dx^\mu(\sigma)/d\sigma = k^\mu(x^\nu)$, where σ is an affine parameter along the ray. For discussion of the higher-order terms in Eqs. (3.56), see MTW, section 22.5.

not alter the fact that the equations of motion are derivable from an action. Consequently, one is led in curved spacetime to an action of the general form

$$\begin{aligned} I_{\rm NG} &= \int \mathcal{L}_{\rm NG} d^4x, \\ \mathcal{L}_{\rm NG} &= \mathcal{L}_{\rm NG}(q_A, q_{A,\mu}, g_{\mu\nu}, g_{\mu\nu,\beta}), \end{aligned} \tag{3.60}$$

where q_A and $q_{A,\mu}$ are the nongravitational fields under consideration and their first partial derivatives (e.g., u^μ, A_μ, $A_{\mu,\nu}$, ...) and $g_{\mu\nu}$ and $g_{\mu\nu,\beta}$ are the metric and its first derivative. (The extension to second and higher derivatives is straightforward). The action principle $\delta I_{\rm NG} = 0$ is covariant, thus, under a coordinate transformation, $\mathcal{L}_{\rm NG}$ must be unchanged in functional form, modulo a divergence (see Trautman (1962) for discussion). Consider the infinitesimal coordinate transformation

$$x^\mu \to x^\mu + \xi^\mu \,. \tag{3.61}$$

Then the metric changes according to [see, for example, Eq. (3.12)],

$$\delta g_{\mu\nu} = -g_{\mu\alpha}\xi^\alpha{}_{,\nu} - g_{\nu\alpha}\xi^\alpha{}_{,\mu} - g_{\mu\nu,\alpha}\xi^\alpha \,. \tag{3.62}$$

Assume the matter and nongravitational field variables change according to

$$\delta q_A = d^\mu_{A\nu}\xi^\nu{}_{,\mu} - q_{A,\nu}\xi^\nu \,, \tag{3.63}$$

where $d^\mu_{A\nu}$ are functions of x^α and encode the tensorial transformation properties of q_A; the second term accounts for the spacetime variation of q_A. Under this transformation, $\mathcal{L}_{\rm NG}$ changes by

$$\sum_A \left(\frac{\partial \mathcal{L}_{\rm NG}}{\partial q_A}\delta q_A + \frac{\partial \mathcal{L}_{\rm NG}}{\partial q_{A,\mu}}\delta q_{A,\mu} \right) + \frac{\partial \mathcal{L}_{\rm NG}}{\partial g_{\mu\nu}}\delta g_{\mu\nu} + \frac{\partial \mathcal{L}_{\rm NG}}{\partial g_{\mu\nu,\beta}}\delta g_{\mu\nu,\beta} \,. \tag{3.64}$$

Substituting Eqs. (3.62) and (3.63), integrating by parts, dropping divergence terms, and demanding that $\mathcal{L}_{\rm NG}$ be unchanged for arbitrary functions ξ^α yields the "Bianchi identities"

$$\sum_A \left[q_{A,\alpha}\frac{\delta \mathcal{L}_{\rm NG}}{\delta q_A} + \left(d^\mu_{A\alpha}\frac{\delta \mathcal{L}_{\rm NG}}{\delta q_A} \right)_{,\mu} \right] + \frac{1}{2} g_{\mu\nu,\alpha}(-g)^{1/2}T^{\mu\nu} - \left[g_{\mu\alpha}(-g)^{1/2}T^{\mu\nu} \right]_{,\nu} = 0 \,, \tag{3.65}$$

where $\delta \mathcal{L}_{\rm NG}/\delta q_A$ is the "variational" derivative of $\mathcal{L}_{\rm NG}$ defined, for any variable ψ, by

$$\frac{\delta \mathcal{L}_{\rm NG}}{\delta \psi} \equiv \frac{\partial \mathcal{L}_{\rm NG}}{\partial \psi} - \frac{\partial}{\partial x^\mu}\left(\frac{\partial \mathcal{L}_{\rm NG}}{\partial \psi_{,\mu}} \right) , \tag{3.66}$$

and $T^{\mu\nu}$ is the "energy-momentum tensor," defined by

$$T^{\mu\nu} \equiv 2(-g)^{-1/2}\frac{\delta \mathcal{L}_{\rm NG}}{\delta g_{\mu\nu}} \,. \tag{3.67}$$

Using the fact that

$$\frac{\partial}{\partial x^\mu}(-g)^{1/2} = (-g)^{1/2}\Gamma^\alpha_{\mu\alpha} \,, \tag{3.68}$$

we can rewrite Eq. (3.65) in the form

$$T^\nu_{\alpha;\nu} = (-g)^{-1/2}\sum_A \left[q_{A,\alpha}\frac{\delta \mathcal{L}_{\rm NG}}{\delta q_A} + \left(d^\mu_{A\alpha}\frac{\delta \mathcal{L}_{\rm NG}}{\delta q_A} \right)_{,\mu} \right] . \tag{3.69}$$

However, the nongravitational field equations and equations of motion are obtained by setting the variational derivative of $\mathcal{L}_{\rm NG}$ with respect to each field variable q_A equal to zero, that is,

$$\delta \mathcal{L}_{\rm NG}/\delta q_A = 0 \,, \tag{3.70}$$

which by Eq. (3.69) is equivalent to

$$T^{\nu}_{\alpha;\nu} = 0 \,. \tag{3.71}$$

Thus, the vanishing of the divergence of the energy-momentum tensor $T^{\mu\nu}$ is a consequence of the nongravitational equations of motion. This result could also have been derived, first working with $\mathcal{L}_{\rm NG}$ in a local Lorentz frame to obtain the equation $T^{\hat\nu}_{\hat\alpha,\hat\nu} = 0$ by the above method, then using EEP to obtain Eq. (3.71). Notice that Equation (3.65) is a consequence purely of universal coupling (EEP) and of the invariance of the nongravitational action, and is valid independent of the field equations for the gravitational fields.

The energy-momentum tensor $T^{\mu\nu}$ for charged particles and electromagnetic fields may be obtained from the action $I_{\rm NG}$, Eq. (3.26), by first rewriting it as an integral over spacetime in the form

$$\begin{aligned} I_{\rm NG} = &-\sum_a m_{0a} \int \left(g_{\mu\nu} \frac{dx^\mu}{d\tau}\frac{dx^\nu}{d\tau} \right)^{1/2} \delta^4[x^\alpha - x^\alpha_a(\tau)] d\tau d^4x \\ &+ \sum_a e_a \int A_\mu(x^\nu) \frac{dx^\mu}{d\tau} \delta^4[x^\alpha - x^\alpha_a(\tau)] d\tau d^4x \\ &- \frac{1}{16\pi} \int g^{\mu\alpha} g^{\mu\beta} F_{\mu\nu} F_{\alpha\beta} (-g)^{1/2} d^4x \,, \end{aligned} \tag{3.72}$$

Since only $g_{\mu\nu}$ (and not its derivatives) appears in $I_{\rm NG}$, we obtain

$$\begin{aligned} T^{\mu\nu} &= 2(-g)^{-1/2} \frac{\partial \mathcal{L}_{\rm NG}}{\partial g_{\mu\nu}} \\ &= \sum_a m_{0a} \frac{u^\mu u^\nu}{u^0 (-g)^{1/2}} \delta^3[\mathbf{x} - \mathbf{x}_a(\tau)] + \frac{1}{4\pi} \left(F^{\mu\alpha} F^{\nu}_{\ \alpha} - \frac{1}{4} g^{\mu\nu} F^{\alpha\beta} F_{\alpha\beta} \right) , \end{aligned} \tag{3.73}$$

where we have used the fact that

$$\frac{\partial (-g)^{1/2}}{\partial g_{\mu\nu}} = \frac{1}{2} (-g)^{1/2} g^{\mu\nu} \,. \tag{3.74}$$

3.2.4 Hydrodynamics

So far, our model for nongravitational physics has been that of point charged particles and their associated electromagnetic fields (we have ignored the weak and strong interactions of the standard model of particle physics). For some applications, this is a useful model, but for most of the applications of this book it is too microscopic. Instead, we will use the perfect fluid as our model for matter. This model is a "coarse graining" of the properties of matter over scales that are large compared to atomic scales, but small compared to the scales over which the bulk properties of the fluid vary. The matter is assumed to be electrically neutral over these scales, so that the direct effects of electromagnetic fields can be ignored. Thus, one can speak of density, pressure, and velocity of fluid elements at a point within the fluid. A perfect fluid is one that has negligible viscosity, heat transport, and shear stresses. It is then possible to show that the energy-momentum tensor for the fluid has the following property: in a local Lorentz frame that is momentarily comoving with a chosen element of the fluid, the energy-momentum tensor for that element has the form

$$T^{\hat{\mu}\hat{\nu}} = \mathrm{diag}[\rho(1+\Pi), p, p, p] , \tag{3.75}$$

where ρ is the rest-mass-energy density of atoms in the fluid element, $\rho\Pi$ is the density of internal kinetic and thermal energy in the fluid element (Π then is the specific energy density), and p is the isotropic pressure, all measured in the locally comoving Lorentz frame. This can also be written in the Lorentz invariant form

$$T^{\hat{\mu}\hat{\nu}} = \rho(1+\Pi)u^{\hat{\mu}}u^{\hat{\nu}} + p(u^{\hat{\mu}}u^{\hat{\nu}} + \eta^{\hat{\mu}\hat{\nu}}) , \tag{3.76}$$

where $u^{\hat{\mu}} = dx^{\hat{\mu}}/d\tau$ is the four-velocity of the fluid element ($= \delta^{\hat{\mu}}_0$ in the comoving frame). Then in curved spacetime, $T^{\mu\nu}$ has the form

$$T^{\mu\nu} = (\rho + \rho\Pi + p)u^{\mu}u^{\nu} + pg^{\mu\nu} , \tag{3.77}$$

and satisfies the equation of energy-momentum conservation

$$T^{\mu\nu}{}_{;\nu} = 0 \tag{3.78}$$

if the nongravitational equations of motion hold. This can also be derived from Eq. (3.73) using suitable techniques in relativistic kinetic theory (see Ehlers 1971).

In our microscopic approach, we treated each particle has having a constant mass m_{0a}. In coarse-graining to obtain a fluid approximation, we assume that the fluid has a chemical composition that is static and homogeneous, and that baryon number is conserved, so that we can treat the rest mass-energy of a fluid element as directly proportional to its total baryon number. In our fluid, therefore, the number of baryons within a given comoving fluid element of volume $\mathcal{V}$ is constant by definition (we obviously are ignoring the effects of a possible nonconservation of baryons, which are utterly negligible on the timescales of interest). If ρ is the total mass energy in the fluid element divided by its volume, then we have the conservation statement

$$d(\rho\mathcal{V})/dt = 0 \tag{3.79}$$

valid in the comoving frame. But because $\mathcal{V}^{-1}d\mathcal{V}/dt = \boldsymbol{\nabla}\cdot\boldsymbol{v}$ where $\boldsymbol{v}$ is the velocity field within the fluid element ($\boldsymbol{v} = 0$ only at a single fiducial point within the comoving element), this conservation statement is equivalent to

$$\frac{d\rho}{dt} + \rho\boldsymbol{\nabla}\cdot\boldsymbol{v} = 0 \Longrightarrow \frac{\partial\rho}{\partial t} + \boldsymbol{\nabla}\cdot(\rho\boldsymbol{v}) = 0 . \tag{3.80}$$

This is the continuity equation for rest mass-energy, expressed in the locally comoving Lorentz frame. The Lorentz invariant version of this equation, valid in any local Lorentz frame, is

$$\frac{\partial}{\partial t}(\rho u^0) + \frac{\partial}{\partial x^j}(\rho u^j) = (\rho u^{\mu})_{,\mu} = 0 . \tag{3.81}$$

In curved spacetime, this then has the form

$$(\rho u^\mu)_{;\mu} = 0 \,. \tag{3.82}$$

Equations (3.77), (3.78) and (3.82) constitute the fundamental equations of hydrodynamics in curved spacetime. To complete the system of equations, one needs to specify an equation of state $p = p(\rho, T, Y, \dots)$ that relates the pressure to the density ρ, temperature T, chemical composition Y and other variables.

Substituting Eq. (3.77) into (3.78) and separating the result into a component along the four-velocity u^μ and components orthogonal to it, we obtain

$$(\mu + p)\frac{Du^\alpha}{D\tau} + (g^{\alpha\beta} + u^\alpha u^\beta)p_{,\beta} = 0 \,, \tag{3.83a}$$

$$\frac{d\mu}{d\tau} + (\mu + p)u^\beta{}_{;\beta} = 0 \,, \tag{3.83b}$$

where $\mu \equiv \rho(1 + \Pi)$, and $Du^\alpha/D\tau = u^\nu u^\alpha{}_{;\nu}$. The first equation is the curved spacetime version of the Euler equation of hydrodynamics, with $\mu + p$ playing the role of inertial mass density, and with the Christoffel symbols embedded within the covariant derivative producing the gravitational accelerations. Inserting the continuity equation (3.82) into Eq. (3.83b) and recalling that $u^\mu{}_{;\mu} = \mathcal{V}^{-1} d\mathcal{V}/d\tau$ (the curved spacetime version of $\boldsymbol{\nabla} \cdot \boldsymbol{v} = \mathcal{V}^{-1} d\mathcal{V}/dt$), we obtain

$$\frac{d}{d\tau}(\rho\Pi\mathcal{V}) + p\frac{d}{d\tau}\mathcal{V} = 0 \,, \tag{3.84}$$

which is nothing but the local first law of thermodynamics, with $E = \rho\Pi\mathcal{V}$, namely,

$$dE + pd\mathcal{V} = 0 \,. \tag{3.85}$$

But since $dE + pd\mathcal{V} = TdS$ in general, where S is the entropy, this implies that the fluid flow is isentropic ($dS = 0$). This property was actually built into the energy-momentum tensor from the start by assuming the perfect fluid form. Entropy changes are associated with heat transport, and to include such effects we would have had to add an additional piece to $T^{\mu\nu}$ given by $2u^{(\mu}q^{\nu)}$, where q^ν is a heat-flux four-vector, along with auxiliary equations relating q^ν to temperature gradients, energy-generation mechanisms, and so on. For further discussion of nonperfect fluids, see MTW, section 22.3, Ehlers (1971), and Andersson and Comer (2007).

The law of conservation of rest mass, Eq. (3.82) permits us to introduce an auxiliary density variable ρ^*, sometimes called the conserved density, which will prove to be very useful throughout this book. By noticing that for any four-vector field A^μ,

$$A^\mu{}_{;\mu} = \frac{1}{\sqrt{-g}}[\sqrt{-g}A^\mu]_{,\mu} \,, \tag{3.86}$$

we can therefore write

$$\sqrt{-g}(\rho u^\mu)_{;\mu} = [\sqrt{-g}\rho u^\mu]_{,\mu} = \frac{\partial}{\partial t}\rho^* + \frac{\partial}{\partial x^j}(\rho^* v^j) = 0 \,, \tag{3.87}$$

where we define

$$\rho^* \equiv \rho\sqrt{-g}u^0 \,. \tag{3.88}$$

Thus ρ^* satisfies an "Eulerian" continuity equation in any $(t,\boldsymbol{x})$ coordinate system. This "conserved" density is useful because for any function $f(t,\boldsymbol{x})$ defined in a volume V whose boundary is outside the matter

$$\frac{\partial}{\partial t}\int_V \rho^*(t,\boldsymbol{x})f(t,\boldsymbol{x})d^3x = \int_V \rho^*\frac{df}{dt}d^3x\,, \tag{3.89}$$

where $df/dt = \partial f/\partial t + \boldsymbol{v}\cdot\boldsymbol{\nabla}f$. To establish this we apply the continuity equation for ρ^* and convert the integral of a divergence into a surface integral that vanishes outside the matter. Notice that Eq. (3.89) implies that

$$\frac{dm}{dt} = 0\,, \tag{3.90}$$

where

$$m \equiv \int_V \rho^* d^3x = \int_V \rho u^0(-g)^{1/2}d^3x = \int_V \rho dV_p \tag{3.91}$$

is the total rest mass energy of the particles in the volume V; $V_p = u^0(-g)^{1/2}d^3x$ is the proper volume element.

3.3 Metric Theories of Gravity and the Strong Equivalence Principle

To obtain a complete metric theory of gravity one must now specify field equations for the metric and for the other possible gravitational fields in the theory. There are two alternatives. The first is to assume that these equations, like the nongravitational equations, can be derived from an invariant action $I_G(\phi_A, \phi_{A,\mu})$, which will be a function of the gravitational fields ϕ_A (which could include $g_{\mu\nu}$) and their derivatives. The complete action is thus

$$I = I_G(\phi_A, \phi_{A,\mu}) + I_{NG}(q_A, q_{A,\mu}, g_{\mu\nu}, g_{\mu\nu,\beta})\,. \tag{3.92}$$

Variation with respect to ϕ_A yields the gravitational field equations

$$\frac{\delta\mathcal{L}_G}{\delta\phi_A} + \frac{\partial g_{\mu\nu}}{\partial\phi_A}\frac{\delta\mathcal{L}_{NG}}{\delta g_{\mu\nu}} = 0\,, \tag{3.93}$$

or, using Eq. (3.67),

$$\frac{\delta\mathcal{L}_G}{\delta\phi_A} = -\frac{1}{2}(-g)^{1/2}T^{\mu\nu}\frac{\partial g_{\mu\nu}}{\partial\phi_A}\,. \tag{3.94}$$

Theories of this type are called Lagrangian-based covariant metric theories of gravity. Many important general properties of such theories are described by Lee et al. (1974). The other alternative is to specify gravitational field equations that are not derivable from an action. These are called non-Lagrangian-based theories. Although many such theories have been devised, they have not met with great success in agreeing with experiment. All the metric theories to be described in Chapter 5 are Lagrangian based.

In any metric theory of gravity, matter and nongravitational fields respond only to the spacetime metric $\boldsymbol{g}$. In principle, however, there could exist other gravitational fields besides the metric, such as scalar fields, vector fields, and so on. If matter does not couple to these fields what can their role in a gravitation theory be? Their role must be that of mediating the method by which matter and nongravitational fields generate gravitational fields and produce the metric. Once determined, however, the metric alone interacts with the matter as prescribed by EEP. What distinguishes one metric theory from another, therefore, is the number and kind of gravitational fields it contains in addition to the metric, and the equations that determine the structure and evolution of these fields. From this viewpoint, one can divide all metric theories of gravity into two fundamental classes: "purely dynamical" and "prior geometric." (This division is independent of whether or not the theory is Lagrangian based.) By "purely dynamical metric theory" we mean any metric theory whose gravitational fields have their structure and evolution determined by coupled partial differential field equations. In other words, the behavior of each field is influenced to some extent by a coupling to at least one of the other fields in the theory. By "prior geometric" theory, we mean any metric theory that contains "absolute elements," fields or equations whose structure and evolution are given *a priori* and are independent of the structure and evolution of the other fields of the theory. These "absolute elements" could include flat background metrics η, cosmic time coordinates T, and algebraic relationships among otherwise dynamical fields, such as

$$g_{\mu\nu} = h_{\mu\nu} + k_\mu k_\nu \,, \tag{3.95}$$

where $h_{\mu\nu}$ and k_μ may be dynamical fields. Note that a field may be absolute even if it is determined by partial differential equations, as long as the equation does not involve any dynamical fields. For instance, a flat background metric is specified by the field equation $\text{Riem}(\eta) = 0$, or a cosmic time function is specified by the field equations $\nabla_\mu \nabla_\nu T = 0$, $\nabla_\mu T \nabla^\mu T = -1$, where the gradients and inner products are taken with respect to a nondynamical background metric, such as η.

General relativity is a purely dynamical theory since it contains only one gravitational field, the metric itself, and its structure and evolution is governed by partial differential equations (Einstein's equations). Brans-Dicke theory and its generalizations are purely dynamical theories; the field equation for the metric involves the scalar field (as well as the matter as source), and the field equation for the scalar field involves the metric. In Chapter 5, we will discuss these and other theories in more detail.

By discussing metric theories of gravity from this broad, "Dicke" point of view, it is possible to draw some general conclusions about the nature of gravity in different metric theories, conclusions that are reminiscent of the Einstein Equivalence Principle, but that will be given a new name: the Strong Equivalence Principle.

Consider a local, freely falling frame in any metric theory of gravity. Let this frame be small enough that inhomogeneities in the external gravitational fields can be neglected throughout its volume. However, let the frame be large enough to encompass a system of gravitating matter and its associated gravitational fields. The system could be a star, a black hole, the solar system, or a Cavendish experiment. Call this frame a "quasilocal Lorentz frame." To determine the behavior of the system we must calculate the metric. The computation proceeds in two stages. First, we determine the external behavior of the metric and gravitational fields, thereby establishing boundary values for the fields generated by the local system, at a boundary of the quasilocal frame "far" from the local system. Second, we solve for the fields generated by the local system. But because the metric is coupled directly or indirectly to the other fields of the theory, its structure and evolution will be influenced by those fields, particularly by the boundary values taken on by those fields far from the local system. This will be true even if we work in a coordinate system in which the asymptotic form of $g_{\mu\nu}$ in the boundary region between the local system and the external world is that of the Minkowski metric. Thus, the gravitational environment in which the local gravitating system resides can influence the metric generated by the local system via the boundary values of the auxiliary fields. Consequently, the results of local gravitational experiments may depend on the location and velocity of the frame relative to the external environment Of course, local *non*gravitational experiments are unaffected since the gravitational fields they generate are assumed to be negligible, and since those experiments couple only to the metric whose form can always be made locally Minkowskian. Local gravitational experiments might include Cavendish experiments, measurements of the acceleration of massive bodies, studies of the structure of stars and planets, and so on. We can now make several statements about different kinds of metric theories (Will and Nordtvedt, 1972).

(a) A theory that contains only the metric $\boldsymbol{g}$ yields local gravitational physics that is independent of the location and velocity of the local system. This follows from the fact that the only field coupling the local system to the environment is $\boldsymbol{g}$, and it is always possible to find a coordinate system in which $\boldsymbol{g}$ takes the Minkowski form at the boundary between the local system and the external environment. Thus, the asymptotic values of $g_{\mu\nu}$ are constants independent of location, and are asymptotically Lorentz invariant, thus independent of velocity. General relativity is an example of such a theory.

(b) A theory that contains the metric $\boldsymbol{g}$ and dynamical scalar fields ϕ_A yields local gravitational physics that may depend on the location of the frame but that is independent of the velocity of the frame. This follows from the asymptotic Lorentz invariance of the Minkowski metric and of the scalar fields, except now the asymptotic values of the scalar fields may depend on the location of the frame. An example is scalar-tensor theory, where the asymptotic scalar field determines the value of the gravitational constant, which can thus vary as ϕ varies.

(c) A theory that contains the metric $\boldsymbol{g}$ and additional dynamical vector or tensor fields or prior-geometric fields yields local gravitational physics that may have both location- and velocity-dependent effects. This will be true, for example, even if the auxiliary field is a flat background metric $\boldsymbol{\eta}$. The background solutions for $\boldsymbol{g}$ and $\boldsymbol{\eta}$ will in general be different, and therefore in a frame in which $g_{\mu\nu}$ takes the asymptotic form diag$(-1, 1, 1, 1)$, $\eta_{\mu\nu}$ will in general have a form that depends on location and that is not Lorentz invariant (although it will still have vanishing curvature). The resulting location and velocity dependence in $\boldsymbol{\eta}$ will act back on the local gravitational problem. Be reminded that these effects are a consequence of the coupling of auxiliary gravitational fields to the metric and to each other, not to the matter and nongravitational fields. For metric theories of gravity, only $g_{\mu\nu}$ couples to the latter.

These ideas can be summarized in the form of a principle called the Strong Equivalence Principle whose statement is given in Box 3.2.

Box 3.2 **The Strong Equivalence Principle (SEP)**

1. WEP is valid for self-gravitating bodies as well as for test bodies. This is known as the Gravitational Weak Equivalence Principle (GWEP).
2. The outcome of any local test experiment is independent of the velocity of the (freely falling) apparatus.
3. The outcome of any local test experiment is independent of where and when in the universe it is performed.

The distinction between SEP and EEP is the inclusion of bodies with self-gravitational interactions (planets, stars) and of experiments involving gravitational forces (Cavendish experiments, gravimeter measurements). Note that SEP contains EEP as the special case in which local gravitational forces are ignored.

It is tempting to ask whether the parallel between SEP and EEP extends as far as a Schiff-type conjecture; for example, "any theory that embodies GWEP also embodies SEP." As in Section 2.5, we can give a plausibility argument in support of this, for the special case of metric theories of gravity with a conservation law for total energy (Haugan, 1979). Generally speaking, this means Lagrangian-based theories. Consider a local self-gravitating system moving slowly in a weak, static, external gravitational potential $U(\boldsymbol{x})$. We assume that the laws governing its motion can be put into a quasi-Newtonian form, with the conserved energy $E_{\rm c}$ given by

$$E_{\rm c} = M_{\rm R}\left[1 + \frac{1}{2}V^2 - U(\boldsymbol{X}) + O(U^2, V^4, UV^2)\right], \tag{3.96}$$

where $M_R = M_0 - E_{\rm B}(\boldsymbol{X}, \boldsymbol{V})$ is the rest mass of the self-gravitating system, with M_0 the total rest-mass of its constituents, and $E_{\rm B}$ the gravitational binding energy, given by

$$E_{\rm B}(\boldsymbol{X}, \boldsymbol{V}) = E_{\rm B}^0 + \delta m_{\rm P}^{jk} U^{jk}(\boldsymbol{X}) - \frac{1}{2}\delta m_{\rm I}^{jk} V^j V^k - O(E_{\rm B} U^2, \ldots), \tag{3.97}$$

(see Section 2.4 for detailed definitions). Here we use units in which the speed of light as measured far from the local system is unity. The position and velocity dependence in $E_{\rm B}$ can manifest itself, for example, as position and velocity dependence in the locally measured gravitational constant. For two bodies in a local Cavendish experiment separated by a distance r, the local gravitational constant is given by

$$G_{\rm L} \equiv \frac{r^2}{m_1 m_2} F_r = \frac{r^2}{m_1 m_2}\frac{dE_{\rm B}}{dr}, \tag{3.98}$$

and thus the anomalous mass tensors will contribute to $G_{\rm L}$ (see Section 6.6 for an explicit calculation in the PPN framework). However, a cyclic *gedanken* experiment identical to that presented in Section 2.4 shows that the anomalous mass tensors $\delta m_{\rm P}^{jk}$ and $\delta m_{\rm I}^{jk}$ also generate violations of GWEP,

$$A^j = g^j + \left(\frac{\delta m_{\rm P}^{k\ell}}{M_{\rm R}}\right) U^{k\ell}{}_{,j} - \left(\frac{\delta m_{\rm I}^{jk}}{M_{\rm R}}\right) g^k, \tag{3.99}$$

where $\mathbf{g} = \nabla U$. Hence the validity of GWEP ($\delta m_{\rm P}^{jk} = \delta m_{\rm I}^{jk} = 0$) implies no preferred-location or preferred-frame effects, thence SEP. In Chapters 4, 5, and 6, we will see specific examples of GWEP and SEP in action in the post-Newtonian limits of arbitrary metric theories of gravity.

This discussion of the coupling of auxiliary fields to local gravitating systems indicates that if SEP is valid, there must be one and only one gravitational field in the universe, the metric $\mathbf{g}$. These arguments are only suggestive however, and it is very unlikely that they can be made more rigorous.

The assumption that there is only one gravitational field is the foundation of many so-called derivations of general relativity. One class of derivations uses a quantum-field-theoretic approach. One begins with the assumption that, in perturbation theory, the gravitational field is associated with the exchange of a single massless particle of spin 2 (corresponding to a single second-rank tensor dynamical field), and by making certain reasonable assumptions that the S-matrix be Lorentz invariant or that the theory be derivable from an action, one can generate the full classical Einstein field equations (Weinberg, 1965; Deser, 1970). Another class of derivations attempts to build the most general field equation for $\mathbf{g}$ out of tensors constructed only from $\mathbf{g}$, subject to certain constraints (no higher than second derivatives, for instance). By demanding that the field equations should be compatible with the matter equations of motion $T^{\mu\nu}{}_{;\nu} = 0$, one is led (except for the possible cosmological constant term) to Einstein's equations. For a review of these and other "derivations" of general relativity the reader is referred to MTW, box 17.2.

However, the implicit use of SEP in all these derivations cannot be emphasized enough. Empirically, it has been found that virtually every metric theory other than general relativity introduces auxiliary gravitational fields, either dynamical or prior geometric, and thus predicts violations of SEP at some level. General relativity seems to be the only metric theory that embodies SEP completely. Thus, the wide variety of derivations of general relativity that assume SEP, plus evidence from alternative theories lends some credence to the conjecture

$$\text{SEP} \Rightarrow [\text{General Relativity}] \,. \tag{3.100}$$

In Chapter 8 we will discuss experimental evidence for the validity of SEP.

4 The Parametrized Post-Newtonian Formalism

We have seen that, despite the possible existence of long-range gravitational fields in addition to the metric in various theories of gravity, the postulates of metric theories demand that matter and nongravitational fields be completely oblivious to them. The only gravitational field that enters the equations of motion is the metric $\boldsymbol{g}$. The role of the other fields that a theory may contain can only be that of helping to generate the spacetime curvature associated with the metric. Matter may create these fields, and they plus the matter may generate the metric, but they cannot act back directly on the matter. Matter responds only to the metric.

Consequently, the metric and the equations of motion for matter become the primary entities for calculating observable effects, and all that distinguishes one metric theory from another is the particular way in which matter and possibly other gravitational fields generate the metric.

The comparison of metric theories of gravity with each other and with experiment becomes particularly simple when one takes the slow-motion, weak-field limit. This approximation, known as the post-Newtonian (PN) limit, is sufficiently accurate to encompass a wide range of past and future solar-system tests. The post-Newtonian approximation as presented in this chapter is intended to be valid within the near-zone of the system, corresponding to a sphere smaller than one gravitational wavelength in size. In a somewhat different guise, it can also be applied to gravitational waves; this will be addressed in Chapter 11. Strictly speaking, the post-Newtonian approximation cannot be applied to systems with compact objects, such as black holes or neutron stars, however, with suitable modifications, a version of the post-Newtonian approximation can be used to describe the orbital motion of compact bodies, provided that the orbital motions are slow and the *interbody* gravitational fields are weak, and that the bodies are sufficiently small compared to the orbital separation that tidal interactions can be ignored (see Chapter 10). The post-Newtonian approximation is not suited to cosmology where rather different assumptions and approximations must be made (see Section 12.4).

In Section 4.1, we discuss the post-Newtonian limit of metric theories of gravity, and devise a general form for the post-Newtonian metric for a system of perfect fluid. This form should be obeyed by most metric theories, with the differences from one theory to the next occurring only in the numerical coefficients that appear in the metric. When the coordinate system is appropriately specialized to a standard "gauge," and arbitrary parameters are used in place of the numerical coefficients, the result, described in Section 4.2 is known as the parametrized post-Newtonian (PPN) formalism, and the parameters are called PPN parameters. In Section 4.3, we discuss the effect of Lorentz transformations on the PPN coordinate system, and show that some theories of gravity may predict gravitational effects that depend on the velocity of the gravitating system relative to a preferred frame, presumably attached to the rest of the universe. In Section 4.4, we analyze the existence of post-Newtonian integral conservation laws for energy, momentum, angular momentum and center-of-mass motion within the PPN formalism, and show that metric theories possess such conserved quantities if and only if their PPN parameters obey certain constraints.

This formalism then provides the framework for a discussion of specific alternative metric theories of gravity (Chapter 5), and for the analysis of solar system tests of relativistic gravitational effects (Chapters 7 – 9).

4.1 The Post-Newtonian Approximation

4.1.1 Newtonian gravity and the Newtonian limit

In the solar system, gravitation is weak enough for Newton's theory of gravity to adequately account for all but the most minute effects. This is also true for almost all stars (excepting neutron stars, and black holes, of course), all galaxies (except in the vicinity of their central massive black holes), and even clusters of galaxies and the large-scale structure of the universe. To an accuracy of at least a part in 10^5, light rays travel through the solar system on straight lines, and test bodies move according to

$$\boldsymbol{a} = \boldsymbol{\nabla} U, \tag{4.1}$$

where $\boldsymbol{a}$ is the body's acceleration and U is the Newtonian gravitational potential produced by rest-mass density ρ according to

$$\nabla^2 U = -4\pi\rho, \qquad U(t,\boldsymbol{x}) = \int \frac{\rho(t,\boldsymbol{x}')}{|\boldsymbol{x}-\boldsymbol{x}'|} d^3x' \,. \tag{4.2}$$

Here and for the remainder of this book we use units in which the speed of light c is unity, and the gravitational constant as measured far from the solar system G is unity. A perfect, nonviscous fluid obeys the usual Eulerian equations of hydrodynamics

$$\frac{\partial \rho}{\partial t} + \boldsymbol{\nabla} \cdot (\rho \boldsymbol{v}) = 0 \,,$$

$$\rho \frac{d\boldsymbol{v}}{dt} = \rho \boldsymbol{\nabla} U - \boldsymbol{\nabla} p \,, \tag{4.3}$$

where $\boldsymbol{v}$ is the velocity of an element of the fluid, ρ is the rest-mass density of matter in the element, p is the total pressure (matter plus radiation) on the element, and $d/dt = \partial/\partial t + \boldsymbol{v} \cdot \boldsymbol{\nabla}$ is the time derivative following the fluid.

From the standpoint of a metric theory of gravity, Newtonian physics may be viewed as a first-order approximation. Consider a test body momentarily at rest in a static external gravitational field. From the geodesic equation (3.44), the body's acceleration $a^j = d^2x^j/dt^2$ in a static $(t,\boldsymbol{x})$ coordinate system is given by

$$a^j = -\Gamma^j_{00} = \frac{1}{2} g^{jk} g_{00,k} \,. \tag{4.4}$$

Far from the Newtonian system, we know that, in an appropriately chosen coordinate system, the metric must reduce to the Minkowski metric (see Subsection 4.1.3)

$$g_{\mu\nu} \to \eta_{\mu\nu} = \mathrm{diag}(-1, 1, 1, 1)\,. \tag{4.5}$$

In the presence of a very weak gravitational field, Eq. (4.4) can yield Newtonian gravitation, Eq. (4.1) only if

$$g^{jk} \sim \delta^{jk}\,, \qquad g_{00} = -1 + 2U\,. \tag{4.6}$$

It is straightforward to show that, with this approximation and an energy-momentum tensor for perfect fluids given by

$$T^{00} = \rho\,, \quad T^{0j} = \rho v^j\,, \quad T^{jk} = \rho v^j v^k + p\delta^{jk}\,, \tag{4.7}$$

the Eulerian equations of Motion 4.3 are equivalent to

$$T^{\mu\nu}{}_{;\nu} \simeq T^{\mu\nu}{}_{,\nu} + \Gamma^{\mu}_{00} T^{00} = 0\,, \tag{4.8}$$

where we retain only terms of lowest order in $v^2 \sim U \sim p/\rho$.

But the Newtonian limit no longer suffices when we begin to demand accuracies greater than a part in 10^5 in the solar system. For example, it cannot account for Mercury's anomalous perihelion advance of $\sim 5 \times 10^{-7}$ radians per orbit. Thus we need a more accurate approximation to the spacetime metric that goes beyond, or "post" Newtonian theory, hence the name post-Newtonian limit.

4.1.2 Post-Newtonian bookkeeping

The key features of the post-Newtonian limit can be better understood if we first develop a "bookkeeping" system for keeping track of "small quantities." In the solar system, the Newtonian gravitational potential U is nowhere larger than 10^{-5} (in geometrized units, U, or more precisely, GU/c^2, is dimensionless). Planetary velocities are related to U by virial relations which yield

$$v^2 \sim U\,. \tag{4.9}$$

The matter making up the Sun and planets is under pressure p, but because the bodies are in hydrostatic equilibrium, this pressure is comparable to ρU (more precisely $\boldsymbol{\nabla} p = \rho \boldsymbol{\nabla} U$), in other words

$$p/\rho \sim U\,. \tag{4.10}$$

From thermodynamics, other forms of energy (compressional energy, radiation, thermal energy, magnetic energy, etc.) are related to pressure by $p\mathcal{V} \sim E \sim \rho\Pi\mathcal{V}$, and thus Π, the ratio of energy density to rest-mass density, is related to U by

$$\Pi \sim p/\rho \sim U\,. \tag{4.11}$$

These four dimensionless quantities are assigned a bookkeeping label ϵ that denotes their "order of smallness":

$$U \sim v^2 \sim p/\rho \sim \Pi \sim O(\epsilon)\,. \tag{4.12}$$

Then single powers of velocity v are $O(\epsilon^{1/2})$, U^2 is $O(\epsilon^2)$, Uv is $O(\epsilon^{3/2})$, $U\Pi$ is $O(\epsilon^2)$, and so on. Also, since the time evolution of the system is governed by the motions of its constituents, we have that

$$\partial/\partial t \sim \boldsymbol{v}\cdot\boldsymbol{\nabla}\,, \tag{4.13}$$

and thus that

$$\frac{|\partial/\partial t|}{|\partial/\partial x|} \sim O(\epsilon^{1/2})\,. \tag{4.14}$$

We can now analyze the "post-Newtonian" metric using this bookkeeping system. The action, Eq. (3.26), from which one can derive the geodesic equation (3.44) for a single neutral particle, may be rewritten, after dropping the electromagnetic terms,

$$\begin{aligned} I_{\rm NG} = I_0 &= -m_0 \int \left(-g_{\mu\nu}\frac{dr^\mu}{dt}\frac{dr^\nu}{dt}\right)^{1/2} dt \\ &= -m_0 \int \left(-g_{00} - 2g_{0j}v^j - g_{jk}v^j v^k\right)^{1/2} dt\,, \end{aligned} \tag{4.15}$$

where $r^\mu = (t, \boldsymbol{r})$ describes the particle's position in spacetime, and $v^j = dr^j/dt$ is its three-dimensional velocity vector. The integrand in Eq. (4.15) may thus be viewed as the Lagrangian L for a single particle in a metric gravitational field. From Eq. (4.6), we see that the Newtonian limit corresponds to

$$L = -m_0(1 - 2U - v^2)^{1/2} \simeq -m_0 + \frac{1}{2}m_0 v^2 + m_0 U\,, \tag{4.16}$$

as can be verified using the Euler-Lagrange equations. In other words, Newtonian physics is given by an approximation for L correct to $O(\epsilon)$. Post-Newtonian physics must therefore involve those terms in L of the next order, $O(\epsilon^2)$.

But what about odd-half-order terms, $O(\epsilon^{1/2})$ or $O(\epsilon^{3/2})$. Odd-half-order terms must contain an odd number of velocities $\boldsymbol{v}$ or of time derivatives $\partial/\partial t$. Since these factors change sign under time reversal, odd-half-order terms must represent dissipative processes in the system. But, roughly speaking, conservation of rest-mass energy prevents terms of $O(\epsilon^{1/2})$ from appearing in L, and conservation of energy in the Newtonian limit prevents terms of $O(\epsilon^{3/2})$. Beyond $O(\epsilon^2)$, different theories may make different predictions. In general relativity, for example, the conservation of post-Newtonian energy prevents terms of $O(\epsilon^{5/2})$. However, terms of $O(\epsilon^{7/2})$ *can* appear; they represent the dissipative effects of gravitational radiation reaction.

In order to express L to $O(\epsilon^2)$, we must know the various metric components to an appropriate order:

$$L = \left(1 - 2U - v^2 - \delta g_{00}[O(\epsilon^2)] - 2g_{0j}[O(\epsilon^{3/2})]v^j - \delta g_{jk}[O(\epsilon)]v^j v^k\right)^{1/2}. \tag{4.17}$$

Thus the post-Newtonian limit of any metric theory of gravity requires a knowledge of

$$\begin{aligned} g_{00} &\ \text{to}\ \ O(\epsilon^2)\,, \\ g_{0j} &\ \text{to}\ \ O(\epsilon^{3/2})\,, \\ g_{jk} &\ \text{to}\ \ O(\epsilon)\,. \end{aligned} \tag{4.18}$$

The post-Newtonian propagation of light rays may also be obtained using the above approximations to the metric. Since light moves along null trajectories ($d\tau = 0$), the Lagrangian L must be formally identical to zero. In the first-order "Newtonian" limit this implies that light must move on straight lines at speed 1, that is,

$$0 = L = (1 - v^2)^{1/2}, \quad v^2 = 1. \tag{4.19}$$

In the next, post-Newtonian order, we must have

$$0 = L = \left(1 - 2U - v^2 - \delta g_{jk}[O(\epsilon)]v^j v^k\right)^{1/2}. \tag{4.20}$$

Thus to obtain post-Newtonian corrections to the propagation of light rays, we need to know

$$\begin{aligned} g_{00} &\quad \text{to} \quad O(\epsilon), \\ g_{jk} &\quad \text{to} \quad O(\epsilon). \end{aligned} \tag{4.21}$$

In a similar manner, we can verify that if we take the perfect-fluid energy-momentum tensor

$$T^{\mu\nu} = (\rho + \rho\Pi + p)u^\mu u^\nu + pg^{\mu\nu}, \tag{4.22}$$

expanded through the following orders of accuracy,

$$\begin{aligned} T^{00} &\quad \text{to} \quad \rho O(\epsilon), \\ T^{0j} &\quad \text{to} \quad \rho O(\epsilon^{3/2}), \\ T^{jk} &\quad \text{to} \quad \rho O(\epsilon^2), \end{aligned} \tag{4.23}$$

and combined with the post-Newtonian metric, then the equation of motion $T^{\mu\nu}{}_{;\nu} = 0$ will yield consistent "post-Eulerian" equations of hydrodynamics.

4.1.3 Isolated post-Newtonian systems

We must now specify the coordinate system to be used for the post-Newtonian limit. We imagine an isolated post-Newtonian system residing in a homogeneous, isotropic universe. We choose a coordinate system whose outer region far from the isolated system is in free fall with respect to the surrounding cosmological model, and is at rest and nonrotating with respect to a frame in which the universe appears isotropic (universe rest frame). In this outer region, we expect the physical metric to have the form

$$ds^2 = -dt^2 + \left(\frac{a(t)}{a_0}\right)^2 \delta_{jk}dx^j dx^k + h_{\mu\nu}dx^\mu dx^\nu, \tag{4.24}$$

where the first two terms comprise the standard spatially flat Friedmann-Robertson-Walker (FRW) line element of the current standard model of big-bang cosmology, and the final term represents the perturbation due to the local system, where $a(t)$ is the cosmological scale factor (a_0 is its present value). Expanding $a(t)$ about the present time, we obtain

$$\frac{a(t)}{a_0} = 1 + H_0 t - \frac{1}{2}H_0^2 q_0 t^2 + \dots, \tag{4.25}$$

where $H_0 \equiv \dot{a}_0/a_0$ and $q_0 \equiv -H_0^{-2}\ddot{a}_0/a_0$ are the present values of the Hubble parameter and deceleration parameter.

We make the coordinate transformation

$$t = t' + H_0 A_1(t', x'^k) + H_0^2 A_2(t', x'^k),$$
$$x^j = x'^j + H_0 B_1^j(t', x'^k) + H_0^2 B_2^j(t', x'^k), \tag{4.26}$$

where we use H_0 as an effective expansion parameter. Substituting into Eq. (4.24), and choosing A_1 and B_1^j to eliminate all terms linear in H_0, we find, after dropping contributions corresponding to trivial boosts, rotations and displacements, that $A_1 = -r'^2/2$ and $B_1^j = -x'^j t'$, and that to second order in H_0, the cosmological part of the metric takes the form

$$g'_{00} = -1 - H_0^2\left(2\dot{A}_2 - r'^2\right),$$
$$g'_{0j} = H_0^2\left(\dot{B}_2^j - A_{2,j} - x'^j t'\right),$$
$$g'_{jk} = \delta_{jk} + H_0^2\left(2\dot{B}^j_{2,k} - t'^2\delta_{jk}(2+q_0) - r'^2\delta_{jk} - x'^j x'^k\right). \tag{4.27}$$

Choosing $B_2^j = \frac{1}{2}t'^2 x'^j(2+q_0) + \frac{1}{3}r'^2 x'^j$ and $A_2 = \frac{1}{2}(1+q_0)r'^2 t'$, we can eliminate the t' dependence from g'_{jk}, set $g'_{0j} = 0$, and obtain finally

$$g'_{00} = -1 - R_{0p0q}x'^p x'^q,$$
$$g'_{0j} = 0,$$
$$g'_{jk} = \delta_{jk} - \frac{1}{3}R_{jpkq}x'^p x'^q, \tag{4.28}$$

where, for the FRW metric, $R_{0p0q} = H_0^2 q_0 \delta_{pq}$, $R_{0jpq} = 0$, and $R_{jpkq} = H_0^2(\delta_{jk}\delta_{pq} - \delta_{jq}\delta_{kp})$. These are called Fermi normal coordinates for describing a local freely falling frame (PW, section 5.2.5). Then, including the contribution from the local system, we find the line element in the form

$$ds^2 = (\eta_{\mu'\nu'} + h_{\mu'\nu'})dx^{\mu'}dx^{\nu'} - H_0^2 r'^2\left[q_0 dt'^2 + \frac{1}{3}r'^2(d\theta'^2 + \sin^2\theta' d\phi'^2)\right]. \tag{4.29}$$

Normally, in finding the metric for an isolated system, we want to impose the boundary condition that the metric tend to the Minkowski metric $\eta_{\mu\nu}$ at infinity, however because of the cosmological background, such a condition will generate an error that grows with distance from the system. But because $H_0^{-1} \sim 10^{23}$ km this will happen *very* far away. To estimate where and how large the error might be, we note that a post-Newtonian potential of order $\epsilon m/r$ will be comparable to $H_0^2 r^2$ at a distance

$$r_0 \sim \left(\frac{\epsilon m}{H_0^2}\right)^{1/3} \sim 3 \times 10^{13}\,\text{km}\left(\frac{\epsilon}{10^{-6}}\frac{m}{m_\odot}\right)^{1/3}, \tag{4.30}$$

and at r_0, the error being made in imposing the normal boundary condition is of order $(r_0 H_0)^2 \sim 5 \times 10^{-20}$ for $\epsilon \sim 10^{-6}$ and $m \sim m_\odot$. This is far smaller than the error $\sim \epsilon^2 \sim 10^{-12}$ made by ignoring 2PN effects within the system itself.

Thus for all solar-system or astrophysical applications of the post-Newtonian approximation, we are justified in ignoring the surrounding cosmological setting when establishing asymptotic boundary conditions on the metric. However, if the theory in question has auxiliary fields, the cosmological boundary conditions on *those* fields may be crucial (see Section 3.3 for further discussion).

We will call the coordinate system thus constructed "local quasi-Cartesian coordinates." In this coordinate system it is useful to define the following conventions and quantities:

1. Unless otherwise noted, spatial vectors are treated as Cartesian vectors, with $x^k \equiv x_k$.
2. Repeated spatial indices or the symbol $|\boldsymbol{x}|$ denotes a Cartesian inner product, for example

$$x^k x_k \equiv x_k x_k \equiv x^k x^k \equiv |\boldsymbol{x}|^2 = x^2 + y^2 + z^2 \,.$$

3. The volume element $d^3x \equiv dxdydz$.

4.2 Building the PPN Formalism

4.2.1 Post-Newtonian potentials

We will assume throughout that the matter composing the system can be idealized as perfect fluid. For the purposes of most solar system or astrophysical situations, this is an adequate assumption. As we will see in more detail in Chapter 5, the post-Newtonian limit for a system of perfect fluid in any metric theory of gravity is best calculated by solving the field equations formally, expressing the metric as a sequence of post-Newtonian functionals of the matter variables, with possible coefficients that may depend on the matching conditions between the local system and the surrounding cosmological model and on other constants of the theory. The evolution of the matter variables, and thence of the metric functionals, is determined by means of the equations of motion $T^{\mu\nu}{}_{;\nu} = 0$ using the matter energy-momentum tensor and the post-Newtonian metric, all evaluated to an order consistent with the post-Newtonian approximation. The evolution of the cosmological matching coefficients is determined by a solution of the appropriate cosmological model. Thus, the most general post-Newtonian metric can be found by simply writing down metric terms composed of all possible post-Newtonian functionals of matter variables, each multiplied by an arbitrary coefficient that may depend on the cosmological matching conditions and on other constants, and adding these terms to the Minkowski metric to obtain the physical metric. Unfortunately, there is an infinite number of such functionals, so that in order to obtain a formalism that is both useful and manageable, we must impose some restrictions on the possible terms to be considered, guided in part by a subjective notion of "reasonableness" and in part by evidence obtained from known gravitation theories. Some of these restrictions are obvious:

1. The metric terms should be of Newtonian or post-Newtonian order; no post-post-Newtonian or higher terms are included.
2. The terms should tend to zero as the distance $|\boldsymbol{x} - \boldsymbol{x}'|$ between the field point $\boldsymbol{x}$ and a typical point $\boldsymbol{x}'$ inside the matter becomes large. This will guarantee that the metric becomes asymptotically Minkowskian in our quasi-Cartesian coordinate system.

3. The coordinates are chosen so that the metric is dimensionless.
4. In our chosen quasi-Cartesian coordinate system, the spatial origin and initial moment of time are completely arbitrary, so the metric should contain no explicit reference to these quantities. This is guaranteed by using functionals in which the field point $\boldsymbol{x}$ always occurs in the combination $\boldsymbol{x}-\boldsymbol{x}'$, where $\boldsymbol{x}'$ is a point associated with the matter distribution, and by making all time dependence in the metric terms implicit via the evolution of the matter variables and of the possible cosmological matching parameters.
5. The metric corrections h_{00}, h_{0j} and h_{jk} should transform under spatial rotations as a scalar, vector, and tensor, respectively, and thus should be constructed out of the appropriate quantities. For variables associated with the matter distribution, examples are: scalars, ρ, $|\boldsymbol{x}-\boldsymbol{x}'|$, v'^2, $\boldsymbol{v}'\cdot(\boldsymbol{x}-\boldsymbol{x}')$, etc.; vectors, v'^j, $(x-x')^j$; and tensors, $(x-x')^j(x-x')^k$, $v'^j v'^k$, etc. For variables associated with the structure of the field equations of the theory or with the cosmological matching conditions, there are only two available quantities in the rest frame of an isotropic universe: scalar cosmological matching parameters or numerical coefficients; and a tensor, δ^{jk}. In the rest frame of an isotropic universe, no vectors or anisotropic tensors can be constructed.
6. The metric functionals should be generated by rest mass, energy, pressure, and velocity, not by their gradients. This restriction is purely subjective, and can be relaxed quite easily if there is ever any reason to do so. No reason has yet arisen.
7. A final constraint is extremely subjective: The functionals should be "simple."

With those restrictions in mind, we can now write down possible terms that may appear in the post-Newtonian metric.

(1) g_{00} to $O(\epsilon)$: We have already determined that $g_{00} = -1+2U+O(\epsilon^2)$, where U is the Newtonian potential. However, we now define a new "Newtonian" potential U using the conserved density ρ^* defined in Section 3.2.4, instead of the locally measured density ρ, in other words, henceforth,

$$U(t,\boldsymbol{x}) \equiv \int \frac{\rho^*(t,\boldsymbol{x}')}{|\boldsymbol{x}-\boldsymbol{x}'|} d^3x' , \quad \nabla^2 U = -4\pi\rho^* . \tag{4.31}$$

We will use ρ^* to define all post-Newtonian potentials.

(2) g_{jk} to $O(\epsilon)$: From condition (5), g_{jk} must behave as a three-dimensional tensor potential under rotations, thus the only terms that can appear are

$$g_{jk}[O(\epsilon)] : \quad U\delta_{jk},\ U_{jk} , \tag{4.32}$$

where U_{jk} is given by

$$U_{jk}(t,\boldsymbol{x}) \equiv \int \rho^*(t,\boldsymbol{x}') \frac{(x-x')_j(x-x')_k}{|\boldsymbol{x}-\boldsymbol{x}'|^3} d^3x' . \tag{4.33}$$

The potential U_{jk} can be expressed more conveniently in terms of the "superpotential" $X(t,\boldsymbol{x})$, given by

$$X(t,\boldsymbol{x}) \equiv \int \rho^*(t,\boldsymbol{x}')|\boldsymbol{x}-\boldsymbol{x}'| d^3x' , \tag{4.34a}$$

$$\nabla^2 X = 2U, \tag{4.34b}$$

$$X_{,jk} = U\delta_{jk} - U_{jk} . \tag{4.34c}$$

Thus, the only terms that we will consider in g_{jk} are

$$g_{jk}[O(\epsilon)]: \quad U\delta_{jk},\ X_{,jk}\,. \tag{4.35}$$

(3) g_{0j} to $O(\epsilon^{3/2})$: These metric components must transform as three-vectors under rotations, and thus contain only the terms

$$g_{0j}[O(\epsilon^{3/2})]: \quad V_j,\ W_j\,, \tag{4.36}$$

where

$$\begin{aligned} V_j &\equiv \int \rho^*(t,\boldsymbol{x}')\frac{v'_j}{|\boldsymbol{x}-\boldsymbol{x}'|}d^3x'\,, \qquad \nabla^2 V_j = -4\pi\rho^* v_j\,, \\ W_j &\equiv \int \rho^*(t,\boldsymbol{x}')\frac{\boldsymbol{v}'\cdot(\boldsymbol{x}-\boldsymbol{x}')(x-x')_j}{|\boldsymbol{x}-\boldsymbol{x}'|^3}d^3x'\,. \end{aligned} \tag{4.37}$$

The potentials V_j and W_j are also related to the superpotential X by

$$X_{,0j} = W_j - V_j\,, \tag{4.38}$$

so that the only terms that we will consider here are

$$g_{0j}[O(\epsilon^{3/2})]: \quad V_j,\ X_{,0j}\,. \tag{4.39}$$

(4) g_{00} to $O(\epsilon^2)$: This component should be a scalar under rotations. The only terms that we will consider are

$$g_{00}[O(\epsilon^2)]: \quad U^2,\ \Phi_1,\ \Phi_2,\ \Phi_3,\ \Phi_4,\ \Phi_5,\ \Phi_6,\ \Phi_W\,, \tag{4.40}$$

where

$$\begin{aligned} \Phi_1 &\equiv \int \frac{\rho^{*\prime} v'^2}{|\boldsymbol{x}-\boldsymbol{x}'|}d^3x'\,, \qquad \nabla^2\Phi_1 = -4\pi\rho^* v^2\,, \\ \Phi_2 &\equiv \int \frac{\rho^{*\prime} U'}{|\boldsymbol{x}-\boldsymbol{x}'|}d^3x'\,, \qquad \nabla^2\Phi_2 = -4\pi\rho^* U\,, \\ \Phi_3 &\equiv \int \frac{\rho^{*\prime} \Pi'}{|\boldsymbol{x}-\boldsymbol{x}'|}d^3x'\,, \qquad \nabla^2\Phi_3 = -4\pi\rho^* \Pi\,, \\ \Phi_4 &\equiv \int \frac{p'}{|\boldsymbol{x}-\boldsymbol{x}'|}d^3x'\,, \qquad \nabla^2\Phi_4 = -4\pi p\,, \\ \Phi_5 &\equiv \int \rho^{*\prime}\boldsymbol{\nabla}' U'\cdot\frac{(\boldsymbol{x}-\boldsymbol{x}')}{|\boldsymbol{x}-\boldsymbol{x}'|}d^3x'\,, \\ \Phi_6 &\equiv \int \rho^{*\prime}\frac{[\boldsymbol{v}'\cdot(\boldsymbol{x}-\boldsymbol{x}')]^2}{|\boldsymbol{x}-\boldsymbol{x}'|^3}d^3x'\,, \\ \Phi_W &\equiv \int\int \rho^{*\prime}\rho^{*\prime\prime}\frac{(\boldsymbol{x}-\boldsymbol{x}')}{|\boldsymbol{x}-\boldsymbol{x}'|^3}\cdot\left[\frac{(\boldsymbol{x}'-\boldsymbol{x}'')}{|\boldsymbol{x}-\boldsymbol{x}''|}-\frac{(\boldsymbol{x}-\boldsymbol{x}'')}{|\boldsymbol{x}'-\boldsymbol{x}''|}\right]d^3x'd^3x''\,. \end{aligned} \tag{4.41}$$

Condition (7) has been used liberally to eliminate otherwise possible metric potentials such as $V_jV_jU^{-1}$, $\Phi_1\Phi_3U^{-2}$, or $U_{ij}U_{ij}$. Should one of these terms ever appear in the post-Newtonian metric of a plausible gravitational theory, the formalism could be modified accordingly. The potentials that we have introduced satisfy a number of useful relationships, including

$$V_{j,j} = -U_{,0}, \tag{4.42a}$$

$$X_{,00} = \Phi_1 + 2\Phi_4 - \Phi_5 - \Phi_6, \tag{4.42b}$$

$$\Phi_W = -U^2 - \Phi_2 - \nabla U \cdot \nabla X + \nabla \cdot \int \frac{\rho^{*\prime}}{|\boldsymbol{x}-\boldsymbol{x}'|} \nabla' X' d^3x' . \tag{4.42c}$$

The first two relationships may be derived by making use of the following identity, obtained using the equation of continuity for the conserved density ρ^*,

$$\frac{\partial}{\partial t} \int \rho^*(t,\boldsymbol{x}') f(\boldsymbol{x},\boldsymbol{x}') d^3x' = \int \rho^*(t,\boldsymbol{x}') \boldsymbol{v}' \cdot \nabla' f(\boldsymbol{x},\boldsymbol{x}') d^3x' . \tag{4.43}$$

The third may be obtained by splitting $\boldsymbol{x}' - \boldsymbol{x}''$ into $(\boldsymbol{x}' - \boldsymbol{x}) + (\boldsymbol{x} - \boldsymbol{x}'')$ and $\boldsymbol{x} - \boldsymbol{x}''$ into $(\boldsymbol{x} - \boldsymbol{x}') + (\boldsymbol{x}' - \boldsymbol{x}'')$ in the numerators of Φ_W and identifying the four resulting terms.

4.2.2 The standard PPN gauge

We can restrict the form of the post-Newtonian metric somewhat by making use of the arbitrariness of coordinates embodied in statement (ii) of the Dicke framework. An infinitesimal coordinate or "gauge" transformation,

$$x^{\bar{\mu}} = x^{\mu} + \xi^{\mu}(x^{\nu}), \tag{4.44}$$

changes the metric to

$$\begin{aligned} \bar{g}_{\bar{\mu}\bar{\nu}}(x^{\bar{\gamma}}) &= \frac{\partial x^{\bar{\mu}}}{\partial x^{\alpha}} \frac{\partial x^{\bar{\nu}}}{\partial x^{\beta}} g_{\alpha\beta} \\ &= g_{\mu\nu}(x^{\gamma}) - \xi_{\mu,\nu} - \xi_{\nu,\mu} + O(\xi^2) . \end{aligned} \tag{4.45}$$

We wish to retain the post-Newtonian character of $g_{\mu\nu}$ and the quasi-Cartesian character of the coordinate system, and to remain in the universe rest frame, thus the functions ξ^{μ} must satisfy: (i) $\xi_{\mu,\nu} + \xi_{\nu,\mu}$ are post-Newtonian functions; (ii) $\xi_{\mu,\nu} + \xi_{\nu,\mu} \rightarrow 0$ far from the system; and (iii) $|\xi^{\mu}|/|x^{\mu}| \rightarrow 0$ far from the system. The only "simple" functional that has this property is the gradient of the superpotential $X_{,\mu}$. Thus, we choose

$$\xi_0 = \lambda_1 X_{,0}, \qquad \xi_j = \lambda_2 X_{,j}, \tag{4.46}$$

and obtain, to 1PN order,

$$\begin{aligned} \bar{g}_{\bar{0}\bar{0}} &= g_{00} - 2\lambda_1 X_{,00}, \\ \bar{g}_{\bar{0}\bar{j}} &= g_{0j} - (\lambda_1 + \lambda_2) X_{,0j}, \\ \bar{g}_{\bar{j}\bar{k}} &= g_{jk} - 2\lambda_2 X_{,jk} . \end{aligned} \tag{4.47}$$

In addition to the changes in the metric components induced by the coordinate change, we must also express the potentials that appear in the original metric $g_{\mu\nu}$ in terms of the new coordinates. Since the coordinate changes are of post-Newtonian order, they will only make a contribution to the post-Newtonian metric via the Newtonian potential U in g_{00}. We recall that $\rho^* d^3x$ is an invariant, thus $\rho^*(t,\boldsymbol{x})d^3x = \bar{\rho}^*(\bar{t},\bar{\boldsymbol{x}})d^3\bar{x}$, and thus we can write that

$$
\begin{aligned}
U(t,\boldsymbol{x}) &= \int \frac{\rho^*(t,\boldsymbol{x}')}{|\boldsymbol{x}-\boldsymbol{x}'|} d^3x' \\
&= \int \frac{\bar{\rho}^*(\bar{t},\bar{\boldsymbol{x}}')}{|\bar{\boldsymbol{x}}-\bar{\boldsymbol{x}}' - \lambda_2 \bar{\boldsymbol{\nabla}}\bar{X} + \lambda_2 \bar{\boldsymbol{\nabla}}'\bar{X}'|} d^3\bar{x}' + O(\epsilon^2) \\
&= \int \frac{\bar{\rho}'^*}{|\bar{\boldsymbol{x}}-\bar{\boldsymbol{x}}'|} d^3\bar{x}' - \lambda_2 \int \bar{\rho}'^* \left(\bar{\boldsymbol{\nabla}}\bar{X} - \bar{\boldsymbol{\nabla}}'\bar{X}'\right) \cdot \bar{\boldsymbol{\nabla}} \frac{1}{|\bar{\boldsymbol{x}}-\bar{\boldsymbol{x}}'|} d^3\bar{x}' .
\end{aligned}
\tag{4.48}
$$

Using Eq. (4.42c), we find that

$$
U(t,\boldsymbol{x}) = \bar{U}(\bar{t},\bar{\boldsymbol{x}}) + \lambda_2(\bar{U}^2 + \bar{\Phi}_2 + \bar{\Phi}_W) , \tag{4.49}
$$

and recalling Eq. (4.42b), we obtain finally, after dropping all bars on the right-hand-side,

$$
\begin{aligned}
g_{\bar{0}\bar{0}} &= g_{00} - 2\lambda_1(\Phi_1 + 2\Phi_4 - \Phi_5 - \Phi_6) + 2\lambda_2(U^2 + \Phi_2 + \Phi_W) , \\
g_{\bar{0}\bar{j}} &= g_{0j} - (\lambda_1 + \lambda_2)X_{,0j} , \\
g_{\bar{j}\bar{k}} &= g_{jk} - 2\lambda_2 X_{,jk} .
\end{aligned}
\tag{4.50}
$$

By an appropriate choice of λ_1 and λ_2, we can eliminate certain terms from the post-Newtonian metric. We will thus adopt a *standard post-Newtonian gauge*, in which the spatial part of the metric is diagonal and isotropic (i.e., $X_{,jk}$ is eliminated) and in which $g_{\bar{0}\bar{0}}$ contains no term Φ_5. There is no physical significance in this gauge choice; it is purely a matter of convenience (see Section 4.5 for other possible gauge choices).

We now have a very general form for the post-Newtonian perfect-fluid metric in any metric theory of gravity, expressed in a local, quasi-cartesian coordinate system at rest with respect to the universe rest frame, and in a standard gauge. The only way that the metric of any one theory can differ from that of any other theory is in the coefficients that multiply each term in the metric. By replacing each coefficient by an arbitrary parameter we obtain a "super post-Newtonian metric theory of gravity" whose special cases (particular values of the parameters) are the post-Newtonian metrics of particular theories of gravity. This "super metric" is called the parametrized post-Newtonian (PPN) metric, and the parameters are called PPN parameters.

This use of parameters to describe the post-Newtonian limit of metric theories of gravity is called the parametrized post-Newtonian (PPN) formalism. Primitive versions of such a formalism were devised and studied by Eddington (1922), Robertson (1962) and Schiff (1967). This Eddington-Robertson-Schiff formalism treated the solar system metric as that of a static, spherical, nonrotating Sun, and idealized the planets as test bodies moving on geodesics of this metric. The metric in this version of the formalism reads

$$
\begin{aligned}
g_{00} &= -1 + 2M/r - 2\beta(M/r)^2 , \\
g_{0j} &= 0 , \\
g_{jk} &= (1 + 2\gamma M/r)\delta_{jk} ,
\end{aligned}
\tag{4.51}
$$

where M is the mass of the Sun, and β and γ are PPN parameters.

These two parameters may be given a physical interpretation. The parameter γ measures the amount of curvature of space produced by a body of mass M at radius r, in the sense that the spatial components of the Riemann curvature tensor are given to PN order by [see Eqs. (3.20) and (3.40)]

$$R_{ijkl} = 3\gamma\frac{M}{r^3}\left(\hat{n}_j\hat{n}_k\delta_{il} + \hat{n}_i\hat{n}_l\delta_{jk} - \hat{n}_i\hat{n}_k\delta_{jl} - \hat{n}_j\hat{n}_l\delta_{ik} - \frac{2}{3}\delta_{jk}\delta_{il} + \frac{2}{3}\delta_{ik}\delta_{jl}\right), \tag{4.52}$$

where $\hat{\boldsymbol{n}} = \boldsymbol{x}/r$, independent of the choice of post-Newtonian gauge. The parameter β is said to measure the amount of nonlinearity $(M/r)^2$ that a given theory puts into the g_{00} component of the metric. However, this statement is valid only in the standard post-Newtonian gauge. The coefficient of $U^2 = (M/r)^2$ depends upon the choice of gauge, as can be seen from Eq. (4.50). In general relativity, for example ($\beta = \gamma = 1$), the $(M/r)^2$ term can be completely eliminated from g_{00} by a gauge transformation that happens to be the post-Newtonian limit of the exact coordinate transformation from isotropic coordinates to Schwarzschild coordinates for the Schwarzschild geometry. Thus, this identification of β should be viewed only as a heuristic one.

Schiff (1967) generalized the metric of Eq. (4.51) to incorporate rotation (Lense-Thirring effect, Section 9.1), and Baierlein (1967) developed a primitive perfect fluid PPN metric. But the pioneering development of the full PPN formalism was initiated by Kenneth Nordtvedt Jr. (1968b), who studied the post-Newtonian metric of a system of gravitating point masses. Will (1971c) generalized the formalism to incorporate matter described by a perfect fluid. A unified version of the PPN formalism was then presented by Will and Nordtvedt (1972). The Whitehead potential Φ_W was added by Will (1973). The version presented here is the canonical version, with conventions chosen to be in accord with standard textbooks such as MTW and PW.

As in the Eddington-Robertson-Schiff version of the PPN formalism, we introduce an arbitrary PPN parameter in front of each post-Newtonian term in the metric. Ten parameters are needed; they are denoted γ, β, ξ, α_1, α_2, α_3, ζ_1, ζ_2, ζ_3, and ζ_4. In terms of them, the PPN metric reads

$$g_{00} = -1 + 2U + 2(\psi - \beta U^2) + O(\epsilon^3), \tag{4.53a}$$

$$g_{0j} = -\frac{1}{2}[4(1+\gamma) + \alpha_1]V_j - \frac{1}{2}[1 + \alpha_2 - \zeta_1 + 2\xi]X_{,0j} + O(\epsilon^{5/2}), \tag{4.53b}$$

$$g_{jk} = (1 + 2\gamma U)\,\delta_{jk} + O(\epsilon^2), \tag{4.53c}$$

where

$$\psi := \frac{1}{2}(2\gamma + 1 + \alpha_3 + \zeta_1 - 2\xi)\Phi_1 - (2\beta - 1 - \zeta_2 - \xi)\Phi_2 + (1 + \zeta_3)\Phi_3 + (3\gamma + 3\zeta_4 - 2\xi)\Phi_4 - \frac{1}{2}(\zeta_1 - 2\xi)\Phi_6 - \xi\Phi_W. \tag{4.54}$$

Although we have used linear combinations of PPN parameters in Eqs. (4.53) and (4.54), it can be seen quite easily that a given set of numerical coefficients for the post-Newtonian terms will yield a unique set of values for the parameters. The linear combinations were chosen in such a way that the parameters α_a and ζ_a will have special physical significance.

Other versions of the PPN formalism have been developed to deal with point masses with charge (Will, 1976b), fluid with anisotropic stresses (MTW, section 39), and bodies with strong internal gravity (Nordtvedt (1985) and Section 10.3). Extensions of the formalism to post-post-Newtonian (2PN) order have been developed for light propagation (Epstein and Shapiro, 1980; Fischbach and Freeman, 1980; Richter and Matzner, 1982), for motion in spherical symmetry (Sarmiento, 1982), and for N-body motion governed by a Lagrangian (Nordtvedt, 1987; Benacquista and Nordtvedt, 1988; Benacquista, 1992). Additional parameters or potentials are needed to deal with some theories, such as theories with massive fields, where Yukawa-type potentials replace Poisson potentials (see Zaglauer (1990) and Helbig (1991) for the case of massive scalar-tensor theory), or theories like Chern-Simons theory, which permit parity violation in gravity (Section 5.6).

4.3 Lorentz Transformations and the PPN Metric

In Section 4.1.3, the PPN metric was devised in a coordinate system whose outer regions were assumed to be at rest with respect to the universe rest frame. For some purposes—for example, the computation of the post-Newtonian metric in a given theory of gravity—this is a useful coordinate system. But for other purposes, such as the computation of observable post-Newtonian effects in systems, such as the solar system, that are in motion relative to the universal rest frame, it is not a convenient coordinate system. In such cases, a better coordinate system would be one in which the center of mass of the system under study is approximately at rest. Again, this is a matter of convenience; the results of experiments cannot be affected by our choice of coordinate system. Because many of our computations will be carried out for such moving systems, it is useful to reexpress the PPN metric in a moving coordinate system. This will also yield some insight into the significance of the PPN parameters α_1, α_2, and α_3 (Will, 1971d).

To do this we make a Lorentz transformation from the original PPN frame to a new frame which moves at velocity $\boldsymbol{w}$ relative to the old frame. In order to preserve the post-Newtonian character of the metric, we assume that $w \equiv |\boldsymbol{w}|$ is small, that is, of $O(\epsilon^{1/2})$. This transformation from rest coordinates $x^\alpha = (t, \boldsymbol{x})$ to moving coordinates $\xi^\mu = (\tau, \boldsymbol{\xi})$ can be expanded in powers of w to the required order: this approximate form of the Lorentz transformation is sometimes called a post-Galilean transformation (Chandrasekhar and Contopoulos, 1967), and has the form

$$\begin{aligned} \boldsymbol{x} &= \boldsymbol{\xi} + \left(1 + \frac{1}{2}w^2\right)\boldsymbol{w}\tau + \frac{1}{2}(\boldsymbol{\xi}\cdot\boldsymbol{w})\boldsymbol{w} + \xi \times O(\epsilon^2), \\ t &= \tau\left(1 + \frac{1}{2}w^2 + \frac{3}{8}w^4\right) + \left(1 + \frac{1}{2}w^2\right)\boldsymbol{\xi}\cdot\boldsymbol{w} + \tau \times O(\epsilon^3), \end{aligned} \tag{4.55}$$

where $\boldsymbol{w}\tau$ is assumed to be $O(\epsilon^0)$.

We use the standard transformation law

$$g_{\mu\nu}(\tau, \boldsymbol{\xi}) = \frac{\partial x^\alpha}{\partial \xi^\mu}\frac{\partial x^\beta}{\partial \xi^\nu} g_{\alpha\beta}(t, \boldsymbol{x}), \tag{4.56}$$

and express the potentials that appear in $g_{\alpha\beta}(t, \boldsymbol{x})$ in terms of the new coordinates. Since ρ, Π, and p are all measured in comoving local Lorentz frames, they are unchanged by the transformation, as is the quantity $\rho^* d^3x$. Thus, for any given element of fluid,

$$\begin{aligned}\rho(t,\boldsymbol{x}) &= \rho(\tau,\boldsymbol{\xi})\,,\\ \Pi(t,\boldsymbol{x}) &= \Pi(\tau,\boldsymbol{\xi})\,,\\ p(t,\boldsymbol{x}) &= p(\tau,\boldsymbol{\xi})\,,\\ \rho^*(t,\boldsymbol{x})d^3x &= \rho^*(\tau,\boldsymbol{\xi})d^3\xi\,.\end{aligned} \tag{4.57}$$

If $\boldsymbol{v}(t,\boldsymbol{x})$ and $\boldsymbol{\nu}(\tau,\boldsymbol{\xi})$ are the matter velocity fields in the two frames, they are related by

$$\boldsymbol{v} = \boldsymbol{\nu} + \boldsymbol{w} + O(\epsilon^{3/2})\,. \tag{4.58}$$

The quantity $\boldsymbol{x}(t) - \boldsymbol{x}'(t)$ that appears in the post-Newtonian potentials transforms according to

$$\begin{aligned}\boldsymbol{x}(t) - \boldsymbol{x}'(t) &= \boldsymbol{\xi}(\tau) - \boldsymbol{\xi}'(\tau') + \left(1 + \frac{1}{2}w^2\right)\boldsymbol{w}(\tau - \tau') + \frac{1}{2}\boldsymbol{w}[\boldsymbol{\xi}(\tau) - \boldsymbol{\xi}'(\tau')]\cdot\boldsymbol{w}\\ &\quad + \xi O(\epsilon^2)\,,\\ 0 &= (\tau - \tau')\left(1 + \frac{1}{2}w^2\right) + [\boldsymbol{\xi}(\tau) - \boldsymbol{\xi}'(\tau')]\cdot\boldsymbol{w} + \tau O(\epsilon^2)\,.\end{aligned} \tag{4.59}$$

But in the $(\tau, \boldsymbol{\xi})$ coordinates, the quantity $\boldsymbol{\xi} - \boldsymbol{\xi}'$ must be evaluated at the common time τ, hence we must use the fact that

$$\boldsymbol{\xi}'(\tau') = \boldsymbol{\xi}'(\tau) + \boldsymbol{\nu}'(\tau' - \tau) + O(\tau' - \tau)^2\,. \tag{4.60}$$

Combining Eqs. (4.59) and (4.60), we obtain

$$\frac{1}{|\boldsymbol{x} - \boldsymbol{x}'|} = \frac{1}{|\boldsymbol{\xi} - \boldsymbol{\xi}'|}\left[1 + \frac{1}{2}(\boldsymbol{w}\cdot\hat{\boldsymbol{n}}')^2 + (\boldsymbol{w}\cdot\hat{\boldsymbol{n}}')(\boldsymbol{\nu}'\cdot\hat{\boldsymbol{n}}') + O(\epsilon^2)\right], \tag{4.61}$$

where $\hat{\boldsymbol{n}}' \equiv (\boldsymbol{\xi} - \boldsymbol{\xi}')/|\boldsymbol{\xi} - \boldsymbol{\xi}'|$. Using Eqs. (4.57), (4.58), and (4.61), along with the definitions (4.31), (4.33), (4.34a), (4.37), and (4.41), we then find that

$$\begin{aligned}U(t,\boldsymbol{x}) &= U(\tau,\boldsymbol{\xi}) + w^jW_j(\tau,\boldsymbol{\xi}) + \frac{1}{2}w^jw^kU_{jk}(\tau,\boldsymbol{\xi}) + O(\epsilon^3)\,,\\ \Phi_1(t,\boldsymbol{x}) &= \Phi_1(\tau,\boldsymbol{\xi}) + 2w^jV_j(\tau,\boldsymbol{\xi}) + w^2U(\tau,\boldsymbol{\xi}) + O(\epsilon^3)\,,\\ \Phi_6(t,\boldsymbol{x}) &= \Phi_6(\tau,\boldsymbol{\xi}) + 2w^jW_j(\tau,\boldsymbol{\xi}) + w^jw^kU_{jk}(\tau,\boldsymbol{\xi}) + O(\epsilon^3)\,,\\ V_j(t,\boldsymbol{x}) &= V_j(\tau,\boldsymbol{\xi}) + w_jU(\tau,\boldsymbol{\xi}) + O(\epsilon^{5/2})\,,\\ X_{,0j}(t,\boldsymbol{x}) &= X_{,0j}(\tau,\boldsymbol{\xi}) - w^kX_{,kj}(\tau,\boldsymbol{\xi}) + O(\epsilon^{5/2})\,,\end{aligned} \tag{4.62}$$

where $_{,0}$ in $X_{,0j}(\tau,\boldsymbol{\xi})$ denotes $\partial/\partial\tau$. The post-Newtonian potentials Φ_2, Φ_3, Φ_4, Φ_5 and Φ_W are unchanged to this order. Applying the post-Galilean transformation Eqs. (4.55) and (4.56) to the PPN metric, Eqs. (4.53) and (4.54), and making use of Eqs. (4.62), we obtain for the metric in the moving $(\tau, \boldsymbol{\xi})$ frame, to PN order,

$$\begin{aligned}g_{00}(\tau,\boldsymbol{\xi}) &= 1 + 2U(\tau,\boldsymbol{\xi}) + 2\big(\psi(\tau,\boldsymbol{\xi}) - \beta U(\tau,\boldsymbol{\xi})^2\big)\\ &\quad - (\alpha_1 - \alpha_3)w^2U(\tau,\boldsymbol{\xi}) + \alpha_2w^jw^kX_{,jk}(\tau,\boldsymbol{\xi}) + (2\alpha_3 - \alpha_1)w^jV_j(\tau,\boldsymbol{\xi})\\ &\quad + (1 - \alpha_2 - \zeta_1 + 2\xi)w^kX_{,0k}(\tau,\boldsymbol{\xi}) + O(\epsilon^3),\end{aligned}$$

$$
\begin{aligned}
g_{0j} = &-\frac{1}{2}[4(1+\gamma)+\alpha_1]V_j(\tau,\boldsymbol{\xi}) - \frac{1}{2}[1+\alpha_2-\zeta_1+2\xi]X_{,0j}(\tau,\boldsymbol{\xi}) \\
&-\frac{1}{2}\alpha_1 w^j U(\tau,\boldsymbol{\xi}) + \alpha_2 w^k X_{,jk}(\tau,\boldsymbol{\xi}) \\
&+\frac{1}{2}(1-\alpha_2-\zeta_1+2\xi)w^k X_{,jk}(\tau,\boldsymbol{\xi}) + O(\epsilon^{5/2}), \\
g_{jk} = &\,(1+2\gamma U(\tau,\boldsymbol{\xi}))\,\delta_{jk} + O(\epsilon^2),
\end{aligned} \tag{4.63}
$$

where

$$
\begin{aligned}
\psi(\tau,\boldsymbol{\xi}) = &\,\frac{1}{2}(2\gamma+1+\alpha_3+\zeta_1-2\xi)\Phi_1(\tau,\boldsymbol{\xi}) - (2\beta-1-\zeta_2-\xi)\Phi_2(\tau,\boldsymbol{\xi}) \\
&+(1+\zeta_3)\Phi_3(\tau,\boldsymbol{\xi}) + (3\gamma+3\zeta_4-2\xi)\Phi_4(\tau,\boldsymbol{\xi}) \\
&-\frac{1}{2}(\zeta_1-2\xi)\Phi_6(\tau,\boldsymbol{\xi}) - \xi\Phi_W(\tau,\boldsymbol{\xi}).
\end{aligned} \tag{4.64}
$$

Because we now have available an additional post-Newtonian variable, $\boldsymbol{w}$, we have an additional gauge freedom that can be employed without altering the standard PPN gauge. This gauge was selected in the frame in which $\boldsymbol{w} = 0$, and was not affected by the post-Galilean transformation (spatial metric diagonal, no Φ_5 in g_{00}). By making the additional transformation

$$
\begin{aligned}
\bar{\tau} &= \tau + \frac{1}{2}(1-\alpha_2-\zeta_1+2\xi)w^k X_{,k}, \\
\bar{\xi}^j &= \xi^j,
\end{aligned} \tag{4.65}
$$

we can eliminate the term $(1-\alpha_2-\zeta_1+2\xi)w^k X_{,0k}$ from g_{00} and the term $(1/2)$ $(1-\alpha_2-\zeta_1+2\xi)w^k X_{,jk}$ from g_{0j}.

This then becomes the standard PPN gauge in a coordinate system moving at velocity $\boldsymbol{w}$ relative to the universal rest frame, namely that gauge in which g_{jk} is diagonal and isotropic, and in which the terms Φ_5 and $w^k X_{,0k}$ are absent from g_{00}. It is then straightforward to show that a further post-Galilean transformation with velocity $\boldsymbol{u}$ (plus a possible gauge transformation to maintain the standard gauge) does not alter the form of the PPN metric, it merely changes the coordinate system velocity $\boldsymbol{w}$ that appears there to $\boldsymbol{w}+\boldsymbol{u}$. At first glance, one might be disturbed by the presence of metric terms that depend on the coordinate system's velocity $\boldsymbol{w}$ relative to the universal rest frame. These terms do not violate the principles of special relativity since they are purely gravitational terms, while special relativity is valid only when the effects of gravitation can be ignored; but they do suggest that the gravitation generated by matter may be affected by its motion relative to the universe (violation of the Strong Equivalence Principle). Nevertheless, the results of physical measurements must not depend on the velocity $\boldsymbol{w}$ (this is a consequence of general covariance). For a system such as the Sun and planets, the only physically measurable velocities are the velocities of elements of matter relative to each other and to the center of mass of the system, and the velocity $\boldsymbol{w}_0$ of the center of mass relative to the universal rest frame (as measured for example by studying the dipole anisotropy in the cosmic microwave radiation induced by Doppler shifts). Thus, the PPN prediction for any physical effect can depend only on these relative velocities and on $\boldsymbol{w}_0$, never on $\boldsymbol{w}$. Therefore, the terms in the PPN metric that depend on $\boldsymbol{w}$ must signal the presence of effects that depend on $\boldsymbol{w}_0$. This can be seen most simply by working in a coordinate system in which the system under study is at rest, i.e., where $\boldsymbol{w} = \boldsymbol{w}_0$. Then, if any one of the set of parameters $\{\alpha_1, \alpha_2, \alpha_3\}$ is nonzero, there may be observable effects which depend on $\boldsymbol{w}_0$. On the other hand, if $\alpha_1 = \alpha_2 = \alpha_3 = 0$, there is no reference to $\boldsymbol{w}$ or $\boldsymbol{w}_0$ in the metric in any coordinate system, and no such effects can occur. Thus, we see that the parameters α_1, α_2 and α_3 measure the extent and manner in which motion relative to the universal rest frame affects the post-Newtonian metric and produces observable effects. These parameters are called "preferred-frame parameters" since they measure the size of post-Newtonian effects produced by motion relative to the "preferred" rest frame of the universe. If all three are zero, no such effects are present, and there is no preferred frame (to post-Newtonian order).

Notice that even if one works in the universal rest frame, where $\boldsymbol{w} = 0$, physical effects will be unchanged, for even though the explicit preferred-frame terms are absent, the velocities of elements of matter $\boldsymbol{v}$ that appear in the PPN metric and in the equations of motion must be decomposed according to

$$\boldsymbol{v} = \boldsymbol{w}_0 + \tilde{\boldsymbol{v}}, \tag{4.66}$$

where $\tilde{\boldsymbol{v}}$ is the velocity of each element relative to the center of mass, and, unless α_1, α_2, and α_3 are all zero, the same effects dependent upon $\boldsymbol{w}_0$ will result. At this point the PPN metric has taken on its standard form. Box 4.1 summarizes the basic definitions and formulae that enter the PPN formalism.

Box 4.1 The parametrized post-Newtonian formalism

A. *Coordinate system:* A quasi-local Lorentz coordinate system (Section 4.1.3) in which the coordinates are (t, x_1, x_2, x_3). Three-dimensional Euclidean vector notation is used. Coordinates are specialized to the standard PPN gauge (Section 4.2.2).

B. *Matter variables:*

1. $\rho =$ density of rest mass of a fluid element as measured in a local, freely falling, momentarily comoving frame.
2. $u^\mu =$ four-velocity of the momentarily comoving frame, with $v^j \equiv u^j/u^0 =$ coordinate velocity of the frame (and thus of the fluid element).
3. $\rho^* = \rho u^0 \sqrt{-g} =$ "conserved" or baryonic mass density. Satisfies the exact continuity equation $\rho^*_{,0} + \boldsymbol{\nabla} \cdot (\rho^* \boldsymbol{v}) = 0$.
4. $p =$ pressure as measured in a local, freely falling, momentarily comoving frame.
5. $\Pi =$ internal energy per unit rest mass. It includes all forms of non-rest-mass, nongravitational energy, such as thermal energy, magnetic energy, etc.
6. $\boldsymbol{w} =$ coordinate velocity of the PPN coordinate system relative to the mean rest frame of the universe.

C. PPN *parameters:*

$\gamma, \beta, \xi, \alpha_1, \alpha_2, \alpha_3, \zeta_1, \zeta_2, \zeta_3, \zeta_4.$

D. *Metric:*

$$g_{00} = -1 + 2U + 2(\psi - \beta U^2) + \Phi^{\rm PF},$$

$$g_{0j} = -\left[2(1+\gamma) + \frac{1}{2}\alpha_1\right]V_j - \frac{1}{2}\left[1 + \alpha_2 - \zeta_1 + 2\xi\right]X_{,0j} + \Phi_j^{\rm PF},$$

$$g_{jk} = (1 + 2\gamma U)\,\delta_{jk},$$

where

$$\psi = \frac{1}{2}(2\gamma + 1 + \alpha_3 + \zeta_1 - 2\xi)\Phi_1 - (2\beta - 1 - \zeta_2 - \xi)\Phi_2 + (1 + \zeta_3)\Phi_3 + (3\gamma + 3\zeta_4 - 2\xi)\Phi_4 - \frac{1}{2}(\zeta_1 - 2\xi)\Phi_6 - \xi\Phi_W.$$

E. *Metric potentials*

$$U = \int \frac{\rho'^*}{|\boldsymbol{x} - \boldsymbol{x}'|} d^3x', \quad V_j = \int \frac{\rho'^* v_j'}{|\boldsymbol{x} - \boldsymbol{x}'|} d^3x', \quad X = \int \rho'^* |\boldsymbol{x} - \boldsymbol{x}'| d^3x',$$

$$\Phi_1 = \int \frac{\rho'^* v'^2}{|\boldsymbol{x} - \boldsymbol{x}'|} d^3x', \quad \Phi_6 = \int \frac{\rho'^* [\boldsymbol{v}' \cdot (\boldsymbol{x} - \boldsymbol{x}')]^2}{|\boldsymbol{x} - \boldsymbol{x}'|^3} d^3x',$$

$$\Phi_2 = \int \frac{\rho'^* U'}{|\boldsymbol{x} - \boldsymbol{x}'|} d^3x', \quad \Phi_3 = \int \frac{\rho'^* \Pi'}{|\boldsymbol{x} - \boldsymbol{x}'|} d^3x', \quad \Phi_4 = \int \frac{p'}{|\boldsymbol{x} - \boldsymbol{x}'|} d^3x',$$

$$\Phi_W = \int \rho^{*\prime} \rho^{*\prime\prime} \frac{(\boldsymbol{x} - \boldsymbol{x}')}{|\boldsymbol{x} - \boldsymbol{x}'|^3} \cdot \left[\frac{(\boldsymbol{x}' - \boldsymbol{x}'')}{|\boldsymbol{x} - \boldsymbol{x}''|} - \frac{(\boldsymbol{x} - \boldsymbol{x}'')}{|\boldsymbol{x}' - \boldsymbol{x}''|}\right] d^3x' d^3x'',$$

F. *Preferred-frame potentials:*

$$\Phi^{\rm PF} = (\alpha_3 - \alpha_1) w^2 U + (2\alpha_3 - \alpha_1) w^j V_j + \alpha_2 w^j w^k X_{,jk},$$

$$\Phi_j^{\rm PF} = -\frac{1}{2}\alpha_1 w_j U + \alpha_2 w^k X_{,jk}.$$

G. *Energy-momentum tensor (perfect fluid):*

$$T^{00} = \rho^* \left[1 + \Pi + \frac{1}{2}v^2 + (2 - 3\gamma)U\right],$$

$$T^{0j} = \rho^* v^j \left[1 + \Pi + \frac{1}{2}v^2 + (2 - 3\gamma)U + p/\rho^*\right],$$

$$T^{jk} = \rho^* v^j v^k \left[1 + \Pi + \frac{1}{2}v^2 + (2 - 3\gamma)U + p/\rho^*\right] + p\delta^{jk}(1 - 2\gamma U),$$

H. *Equations of motion:*

1. Hydrodynamics:

$$T^{\mu\nu}{}_{;\nu} = 0,$$

2. Test bodies

$$\frac{d^2x^\mu}{d\lambda^2} + \Gamma^\mu_{\nu\lambda} \frac{dx^\nu}{d\lambda} \frac{dx^\lambda}{d\lambda} = 0,$$

3. Maxwell's equations

$$F^{\mu\nu}{}_{;\nu} = 4\pi J^\mu, \quad F_{\mu\nu} \equiv A_{\nu;\mu} - A_{\mu;\nu}.$$

4.4 Global Conservation Laws

Conservation laws in Newtonian gravitation theory are familiar: for isolated gravitating systems, mass is conserved, energy is conserved, linear and angular momenta are conserved, and the center of mass of the system moves uniformly. This does not apply to every metric theory of gravity, however. Some theories violate some of these conservation laws at the post-Newtonian level, and it is the purpose of this section to explore such violations using the PPN formalism.

One can distinguish two kinds of conservation laws: local and global. Local conservation laws are laws that are valid in any local Lorentz frame, and are independent of the metric theory of gravity. They depend rather, upon the structure of matter that one assumes. They include local baryon number conservation and the local conservation of energy or the first law of thermodynamics. These have been discussed in Chapter 3.

Global conservation laws, however, are statements about gravitating systems in asymptotically flat spacetime. Because they incorporate the structure of both the matter and the gravitational fields, they depend on the metric theory in question.

When we attempt to devise more general integral conservation laws, such as for total energy, total momentum, or total angular momentum, we run into difficulties. It is well known that integral conservation laws cannot be obtained directly from the equation of hydrodynamics $T^{\mu\nu}{}_{;\nu} = 0$ because of the presence of the Christoffel symbols in the covariant derivative. Rather, we search for a quantity $\tau^{\mu\nu}$, which reduces to $T^{\mu\nu}$ in flat spacetime and whose *ordinary* divergence in a coordinate basis vanishes, that is,

$$\tau^{\mu\nu}{}_{,\nu} = 0\,. \tag{4.67}$$

Then, provided that $\tau^{\mu\nu}$ is symmetric, we find that the quantities

$$P^\mu \equiv \int_\Sigma \tau^{\mu\nu} d\Sigma_\nu\,, \quad J^{\mu\nu} \equiv 2\int_\Sigma x^{[\mu}\tau^{\nu]\lambda} d\Sigma_\lambda\,, \tag{4.68}$$

are conserved, i.e., the integrals in Eq. (4.68) vanish when taken over a closed three-dimensional hypersurface Σ. If we choose a coordinate system $(t, \boldsymbol{x})$ in which Σ is a constant-time hypersurface that extends infinitely far in all spatial directions, then, provided that $\tau^{\mu\nu}$ vanishes sufficiently rapidly with spatial distance from the matter (thereby excluding for the moment a flux of gravitational waves), then P^μ and $J^{\mu\nu}$ are independent of time and are given by

$$P^\mu \equiv \int \tau^{\mu 0} d^3x\,, \quad J^{\mu\nu} \equiv 2\int x^{[\mu}\tau^{\nu]0} d^3x\,. \tag{4.69}$$

An appropriate choice of $\tau^{\mu\nu}$ allows one to interpret the components of P^μ and $J^{\mu\nu}$ in the usual way: as measured in the asymptotically flat spacetime far from the matter, P^0 is the total energy, P^j is the total momentum, J^{jk} is the total angular momentum, and J^{0j} determines the motion of the center of mass of the matter. If a suitable $\tau^{\mu\nu}$ exists but is not symmetric, then P^μ is conserved but $J^{\mu\nu}$ varies according to

$$\frac{dJ^{\mu\nu}}{dt} = -2\int \tau^{[\mu\nu]} d^3x\,. \tag{4.70}$$

The quantity $\tau^{\mu\nu}$, often called the energy-momentum pseudotensor, has been found for the exact versions of general relativity and scalar-tensor theories (see Chapter 5); for a general discussion of the existence of such objects see Lee et al. (1974). A wide variety of nonsymmetric stress-energy pseudotensors have been devised and discussed within general relativity, but only the symmetric version guarantees conservation of angular momentum.

There is a close connection between integral conservation laws and covariant Lagrangian formulations of metric theories. It is known (Lee et al., 1974) that every Lagrangian-based, generally covariant metric theory of gravity that either (i) is purely dynamical (possesses no absolute variables) or (ii) contains prior geometry, with a simple constraint on the symmetry group of its absolute variables (a constraint satisfied by all specific metric theories known), possesses conservation laws of the form $\tau^{\mu\nu}{}_{,\nu} = 0$, where $\tau^{\mu\nu}$ is a function of certain variational derivatives of the Lagrangian of the theory that reduces to $T^{\mu\nu}$ in the absence of gravity. When there are no absolute variables, the conservation laws are the result of invariance under coordinate transformations, and the $\tau^{\mu\nu}$ are not tensors (or tensor densities), hence the name pseudotensors: moreover, there may be infinitely many of them. When absolute variables are present, their symmetry group produces the conservation laws and $\tau^{\mu\nu}$ typically are true tensors (or tensor densities). Although $\tau^{\mu\nu}$ is guaranteed to exist for any Lagrangian-based metric theory, there is no guarantee that it will be symmetric, and no general argument is known to determine the conditions under which it will be symmetric.

In the post-Newtonian limit, the existence of conservation laws of the form of Eq. (4.67) can be translated into a condition on values of some of the PPN parameters. We will attempt to construct a $\tau^{\mu\nu}$ by adopting a form, valid to PN order,

$$\tau^{\mu\nu} \equiv (1 - aU)\,(T^{\mu\nu} + t^{\mu\nu}) \tag{4.71}$$

where a is a constant to be determined and $t^{\mu\nu}$ is a quantity (which may be interpreted under some circumstances as "gravitational energy-momentum") that vanishes in flat spacetime, and that is a function of the fields $U, X, \Phi_1, \Phi_W, V_j, \ldots$ and their derivatives, the velocity $\mathbf{w}$, (and may also contain the matter variables ρ, p, Π and $\mathbf{v}$). We reject PN terms in $\tau^{\mu\nu}$ of the form $v^2 T^{\mu\nu}$, $\Pi T^{\mu\nu}$, $(p/\rho)T^{\mu\nu}$, or $w^2 T^{\mu\nu}$, since such terms do not vanish in general in regions of negligible gravitational field.

By imposing the conditions $\tau^{\mu\nu}{}_{,\nu} = 0$, and $T^{\mu\nu}{}_{;\nu} = 0$ we find that, to post-Newtonian order, $t^{\mu\nu}$ must satisfy the equation

$$t^{\mu\nu}{}_{,\nu} - aU_{,\nu}t^{\mu\nu} = \Gamma^{\mu}_{\nu\lambda}T^{\nu\lambda} + \Gamma^{\lambda}_{\nu\lambda}T^{\mu\nu} + aU_{,\nu}T^{\mu\nu} \,. \tag{4.72}$$

In our attempt to integrate Eq. (4.72) we will make use of Box 4.1 and the definitions of the post-Newtonian potentials given in Section 4.2.1, together with formulae for the Christoffel symbols displayed in Box 6.1. We will also repeatedly exploit the following identity, which is valid for any function f:

$$4\pi\rho f_{,j} = -2\frac{\partial}{\partial x^k}\Gamma_{jk}(f) + U_{,j}\nabla^2 f, \tag{4.73}$$

where

$$\Gamma_{jk}(f) \equiv U_{,(j}f_{,k)} - \frac{1}{2}\delta_{jk}\boldsymbol{\nabla} U \cdot \boldsymbol{\nabla} f. \tag{4.74}$$

Two additional identities will be useful. To lowest order, it can be shown that $(UT^{0\nu})_{,\nu} = (\rho^* U)_{,0} + (\rho^* v^j U)_{,j}$; inserting the equations $\nabla^2 U = -4\pi\rho^*$ and $\nabla^2 V_j = -4\pi\rho^* v_j$ and manipulating the derivatives suitably, we obtain the identity

$$4\pi(UT^{0\nu})_{,\nu} - \frac{\partial}{\partial t}|\nabla U|^2 + \frac{\partial}{\partial x^k}\left(U_{,0}U_{,k} + 2U_{,j}V_{[j,k]}\right) = 0\,. \tag{4.75}$$

A second identity, obtained after tedious calculations, is given by

$$\begin{aligned}\rho^*\Phi_{W,j} = &-\frac{1}{4\pi}\frac{\partial}{\partial x^k}\Big[2\Gamma_{jk}(\Phi_W) + 3U\Gamma_{jk}(U) + 2\Gamma_{jk}(\nabla X\cdot\nabla U)\\ &- \Sigma_{\ell,\ell}X_{,jk} - X\Sigma_{\ell,\ell jk} + \delta_{jk}(X\Sigma_{\ell,\ell m})_{,m}\\ &+ 8\pi\rho^* X_{,(j}U_{,k)} + \delta_{jk}(U_{,0})^2\Big] + \frac{1}{2\pi}\frac{\partial}{\partial t}(U_{,j}U_{,0})\\ &+ U_{,j}\left[\rho^* v^2 + 2p - \frac{1}{8\pi}|\nabla U|^2 + \frac{1}{4\pi}\nabla^2\Phi_6\right],\end{aligned} \tag{4.76}$$

where Σ_ℓ is the solution of the equation

$$\nabla^2\Sigma_\ell = -4\pi\rho^* U_{,\ell}\,. \tag{4.77}$$

Then, Eq. (4.72) can be put into the form

$$\begin{aligned}4\pi t^{0\nu}{}_{,\nu} &= 4\pi\left(t^{00}{}_{,0} + t^{0j}{}_{,j}\right)\\ &= \frac{\partial}{\partial t}\left[\frac{1}{2}(6\gamma - 5 + 2a - 2b)|\nabla U|^2\right]\\ &\quad - \frac{\partial}{\partial x^j}\Big[(3\gamma - 2 + a - b)U_{,0}U_{,j} + 2(3\gamma - 3 + a - b)V_{[k,j]}U_{,k}\Big]\\ &\quad + 4\pi b\frac{\partial}{\partial x^\nu}(UT^{0\nu}),\\ 4\pi t^{j\nu}{}_{,\nu} &= 4\pi\left(t^{j0}{}_{,0} + t^{jk}{}_{,k}\right)\\ &= \frac{\partial}{\partial t}\Big\{\frac{1}{2}(4\gamma + 2 + \alpha_1 - 2\alpha_2 + 2\zeta_1)U_{,0}U_{,j} + (4\gamma + 4 + \alpha_1)V_{[k,j]}U_{,k}\\ &\quad + \frac{1}{2}\alpha_1 w^j U\nabla^2 U + \alpha_2 U_{,j}(\boldsymbol{w}\cdot\nabla U)\Big\}\\ &\quad \frac{\partial}{\partial x^k}\Big\{\Big[1 - (\zeta_2 + 4\xi - a)U + \frac{1}{2}(\alpha_3 - \alpha_1)w^2\Big]\Gamma_{jk}(U)\\ &\quad + 2\Gamma_{jk}(\psi) - 2\xi\Gamma_{jk}(\nabla X\cdot\nabla U) + (1 + \alpha_2 - \zeta_1 + 2\xi)\Gamma_{jk}(\ddot{X})\\ &\quad - 2(4\gamma + 4 + \alpha_1)\left(V_{[j,\ell]}V_{[k,\ell]} - \frac{1}{4}\delta_{jk}V_{[m,\ell]}V_{[m,\ell]}\right)\\ &\quad + (4\gamma + 4 + \alpha_1)\left(U_{,(j}V_{k),0} - \frac{1}{2}\delta_{jk}U_{,\ell}V_{\ell,0}\right)\\ &\quad - \frac{1}{4}(4\gamma + 2 + \alpha_1 - 2\alpha_2 + 2\zeta_1)\delta_{jk}\dot{U}^2\\ &\quad + \xi\left[2\nabla^2 UX_{,(j}U_{,k)} + X_{,jk}\Sigma_{\ell,\ell} + X\Sigma_{\ell,\ell jk} - \delta_{jk}(X\Sigma_{\ell,\ell m})_{,m}\right]\\ &\quad + (2\alpha_3 - \alpha_1)w^\ell\Gamma_{jk}(V_\ell) - 2\alpha_2 w^\ell\Gamma_{jk}(\dot{X}_{,\ell}) + \alpha_2 w^\ell w^m\Gamma_{jk}(X_{,\ell m})\end{aligned} \tag{4.78}$$

$$+ \alpha_2 \delta_{jk} \left[\frac{1}{2}(\boldsymbol{w} \cdot \boldsymbol{\nabla} U)^2 - \dot{U}(\boldsymbol{w} \cdot \boldsymbol{\nabla} U) \right] + B^{jk} \bigg\}$$

$$+ 4\pi(5\gamma + a - 1)\frac{\partial}{\partial x^\nu}(UT^{j\nu}) + 4\pi Q^j \,, \tag{4.79}$$

where ψ is given by Eq. (4.54),

$$B^{jk} = \frac{1}{2}\alpha_1 U w_j \nabla^2 V_k + \alpha_2 w_k U_{,j} \dot{U} - \alpha_2 w_k U_{,j} (\boldsymbol{w} \cdot \boldsymbol{\nabla} U) \,, \tag{4.80}$$

and the residual term Q^j is given by

$$Q^j = U_{,j} \left[\frac{1}{2}(\alpha_3 + \zeta_1)\rho^* v^2 + \frac{1}{8\pi}\zeta_1 \nabla^2 \Phi_6 + \frac{1}{8\pi}\zeta_2 |\boldsymbol{\nabla} U|^2 + \zeta_3 \rho^* \Pi \right.$$

$$\left. + 3\zeta_4 p + \alpha_3 \rho^* \boldsymbol{w} \cdot \boldsymbol{v} \right] . \tag{4.81}$$

We have also taken the liberty of adding b times the identity (4.75) to the right-hand-side of the time component $t^{0\nu}{}_{,\nu}$. The reason for this will be clear shortly.

We first notice that Eq. (4.78) is automatically integrable, but Eq. (4.79) is not, because of the residual term Q^j. Any attempt to write Q^j purely as a combination of gradients and time derivatives of gravitational fields and matter variables simply generates other residual terms of similar forms. Thus the integrability of Eq. (4.79) for arbitrary systems requires that each of the terms in Q^i vanish identically, and therefore that

$$\alpha_3 \equiv \zeta_1 \equiv \zeta_2 \equiv \zeta_3 \equiv \zeta_4 \equiv 0 \,. \tag{4.82}$$

These constraints must be satisfied by any metric theory in order that there be conservation laws of the form of Eq. (4.67). If these conditions hold, then expressions for the conserved energy and momentum can be obtained using Eqs. (4.69) and (4.71), combined with expressions for T^{00} and T^{j0} from Box 4.1, and for t^{00} and t^{j0} read off from Eqs. (4.78) and (4.79), respectively. The results are (after integrations by parts):

$$P^0 = \int \rho^* \left(1 + \frac{1}{2}v^2 - \frac{1}{2}U + \Pi\right) d^3x \,,$$

$$P^j = \int \rho^* \left[v^j \left(1 + \frac{1}{2}v^2 - \frac{1}{2}U + \Pi + p/\rho\right) - \frac{1}{2}(1 + \alpha_1)V_j - \frac{1}{2}(1 + \alpha_2)X_{,0j} \right.$$

$$\left. - \frac{1}{2}\left(\alpha_1 w_j U - \alpha_2 w^k X_{,jk}\right) \right] d^3x \,. \tag{4.83}$$

These results are independent of the parameter a. In the expression for P^0, the first term is the total conserved rest mass of particles in the fluid. The other terms are the total kinetic, gravitational, and internal energies in the fluid, whose sum is conserved according to Newtonian theory (which can be used in any post-Newtonian terms). Thus, P^0 is simply the total mass-energy of the fluid, accurate to $O(\epsilon)$ beyond the rest mass, and is conserved irrespective of the validity of the conditions in Eqs. (4.82). This is related to the automatic integrability of Eq. (4.78). However, if those conditions were violated, one would expect violations of the conservation of P^0 at $O(\epsilon^2)$.

An alternative derivation of the conserved momentum uses the technique of integrating the hydrodynamic equations of motion $T^{\mu\nu}{}_{;\nu} = 0$ over all space, and searching for a quantity $\tilde{P}^j$ whose total time derivative vanishes. This procedure is blocked by a term $\int Q^j d^3x$, where Q^j, is given by Eq. (4.81). This integral can be written as a total time derivative only if Q^j can be written as a combination of time derivatives and spatial divergences (which lead to surface integrals at infinity that vanish). But according to the reasoning given earlier, this can be true only if the five parameter constraints of Eq. (4.82) are satisfied. Then Q^j in fact vanishes, and the conserved $\tilde{P}^j$ derived by this method agrees with Eq. (4.83).

We now see the physical significance of the parameters α_3, ζ_1, ζ_2, ζ_3 and ζ_4: they measure the extent and manner in which a given metric theory of gravity predicts violations of conservation of total momentum at PN order. If all five are zero in any given theory, then momentum is conserved; if some are nonzero, then momentum may not be conserved. According to the theorem of Lee et al. (1974), every Lagrangian-based metric theory of gravity must have all five conservation law parameters vanish. Notice that the parameter α_3 plays a dual role in the PPN formalism, both as a conservation-law parameter and as a preferred-frame parameter.

In order to guarantee conservation of the tensor $J^{\mu\nu}$, the object $t^{\mu\nu}$ must be symmetric. Eqs. (4.78) and (4.79) show that $t^{0j} \neq t^{j0}$ and that there are nonsymmetric terms, B^{jk} in t^{jk}. However, in integrating Eqs. (4.78) and (4.79), we have the freedom to add to the nominal solutions for $t^{\mu\nu}$ any quantity $S^{\mu\nu}$ that satisfies the identity $S^{\mu\nu}_{,\nu} = 0$. We have already used such an identity (4.75) in $t^{0\nu}$, but we have been utterly unable to find an identity that will eliminate or symmetrize the offending terms B^{jk} in t^{jk}. As for the t^{0j} and t^{j0} components, we can select the parameter $b = 5\gamma + a - 1$ so that the term UT^{0j} in t^{0j} matches the term UT^{j0} in t^{j0}. The result is that

$$\begin{aligned} 8\pi t^{[0j]} &= -\frac{1}{2}(\alpha_1 - 2\alpha_2)U_{,0}U_{,j} - \alpha_1 U_{,k}V_{[k,j]} - \frac{1}{2}\alpha_1 w_j U\nabla^2 U - \alpha_2 U_{,j}(\boldsymbol{w}\cdot\boldsymbol{\nabla}U)\,, \\ 8\pi t^{[jk]} &= 2B^{[jk]} = \alpha_1 U w_{[j}\nabla^2 V_{k]} - 2\alpha_2 w_{[j}U_{,k]}(\dot{U} - \boldsymbol{w}\cdot\boldsymbol{\nabla}U)\,. \end{aligned} \tag{4.84}$$

Symmetry of $t^{\mu\nu}$ requires that each of the terms in Eqs. (4.84) vanishes identically, that is,

$$\alpha_1 \equiv \alpha_2 \equiv 0\,. \tag{4.85}$$

Any other choice of b will lead to the same constraints.

We apply the name Fully Conservative Theory to any theory of gravity that possesses a full complement of post-Newtonian conservation laws: energy, momentum, angular momentum, and center-of-mass motion, that is, whose PPN parameters satisfy

$$\alpha_1 \equiv \alpha_2 \equiv \alpha_3 \equiv \zeta_1 \equiv \zeta_2 \equiv \zeta_3 \equiv \zeta_4 \equiv 0\,. \tag{4.86}$$

A fully conservative theory cannot be a preferred frame theory to post-Newtonian order since $\alpha_1 = \alpha_2 = \alpha_3 = 0$. For such theories, only three PPN parameters, γ, β and ξ may vary from theory to theory; $\tau^{\mu\nu}$ and $t^{\mu\nu}$ in this case take the form

$$\tau^{\mu\nu} = [1 + (5\gamma - 1)U]\,T^{\mu\nu} + \tilde{t}^{\mu\nu}\,, \tag{4.87}$$

where

$$\tilde{t}^{00} = -\frac{1}{8\pi}(4\gamma + 3)|\boldsymbol{\nabla}U|^2\,,$$

$$
\begin{aligned}
\tilde{t}^{0j} = \tilde{t}^{j0} &= \frac{1}{4\pi}\left[(2\gamma+1)U_{,j}U_{,0} + 4(\gamma+1)U_{,k}V_{[k,j]}\right], \\
\tilde{t}^{jk} &= (1-4\xi U)\Gamma_{jk}(U) + 2\Gamma_{jk}(\psi) - 2\xi\Gamma_{jk}(\nabla X\cdot\nabla U) + (1+2\xi)\Gamma_{jk}(\ddot{X}) \\
&\quad - 8(\gamma+1)\left(V_{[j,\ell]}V_{[k,\ell]} - \frac{1}{4}\delta_{jk}V_{[m,\ell]}V_{[m,\ell]}\right) \\
&\quad + 4(\gamma+1)\left(U_{,(j}V_{k),0} - \frac{1}{2}\delta_{jk}U_{,\ell}V_{\ell,0}\right) - \frac{1}{2}(2\gamma+1)\delta_{jk}\dot{U}^2 \\
&\quad + \xi\left[2\nabla^2 U X_{,(j}U_{,k)} + X_{,jk}\Sigma_{\ell,\ell} + X\Sigma_{\ell,\ell jk} - \delta_{jk}(X\Sigma_{\ell,\ell m})_{,m}\right],
\end{aligned}
\tag{4.88}
$$

where

$$
\psi = \frac{1}{2}(2\gamma+1-2\xi)\Phi_1 - (2\beta-1-\xi)\Phi_2 + \Phi_3 + (3\gamma-2\xi)\Phi_4 + \xi\Phi_6 - \xi\Phi_W. \tag{4.89}
$$

The conserved quantities are

$$
\begin{aligned}
P^0 &= \int \rho^*\left(1 + \frac{1}{2}v^2 - \frac{1}{2}U + \Pi\right)d^3x, \\
P^j &= \int \rho^*\left[v^j\left(1 + \frac{1}{2}v^2 - U + \Pi + p/\rho^*\right) - \frac{1}{2}X_{,0j}\right]d^3x, \\
J^{jk} &= \int \rho^* x^{[j}\left\{v^{k]}\left[1 + \frac{1}{2}v^2 - (2\gamma+1)U + \Pi + p/\rho^*\right]\right. \\
&\quad \left. - (2\gamma+2)V^{k]} - \frac{1}{2}\dot{X}^{,k]}\right\}d^3x, \\
J^{0j} &= \int \rho^* x^j\left(1 + \frac{1}{2}v^2 - \frac{1}{2}U + \Pi\right)d^3x - P^j t.
\end{aligned}
\tag{4.90}
$$

By defining a center of mass X^j given by

$$
X^j = \frac{\int \rho^* x^j\left(1 + \frac{1}{2}v^2 - \frac{1}{2}U + \Pi\right)d^3x}{\int \rho^*\left(1 + \frac{1}{2}v^2 - \frac{1}{2}U + \Pi\right)d^3x}, \tag{4.91}
$$

we find from Eqs. (4.90) and the constancy of J^{0j} that

$$
\frac{dX^j}{dt} = \frac{P^j}{P^0}, \tag{4.92}
$$

that is, the center of mass moves uniformly with velocity P^j/P^0.

Some theories of gravity may possess only energy and momentum conservation laws, that is, their parameters may satisfy

$$
\alpha_3 \equiv \zeta_1 \equiv \zeta_2 \equiv \zeta_3 \equiv \zeta_4 \equiv 0, \quad \text{one of}\,\{\alpha_1, \alpha_2\} \neq 0. \tag{4.93}
$$

We call such theories Semiconservative Theories. Their conserved four-momentum P^μ may be obtained from Eqs. (4.83), but their nonconserved $J^{\mu\nu}$ are ambiguous. A peculiar feature of the semiconservative case is that in a coordinate system at rest with respect to the universe, $\boldsymbol{w} = 0$, and the spatial components t^{jk} are automatically symmetric, irrespective of the values of α_1 and α_2, (since $B^{jk} = 0$ if $\boldsymbol{w} = 0$; see Eq. (4.80)). Thus, spatial angular momentum J^{jk} is a conserved quantity in this frame, whereas it is not in a moving frame. The center-of-mass component J^{0j}, however, is not conserved in any frame, since $t^{0j} \neq t^{j0}$ for any $\boldsymbol{w}$. This discrepancy can be understood by noting that the distinction between J^{jk} and J^{0j} is not a Lorentz-invariant distinction. Because the PPN metric is post-Galilean invariant, the quantities P^μ and $J^{\mu\nu}$ should transform as a vector and antisymmetric tensor respectively under post-Galilean transformations. This can be verified explicitly by applying the transformation Eq. (4.55) to the integrals that comprise P^μ and $J^{\mu\nu}$, with the result, valid to post-Newtonian order

$$\begin{aligned} P^{0'} &= P^0\left(1+\frac{1}{2}w^2\right) - \boldsymbol{w}\cdot\boldsymbol{P}, \\ \boldsymbol{P}' &= \boldsymbol{P} - \left(1+\frac{1}{2}w^2\right)\boldsymbol{w}P^0 + \frac{1}{2}\boldsymbol{w}(\boldsymbol{w}\cdot\boldsymbol{P}), \\ J^{jk'} &= J^{jk} - J^{\ell[j}w^{k]}w^\ell + 2\left(1+\frac{1}{2}w^2\right)J^{0[j}w^{k]}, \\ J^{0j'} &= \left(1+\frac{1}{2}w^2\right)J^{0j} - w^k J^{kj} - \frac{1}{2}w^j w^k J^{0k}, \end{aligned} \tag{4.94}$$

where $\boldsymbol{w}$ is the velocity of the boost. Thus a boost from the universe rest frame where $dJ^{jk}/dt = 0$ to a frame moving with velocity $\boldsymbol{w}$ yields

$$\dot{J}^{jk'} = 2\dot{J}^{0[j}w^{k]}\left[1 + O(w^2)\right]. \tag{4.95}$$

Thus, the violation of angular momentum conservation is intimately connected with the violation of uniform center-of-mass motion. This is our reason for stating that semiconservative theories of gravity possess *only* energy and momentum conservation laws. Eq. (4.95) may be verified explicitly using Eq. (4.84), and the fact that to PN order,

$$\dot{J}^{jk} = -2\int t^{[jk]}d^3x, \quad \dot{J}^{0j} = -2\int t^{[0j]}d^3x. \tag{4.96}$$

Every Lagrangian-based theory of gravity is at least semiconservative.

Nonconservative theories possess no conservation laws (other than the trivial one for P^0); their parameters satisfy

$$\text{one of}\{\alpha_3, \zeta_1, \zeta_2, \zeta_3, \zeta_4\} \neq 0. \tag{4.97}$$

Table 4.1 summarizes the significance of the various PPN parameters.

4.5 Other Post-Newtonian Gauges

The choice of the standard PPN gauge has no physical implications; it is merely a choice of coordinates. It was made mainly for historical reasons: it was the gauge used by Chandrasekhar (1965) in his seminal papers on post-Newtonian hydrodynamics, and it was these papers that started the author along the path of post-Newtonian theory. In general relativity, another coordinate choice, called "harmonic" coordinates has proven to be extremely useful in developing systematic approximation methods to solutions of Einstein's equation, both for post-Newtonian theory and for gravitational radiation theory.

Table 4.1 PPN parameters and their physical significance.

Parameter	What it measures relative to GR	GR value	Semi-conservative theories	Fully-conservative theories
γ	How much spatial curvature produced by mass?	1	γ	γ
β	How much nonlinearity in superposition of gravity?	1	β	β
ξ	Preferred-location effects?	0	ξ	ξ
α_1	Preferred-frame effects?	0	α_1	0
α_2		0	α_2	0
α_3		0	0	0
α_3	Is total momentum conserved?	0	0	0
ζ_1		0	0	0
ζ_2		0	0	0
ζ_3		0	0	0
ζ_4		0	0	0

Note that α_3 is listed twice to indicate that it is a measure of two separate effects.

This choice is the basis for the development of post-Minkowskian and post-Newtonian theory described in detail in PW, for example. In fact, we will use harmonic coordinates in Chapter 5 to derive the post-Newtonian limits of general relativity and scalar-tensor theory, and will use them again in Chapter 11 to analyze gravitational radiation in those theories.

Accordingly, it would be useful to express the PPN metric in a coordinate system that is a generalization of harmonic coordinates that might occur in a range of alternative theories. In general relativity, harmonic coordinates are established via the condition

$$\mathfrak{g}^{\mu\nu}{}_{,\nu} = 0 \,, \tag{4.98}$$

where the "gothic inverse metric" $\mathfrak{g}^{\mu\nu}$ is defined by

$$\mathfrak{g}^{\mu\nu} \equiv \sqrt{-g} g^{\mu\nu} \,, \tag{4.99}$$

where g is the determinant of the metric. In an alternative theory of gravity, this might not be the most appropriate choice, but we could imagine that a generalized post-Newtonian "harmonic" condition,

$$\left[(1 - aU)\sqrt{-g}\, g^{\mu\nu}\right]_{,\nu} = 0 \,, \tag{4.100}$$

might be appropriate, where a is a parameter to be determined within the given theory, and U is the Newtonian potential. In general relativity, $a = 0$, and the gauge is the standard harmonic gauge. In Chapter 5, we will see that in scalar-tensor theories, the factor $1 - aU$ is related to a function of the scalar field. In what follows we will see that only the $O(\epsilon)$ term in the factor is needed for the gauge condition at the first PN order. We will confine our attention to semi-conservative theories of gravity, since the overwhelming majority of currently viable theories are Lagrangian based.

To see what the condition (4.100) implies for the PPN metric, we first make a general gauge transformation of the form $x^{\bar{\mu}} = x^\mu + \xi^\mu$, where

$$\xi_0 = \lambda_1 X_{,0} + \lambda_3 w^j X_{,j}\,, \quad \xi_j = \lambda_2 X_{,j}\,. \tag{4.101}$$

The metric in the new coordinates takes the form

$$\begin{aligned}
g_{\bar{0}\bar{0}} &= g_{00} - 2\lambda_1 X_{,00} + 2\lambda_2 (U^2 + \Phi_2 + \Phi_W) - 2\lambda_3 w^j X_{,0j}\,,\\
g_{\bar{0}\bar{j}} &= g_{0j} - (\lambda_1 + \lambda_2) X_{,0j} - \lambda_3 w^k X_{,jk}\,,\\
g_{\bar{j}\bar{k}} &= g_{jk} - 2\lambda_2 X_{,jk}\,,
\end{aligned} \tag{4.102}$$

and $-\bar{g} = 1 + (6\gamma - 4\lambda_2 - 2)U$. For the "generalized gothic inverse metric" $\bar{\mathfrak{g}}^{\bar{\alpha}\bar{\beta}} \equiv (1 - aU)\sqrt{-\bar{g}}\bar{g}^{\bar{\alpha}\bar{\beta}}$, we obtain

$$\begin{aligned}
\bar{\mathfrak{g}}^{\bar{0}\bar{0}} &= -1 - (3\gamma - 2\lambda_2 - a + 1)\, U + O(\epsilon^2)\,,\\
\bar{\mathfrak{g}}^{\bar{0}\bar{j}} &= -\left[2(\gamma+1) + \frac{1}{2}\alpha_1\right] V_j - \frac{1}{2}(1 + \alpha_2 + 2\xi + 2\lambda_1 + 2\lambda_2) X_{,0j}\\
&\quad - \frac{1}{2}\alpha_1 w^j U + (\alpha_2 - \lambda_3) w^k X_{,jk}\,,\\
\bar{\mathfrak{g}}^{\bar{j}\bar{k}} &= [1 + (\gamma - 2\lambda_2 - 1 - a)U]\,\delta_{jk} + 2\lambda_2 X_{,jk}\,,
\end{aligned} \tag{4.103}$$

where the potentials are expressed in the new coordinates. Imposing the four conditions $\bar{\mathfrak{g}}^{\bar{\alpha}\bar{\beta}}{}_{,\bar{\beta}} = 0$ we obtain the conditions

$$\begin{aligned}
0 &= \left(a - \gamma + \frac{1}{2}\alpha_1 - \alpha_2 - 2\xi - 2\lambda_1\right)\dot{U} + \left(2\alpha_2 - \frac{1}{2}\alpha_1 - 2\lambda_3\right)\boldsymbol{w}\cdot\boldsymbol{\nabla} U\,,\\
0 &= (\gamma - 1 - a + 2\lambda_2) U_{,j}\,,
\end{aligned} \tag{4.104}$$

which fix the values of λ_1, λ_2 and λ_3. Substituting these values back into Eq. (4.102), we obtain for the PPN metric in generalized harmonic gauge (we drop the overbars):

$$\begin{aligned}
g_{00} &= -1 + 2U + \left[2\psi_{\rm Harm} - (2\beta + \gamma - 1 - a)\, U^2\right]\\
&\quad + \left(\gamma - a - \frac{1}{2}\alpha_1 + \alpha_2 + 2\xi\right)\ddot{X} + \Phi^{\rm PF}_{\rm Harm}\,,\\
g_{0j} &= -\left[2(1+\gamma) + \frac{1}{2}\alpha_1\right] V_j - \left[a + 1 - \gamma + \frac{1}{4}\alpha_1\right] X_{,0j} + \Phi^{\rm PF}_{j\,\rm Harm}\,,\\
g_{jk} &= (1 + 2\gamma U)\,\delta_{jk} - (a + 1 - \gamma) X_{,jk}\,,
\end{aligned} \tag{4.105}$$

where

$$\begin{aligned}
\psi_{\rm Harm} &= \frac{1}{2}(2\gamma + 1 - 2\xi)\Phi_1 - \left(2\beta - \frac{3}{2} + \frac{1}{2}\gamma - \frac{1}{2}a - \xi\right)\Phi_2 + \Phi_3\\
&\quad + (3\gamma - 2\xi)\Phi_4 + \xi\Phi_6 + \frac{1}{2}(a - \gamma + 1 - 2\xi)\,\Phi_W\,,\\
\Phi^{\rm PF}_{\rm Harm} &= -\alpha_1 w^2 U - \alpha_1 w^j V_j + \alpha_2 w^j w^k X_{,jk} - \left(2\alpha_2 - \frac{1}{2}\alpha_1\right) w^j X_{,0j}\,,\\
\Phi^{\rm PF}_{j\,\rm Harm} &= -\frac{1}{2}\alpha_1 w^j U + \frac{1}{4}\alpha_1 w^k X_{,jk}\,.
\end{aligned} \tag{4.106}$$

We can choose $a = \gamma - 1$, and put the metric in the form

$$
\begin{aligned}
g_{00} &= -1 + 2U + \left[2\psi - 2\beta U^2\right] + \left(1 - \frac{1}{2}\alpha_1 + \alpha_2 + 2\xi\right)\ddot{X} + \Phi^{\rm PF}_{\rm Harm} \,, \\
g_{0j} &= -\left[2(1+\gamma) + \frac{1}{2}\alpha_1\right] V_j - \frac{1}{4}\alpha_1 X_{,0j} + \Phi^{\rm PF}_{j\,{\rm Harm}} \,, \\
g_{jk} &= (1 + 2\gamma U)\,\delta_{jk} \,,
\end{aligned}
\tag{4.107}
$$

where ψ is given by Eq. (4.89). Thus in a generalized harmonic gauge, the PPN parameters of the semiconservative theory can equally well be read off from the metric.

5 Metric Theories of Gravity and Their Post-Newtonian Limits

Despite the fact that today general relativity is widely considered to be the "standard model" for gravity, there has never been a period in its history in which serious alternative theories were not being studied (here we exclude the many "crackpot" theories, which are not serious and which inhabit a world, and preprint archive, of their own). When the first edition of this book was published in 1981, the list of alternative theories looked entirely different from the list that we will present in this chapter. Many of the theories discussed then – Whitehead's theory, Rosen's theory – are no longer viable or of interest, and many of the theories discussed here – Einstein-Æther, TeVeS, Chern-Simons, massive gravity, and so on – did not exist in 1981. On the other hand, there has been one alternative theory that, like the zombies of the movies, keeps recovering from near death and returning for more action: the scalar-tensor theory.

The history of alternative theories of gravity can be divided roughly into three periods. The first period, from the publication of general relativity until the development of the Brans-Dicke scalar-tensor theory around 1960, was one in which alternative theories were invented in response to perceived difficulties with general relativity in the minds of their inventors. These difficulties included the relative lack of empirical support, problems with some of the concepts or consequences of curved spacetime, or the apparent absence of certain features, such as Mach's principle. In addition to the Brans-Dicke theory, examples include Whitehead's 1922 theory, Birkhoff's 1943 theory, and Belinfante and Swihart's 1957 theory. Many of these early alternative theories were examined in detail in the review by Whitrow and Morduch (1965).

The second period, from around 1960 to the mid 1980s, might be called the "straw-man" period. Armed with the deep understanding of the nature of gravitational theory provided by the Dicke framework (see Chapter 2) together with the PPN framework, many investigators invented alternative theories, not because they believed in them necessarily, but to illustrate the kinds of effects that could be predicted and to explore and sharpen their differences with general relativity. They posed questions like "what does it take to have a theory with a non-zero value of α_1 or α_2?," or "what if we turned the coupling constant ω of Brans-Dicke theory into a function $\omega(\phi)$? Even the author of this book, in collaboration with Kenneth Nordtvedt Jr. was guilty of inventing an alternative theory in this manner (see Section 5.4.2). To be fair, some theories, notably Nathan Rosen's 1973 bimetric theory, *were* invented for a real purpose and to be in competition with general relativity.

The third period, beginning in the middle 1980s might be termed the "beyond-Einstein" period. This period opened with string theory, which revealed the potentially deep connection between the microscopic world of high-energy particles and the large-scale world of gravity. Although developed in part in an attempt understand the strong interactions, string theory had the unexpected property that it seemed to predict general relativity in the small string-coupling limit. Yet the limit wasn't pure general relativity, instead it could be interpreted as a scalar tensor theory with $\omega = -1$; this would strongly violate experiment (see Chapter 7), were it not for the expectation that the scalar field would acquire a large mass via spontaneous symmetry breaking, and thus its effects would be exponentially suppressed on any macroscopic scale, restoring an effective theory equivalent to general relativity to high accuracy. This and other connections with particle physics then provided the motivation for many of the alternative theories discussed in this chapter. Another motivation was the exploration of phenomena, such as violations of Lorentz symmetry, that might be relics of quantum gravity. A third motivation was provided by the 1998 discovery of the accelerated expansion of the universe. While the nominal "ΛCDM" cosmological model treated this as the result either of the presence of Einstein's old cosmological constant Λ, or of a new form of matter, dubbed "dark energy," the alternative possibility was a modification of general relativity itself at cosmological scales.

We call this the beyond-Einstein period because, for the most part, the alternatives studied were not meant to replace or compete with general relativity, but rather to extend it into regimes far beyond those where it had been well tested, such as the solar system and binary pulsars, but with modifications motivated by particle physics, quantum gravity or cosmology.

In this chapter, we will focus in some detail on general relativity and a variety of modifications of it generated by introducing additional scalar fields (the Brans-Dicke theory being the iconic example), vector fields or tensor fields. We will also briefly review a variety of theories that have been motivated either by the discovery of the acceleration of the universe, by considerations of how quantum gravity might change general relativity in the low-energy limit, or by efforts to make gravity a massive field. The number and variety of alternative theories of gravity has become so large that it is impossible to cover them all here. For a comprehensive review, see Berti et al. (2015).

We begin in Section 5.1 with a discussion of the general method of calculating post-Newtonian limits of metric theories of gravity. We then discuss in turn, general relativity (Section 5.2), scalar-tensor theories (Section 5.3), vector-tensor theories (Section 5.4), and tensor-vector-scalar (TeVeS) theories (Section 5.5). Sections 5.6 and 5.7 briefly cover quadratic gravity and Chern-Simons theories, and massive gravity theories. Section 5.8 will conclude this chapter with brief historical remarks about the rise and fall of some notable alternative theories.

5.1 Method of Calculation

Despite the large differences in structure between different metric theories of gravity, the calculation of the post-Newtonian limit possesses a number of universal features that are worth summarizing. It is just these common features that cause the post-Newtonian limit to have a nearly universal form, except for the values of the PPN parameters. Thus, the computation of the post-Newtonian limits of various theories tends to have a repetitive character, the major variable usually being the amount algebraic complexity involved. In order to streamline the presentation of specific theories in the following sections, and to establish a uniform notation, we present a "cookbook" for calculating post-Newtonian limits of any metric theory of gravity.

Step 1: Identify the variables: (a) dynamical gravitational variables such as the metric $g_{\mu\nu}$, scalar field ϕ, vector field K^μ (generally assumed to be timelike), tensor field $B_{\mu\nu}$, and so on; (b) prior-geometrical variables such as a flat or de Sitter background metric $\eta_{\mu\nu}$, a cosmic time function T, and so on; and (c) matter and nongravitational field variables.

Step 2: Set the cosmological boundary conditions. Assume a homogeneous isotropic cosmology, and at a chosen moment of time and in an asymptotic coordinate system define the values of the variables far from the post-Newtonian system. With isotropic coordinates in the rest frame of the universe, a convenient choice that is compatible with the symmetry of the situation is, for the dynamical variables,

$$\begin{aligned} g_{\mu\nu} &\to g^{(0)}_{\mu\nu} = \mathrm{diag}(-c_0,\, c_1,\, c_1,\, c_1)\,, \\ \phi &\to \phi_0\,, \\ K^\mu &\to (K,\, 0,\, 0,\, 0)\,, \\ B_{\mu\nu} &\to B^{(0)}_{\mu\nu} = \mathrm{diag}(-b_0,\, b_1,\, b_1,\, b_1)\,, \end{aligned} \tag{5.1}$$

and for the prior-geometric variables (these values are valid everywhere since these variables are independent of the local system),

$$\begin{aligned} \eta_{\mu\nu} &= \mathrm{diag}(-1,\, 1,\, 1,\, 1)\,, \\ T &= T, \text{ with } \boldsymbol{\nabla} T = (1,\, 0,\, 0,\, 0)\,. \end{aligned} \tag{5.2}$$

The relationships among and the evolution of these asymptotic values will be set by a solution of the cosmological problem. Because these asymptotic values may affect the values of the PPN parameters, a complete determination of the post-Newtonian limit may in fact require a complete cosmological solution. This can be very complicated in some theories. For the present, we will avoid these complications by simply assuming that the cosmological matching parameters are arbitrary constants (or more precisely, arbitrary slowly varying functions of time). Notice that if a flat background metric $\boldsymbol{\eta}$ is present, it is almost always most convenient to work in a coordinate system in which it has the Minkowski form, for in many theories the resulting field equations involve flat-spacetime wave equations, which are easy to solve. Then the asymptotic form of $\boldsymbol{g}$ shown is determined by the cosmological solution. If $\boldsymbol{\eta}$ is present it is *not* generally possible (unless in a special cosmology or at a special cosmological epoch) to make both it and $\boldsymbol{g}$ have the asymptotic Minkowski form simultaneously. Of course, once the post-Newtonian metric $\boldsymbol{g}$ has been determined, one can always choose a local quasi-Cartesian coordinate system (see Section 4.1.3) in which it takes the asymptotic Minkowski form. The form that $\boldsymbol{\eta}$ now takes is irrelevant since, unlike $\boldsymbol{g}$, it does not couple to matter. In theories without $\boldsymbol{\eta}$, it is usually convenient to choose asymptotically Minkowski coordinates right away.

Step 3: Expand in a post-Newtonian series about the asymptotic values:

$$\begin{aligned} g_{\mu\nu} &= g^{(0)}_{\mu\nu} + p_{\mu\nu}\,, \\ \phi &= \phi_0(1+\Psi)\,, \\ K^\mu &= (K+k^0,\, k^1,\, k^2,\, k^3)\,, \\ B_{\mu\nu} &= B^{(0)}_{\mu\nu} + b_{\mu\nu}\,. \end{aligned} \tag{5.3}$$

Generally, the post-Newtonian orders of these perturbations are given by

$$\begin{aligned} &p_{00} \sim O(\epsilon) + O(\epsilon^2)\,, \quad p_{0j} \sim O(\epsilon^{3/2})\,, \quad p_{jk} \sim O(\epsilon)\,, \\ &\Psi \sim O(\epsilon) + O(\epsilon^2)\,, \\ &k^0 \sim O(\epsilon) + O(\epsilon^2)\,, \quad k^j \sim O(\epsilon^{3/2})\,, \\ &b_{00} \sim O(\epsilon) + O(\epsilon^2)\,, \quad b_{0j} \sim O(\epsilon^{3/2})\,, \quad b_{jk} \sim O(\epsilon)\,. \end{aligned} \tag{5.4}$$

Step 4: Substitute these forms into the field equations, keeping only such terms as are necessary to obtain a final, consistent post-Newtonian solution for $p_{\mu\nu}$. Make use of all the bookkeeping tools of the post-Newtonian limit (Section 4.1), including the relation $(\partial/\partial t)/(\partial/\partial x) \sim O(\epsilon^{1/2})$. For the matter sources, substitute the perfect-fluid stress-energy tensor $T^{\mu\nu}$ and associated fluid variables.

Step 5 : Solve for p_{00} to $O(\epsilon)$. Only the lowest post-Newtonian order equations are needed. Assuming that $p_{00} \to 0$ far from the system, one obtains the form

$$p_{00} = 2\alpha U\,, \tag{5.5}$$

where U is the Newtonian gravitational potential, defined in terms of the "conserved" density ρ^* [Eq. (4.31)], and where α may be a complicated function of cosmological matching parameters and of other coupling constants that may appear in the theory's field equations (such as a "gravitational coupling constant"). To Newtonian order, the metric thus has the form

$$\begin{aligned} g_{00} &= -c_0 + 2\alpha U + O(\epsilon^2)\,, \\ g_{0j} &= O(\epsilon^{3/2})\,, \\ g_{jk} &= c_1\delta_{jk} + O(\epsilon)\,. \end{aligned} \tag{5.6}$$

To put the metric into standard Newtonian and post-Newtonian form in local quasi-Cartesian coordinates, we must make the coordinate transformation

$$x^{\bar 0} = (c_0)^{1/2}x^0\,, \quad x^{\bar j} = (c_1)^{1/2}x^j\,, \tag{5.7}$$

which results in $g_{\bar 0\bar 0} = c_0^{-1}g_{00}$, $g_{\bar 0\bar j} = (c_0c_1)^{-1/2}g_{0j}$, and $g_{\bar j\bar k} = c_1^{-1}g_{jk}$, and $\bar U = (c_1)^{-1/2}U$ (recall that $\rho^* d^3x$ is invariant). We arrive finally at

$$\begin{aligned} g_{\bar 0\bar 0} &= -1 + 2(\alpha c_0^{-1}c_1^{1/2})U + O(\epsilon^2)\,, \\ g_{\bar 0\bar j} &= O(\epsilon^{3/2})\,, \\ g_{\bar j\bar k} &= \delta_{jk} + O(\epsilon)\,. \end{aligned} \tag{5.8}$$

Because we work in units in which the gravitational constant measured today far from gravitating matter is unity, we must set

$$G_{\text{today}} \equiv \alpha c_0^{-1} c_1^{1/2} = 1 \,. \tag{5.9}$$

The constraint provided by this equation often simplifies other calculations, however there is no physical constraint implied; it is merely a definition of units.

Step 6: Solve for p_{jk} to $O(\epsilon)$ and p_{0j} to $O(\epsilon^{3/2})$. These solutions can be obtained from the linearized versions of the field equations. The field equations of some theories have a gauge freedom, and a certain choice of gauge often simplifies solution of the equations. However, the gauge so chosen need not be the standard PPN gauge (Section 4.2.2), and a gauge (coordinate) transformation into the standard gauge (Box 4.1) or the generalized harmonic gauge [Eq. (4.105)] may be necessary once the complete solution has been obtained.

Step 7: Solve for p_{00} to $O(\epsilon^2)$. This is the messiest step, involving all the nonlinearities in the field equations and many of the lower-order solutions for the gravitational variables. For example, if p_{jk} from Step 6 has the form $p_{jk} = 2a_1 U\delta_{jk} + a_2 X_{,jk}$, then the determinant of the metric has the form

$$-g = c_0 c_1^3 \left[1 - 2\left(\frac{\alpha}{c_0} - \frac{3a_1 + a_2}{c_1} \right) U \right] . \tag{5.10}$$

The energy-momentum tensor $T^{\mu\nu}$ must also be expanded to post-Newtonian order. Using Eqs. (3.77), (5.6), (5.9), and (5.10), we obtain

$$\begin{aligned} T^{00} &= \frac{\rho^*}{c_0 c_1^{3/2}} \left[1 + \Pi + \frac{c_1}{2c_0} v^2 + \left(\frac{2\alpha}{c_0} - \frac{3a_1 + a_2}{c_1} \right) U + O(\epsilon^2) \right] , \\ T^{0j} &= \frac{\rho^*}{c_0 c_1^{3/2}} \left[v^j + O(\epsilon^{3/2}) \right] , \\ T^{jk} &= \frac{\rho^*}{c_0 c_1^{3/2}} v^j v^k + \frac{p}{c_1} \delta^{jk} + \rho^* O(\epsilon^2) \,. \end{aligned} \tag{5.11}$$

Step 8: Convert to local quasi-Cartesian coordinates (Eq. (5.7)) and to the standard PPN gauge (Section 4.2.2).

Step 9: By comparing the result for $g_{\mu\nu}$ with Eq. (4.53) or with the PPN metric in Box 4.1 (with $\boldsymbol{w} = 0$), read off the PPN parameter values.

In general relativity, it is convenient to expand not $g_{\mu\nu}$, but the gothic inverse metric, $\mathfrak{g}^{\mu\nu} \equiv \sqrt{-g} g^{\mu\nu} = \eta^{\mu\nu} - h^{\mu\nu}$. This leads to what is called the Landau-Lifshitz formulation of general relativity, an approach that is very useful in post-Newtonian theory and gravitational radiation (for a complete exposition, see PW). We will use this method to calculate the PPN parameters for general relativity in the next section. In others, such as scalar-tensor theories, it is convenient first to make a conformal transformation to an auxiliary metric $\tilde{g}_{\mu\nu} = f(\phi) g_{\mu\nu}$ and to rewrite the field equations in term of that metric. It turns out that the same Landau-Lifshitz approach that works in general relativity can then be applied using a gothic inverse metric defined using $\tilde{g}_{\mu\nu}$.

5.2 General Relativity

5.2.1 Statement of the theory

We begin with the "standard model" of gravitation, general relativity.

(a) Gravitational fields present: the metric $\boldsymbol{g}$.

(b) Arbitrary parameters and functions: None. We will ignore the effects of either dark energy or a cosmological constant, which are too small to be directly measurable in the solar system or binary pulsars.

(c) Cosmological matching parameters: None.

(d) Field Equations: The field equations are derivable from an invariant action principle $\delta I = 0$, where

$$I = \frac{1}{16\pi G}\int R\sqrt{-g}\,d^4x + I_{\mathrm{NG}}(q_A, g_{\mu\nu}), \tag{5.12}$$

where $R = g^{\mu\nu}R_{\mu\nu}$ is the Ricci scalar constructed from the Ricci tensor [Eq. (3.42)], G is the gravitational coupling constant, and I_{NG} is the universally coupled nongravitational action. By varying the action with respect to $g_{\mu\nu}$, we obtain the field equations

$$G_{\mu\nu} = 8\pi G T_{\mu\nu}, \tag{5.13}$$

where $G_{\mu\nu} = R_{\mu\nu} - \frac{1}{2}g_{\mu\nu}R$ is the Einstein tensor. Varying I_{NG} with respect to the matter variables q_A yields the matter field equations, while the general covariance of I_{NG} yields the energy-momentum conservation equation $T^{\mu\nu}{}_{;\nu} = 0$ if the matter field equations are satisfied (see Eq. (3.69)). This is compatible with the gravitational field equations (5.13) and the Bianchi identity $G^{\mu\nu}{}_{;\nu} = 0$ satisfied by the Einstein tensor. This compatibility is not unique to general relativity. It holds in any Lagrangian-based theory that has no "absolute elements" (Lee et al., 1974).

5.2.2 The post-Newtonian limit and the PPN parameters

We will derive the post-Newtonian limit using the so-called post-Minkowskian theory, which is based on the Landau-Lifshitz formulation of general relativity. The reader is referred to chapter 6 of PW for a pedagogical development of this method. Because $\boldsymbol{g}$ is the only gravitational field present, we can choose it to be asymptotically Minkowskian without affecting any other fields. Thus, we can set $c_1 = c_2 = 1$

We first define the gothic inverse metric (really a metric density),

$$\mathfrak{g}^{\mu\nu} \equiv \sqrt{-g}g^{\mu\nu}, \tag{5.14}$$

along with the object $H^{\alpha\mu\beta\nu} \equiv \mathfrak{g}^{\alpha\beta}\mathfrak{g}^{\mu\nu} - \mathfrak{g}^{\alpha\nu}\mathfrak{g}^{\beta\mu}$, and make use of the identity

$$H^{\alpha\mu\beta\nu}{}_{,\mu\nu} = 2(-g)G^{\alpha\beta} + 16\pi G(-g)t^{\alpha\beta}_{\mathrm{LL}}, \tag{5.15}$$

where $t^{\mu\nu}_{\rm LL}$ is the Landau-Lifshitz pseudotensor, given by

$$\begin{aligned}(-g)t^{\alpha\beta}_{\rm LL} \equiv \frac{1}{16\pi G}\Big\{ & \mathfrak{g}^{\alpha\beta}_{,\lambda}\mathfrak{g}^{\lambda\mu}_{,\mu} - \mathfrak{g}^{\alpha\lambda}_{,\lambda}\mathfrak{g}^{\beta\mu}_{,\mu} + \frac{1}{2}\mathfrak{g}^{\alpha\beta}\mathfrak{g}_{\lambda\mu}\mathfrak{g}^{\lambda\nu}_{,\rho}\mathfrak{g}^{\mu\rho}_{,\nu} \\ & - \mathfrak{g}^{\alpha\lambda}\mathfrak{g}_{\mu\nu}\mathfrak{g}^{\beta\nu}_{,\rho}\mathfrak{g}^{\mu\rho}_{,\lambda} - \mathfrak{g}^{\beta\lambda}\mathfrak{g}_{\mu\nu}\mathfrak{g}^{\alpha\nu}_{,\rho}\mathfrak{g}^{\mu\rho}_{,\lambda} + \mathfrak{g}_{\lambda\mu}\mathfrak{g}^{\nu\rho}\mathfrak{g}^{\alpha\lambda}_{,\nu}\mathfrak{g}^{\beta\mu}_{,\rho} \\ & + \frac{1}{8}(2\mathfrak{g}^{\alpha\lambda}\mathfrak{g}^{\beta\mu} - \mathfrak{g}^{\alpha\beta}\mathfrak{g}^{\lambda\mu})(2\mathfrak{g}_{\nu\rho}\mathfrak{g}_{\sigma\tau} - \mathfrak{g}_{\rho\sigma}\mathfrak{g}_{\nu\tau})\mathfrak{g}^{\nu\tau}_{,\lambda}\mathfrak{g}^{\rho\sigma}_{,\mu}\Big\}. \end{aligned} \tag{5.16}$$

We then impose a "harmonic" coordinate condition

$$\mathfrak{g}^{\alpha\beta}{}_{,\beta} = 0\,, \tag{5.17}$$

and introduce the potentials

$$h^{\alpha\beta} \equiv \eta^{\alpha\beta} - \mathfrak{g}^{\alpha\beta}\,, \tag{5.18}$$

where $\eta^{\alpha\beta} = \mathrm{diag}(-1, 1, 1, 1)$ is the asymptotic Minkowski metric expressed in Cartesian coordinates, and where $h^{\alpha\beta}$ also satisfies the harmonic condition $h^{\alpha\beta}_{,\beta} = 0$. The Einstein equations can then be expressed equivalently as a wave equation for $h^{\alpha\beta}$,

$$\Box h^{\alpha\beta} = -16\pi G(-g)\left(T^{\alpha\beta} + t^{\alpha\beta}_{\rm LL} + t^{\alpha\beta}_{\rm H}\right), \tag{5.19}$$

where $\Box$ is the flat spacetime d'Alembertian, $-(\partial/\partial t)^2 + \nabla^2$, and where $t^{\alpha\beta}_{\rm H}$ is a "harmonic" pseudotensor, given by

$$(-g)t^{\alpha\beta}_{\rm H} \equiv \frac{c^4}{16\pi G}\left(h^{\alpha\nu}_{,\mu}h^{\beta\mu}_{,\nu} - h^{\mu\nu}h^{\alpha\beta}_{,\mu\nu}\right). \tag{5.20}$$

Because $h^{\alpha\beta}$ appears on the right-hand side of Eq. (5.19), either embedded within the matter source term, or quadratically in the Landau-Lifshitz and harmonic pseudotensors, the wave equation (Eq. 5.19) lends itself to iteration. Substitute the zeroth-order solution $h^{\alpha\beta}_0 = 0$ into the right-hand side, solve the wave equation for the first-order solution $h^{\alpha\beta}_1$, substitute that into the right-hand side, solve, and so on (see PW for detailed discussion).

To obtain the first iteration, we have that $T^{00} = \rho^*$, $T^{0j} = \rho^* v^j$, $T^{jk} = \rho^* v^j v^k + p\delta^{jk}$, $-g = 1$, modulo corrections of order ϵ. Substituting $h^{\mu\nu}_0 = 0$ into the right-hand side of the wave equation (Eq. 5.19), it is straightforward to obtain

$$\begin{aligned} h^{00}_1 &= 4GU + O(\epsilon^2)\,, \\ h^{0j}_1 &= 4GV^j + O(\epsilon^{5/2})\,, \\ h^{jk}_1 &= O(\epsilon^2)\,. \end{aligned} \tag{5.21}$$

With this starting point, we can use Eqs. (5.14) and (5.18) to express the metric as an expansion in terms of the potentials $h^{\alpha\beta}$, keeping only the terms needed to obtain the post-Newtonian solution:

$$g_{00} = -1 + \frac{1}{2}h^{00} - \frac{3}{8}(h^{00})^2 + \frac{1}{2}h^{kk} + O(\epsilon^3)\,, \tag{5.22}$$

$$g_{0j} = -h^{0j} + O(\epsilon^{5/2})\,, \tag{5.23}$$

$$g_{jk} = \delta^{ij}\left(1 + \frac{1}{2}h^{00}\right) + O(\epsilon^2), \tag{5.24}$$

$$(-g) = 1 + h^{00} + O(\epsilon^2). \tag{5.25}$$

We now choose units in which $G = 1$, so that $g_{00} = -1 + 2U + O(\epsilon^2)$. To complete the post-Newtonian metric, we need to obtain the second-iterated solution only for h^{00} and for the trace h^{kk}. Substituting $h_1^{\mu\nu}$ into the right-hand side of the wave equation, making use of Eq. (5.11) with $c_0 = c_1 = \alpha = 1$ and $3a_1 + a_2 = 3$, and keeping only terms that will contribute to the required PN order, we obtain the wave equations for the second iteration,

$$\begin{aligned}\Box h_2^{00} &= -16\pi\rho^*\left[1 + \frac{1}{2}v^2 + 3U + \Pi\right] + 14|\nabla U|^2, \\ \Box h_2^{kk} &= -16\pi\left(\rho^* v^2 + 3p\right) + 2|\nabla U|^2.\end{aligned} \tag{5.26}$$

Noting that

$$|\nabla U|^2 = \frac{1}{2}\nabla^2 U^2 - \nabla^2\Phi_2, \tag{5.27}$$

and that the solution of the wave equation $\Box V = -4\pi\rho^*$ is the retarded solution

$$V = \int \frac{\rho^*(t - |\boldsymbol{x} - \boldsymbol{x}'|, \boldsymbol{x}')}{|\boldsymbol{x} - \boldsymbol{x}'|} d^3x', \tag{5.28}$$

which can be approximated within the near zone by expanding the retardation about time t,

$$\begin{aligned}V &= \int \frac{\rho^*(t, \boldsymbol{x}')}{|\boldsymbol{x} - \boldsymbol{x}'|} d^3x' - \frac{\partial}{\partial t}\int \rho^*(t, \boldsymbol{x}') d^3x' + \frac{1}{2}\frac{\partial^2}{\partial t^2}\int \rho^*(t, \boldsymbol{x}')|\boldsymbol{x} - \boldsymbol{x}'| d^3x' + \dots \\ &= U + \frac{1}{2}\ddot{X} + O(\epsilon^{5/2}),\end{aligned} \tag{5.29}$$

we obtain the solutions

$$\begin{aligned}h_2^{00} &= 4U + 7U^2 + 2\Phi_1 - 2\Phi_2 + 4\Phi_3 + 2\ddot{X} + O(\epsilon^{5/2}), \\ h_2^{kk} &= U^2 + 4\Phi_1 - 2\Phi_2 + 12\Phi_4 + O(\epsilon^{5/2}).\end{aligned} \tag{5.30}$$

The $O(\epsilon^{5/2})$ terms in Eq. (5.30) turn out to be trivial gauge effects, and thus the errors are actually of $O(\epsilon^3)$. Substituting into Eq. (5.25) we obtain the post-Newtonian metric

$$\begin{aligned}g_{00} &= -1 + 2U + 2\left(\psi - U^2\right) + \ddot{X} + O(\epsilon^3), \\ g_{0j} &= -4V_j + O(\epsilon^{5/2}), \\ g_{jk} &= \delta_{jk}(1 + 2U) + O(\epsilon^2).\end{aligned} \tag{5.31}$$

where

$$\psi \equiv \frac{3}{2}\Phi_1 - \Phi_2 + \Phi_3 + 3\Phi_4. \tag{5.32}$$

We can put this into the standard PPN gauge by making the coordinate transformation $\bar{t} = t - \frac{1}{2}\dot{X}$. By comparing the result with the PPN metric displayed in Box 4.1 it is easy to read off the PPN parameter values

$$\begin{aligned}&\gamma = \beta = 1, \quad \xi = 0, \\ &\alpha_1 = \alpha_2 = \alpha_3 = \zeta_1 = \zeta_2 = \zeta_3 = \zeta_4 = 0.\end{aligned} \tag{5.33}$$

Alternatively, we can remain in harmonic coordinates and compare the metric (5.31) directly with the version given in Eq. (4.105), with the parameter $a = 0$. Again, it is simple to read off the same PPN parameter values. Note that general relativity is a fully conservative theory with no preferred-frame effects.

5.3 Scalar-Tensor Theories

5.3.1 Statement of the theory

Scalar-tensor theories have proven to be the most interesting, compelling and resilient of alternatives to general relativity. The first and most famous such theory was formulated by Brans and Dicke (1961), based on previous work by Fierz and Jordan. For a time during the late 1970s and early 1980s, mounting experimental evidence in strong support of general relativity caused interest in the Brans-Dicke theory to wane, but soon the theory returned, albeit in a generalized form, inspired by ideas from string theory, inflation, extra-dimensions, and so on [early generalizations were studied by Wagoner (1970) and Nordtvedt (1970)]. The theories are simple, adding a single scalar field to the metric tensor.

(a) Gravitational fields present: the metric $\mathbf{g}$ and a dynamical scalar field ϕ.

(b) Arbitrary parameters and functions: A coupling function $\omega(\phi)$ and a potential $U(\phi)$.

(c) Cosmological matching parameters: ϕ_0, the asymptotic value of the scalar field.

(d) Field Equations: The field equations are derived from the action

$$I = \frac{1}{16\pi G}\int\left[\phi R - \frac{\omega(\phi)}{\phi}g^{\mu\nu}\phi_{,\mu}\phi_{,\nu} - U(\phi)\right]\sqrt{-g}d^4x + I_{\rm NG}(q_A, g_{\mu\nu})\,. \tag{5.34}$$

It is straightforward to vary the action with respect to $g_{\mu\nu}$ and ϕ to obtain the field equations,

$$G_{\mu\nu} = \frac{8\pi G}{\phi}T_{\mu\nu} + \frac{\omega(\phi)}{\phi^2}\left(\phi_{,\mu}\phi_{,\nu} - \frac{1}{2}g_{\mu\nu}\phi_{,\lambda}\phi^{,\lambda}\right) + \frac{1}{\phi}\left(\phi_{;\mu\nu} - g_{\mu\nu}\Box_g\phi\right)\,, \tag{5.35a}$$

$$\Box_g\phi = \frac{1}{3+2\omega(\phi)}\left(8\pi G T - \frac{d\omega}{d\phi}\phi_{,\lambda}\phi^{,\lambda} + \frac{d}{d\phi}(\phi^2 U)\right)\,, \tag{5.35b}$$

where $T = g^{\mu\nu}T_{\mu\nu}$ and $\Box_g$ is the scalar d'Alembertian with respect to the metric.

There is an alternative formulation of the theory in which one introduces an auxiliary metric $\tilde{g}_{\mu\nu}$ related to the physical metric $g_{\mu\nu}$ by the conformal transformation

$$\tilde{g}_{\mu\nu} = \frac{\phi}{\phi_0}g_{\mu\nu}\,, \tag{5.36}$$

where ϕ_0 is a constant to be chosen later. This recasts the action of the theory into the form

$$I = \frac{1}{16\pi\tilde{G}}\int\left[\tilde{R} - \frac{3+2\omega(\phi)}{2\phi^2}\tilde{g}^{\mu\nu}\phi_{,\mu}\phi_{,\nu} - V(\phi)\right]\sqrt{-\tilde{g}}d^4x + I_{\rm NG}(q_A, \phi^{-1}\tilde{g}_{\mu\nu})\,, \tag{5.37}$$

where $\tilde{R}$ is the Ricci scalar constructed from $\tilde{g}_{\mu\nu}$, $\tilde{G} = G/\phi_0$ and $V(\phi) = \phi_0 U(\phi)/\phi^2$. This formulation looks like a standard general relativistic action $\tilde{R}\sqrt{-\tilde{g}}$ plus a kinetic and a potential term for the scalar field. The appearance of ϕ in the nongravitational action would appear to violate the metric hypothesis, but this is an artifact of the redefinition of $g_{\mu\nu}$. In terms of the original metric $g_{\mu\nu}$, this is still a metric theory. The representation of the theory in Eq. (5.34) is often called the "Jordan frame," while the representation in (5.37) is called the "Einstein frame" (why this convention is the opposite of what one might have guessed *a priori* is a mystery). The Einstein frame is useful for discussing general characteristics of such theories, for certain numerical relativity calculations, and for some cosmological applications, while the metric representation or Jordan frame is most useful for calculating observable effects. The following redefinition of the scalar field,

$$A(\varphi) \equiv \phi^{-1/2}\,,$$
$$\frac{dA(\varphi)}{d\varphi} \equiv \phi^{-1/2}[3 + 2\omega(\phi)]^{-1/2}\,, \tag{5.38}$$

puts the action into the even simpler form

$$I = \frac{1}{16\pi G}\int \left[\tilde{R} - 2\tilde{g}^{\mu\nu}\varphi_{,\mu}\varphi_{,\nu} - \bar{V}(\varphi)\right]\sqrt{-\tilde{g}}d^4x + I_{\rm NG}(q_A, A(\varphi)^2\tilde{g}_{\mu\nu})\,, \tag{5.39}$$

where $\bar{V}(\varphi) = V(\phi)$. This version was popularized by Damour and Esposito-Farèse (1992a), who also studied multiscalar tensor theories.

5.3.2 Post-Newtonian limit and PPN parameters

We follow the general procedure described for general relativity in Section 5.2.2, but working in terms of the auxiliary metric $\tilde{g}_{\alpha\beta}$ instead of the physical metric $g_{\alpha\beta}$. We introduce the gothic inverse metric

$$\tilde{\mathfrak{g}}^{\alpha\beta} \equiv \sqrt{-\tilde{g}}\tilde{g}^{\alpha\beta}\,, \tag{5.40}$$

the object $\tilde{H}^{\alpha\mu\beta\nu} = \tilde{\mathfrak{g}}^{\alpha\beta}\tilde{\mathfrak{g}}^{\mu\nu} - \tilde{\mathfrak{g}}^{\alpha\nu}\tilde{\mathfrak{g}}^{\beta\mu}$, and rely on the same identity (5.15) and Landau-Lifshitz pseudotensor (5.16), but with everything now expressed in terms of "tilde" quantities. We introduce the potentials

$$\tilde{h}^{\alpha\beta} \equiv \eta^{\alpha\beta} - \tilde{\mathfrak{g}}^{\alpha\beta}\,, \tag{5.41}$$

impose the harmonic gauge condition

$$\tilde{h}^{\alpha\beta}_{\ ,\beta} = 0\,, \tag{5.42}$$

and arrive at the wave equation

$$\Box\tilde{h}^{\alpha\beta} = -16\pi\tilde{G}(-\tilde{g})\left(\tilde{T}^{\alpha\beta} + \tilde{t}^{\alpha\beta}_{\phi} + \tilde{t}^{\alpha\beta}_{\rm LL} + \tilde{t}^{\alpha\beta}_{\rm H}\right), \tag{5.43}$$

where $\tilde{T}^{\alpha\beta} = (\phi_0/\phi)^3 T^{\alpha\beta}$, $\tilde{t}^{\alpha\beta}_{\rm LL}$ and $\tilde{t}^{\alpha\beta}_{\rm H}$ are the Landau-Lifshitz and harmonic pseudotensors of Eq. (5.20), but defined using $\tilde{h}^{\alpha\beta}$, and

$$(-\tilde{g})\tilde{t}^{\alpha\beta}_{\phi} \equiv \frac{1}{16\pi\tilde{G}}\left[\frac{3+2\omega}{\phi^2}\left(\tilde{g}^{\alpha\mu}\tilde{g}^{\beta\nu} - \frac{1}{2}\tilde{g}^{\alpha\beta}\tilde{g}^{\mu\nu}\right)\phi_{,\mu}\phi_{,\nu} - V(\phi)\tilde{g}_{\alpha\beta}\right]. \tag{5.44}$$

The scalar field equation (5.35b) can also be expressed in the form of a wave equation in flat spacetime:

$$\Box\phi = -8\pi\tilde{G}\tau_s\,, \tag{5.45}$$

where

$$\begin{aligned}\tau_s = &-\sqrt{-\tilde{g}}\frac{\phi}{3+2\omega}\left(\tilde{T}+\frac{\phi}{8\pi\tilde{G}}\frac{dV}{d\phi}\right)\\ &+\frac{1}{16\pi\tilde{G}}\left\{\frac{d}{d\phi}\left[\ln\left(\frac{3+2\omega}{\phi^2}\right)\right]\tilde{\mathfrak{g}}^{\alpha\beta}\phi_{,\alpha}\phi_{,\beta}-2\tilde{h}^{\alpha\beta}\phi_{,\alpha\beta}\right\},\end{aligned} \tag{5.46}$$

where $\tilde{T}\equiv\tilde{T}^{\mu\nu}\tilde{g}_{\mu\nu}=(\phi_0/\phi)^2T$. Henceforth we will ignore the scalar field potential $V(\phi)$, which, among other effects, can give the scalar field an effective mass, and can generate self-interactions or inflationary behavior.

We now expand $\phi=\phi_0(1+\Psi)$, where ϕ_0 is identified as the asymptotic value of ϕ far from the system, and $\Psi\sim O(\epsilon)$, and anticipating that $\tilde{h}^{00}\sim O(\epsilon)$, $\tilde{h}^{0j}\sim O(\epsilon^{3/2})$, and $\tilde{h}^{jk}\sim O(\epsilon^2)$, we can use Eqs. (5.36) and (5.40) to express the physical metric to PN order in the form

$$g_{00} = -1+\frac{1}{2}\tilde{h}^{00}+\Psi-\frac{3}{8}\left(\tilde{h}^{00}\right)^2+\frac{1}{2}\tilde{h}^{kk}-\frac{1}{2}\tilde{h}\Psi-\Psi^2+O(\epsilon^3)\,, \tag{5.47}$$

$$g_{0j} = -\tilde{h}^{0j}+O(\epsilon^{5/2})\,, \tag{5.48}$$

$$g_{jk} = \delta^{ij}\left(1+\frac{1}{2}\tilde{h}^{00}-\Psi\right)+O(\epsilon^2)\,, \tag{5.49}$$

$$(-g) = 1+\tilde{h}^{00}-4\Psi+O(\epsilon^2)\,. \tag{5.50}$$

Then from the first iteration of the field equations (5.43) and (5.45), we obtain

$$\begin{aligned}\tilde{h}_1^{00} &= 4\tilde{G}U+O(\epsilon^2)\,,\\ \tilde{h}_1^{0j} &= 4\tilde{G}V^j+O(\epsilon^{5/2})\,,\\ \tilde{h}_1^{jk} &= O(\epsilon^2)\,,\end{aligned} \tag{5.51}$$

and

$$\Psi = \frac{2}{3+2\omega_0}\tilde{G}U+O(\epsilon^2)\,, \tag{5.52}$$

where $\omega_0\equiv\omega(\phi_0)$. Then, to Newtonian order, we have

$$g_{00} = -1+2\tilde{G}\frac{4+2\omega_0}{3+2\omega_0}U\equiv -1+2G_{\text{today}}U\,, \tag{5.53}$$

where we define the measured present value G_{today} by

$$G_{\text{today}}\equiv\tilde{G}\frac{4+2\omega_0}{3+2\omega_0}\,. \tag{5.54}$$

We then choose units in which $G_{\text{today}}=1$. Note that if ϕ_0 changes as a result of cosmic evolution, then G_{today} may change from its present value of unity. We will discuss experimental bounds on any time variation of G_{today} in Chapter 8.

Proceeding to the second iteration of the field equations, noting that we must expand $\omega(\phi) = \omega_0 + \omega_0'(\phi_0\Psi) + \ldots$, where $\omega_0' \equiv d\omega/d\phi|_0$, and defining the parameters

$$\zeta \equiv \frac{1}{4+2\omega_0}, \qquad \lambda \equiv \frac{\phi_0\omega_0'}{(3+2\omega_0)(4+2\omega_0)}, \tag{5.55}$$

we obtain finally

$$\begin{aligned} g_{00} &= -1 + 2U + 2\left[\psi - (1+\zeta\lambda)U^2 + \frac{1}{2}X_{,00}\right] + O(\epsilon^3), \\ g_{0j} &= -\frac{4}{c^3}(1-\zeta)V^j + O(\epsilon^{5/2}), \\ g_{jk} &= \delta_{jk}\left[1 + 2(1-2\zeta)U\right] + O(\epsilon^2), \end{aligned} \tag{5.56}$$

where

$$\psi = \frac{1}{2}(3-4\zeta)\Phi_1 - (1+2\zeta\lambda)\Phi_2 + \Phi_3 + 3(1-2\zeta)\Phi_4 . \tag{5.57}$$

Since $\phi/\phi_0 = 1+\Psi = 1+2\zeta U+O(\epsilon^2)$, it is simple to show that the gauge condition (5.42) corresponds to our generalized harmonic gauge of Chapter 4, Eq. (4.100), with $a = -2\zeta$, and so comparing Eqs. (5.56) and (5.57) with (4.105) and (4.106) (with $\boldsymbol{w} = 0$, since we are working in the universal rest frame) we can read off the PPN parameters:

$$\begin{aligned} \gamma &= 1 - 2\zeta = \frac{1+\omega_0}{2+\omega_0}, \\ \beta &= 1 + \zeta\lambda = 1 + \frac{\phi_0\omega_0'}{(3+2\omega_0)(4+2\omega_0)^2}, \end{aligned} \tag{5.58}$$

and $\xi = \alpha_i = \zeta_i = 0$. Alternatively, we can bring Eq. (5.56) to the standard PPN gauge of Box 4.1 by making the coordinate transformation $t = \bar{t} + \frac{1}{2}X_{,0}$ and $x^j = \bar{x}^j$, and comparing the two metrics.

5.3.3 General remarks

In Brans-Dicke theory, $\omega = \omega_{BD}$ is strictly constant; the larger the value of ω_{BD}, the smaller the effects of the scalar field, and in the limit $\omega_{BD} \to \infty$, the theory becomes indistinguishable from GR in all its predictions. This is largely why the fortunes of Brans-Dicke theory declined by the late 1970s, as experimental data pushed the allowed value of ω_{BD} higher and higher (into the thousands). As long as observations continue to support general relativity, Brans-Dicke theory with a large enough ω_{BD} can never be ruled out; the principle of Occam's razor would have to be invoked to select the simpler theory, in this case general relativity.

On the other hand, generalized scalar-tensor theories possess a richer phenomenology. The function $\omega(\phi)$ could have the property that, at the present epoch, and in weak-field situations, the value of the scalar field ϕ_0 forces ω to be very large, in agreement with the same observations that support general relativity. However, for past or future values of ϕ, or in strong-field regions such as the interiors of neutron stars, ω and λ could take on values leading to significant differences from general relativity.

Damour and collaborators carried out in-depth studies of the possible differences. They first expanded $\ln A(\varphi)$ about a cosmological background field value φ_0:

$$\ln A(\varphi) = \alpha_0(\varphi - \varphi_0) + \frac{1}{2}\beta_0(\varphi - \varphi_0)^2 + \dots \tag{5.59}$$

A precisely linear function produces pure Brans-Dicke theory, with $\alpha_0^2 = 1/(2\omega_{\rm BD} + 3)$, or $1/(2 + \omega_{\rm BD}) = 2\alpha_0^2/(1 + \alpha_0^2)$. The function $\ln A(\varphi)$ acts as a potential for the scalar field φ within matter, and, if $\beta_0 > 0$, then it turns out that during cosmological evolution, the scalar field naturally evolves toward the minimum of the potential where $\phi \sim \phi_0$, that is, toward $dA(\varphi)/d\varphi \sim 0$, which implies $\omega \to \infty$. The result is a theory close to, although not precisely GR (Damour and Nordtvedt, 1993a, 1993b). Estimates of the expected relic deviations from GR today in such theories depend on the cosmological model, but in some models range from 10^{-5} to a few times 10^{-7} for $|\gamma - 1|$.

However, negative values of β_0 correspond to a "locally unstable" scalar potential (the overall theory is still stable in the sense of having no tachyons or ghosts). In this case, objects such as neutron stars can experience a "spontaneous scalarization," whereby the interior values of φ can be very different from the exterior values, through nonlinear interactions between strong gravity and the scalar field, dramatically affecting the stars' internal structure and leading to strong violations of SEP (Damour and Esposito-Farèse, 1993, 1996). There is evidence from recent numerical simulations of the occurrence of a dynamically induced scalarization during the inspirals of compact binary systems containing neutron stars, which can strongly affect both the final motion and the gravitational-wave emission (Barausse et al., 2013; Palenzuela et al., 2014; Shibata et al., 2014). In Chapters 7 and 12, we will discuss the bounds that solar-system and binary-pulsar measurements place on the parameters α_0 and β_0.

On the other hand, in the case $\beta_0 < 0$, one must confront that fact that, with an unstable φ potential, cosmological evolution can drive the system away from the peak where $dA(\varphi)/d\varphi \sim 0$, toward parameter values that can be excluded by solar system experiments (Anderson et al., 2016; Anderson and Yunes, 2017).

Scalar fields coupled to gravity or matter are ubiquitous in particle-physics-inspired models of unification, such as string theory (Taylor and Veneziano, 1988; Maeda, 1988; Damour and Polyakov, 1994; Damour et al., 2002a, 2002b). In some models, the coupling to matter may lead to violations of EEP, which could be tested or bounded by the experiments described in Section 2.3. In many models the scalar field could be massive; if the Compton wavelength is of macroscopic scale, its effects are those of a "fifth force." Only if the theory can be cast as a metric theory with a scalar field of infinite range or of range long compared to the scale of the system in question (solar system) can the PPN framework be strictly applied. Alsing et al. (2012) worked out the post-Newtonian limit for finite scalar field mass, showing how the PPN parameters are effectively modified by Yukawa-type factors $e^{-r/\lambda}$, where r is the characteristic scale of the problem, and $\lambda \propto m^{-1}$ is the Compton wavelength of the scalar field. If the mass of the scalar field is sufficiently large that its range is much smaller than solar-system scales, the scalar field is suppressed, and the theory is essentially equivalent to general relativity.

For a review of scalar-tensor theories, see Fujii and Maeda (2007).

5.3.4 *f(R)* theories

These are theories whose action has the form

$$I = \frac{1}{16\pi G}\int f(R)(-g)^{1/2}\, d^4x + I_{\rm NG}(q_A, g_{\mu\nu}), \tag{5.60}$$

where f is a function chosen so that at cosmological scales, the universe will experience accelerated expansion without resorting to either a cosmological constant or dark energy. However, it turns out that such theories are equivalent to scalar-tensor theories: replace $f(R)$ by $f(\chi)-f_{,\chi}(\chi)(R-\chi)$, where χ is a dynamical scalar field. Varying the action with respect to χ yields $f_{,\chi\chi}(R-\chi)=0$, which implies that $\chi = R$ as long as $f_{,\chi\chi}\neq 0$. Then defining a scalar field $\phi \equiv -f_{,\chi}(\chi)$ one puts the action into the form of a scalar-tensor theory given by Eq. (5.34), with $\omega(\phi)=0$ and $\phi^2 V = \phi\chi(\phi) - f(\chi(\phi))$. As we will see, this value of ω would ordinarily strongly violate solar-system experiments, but it turns out that in many models, the potential $V(\phi)$ has the effect of giving the scalar field a large effective mass in the presence of matter, via the so-called "chameleon mechanism" (Khoury and Weltman, 2004). As a result, the scalar field is suppressed or screened at distances that extend outside bodies like the Sun and Earth. In this way, with only modest fine tuning, $f(R)$ theories can claim to obey standard tests, while providing interesting, non general-relativistic behavior on cosmic scales. For detailed reviews of this class of theories, see Sotiriou and Faraoni (2010) and De Felice and Tsujikawa (2010), and for a review of chameleon and other screening mechanisms, see Burrage and Sakstein (2017).

5.4 Vector-Tensor Theories

These theories contain the metric $\mathbf{g}$ and a dynamical, timelike, four-vector field K^μ. In some models, the four-vector is unconstrained, while in others, called Einstein-Æther theories it is constrained to have unit norm.

(a) Gravitational fields present: the metric $\mathbf{g}$ and a dynamical four-vector field K^α.

(b) Arbitrary parameters and functions: Five arbitrary coupling constants ω, c_1, c_2, c_3, and c_4, and a constraint parameter λ.

(c) Cosmological matching parameters: In unconstrained versions, K^0_0, the asymptotic value of the vector field.

(d) Field Equations: The field equations are derived from the action

$$I = \frac{1}{16\pi G}\int \Big[(1+\omega K_\mu K^\mu)R - E^{\mu\nu}_{\alpha\beta}K^\alpha{}_{;\mu}K^\beta{}_{;\nu} + \lambda(K_\mu K^\mu + 1)\Big](-g)^{1/2}\, d^4x$$
$$+ I_{\rm NG}(q_A, g_{\mu\nu}), \tag{5.61}$$

where

$$E^{\mu\nu}_{\alpha\beta} = c_1 g^{\mu\nu}g_{\alpha\beta} + c_2\delta^\mu_\alpha\delta^\nu_\beta + c_3\delta^\mu_\beta\delta^\nu_\alpha - c_4 K^\mu K^\nu g_{\alpha\beta}. \tag{5.62}$$

Note that we have not included the possible term $K^\mu K^\nu R_{\mu\nu}$, as it can be shown under integration by parts to be equivalent to a linear combination of the terms involving c_2 and c_3.

5.4.1 Constrained Theories

In the constrained theories, λ is a Lagrange multiplier, and by virtue of the constraint $K_\mu K^\mu = -1$, the factor $\omega K_\mu K^\mu$ in front of the Ricci scalar can be absorbed into a rescaling of G; equivalently, in the constrained theories, we can set $\omega = 0$. In a coordinate system in which the metric is asymptotically Minkowskian and the vector field has only a time coordinate (presumably the universal rest frame), the asymptotic value of K^0 is unity.

Einstein-Æther theory

The Einstein-Æther theories were motivated in part by a desire to explore possibilities for violations of Lorentz invariance in gravity, in parallel with similar studies in matter interactions, such as the SME. The general class of theories was analyzed by Jacobson and collaborators (Jacobson and Mattingly, 2001; Mattingly and Jacobson, 2002; Jacobson and Mattingly, 2004; Eling and Jacobson, 2004; Foster and Jacobson, 2006), motivated in part by Kostelecký and Samuel (1989). Analyzing the post-Newtonian limit,[1] they were able to infer values of the PPN parameters γ and β as follows (Foster and Jacobson, 2006):

$$\begin{aligned}
&\gamma = 1, \quad \beta = 1, \\
&\alpha_1 = -\frac{8(c_3^2 + c_1 c_4)}{2c_1 - c_1^2 + c_3^2}, \\
&\alpha_2 = \frac{1}{2}\alpha_1 - \frac{(2c_{13} - c_{14})(c_{13} + c_{14} + 3c_2)}{c_{123}(2 - c_{14})},
\end{aligned} \tag{5.63}$$

with $\xi = \alpha_3 = \zeta_1 = \zeta_2 = \zeta_3 = \zeta_4 = 0$, where $c_{123} = c_1 + c_2 + c_3$, $c_{13} = c_1 + c_3$, $c_{14} = c_1 + c_4$, subject to the constraints $c_{123} \neq 0$, $c_{14} \neq 2$, $2c_1 - c_1^2 + c_3^2 \neq 0$. The present value of the gravitational constant is given by

$$G_{\text{today}} = G\left(1 - \frac{c_{14}}{2}\right)^{-1} = 1\,. \tag{5.64}$$

By requiring that gravitational wave modes have real (as opposed to imaginary) frequencies, one can impose the bounds $c_1/c_{14} \geq 0$ and $c_{123}/c_{14} \geq 0$. Considerations of positivity of energy impose the constraints $c_1 > 0$, $c_{14} > 0$ and $c_{123} > 0$. Given these constraints, the conditions

$$\begin{aligned}
&c_4 = -\frac{c_3^2}{c_1}, \quad \text{and} \\
&\text{either } c_{13} = 0, \text{ or } c_2 = \frac{c_{13}}{3c_1}(c_3 - 2c_1)\,,
\end{aligned} \tag{5.65}$$

are necessary and sufficient to guarantee that $\alpha_1 = \alpha_2 = 0$, so that the PPN parameters can be made to agree completely with those of general relativity. In many applications of Einstein-Æther theory, these constraints are imposed *a priori*, and one defines the two free parameters

$$c_\pm = c_1 \pm c_3\,, \tag{5.66}$$

[1] Note that the minus sign in front the c_4 term in Eq. (5.62) compared to that in the references is a result of our convention for the signature of the metric

so that

$$c_2 = -\frac{c_+(c_+ + 3c_-)}{3(c_+ + c_-)},$$
$$c_4 = -\frac{(c_+ - c_-)^2}{4(c_+ + c_-)},$$
$$c_{14} = \frac{2c_+c_-}{(c_+ + c_-)}. \tag{5.67}$$

Another constraint on the parameters of Einstein-Æther theory comes from the requirement that the speeds of propagation of the various gravitational-wave modes (see Chapter 11) exceed the speed of light. Otherwise, a particle moving at nearly the speed of light could emit gravitational Čerenkov radiation and thereby lose energy (Caves, 1980; Elliott et al., 2005). The observation of ultrahigh energy cosmic rays places such a strong constraint on this effect that it essentially forces all such speeds to exceed that of light, leading (in the limit of small values of α_1 and α_2) to the constraints $0 \leq c_+ \leq 1$ and $0 \leq c_- \leq c_+/3(1 - c_+)$.

Khronometric theory

This is the low-energy limit of "Hořava gravity," a proposal for a gravity theory that is power-counting renormalizable (Hořava, 2009). The vector field is required to be hypersurface orthogonal, which implies that $K^\alpha \propto T^{;\alpha}$, where T is a scalar field related to a preferred time direction. Equivalently the twist $\omega^{\alpha\beta} = K^{[\beta;\alpha]} + K^{[\alpha}A^{\beta]}$ must vanish, where $A^\beta = K^\mu K^\beta{}_{;\mu}$, so that higher spatial derivative terms could be introduced to effectuate renormalizability. A "healthy" version of the theory (Blas et al., 2010, 2011) can be shown to correspond to the values $c_1 = -\epsilon$, $c_2 = \lambda_K$, $c_3 = \beta_K + \epsilon$, and $c_4 = \alpha_K + \epsilon$, where the limit $\epsilon \to \infty$ is to be taken. The idea is to extract ϵ times $\omega_{\alpha\beta}\omega^{\alpha\beta}$ from the Einstein-Æther action and let $\epsilon \to \infty$ to enforce the twist-free condition (Jacobson, 2014). In this case $\gamma = \beta = 1$, α_1 and α_2 are given by

$$\alpha_1 = \frac{4(\alpha_K - 2\beta_K)}{\beta_K - 1},$$
$$\alpha_2 = \frac{1}{2}\alpha_1 + \frac{(\alpha_K - 2\beta_K)(\alpha_K + \beta_K + 3\lambda_K)}{(2 - \alpha_K)(\beta_K + \lambda_K)}, \tag{5.68}$$

and the remaining PPN parameters vanish.

5.4.2 Unconstrained Theories

In the unconstrained theories, $\lambda \equiv 0$ and ω is arbitrary. Unconstrained theories were studied during the 1970s as "straw-man" alternatives to GR. In addition to having up to four arbitrary parameters, they also left the magnitude of the vector field arbitrary, since it satisfies a linear homogenous vacuum field equation of the form $\mathcal{L}K^\mu = 0$ ($c_4 = 0$ in all such cases studied). Each unconstrained theory studied corresponds to a special case of the action (5.61), all with $\lambda \equiv 0$.

General vector-tensor theory

The gravitational Lagrangian for this class of theories had the form $R + \omega K_\mu K^\mu R + \eta K^\mu K^\nu R_{\mu\nu} - \epsilon F_{\mu\nu} F^{\mu\nu} + \tau K_{\nu;\mu} K^{\nu;\mu}$, where $F_{\mu\nu} = K_{\nu;\mu} - K_{\mu;\nu}$, corresponding to the values $c_1 = 2\epsilon - \tau$, $c_2 = -\eta$, $c_1 + c_2 + c_3 = -\tau$, $c_4 = 0$. In these theories γ, β, α_1, and α_2 are complicated functions of the parameters and of the asymptotic value of $K^2 \equiv -K^\mu K_\mu$, while the rest vanish. This class of theories was first analyzed in the first edition of this book.

Will-Nordtvedt theory

This is the special case $c_1 = -1$, $c_2 = c_3 = c_4 = 0$. In this theory, the PPN parameters are given by $\gamma = \beta = 1$, $\alpha_2 = K^2/(1 + K^2/2)$, and zero for the rest (Will and Nordtvedt, 1972).

Hellings-Nordtvedt theory

This is the special case $c_1 = 2$, $c_2 = 2\omega$, $c_1 + c_2 + c_3 = 0 = c_4$. Here γ, β, α_1 and α_2 are complicated functions of the parameters and of K^2, while the rest vanish (Hellings and Nordtvedt, 1973).

5.5 Tensor-Vector-Scalar (TeVeS) Theories

This class of theories was invented to provide a fully relativistic theory of gravity that could mimic the behavior of so-called Modified Newtonian Dynamics (MOND). MOND is a phenomenological mechanism (Milgrom, 1983) whereby Newton's equation of motion $a = Gm/r^2$ holds as long as a is large compared to some fundamental scale a_0, but in a regime where $a < a_0$, the equation of motion takes the form $a^2/a_0 = Gm/r^2$. With such a behavior, the rotational velocity of a particle far from a central mass where $a < a_0$ would have the form $v \sim \sqrt{ar} \sim (Gma_0)^{1/4}$, thus reproducing the flat rotation curves observed for spiral galaxies, without invoking a distribution of dark matter.

Devising a relativistic theory that would embody the MOND phenomenology turned out to be no simple matter, and the final result, TeVeS was rather complicated (Bekenstein, 2004). Furthermore, it was shown to have unexpected singular behavior that was most simply cured by incorporating features of the Einstein-Æther theory (Skordis, 2008).

(a) Gravitational fields present: the metric $\mathbf{g}$, a dynamical four-vector field K^α and a dynamical scalar field ϕ.

(b) Arbitrary parameters and functions: four arbitrary coupling constants c_1, c_2, c_3, and c_4, a constraint parameter λ for the vector field, and a coupling constant k, a length scale ℓ and an interpolating function $\mathcal{F}$ for the scalar field.

(c) Cosmological matching parameters: ϕ_0, the asymptotic value of the scalar field.

(d) Field equations: the field equations are derived from the action

$$I = \frac{1}{16\pi G}\int \left[\tilde{R} - E^{\mu\nu}_{\alpha\beta}K^{\alpha}{}_{;\mu}K^{\beta}{}_{;\nu} + \lambda(K_\mu K^\mu + 1)\right](-\tilde{g})^{1/2}\,d^4x$$
$$-\frac{1}{2k^2\ell^2 G}\int \mathcal{F}(k\ell^2 h^{\mu\nu}\phi_{,\mu}\phi_{,\nu})(-\tilde{g})^{1/2}\,d^4x + I_{\rm NG}(q_A, g_{\mu\nu}), \tag{5.69}$$

where $\tilde{g}_{\mu\nu}$ is an auxiliary metric related to the physical metric $g_{\mu\nu}$ by

$$g_{\mu\nu} \equiv e^{-2\phi}\tilde{g}_{\mu\nu} - 2K_\mu K_\nu \sinh(2\phi)\,, \tag{5.70}$$

$\tilde{R}$ is the Ricci scalar constructed from the auxiliary metric, and

$$E^{\mu\nu}_{\alpha\beta} \equiv c_1\tilde{g}^{\mu\nu}\tilde{g}_{\alpha\beta} + c_2\delta^\mu_\alpha\delta^\nu_\beta + c_3\delta^\mu_\beta\delta^\nu_\alpha - c_4K^\mu K^\nu\tilde{g}_{\alpha\beta}\,,$$
$$h^{\mu\nu} \equiv \tilde{g}^{\mu\nu} - K^\mu K^\nu\,. \tag{5.71}$$

In the gravitational part of the action, indices are raised and lowered using the auxiliary metric. The interpolating function $\mathcal{F}(y)$ is chosen so that $\mu(y) \equiv d\mathcal{F}/dy$ is unity in the high-acceleration, or normal Newtonian and post-Newtonian regimes, and nearly zero in the MOND regime.

The PPN parameters have been computed in the limit where the function $\mathcal{F}(y)$ is a linear function of its argument $y = k\ell^2 h^{\mu\nu}\phi_{,\mu}\phi_{,\nu}$. In this limit, the scale ℓ drops out of the problem, and the PPN parameters (Sagi, 2009) have the values $\gamma = \beta = 1$ and $\xi = \alpha_3 = \zeta_i = 0$, while the parameters α_1 and α_2 are given by

$$\alpha_1 = (\alpha_1)_{Æ} - 16G\frac{\kappa c_1(2-c_{14}) - c_3\sinh 4\phi_0 + 2(1-c_1)\sinh^2 2\phi_0}{2c_1 - c_1^2 + c_3^2}\,,$$
$$\alpha_2 = (\alpha_2)_{Æ} - 2G\left(A_1\kappa - 2A_2\sinh 4\phi_0 - A_3\sinh^2 2\phi_0\right)\,, \tag{5.72}$$

where $(\alpha_1)_{Æ}$ and $(\alpha_2)_{Æ}$ are given by their Einstein-Æther values in Eq. (5.63), and

$$A_1 \equiv \frac{(2c_{13}-c_{14})^2}{c_{123}(2-c_{14})} + \frac{4c_1(2-c_{14})}{2c_1 - c_1^2 + c_3^2} - \frac{6(1+c_{13}-c_{14})}{2-c_{14}}\,, \tag{5.73}$$

$$A_2 \equiv \frac{(2c_{13}-c_{14})^2}{c_{123}(2-c_{14})^2} - \frac{4(1-c_1)}{2c_1 - c_1^2 + c_3^2} + \frac{2(1-c_{13})}{2-c_{14}}\left(\frac{2}{c_{123}} + \frac{3}{2-c_{14}}\right)\,, \tag{5.74}$$

$$A_3 \equiv \frac{(2c_{13}-c_{14})^2}{c_{123}(2-c_{14})^2} + \frac{4c_3}{2c_1 - c_1^2 + c_3^2} + \frac{2}{(2-c_{14})}\left(\frac{3(1-c_{13})}{c_{123}} - \frac{2c_{13}-c_{14}}{2-c_{14}}\right)\,, \tag{5.75}$$

where $\kappa \equiv k/8\pi$,

$$G_{\rm today} \equiv 2G\left(\frac{1+\kappa(2-c_{14})}{2-c_{14}}\right) = 1\,, \tag{5.76}$$

and ϕ_0 is the asymptotic value of the scalar field. In the limit $\kappa \to 0$ and $\phi_0 \to 0$, α_1, α_2 and $G_{\rm today}$ reduce to their Einstein-Æther forms.

When one takes into account the fact that the function $\mu(y) = d\mathcal{F}/dy$ must interpolate between unity and zero to reach the MOND regime, it has been found that the dynamics of local systems is more strongly affected by the fields of surrounding matter than was anticipated. This "external field effect" (EFE) (Milgrom, 2009; Blanchet and Novak,

2011a, 2011b) produces a quadrupolar contribution to the local Newtonian gravitational potential that depends on the external distribution of matter and on the shape of the function $\mu(y)$. For the solar system, this effect can be significantly larger than the galactic tidal contribution. Although the calculations of EFE have been carried out using phenomenological MOND equations rather than with TeVeS itself (for which the calculations would be very difficult), it is expected to be a generic phenomenon, applicable to TeVeS as well. Analysis of the orbit of Saturn using Cassini data has placed interesting constraints on the MOND interpolating function $\mu(y)$ (Hees et al., 2014).

For thorough reviews of MOND and TeVeS, and their confrontation with the dark-matter paradigm, see Skordis (2009) and Famaey and McGaugh (2012).

An alternative tensor-vector-scalar theory, containing a vector field and three scalar fields, in addition to the metric, was devised by Moffat (2006).

5.6 Quadratic Gravity and Chern-Simons Theories

Quadratic gravity is a recent incarnation of an old idea of adding to the action of GR terms quadratic in the Riemann and Ricci tensors or the Ricci scalar, as "effective field theory" models for more fundamental string or quantum gravity theories. The general action for such theories can be written as

$$\begin{aligned} I = \int \Big[& \kappa R + \alpha_1 f_1(\phi) R^2 + \alpha_2 f_2(\phi) R_{\alpha\beta} R^{\alpha\beta} + \alpha_3 f_3(\phi) R_{\alpha\beta\gamma\delta} R^{\alpha\beta\gamma\delta} \\ & + \alpha_4 f_4(\phi)\, {}^*RR - \frac{\beta}{2}\Big(g^{\mu\nu}\partial_\mu\phi\partial_\nu\phi + 2V(\phi)\Big)\Big] (-g)^{1/2} d^4x \\ & + I_{\rm NG}(q_A, g_{\mu\nu}) \,, \end{aligned} \tag{5.77}$$

where $\kappa = (16\pi G)^{-1}$, ϕ is a scalar field, α_i are dimensionless coupling constants (if the functions $f_i(\phi)$ are dimensionless), β is a constant whose dimension depends on that of ϕ, and ${}^*RR \equiv \frac{1}{2}\epsilon^{\gamma\delta\rho\sigma} R^\alpha{}_{\beta\rho\sigma} R^\beta{}_{\alpha\gamma\delta}$.

One challenge inherent in these theories is to find an argument or a mechanism that evades making the natural choice for each of the α parameters to be of order unity. Such a choice makes the effects of the additional terms essentially unobservable in most laboratory or astrophysical situations because of the enormous scale of $\kappa \propto 1/\ell^2_{\rm Planck}$ in the leading term. This class of theories is too vast and diffuse to cover in this book, and no comprehensive review is available, to our knowledge.

Chern-Simons gravity is the special case of this class of theories in which only the parity-violating term ${}^*RR$ is present ($\alpha_1 = \alpha_2 = \alpha_3 = 0$) (Jackiw and Pi, 2003). It can arise in various anomaly cancellation schemes in the standard model of particle physics, in cancelling the Green-Schwarz anomaly in string theory, or in effective field theories of inflation (Weinberg, 2008). It can also arise in loop quantum gravity (Taveras and Yunes, 2008; Mercuri and Taveras, 2009). The action in this case is given by

$$I = \int \left[\kappa R + \frac{\alpha}{4}\phi \, {}^*RR - \frac{\beta}{2}\left(g^{\mu\nu}\partial_\mu\phi\partial_\nu\phi + 2V(\phi) \right) \right] (-g)^{1/2} d^4x + I_{\rm NG}(q_A, g_{\mu\nu}) , \quad (5.78)$$

where α and β are coupling constants with dimensions ℓ^A, and ℓ^{2A-2}, assuming that the scalar field has dimensions ℓ^{-A}.

There are two different versions of Chern-Simons theory, a non-dynamical version in which $\beta = 0$, so that ϕ, given *a priori* as some specified function of spacetime, plays the role of a Lagrange multiplier enforcing the constraint ${}^*RR = 0$, and a dynamical version, in which $\beta \neq 0$.

The PPN parameters for a nondynamical version of the theory with $\alpha = \kappa$ and $\beta = 0$ are identical to those of GR; however, there is an additional, parity-even potential in the g_{0i} component of the metric that does not appear in the standard PPN framework, given by

$$\delta g_{0i} = 2\frac{d\phi}{dt} \left(\nabla \times \mathbf{V} \right)_i . \quad (5.79)$$

Unfortunately, the nondynamical version has been shown to be unstable (Dyda et al., 2012), while the dynamical version is sufficiently complex that its observable consequences have been analyzed for only special situations (Ali-Haïmoud and Chen, 2011; Yagi et al., 2013). Alexander and Yunes (2009) give a thorough review of Chern-Simons gravity.

Einstein-Dilaton-Gauss-Bonnet gravity is another special case, in which the Chern-Simons term is neglected ($\alpha_4 = 0$), and the three other curvature-squared terms collapse to the Gauss-Bonnet invariant, $R^2 - 4R_{\alpha\beta}R^{\alpha\beta} + R_{\alpha\beta\gamma\delta}R^{\alpha\beta\gamma\delta}$, i.e. $f_1(\phi) = f_2(\phi) = f_3(\phi)$ and $\alpha_1 = -\alpha_2/4 = \alpha_3$ [see Moura and Schiappa (2007) and Pani and Cardoso (2009)].

Berry and Gair (2011) studied $f(R)$ theories in the small R limit, using the expansion $f(R) = R + a_2R^2/2 + \ldots$, making this another special case of quadratic gravity; they used a variety of observations to place limits on the parameter a_2.

5.7 Massive Gravity

Massive gravity theories attempt to give the field carrying the gravitational interaction a mass. In a quantum context, such a field is often called the "graviton," but the realm being discussed here is entirely classical; nevertheless, the term graviton is a convenient shorthand. The simplest attempt to implement this in a ghost-free manner suffers from the so-called van Dam-Veltman-Zakharov (vDVZ) discontinuity (van Dam and Veltman, 1970; Zakharov, 1970). Because of the three additional helicity states available to the massive spin-2 graviton, the limit of small graviton mass does not coincide with pure GR, and the predicted perihelion advance, for example, violates experiment. A model theory by Visser (1998) attempts to circumvent the vDVZ problem by introducing a nondynamical flat-background metric. This theory is truly continuous with GR in the limit of vanishing graviton mass; on the other hand, its observational implications have been only partially explored. Braneworld scenarios predict a tower or a continuum of massive gravitons, and may avoid the vDVZ discontinuity, although the full details are still a work in progress

(Deffayet et al., 2002; Creminelli et al., 2005). Attempts to avert the vDVZ problem involve treating nonlinear aspects of the theory at the fundamental level (Vainshtein, 1972); many models incorporate a second tensor field in addition to the metric. For recent reviews, see Hinterbichler (2012), de Rham (2014), and a focus issue in *Classical and Quantum Gravity* (Vol. 30, No. 18).

5.8 The Rise and Fall of Alternative Theories of Gravity

In this section, we will take a selection of theories from the first two periods of alternative theories and answer the question "Whatever happened to . . . ?".

5.8.1 Whitehead's theory

In 1922, the great mathematician and philosopher Alfred North Whitehead found himself uncomfortable with the fact that, in general relativity, the causal relationships among events in spacetime are not known *a priori*, but are only revealed *after* the field equations have been solved. In response, he proposed a Lorentz-invariant theory of gravity in which the physical metric $\boldsymbol{g}$ was constructed algebraically from a flat background Minkowski metric $\boldsymbol{\eta}$ and matter variables, according to

$$
\begin{aligned}
g_{\mu\nu}(x^\alpha) &\equiv \eta_{\mu\nu} - 2\int_{\Sigma^-} \frac{y_\mu^- y_\nu^-}{(w^-)^3}\left[\sqrt{-g}\rho u^\alpha d\Sigma_\alpha\right]^- , \\
(y^\mu)^- &\equiv x^\mu - (x^\mu)^- , \quad \eta_{\mu\nu}(y^\mu)^-(y^\nu)^- = 0 , \\
w^- &\equiv \eta_{\mu\nu}(y^\mu)^-(u^\nu)^- , \quad u^\mu \equiv dx^\mu/d\sigma , \\
d\sigma^2 &\equiv \eta_{\mu\nu}dx^\mu dx^\nu ,
\end{aligned}
\tag{5.80}
$$

where the superscript $^-$ indicates quantities to be evaluated along the past flat spacetime null cone of the field point x^α. Eq. (5.80) are a modern version of the original theory. Whitehead required that clocks respond only to $\boldsymbol{\eta}$, thereby satisfying his concerns about causal relationships, but this meant that there would be no gravitational redshift effect, in disagreement with general relativity. But in 1922 this was not a problem empirically because attempts to measure the shift of spectral lines from the Sun had thus far failed to detect the effect predicted by general relativity. The full metric in Whitehead's theory was assumed to govern the motion of light and matter via the geodesic equation. Furthermore, the theory had the property that, for a point mass at rest, the metric was equivalent via a coordinate transformation to the Schwarzschild metric. And once the gravitational redshift had been measured, it was natural to reinterpret Whitehead's theory as a two-tensor theory with a flat background metric $\boldsymbol{\eta}$, and with the physical metric $\boldsymbol{g}$ governing all physics, including clocks. As a consequence, Whitehead's theory could not be experimentally distinguished from general relativity using the gravitational redshift, light deflection, and perihelion advance, the main effects that supported Einstein's theory. This led to a conundrum discussed extensively by philosophers of science at the time of how to select among competing theories that equally satisfy experimental observations.

But Will (1971b) pointed out that, when the theory was extended in a natural way to more than one gravitating body or to extended bodies, then the gravitational attraction between any pair of masses in the presence of a third body would be anisotropic, that is, dependent upon the orientation of the pair relative to the distant body (an effect additional to the normal tidal gravitational effects). This effective "anisotropy in Newton's constant G" (see Section 6.6) would result in anomalous tide-like distortions of the Earth in the presence of the mass of the galaxy, that were ruled out by precise measurements made with gravimeters. This however did not totally end the fascination with Whitehead's theory, especially among philosophers, and so Gibbons and Will (2008) embarked on a "serial killing" of Whitehead's theory, pointing out that it actually fails the test of experiment in five different ways:

- *Anisotropy in G.* A reanalysis of Will's 1971 result verified that the theory violates gravimeter data on anomalous Earth tides by a factor of at least 100 (see Section 8.4).
- *Nordtvedt effect.* The theory predicts a violation of the equivalence principle for gravitating bodies, leading to a Nordtvedt effect in lunar laser ranging 400 times larger than the data will permit (see Section 8.1).
- *Birkhoff's theorem and LAGEOS data.* It was already known in the 1950s that the theory predicts that the metric of a static, spherically symmetric *finite-sized* body has an additional size-dependent contribution. This contributes an additional advance of the perigee of the LAGEOS II satellite, in disagreement with observations by a factor of 10 (see Section 9.1.3).
- *Momentum non-conservation.* The theory predicts an acceleration of the center of mass of a binary system, an effect ruled out by binary pulsar data by a factor of a million (see Section 9.3).
- *Gravitational-radiation damping.* The theory predicts anti-damping in binary orbits due to gravitational radiation, in violation of binary pulsar data by four orders of magnitude (see Section 12.1.2).

The purpose of this list of failures is not to gang up on Whitehead, but rather to illustrate that matching the Schwarzschild geometry is no longer sufficient to match experimental tests, and that the current generation of empirical data strongly and deeply constrains the theoretical possibilities.

Whitehead's theory is an example of a class of theories known as "quasilinear." Quasilinear theories of gravity are theories whose post-Newtonian metric for a system of multiple bodies, in a particular post-Newtonian gauge, contains only potentials linear in the masses of the source bodies. In particular, in such a gauge, g_{00} lacks the potentials U^2 and Φ_W. This is a property of many theories that attempt to describe gravity by means of a linear field theory on a flat spacetime background. If the gauge in which this occurs is not the standard PPN gauge, then the gauge transformation (4.47) will yield

$$\bar{g}_{\bar{0}\bar{0}} = g_{00} + 2\lambda_2(U^2 + \Phi_2 + \Phi_W) - 2\lambda_1 X_{,00} \,. \tag{5.81}$$

Since g_{00} did not initially contain either U^2 or Φ_W, we can conclude immediately that $\xi = \beta$ in this class of theories, in severe violation of experiment.

Another group of theories in this class is known as Linear Fixed-Gauge (LFG) theories. The standard field theoretic approach to the construction of a tensor gravitation theory on a flat spacetime background is to use the gauge-invariant action for a spin-two tensor field $p_{\mu\nu}$, combined with the universally coupled nongravitational action to yield

$$I = \frac{1}{16\pi G}\int \left[2p^{\mu\nu}{}_{|\nu}p_{\mu\lambda}{}^{|\lambda} - 2p^{\mu\nu}{}_{|\nu}p^{\lambda}_{\lambda|\mu} + p^{\nu}_{\nu|\mu}p^{\lambda|\mu}_{\lambda} - p^{\mu\nu|\lambda}p_{\mu\nu|\lambda}\right](-\eta)^{1/2}d^4x$$
$$+ I_{\rm NG}(q_A, g_{\mu\nu}), \tag{5.82}$$

where $g_{\mu\nu} = \eta_{\mu\nu} + p_{\mu\nu}$, and "|" denotes a covariant derivative with respect to a Riemann-flat background metric $\boldsymbol{\eta}$. However, the action is singular: the gravitational part is invariant under the gauge transformation $p_{\mu\nu} \to p_{\mu\nu} - \xi_{(\mu|\nu)}$ while $I_{\rm NG}$ is not. The Bianchi identity associated with the gauge invariance of the gravitational action $T^{\mu\nu}{}_{|\nu} = 0$ is incompatible with that associated with the general covariance of $I_{\rm NG}$, namely, $T^{\mu\nu}{}_{;\nu} = 0$. (This incompatibility is the same one that many standard textbooks emphasize when discussing linearized general relativity.) LFG theories seek to remedy this by breaking the gauge invariance of the gravitational action through the introduction of auxiliary gravitational fields that couple to $p_{\mu\nu}$ in such a way as to fix the gauge of $p_{\mu\nu}$. Nevertheless, these theories, devised by Deser and Laurent (1968) and Bollini et al. (1970), turn out to be quasilinear in the sense defined above, and predict $\xi = \beta$, in violation of experiment (Will, 1973).

5.8.2 Rosen's bimetric theory

Nathan Rosen was an assistant of Einstein during the late 1930s and collaborated with him on several seminal papers (the Einstein-Podolsky-Rosen paper, the Einstein-Rosen bridge, and a notorious paper on gravitational waves), but in the 1960s he convinced himself that black holes should not exist. Accordingly, he developed an alternative theory of gravity that would permit compact objects but without the event horizon. Rosen's bimetric theory (Rosen, 1973, 1974, 1977) was the result. It is a metric theory based on the action

$$I = \frac{1}{64\pi G}\int \eta^{\mu\nu}g^{\alpha\beta}g^{\gamma\delta}\left(g_{\alpha\gamma|\mu}g_{\beta\delta|\nu} - \frac{1}{2}g_{\alpha\beta|\mu}g_{\gamma\delta|\nu}\right)(-\eta)^{1/2}d^4x$$
$$+ I_{\rm NG}(q_A, g_{\mu\nu}), \tag{5.83}$$

where $\boldsymbol{\eta}$ is a flat background metric, and "|" denotes covariant derivative with respect to $\boldsymbol{\eta}$. The PPN parameters turn out to be identical to those of general relativity Lee et al. (1976), except for α_2, which is given by

$$\alpha_2 = \frac{c_0}{c_1} - 1, \tag{5.84}$$

where c_0 and c_1 are cosmological matching parameters defined by

$$c_0 + 3c_1 = \eta^{\mu\nu}g^{(0)}_{\mu\nu}, \quad c_0^{-1} + 3c_1^{-1} = \eta_{\mu\nu}g^{(0)\mu\nu}. \tag{5.85}$$

These parameters express the mismatch between the flat metric $\eta_{\mu\nu}$ and the asymptotic form of the physical metric $g^{(0)}_{\mu\nu}$ determined by the cosmological model. Depending on the cosmological model, it was generally possible to choose c_0 and c_1 so as to make α_2 small enough to conform to solar system bounds (see Section 8.4).

Death to Rosen's bimetric theory came with the Hulse-Taylor binary pulsar. Despite the ability to agree with general relativity at post-Newtonian order, the theory predicted violations of the Strong Equivalence Principle in the presence of strongly gravitating neutron stars (Will and Eardley, 1977), and also predicted significant differences from general relativity for gravitational radiation. In particular, for the binary pulsar, it predicted dipole gravitational radiation with a *negative* energy flux (Will, 1977; Will and Eardley, 1977), leading to a predicted *increase* in the binary pulsar's orbital period, in strong contradiction with the observed decrease.

5.8.3 Other theories

In the years prior to general relativity, Nordström (1913), toyed with a conformally flat theory of gravity, in which the metric has the form $\boldsymbol{g} \equiv f(\phi)\boldsymbol{\eta}$, where $\boldsymbol{\eta}$ is the Minkowski metric, and ϕ is a scalar field. Irrespective of the field equations for ϕ, the conformal flatness and a proper Newtonian limit require that $g_{ij} = -\delta_{ij}g_{00} = \delta_{ij}(1-2U)$. Therefore $\gamma = -1$, in strong disagreement with observation. This result can also be deduced from the conformal invariance of Maxwell's equations (i.e., invariance under the transformation $g_{\mu\nu} \to \phi g_{\mu\nu}$), which implies that propagation of light rays in the metric $f(\phi)\boldsymbol{\eta}$ is identical to propagation in the flat spacetime metric $\boldsymbol{\eta}$, namely, straight-line propagation at constant speed.

Another class of alternative theories asserted that, in a certain coordinate system, the metric should always have the form $ds^2 = -A(\phi)dt^2 + B(\phi)(dx^2+dy^2+dz^2)$, independent of the nature of the source, with ϕ determined by some field equations. Such theories can actually be made generally covariant by introducing a nondynamical background Minkowski metric $\boldsymbol{\eta}$, and a nondynamical scalar field T with the properties $T_{|\mu\nu} = 0$ and $T_{,\mu}T_{,\nu}\eta^{\mu\nu} = -1$. This scalar field acts like a cosmic time coordinate, selecting preferred spatial sections or "strata" that are orthogonal to $\boldsymbol{\nabla}T$. With these definitions, the metric in these "stratified theories with time-orthogonal space slices" takes the covariant form $\boldsymbol{g} = (B-A)\mathbf{d}T\otimes\mathbf{d}T + B\boldsymbol{\eta}$. Since the metric is diagonal in a coordinate system in which $T_{,\mu} = \delta^0_\mu$ (the preferred cosmic rest frame), independently of the nature of the source, then in the post-Newtonian limit and in the standard PPN gauge, we must have that $g_{0j} = -\epsilon X_{,0j}$ for some ϵ. From Box 4.1, we then conclude that $\alpha_1 = -4(\gamma+1)$ in gross violation of observation. Some "straw-man" generalizations of this class of theories circumvented this problem by introducing auxiliary dynamical vector and tensor fields, in addition to the scalar field (Lee et al., 1974; Ni, 1973). It turned out however, that when the parameters were adjusted to conform to solar-system experiments, most of these theories predicted negative gravitational energy flux, in strong disagreement with binary pulsar data (Will, 1977).

6 Equations of Motion in the PPN Formalism

One of the consequences of the fundamental postulates of metric theories of gravity is that matter and nongravitational fields couple only to the metric, in a manner dictated by EEP. The resulting equations of motion include

$$\begin{aligned}
T^{\mu\nu}{}_{;\nu} &= 0 && \text{[hydrodynamics and nongravitational fields]}\,,\\
u^{\nu}u^{\mu}{}_{;\nu} &= 0 && \text{[neutral test body: geodesics]}\,,\\
F^{\mu\nu}{}_{;\nu} &= 4\pi J^{\mu} && \text{[Maxwell's equations]}\,,\\
k^{\nu}k^{\mu}{}_{;\nu} &= 0 && \text{[light rays: geodesics]}\,,
\end{aligned} \tag{6.1}$$

(see Section 3.2 for discussion). In Chapter 4, we developed the general spacetime metric through post-Newtonian order as a functional of matter variables and as a function of ten PPN parameters. If this metric is substituted into these equations of motion, we obtain coupled sets of equations of motion for matter and nongravitational field variables in terms of other matter and nongravitational field variables. For specific problems, these equations can be solved using standard techniques to obtain predictions for the behavior of matter in terms of the PPN parameters. These predictions can then be compared with experiment. It is the purpose of this chapter to cast the above equations of motion into a form that can be applied to specific situations and experiments. That application will be made in Chapters 7, 8 and 9. In Section 6.1, we carry out this procedure for light rays. Section 6.2 lays out the equations of hydrodynamics in the PPN framework, and Section 6.3 deals with massive, self-gravitating bodies and presents appropriate N-body equations of motion. Section 6.5 specializes to semiconservative theories of gravity and presents an N-body Lagrangian and Hamilton from which the semiconservative N-body equations of motion can be derived. In Section 6.6, we derive the relative acceleration between two bodies, including the effects of nearby gravitating bodies and of motion with respect to the universe rest frame, and put it into a form from which one can identify a "locally measured" Newtonian gravitational constant. In Section 6.7, we derive post-Newtonian equations of motion for spinning bodies.

6.1 Equations of Motion for Photons

We begin with the geodesic equation obtained from Maxwell's equations in the geometrical-optics limit, Eq. (6.1):

$$k^{\nu}k^{\mu}{}_{;\nu} = 0\,, \tag{6.2}$$

where k^μ is the wave vector tangent to the "photon" trajectory, with

$$k^\mu k_\mu = 0\,. \tag{6.3}$$

Substituting $k^\mu = dx^\mu/d\sigma$ where σ is an "affine" parameter measured along the trajectory, we obtain

$$\frac{d^2x^\mu}{d\sigma^2} + \Gamma^\mu_{\nu\lambda}\frac{dx^\nu}{d\sigma}\frac{dx^\lambda}{d\sigma} = 0\,. \tag{6.4}$$

As it stands, this equation is not particularly useful, since we have no experimental access to the parameter σ. On the other hand, we do have access to coordinate time t. For example, the arrival of photons at Earth can be recorded in terms of atomic time $t_{\rm atomic}$ on Earth; this time can be directly related to PPN coordinate time $t = x^0$ by well known transformations (see, e.g., PW, chapter 10). We can rewrite Eq. (6.4) using t rather than σ by noticing that

$$\frac{d^2t}{d\sigma^2} + \Gamma^0_{\nu\lambda}\frac{dx^\nu}{d\sigma}\frac{dx^\lambda}{d\sigma} = 0\,. \tag{6.5}$$

Then the spatial components of Eq. (6.4) can be rewritten in the form

$$\frac{d^2x^j}{dt^2} + \left(\Gamma^j_{\nu\lambda} - \Gamma^0_{\nu\lambda}\frac{dx^j}{dt}\right)\frac{dx^\nu}{dt}\frac{dx^\lambda}{dt} = 0\,. \tag{6.6}$$

Eq. (6.3) can be written

$$g_{\mu\nu}\frac{dx^\mu}{dt}\frac{dx^\nu}{dt} = 0\,. \tag{6.7}$$

To post-Newtonian accuracy, Eqs. (6.6) and (6.7) take the form (see Box 6.1 for expressions for the Christoffel symbols $\Gamma^\mu_{\nu\lambda}$)

$$\frac{d\boldsymbol{v}}{dt} = \boldsymbol{\nabla}U\left(1+\gamma v^2\right) - 2(1+\gamma)\boldsymbol{v}(\boldsymbol{v}\cdot\boldsymbol{\nabla}U)\,, \tag{6.8a}$$

$$0 = 1 - 2U - v^2(1+2\gamma U)\,, \tag{6.8b}$$

where $\boldsymbol{v} = d\boldsymbol{x}/dt$, with $|\boldsymbol{v}| \sim 1$. Eq. (6.8b) implies that $\boldsymbol{v}$ can be written as

$$\boldsymbol{v} = [1-(1+\gamma)U]\,\boldsymbol{n}\,, \tag{6.9}$$

where $\boldsymbol{n}$ is a unit vector, satisfying $\boldsymbol{n}\cdot\boldsymbol{n} = 1$. Substituting Eq. (6.9) back into (6.8a) yields an equation for the unit vector $\boldsymbol{n}$,

$$\frac{dn^j}{dt} = (1+\gamma)(\delta^{jk} - n^jn^k)U_{,k}\,. \tag{6.10}$$

Notice that the right-hand-side of Eq. (6.10) is orthogonal to $\boldsymbol{n}$, so that $\boldsymbol{n}$ remains a unit vector; only its direction changes.

Eqs. (6.9) and (6.10), can be solved in a straightforward way. Consider a light signal emitted at PPN coordinate time t_e at a point $\boldsymbol{x}_e$ in an initial direction described by the unit vector $\boldsymbol{k}$, where $\boldsymbol{k}\cdot\boldsymbol{k} = 1$. The zeroth-order solution to Eq. (6.10) is $\boldsymbol{n} = \boldsymbol{k}$, so that the light trajectory is given to zeroth order by

$$\boldsymbol{x}(t) = \boldsymbol{x}_e + \boldsymbol{k}(t-t_e)\,, \tag{6.11}$$

where $\boldsymbol{x}_e = \boldsymbol{x}(t_e)$. At the first PN order, we define

$$\boldsymbol{n} = \boldsymbol{k} + \boldsymbol{\alpha}\,, \tag{6.12}$$

where to post-Newtonian order, $\boldsymbol{\alpha}$ satisfies

$$\frac{d\alpha^j}{dt} = (1+\gamma)(\delta^{jk} - k^j k^k)U_{,k}\,, \tag{6.13}$$

with the initial condition $\alpha(t_e) = 0$, where $U_{,k}$ is to be evaluated at $\boldsymbol{x}(t)$, and U is given by Eq. (6.11),

$$U(\boldsymbol{x}(t)) = \int \frac{\rho^{*\prime}}{s} d^3x'\,, \tag{6.14}$$

where $\boldsymbol{s} = \boldsymbol{x}(t) - \boldsymbol{x}'$ and $s = |\boldsymbol{s}|$. This gives

$$\frac{d\boldsymbol{\alpha}}{dt} = -(1+\gamma)\int \rho^{*\prime} \frac{\boldsymbol{b}}{s^3} d^3x'\,, \tag{6.15}$$

where

$$\boldsymbol{b} \equiv \boldsymbol{s}_e - (\boldsymbol{s}_e \cdot \boldsymbol{k})\boldsymbol{k}\,, \quad \boldsymbol{s}_e \equiv \boldsymbol{x}_e - \boldsymbol{x}'\,. \tag{6.16}$$

The vector $\boldsymbol{b}$ is directed from a point $\boldsymbol{x}'$ within the source to the point of closest approach of the light ray relative to that point. To post-Newtonian order, the velocity $\boldsymbol{v}$ of the light ray is now given by

$$\boldsymbol{v} = [1 - (1+\gamma)U]\,\boldsymbol{k} + \boldsymbol{\alpha}\,. \tag{6.17}$$

To solve for $\boldsymbol{\alpha}$ and to integrate Eq. (6.17) to obtain the trajectory, we employ the useful identities, valid to lowest PN order:

$$\frac{ds}{dt} = \frac{\boldsymbol{s}\cdot\boldsymbol{k}}{s}\,, \tag{6.18a}$$

$$\frac{d}{dt}\left(\frac{\boldsymbol{s}\cdot\boldsymbol{k}}{s}\right) = \frac{b^2}{s^3}\,, \tag{6.18b}$$

$$\frac{d}{dt}\ln(s + \boldsymbol{s}\cdot\boldsymbol{k}) = \frac{1}{s}\,. \tag{6.18c}$$

We can then use the identity (6.18b) to express $d\boldsymbol{\alpha}/dt$ as

$$\frac{d\boldsymbol{\alpha}}{dt} = -(1+\gamma)\frac{d}{dt}\int \rho^{*\prime} \frac{\boldsymbol{b}}{b^2}\frac{\boldsymbol{s}\cdot\boldsymbol{k}}{s} d^3x'\,, \tag{6.19}$$

which can be integrated to yield

$$\boldsymbol{\alpha}(t) = -(1+\gamma)\int \rho^{*\prime} \frac{\boldsymbol{b}}{b^2}\left(\frac{\boldsymbol{s}\cdot\boldsymbol{k}}{s} - \frac{\boldsymbol{s}_e\cdot\boldsymbol{k}}{s_e}\right) d^3x'\,. \tag{6.20}$$

Integrating Eq. (6.17) making use of the identities (6.18a) and (6.18c), we find the photon's trajectory in the form

$$\boldsymbol{x}(t) = \boldsymbol{x}_e + \boldsymbol{k}(t - t_e) + \boldsymbol{k}\,\delta x_{\parallel}(t) + \delta\boldsymbol{x}_{\perp}(t)\,, \tag{6.21}$$

where

$$\delta x_{\parallel}(t) = -(1+\gamma)\int \rho^{*\prime} \ln\left[\frac{(s+\boldsymbol{s}\cdot\boldsymbol{k})(s_e - \boldsymbol{s}_e\cdot\boldsymbol{k})}{b^2}\right] d^3x', \tag{6.22a}$$

$$\delta \boldsymbol{x}_{\perp}(t) = -(1+\gamma)\int \rho^{*\prime} \frac{\boldsymbol{b}}{b^2}\left(s - \frac{\boldsymbol{s}\cdot\boldsymbol{s}_e}{s_e}\right) d^3x', \tag{6.22b}$$

assuming the initial conditions $\delta x_{\parallel}(t_e) = \delta \boldsymbol{x}_{\perp}(t_e) = 0$. In the expression for $\delta x_{\parallel}$, we have made use of the identity $b^2 = (s_e - \boldsymbol{s}_e\cdot\boldsymbol{k})(s_e + \boldsymbol{s}_e\cdot\boldsymbol{k})$. Equations (6.21) and (6.22) give a complete post-Newtonian solution for the trajectory of light around any mass distribution.

6.2 PPN Hydrodynamics

The PPN equations of hydrodynamics can be obtained by substituting the Christoffel symbols, displayed to the appropriate order in Box 6.1, into the equation of motion $T^{\mu\nu}_{\ ;\nu} = 0$, using the post-Newtonian form of $T^{\mu\nu}$ given in Box 4.1. Using coordinate time t

Box 6.1 Christoffel symbols for the PPN metric

To the post-Newtonian order sufficient for use in the equations of hydrodynamics or the geodesic equation for test bodies, the Christoffel symbols are given by

$$\begin{aligned}
\Gamma^0_{00} &= -U_{,0}\,, \\
\Gamma^0_{0j} &= -U_{,j}\,, \\
\Gamma^0_{jk} &= \gamma\delta_{jk}U_{,0} + \frac{1}{2}(4\gamma+4+\alpha_1)V_{(j,k)} + \frac{1}{2}(1+\alpha_2-\zeta_1+2\xi)X_{,0jk} \\
&\quad + \frac{1}{2}\alpha_1 w_{(j}U_{,k)} - \alpha_2 w^\ell X_{,\ell jk}\,, \\
\Gamma^j_{00} &= -U_{,j} + 2(\beta+\gamma)UU_{,j} - \psi_{,j} - \frac{1}{2}(4\gamma+4+\alpha_1)V_{j,0} \\
&\quad - \frac{1}{2}(1+\alpha_2-\zeta_1+2\xi)X_{,j00} + \frac{1}{2}(\alpha_1-\alpha_3)w^2U_{,j} - \frac{1}{2}\alpha_2 w^k w^\ell X_{,k\ell j} \\
&\quad - \frac{1}{2}\alpha_1 w^j U_{,0} + \frac{1}{2}(\alpha_1 - 2\alpha_3)w^k V_{k,j} + \alpha_2 w^k X_{,jk0}\,, \\
\Gamma^j_{0k} &= \gamma\delta_{jk}U_{,0} - \frac{1}{2}(4\gamma+4+\alpha_1)V_{[j,k]} - \frac{1}{2}\alpha_1 w_{[j}U_{,k]}\,, \\
\Gamma^j_{k\ell} &= \gamma(\delta_{jk}U_{,\ell} + \delta_{j\ell}U_{,k} - \delta_{k\ell}U_{,j})\,,
\end{aligned} \tag{6.23}$$

where

$$\begin{aligned}
\psi &= \frac{1}{2}(2\gamma+1+\alpha_3+\zeta_1-2\xi)\Phi_1 - (2\beta-1-\zeta_2-\xi)\Phi_2 \\
&\quad + (1+\zeta_3)\Phi_3 + (3\gamma+3\zeta_4-2\xi)\Phi_4 - \frac{1}{2}(\zeta_1-2\xi)\Phi_6 - \xi\Phi_W\,.
\end{aligned} \tag{6.24}$$

to parametrize the time evolution, and organizing terms to put the resulting equation into a form that matches the Euler equation (4.3) in the Newtonian limit, we obtain

$$\begin{aligned}
\rho^* \frac{dv^j}{dt} &= \rho^* U_{,j} - p_{,j} \\
&+ \left[\frac{1}{2} v^2 + (2-\gamma) U + \Pi + \frac{p}{\rho^*} \right] p_{,j} - v^j p_{,0} \\
&+ \rho^* \Big\{ \left[\gamma v^2 - 2(\gamma + \beta) U \right] U_{,j} + \psi_{,j} - v^j \left[(2\gamma + 1) U_{,0} + 2(\gamma + 1) v^k U_{,k} \right] \\
&\qquad + \frac{1}{2}(4\gamma + 4 + \alpha_1) \left[V_{j,0} + v^k (V_{j,k} - V_{k,j}) \right] + \frac{1}{2}(1 + \alpha_2 - \zeta_1 + 2\xi) X_{,00j} \Big\} \\
&+ \rho^* \left[\frac{1}{2} \Phi^{\rm PF}_{,j} - \Phi^{\rm PF}_{j,0} - v^k (\Phi^{\rm PF}_{j,k} - \Phi^{\rm PF}_{k,j}) \right] , \qquad (6.25)
\end{aligned}$$

where ψ, $\Phi^{\rm PF}$ and $\Phi^{\rm PF}_j$ are displayed in Box 4.1.

6.3 Equations of Motion for Massive Bodies

One method of obtaining equations of motion for massive bodies is to assume that each body moves on a test-body geodesic in a spacetime whose PPN metric is produced by the other bodies in the system as well as by the body itself (with proper care taken of infinite self-field terms). However, the resulting equations of motion *cannot* be applied to massive self-gravitating bodies, such as planets, stars, or the Sun (except in general relativity, as it turns out), because such bodies do not necessarily follow geodesics of any PPN metric. Instead, their motion may depend upon their internal structure (a violation of GWEP). This was first demonstrated by Nordtvedt (1968b). Therefore, one must treat each body realistically, as a finite, self-gravitating "ball" of matter and solve the hydrodynamic equations of motion (6.25) to obtain equations of motion for a suitably chosen center of mass of each body. For the purposes of solar-system and stellar-system experiments, it is adequate to treat the matter composing each body as perfect fluid.

In Newtonian gravitation theory, this program is straightforward. By defining an inertial mass and a center of mass for body a according to

$$\begin{aligned}
m_a &\equiv \int_a \rho\, d^3x , \\
\boldsymbol{x}_a &\equiv m_a^{-1} \int_a \rho \boldsymbol{x}\, d^3x , \qquad (6.26)
\end{aligned}$$

one can show, using the Newtonian equation of continuity (4.3), that

$$\begin{aligned}
\frac{dm_a}{dt} &= 0 , \\
\boldsymbol{v}_a &\equiv \frac{d\boldsymbol{x}_a}{dt} = m_a^{-1} \int_a \rho \boldsymbol{v}\, d^3x , \\
\boldsymbol{a}_a &\equiv \frac{d\boldsymbol{v}_a}{dt} = m_a^{-1} \int_a \rho \frac{d\boldsymbol{v}}{dt} d^3x , \qquad (6.27)
\end{aligned}$$

as long as there is no flux of matter from the body, such as a stellar wind, or mass ejection. Inserting the Newtonian Euler equation (4.3), we obtain the equation of motion for body a,

$$\boldsymbol{a}_a = \boldsymbol{\nabla}\mathfrak{U}, \tag{6.28}$$

where

$$\mathfrak{U} = \sum_{b \neq a} \left[\frac{m_b}{r_{ab}} + \frac{1}{2} \left(I_b^{\langle ij \rangle} + \frac{m_b}{m_a} I_a^{\langle ij \rangle} \right) \nabla_{ai} \nabla_{aj} \frac{1}{r_{ab}} + O(r_{ab}^{-4}) \right], \tag{6.29}$$

where m_b is the inertial mass of the bth body, $I_b^{\langle ij \rangle}$ is its *symmetric, tracefree* quadrupole moment tensor, given by

$$I_b^{\langle ij \rangle} \equiv \int_b \rho \left(\bar{x}^i \bar{x}^j - \frac{1}{3} |\bar{x}|^2 \delta^{ij} \right) d^3x, \quad \bar{\boldsymbol{x}} \equiv \boldsymbol{x} - \boldsymbol{x}_b, \tag{6.30}$$

with analogous expressions for body a, and $\boldsymbol{x}_{ab}$ and r_{ab} are given by

$$\boldsymbol{x}_{ab} \equiv \boldsymbol{x}_a - \boldsymbol{x}_b, \quad r_{ab} \equiv |\boldsymbol{x}_{ab}|. \tag{6.31}$$

Here we display the leading finite-size effects on the motion, proportional to the quadrupole moment of each body; a complete expression including all multipole moments may be found in PW, section 1.6.5. Because the gravitational potential is not uniform, the distribution of mass within a finite-size body can affect its own motion and that of other bodies in the system. These effects can have real-world consequences, such as tidal locking, orbital perturbations and the precession of the Earth's angular momentum. In the limit in which the bodies are small compared to their separations, such effects decrease as $(R/r_{ab})^\ell$, with $\ell \geq 2$, where R is a characteristic size of a body.

We now wish to generalize these equations to the post-Newtonian approximation, using the PPN formalism. We will ignore all finite-size effects that vanish in the limit of small bodies, such as the multipole moment terms we just described. Because there are many different "mass densities" in the post-Newtonian limit – locally measured rest-mass ρ, mass-energy density $\rho(1+\Pi)$, conserved density ρ^*, and so on – there is a variety of possible definitions for inertial mass and center of mass. The main requirements for a sensible definition of center of mass are that it be located somewhere inside the body (it should not wander too far off), that it be useful and convenient, and that it be used consistently in all developments. Once these requirements are satisfied, the freedom of choice is unlimited, and ultimately the most important requirement is one of convenience. Since fluid bodies in nature tend to be spherically symmetric to a high degree, except under extreme conditions such as tidal disruption, any reasonable definition of the center of mass will place it close to the geometrical center of the body.

The choice we will make here has the advantage that the final equations of motion are simple (although "simple" is clearly a subjective statement). We define the inertial mass and the center of mass of the ath body to be

$$\begin{aligned} m_a &\equiv \int_a \rho^* \left(1 + \frac{1}{2}\bar{v}^2 - \frac{1}{2}\bar{U} + \Pi \right) d^3x, \\ \boldsymbol{x}_a &\equiv \frac{1}{m_a} \int_a \rho^* \left(1 + \frac{1}{2}\bar{v}^2 - \frac{1}{2}\bar{U} + \Pi \right) \boldsymbol{x}\, d^3x, \end{aligned} \tag{6.32}$$

where $\bar{\boldsymbol{v}} \equiv \boldsymbol{v} - \boldsymbol{v}_{a(0)}$, where, to lowest order,

$$\boldsymbol{v}_{a(0)} \equiv \frac{1}{m_a} \int_a \rho^* \boldsymbol{v} d^3x , \tag{6.33}$$

and where

$$\bar{U} \equiv \int_a \frac{\rho^{*\prime}}{|\boldsymbol{x} - \boldsymbol{x}'|} d^3x' . \tag{6.34}$$

In many post-Newtonian expressions, it will also be useful to define $\bar{\boldsymbol{x}} \equiv \boldsymbol{x} - \boldsymbol{x}_{a(0)}$, where, to lowest order,

$$\boldsymbol{x}_{a(0)} \equiv \frac{1}{m_a} \int_a \rho^* \boldsymbol{x} d^3x . \tag{6.35}$$

Note that, roughly speaking, m_a is the total mass-energy of the body – rest mass of particles plus kinetic, gravitational, and internal energies – as measured in a local, comoving, nearly inertial frame surrounding the body. As long as we ignore tidal forces on the ath body, which would be proportional to its finite size, then according to our discussion of conservation laws in the PPN formalism (Section 3.2.4), m_a is conserved to post-Newtonian accuracy, that is,

$$\frac{dm_a}{dt} = 0 . \tag{6.36}$$

Making use of the equation of continuity for ρ^*, and using Newtonian equations of motion in any post-Newtonian terms, we obtain

$$\boldsymbol{v}_a \equiv \frac{d\boldsymbol{x}_a}{dt} = \frac{1}{m_a} \int_a \left[\rho^* \left(1 - \frac{1}{2}\bar{v}^2 - \frac{1}{2}\bar{U} + \Pi \right) \boldsymbol{v} + p\bar{\boldsymbol{v}} - \frac{1}{2}\rho^* \bar{\boldsymbol{W}}_a \right] d^3x , \tag{6.37}$$

where here and for future use we define

$$\bar{\boldsymbol{V}}_a \equiv \int_a \rho^{*\prime} \frac{\bar{\boldsymbol{v}}}{|\boldsymbol{x} - \boldsymbol{x}'|} d^3x' ,$$
$$\bar{\boldsymbol{W}}_a \equiv \int_a \rho^{*\prime} \frac{\bar{\boldsymbol{v}}' \cdot (\boldsymbol{x} - \boldsymbol{x}')(\boldsymbol{x} - \boldsymbol{x}')}{|\boldsymbol{x} - \boldsymbol{x}'|^3} d^3x' . \tag{6.38}$$

The acceleration $\boldsymbol{a}_a$ is then given by

$$\begin{aligned} \boldsymbol{a}_a &\equiv \frac{d\boldsymbol{v}_a}{dt} \\ &= \frac{1}{m_a} \left\{ \int_a \rho^* \left(1 - \frac{1}{2}\bar{v}^2 - \frac{1}{2}\bar{U} + \Pi \right) \frac{d\boldsymbol{v}}{dt} d^3x + \int_a (\boldsymbol{v}_a \cdot \boldsymbol{\nabla} p) \bar{\boldsymbol{v}} d^3x \right. \\ &\quad \left. + \int_a \left(p_{,0}\bar{\boldsymbol{v}} - \frac{p}{\rho^*} \boldsymbol{\nabla} p \right) d^3x - \frac{1}{2}\frac{d}{dt} \int_a \rho^* \bar{\boldsymbol{W}}_a d^3x + \frac{1}{2}\mathcal{T}_a - \frac{1}{2}\mathcal{T}_a^* + \mathcal{P}_a \right\} , \end{aligned} \tag{6.39}$$

where $\mathcal{T}_a$, $\mathcal{T}_a^*$ and $\mathcal{P}_a$ are determined purely by the internal structure of the ath body. Formulae for these and other "internal" terms are given in Box 6.2. Notice that the acceleration of our chosen center of mass is more than just the weighted average of the accelerations of individual fluid elements, as it is in Newtonian theory. We now evaluate the first integral in Eq. (6.39) using the PPN equations of hydrodynamics (6.25). We make use of the Newtonian equations of motion where necessary to simplify post-Newtonian terms. For more details on how individual terms are evaluated, see PW, section 9.3.

Box 6.2 Integrals for massive bodies in the PPN equations of motion

Scalar integrals

$$
\begin{aligned}
I_a &\equiv \int_a \rho^* |\bar{\boldsymbol{x}}|^2 \, d^3x , \\
\mathcal{T}_a &\equiv \frac{1}{2} \int_a \rho^* \bar{v}^2 \, d^3x , \\
\Omega_a &\equiv -\frac{1}{2} \int_a \frac{\rho^* \rho^{*\prime}}{|\boldsymbol{x} - \boldsymbol{x}'|} \, d^3x' d^3x , \\
P_a &\equiv \int_a p \, d^3x , \\
E_a &\equiv \int_a \rho^* \Pi \, d^3x .
\end{aligned}
\tag{6.40}
$$

Tensor integrals

$$
\begin{aligned}
I_a^{jk} &\equiv \int_a \rho^* \bar{x}^j \bar{x}^k \, d^3x , \\
S_a^{jk} &\equiv \int_a \rho^* (\bar{x}^j \bar{v}^k - \bar{x}^k \bar{v}^j) \, d^3x , \\
\mathcal{T}_a^{jk} &\equiv \frac{1}{2} \int_a \rho^* \bar{v}^j \bar{v}^k \, d^3x , \\
L_a^{jk} &\equiv \int_a \bar{v}^j \partial_k p \, d^3x , \\
\Omega_a^{jk} &\equiv -\frac{1}{2} \int_a \rho^* \rho^{*\prime} \frac{(x - x')^j (x - x')^k}{|\boldsymbol{x} - \boldsymbol{x}'|^3} \, d^3x' d^3x , \\
H_a^{jk} &\equiv \int_a \rho^* \rho^{*\prime} \frac{\bar{v}'^j (x - x')^k}{|\boldsymbol{x} - \boldsymbol{x}'|^3} \, d^3x' d^3x , \\
K_a^{jk} &\equiv \int_a \rho^* \rho^{*\prime} \frac{\bar{\boldsymbol{v}}' \cdot (\boldsymbol{x} - \boldsymbol{x}') (x - x')^j (x - x')^k}{|\boldsymbol{x} - \boldsymbol{x}'|^5} \, d^3x' d^3x .
\end{aligned}
\tag{6.41}
$$

Vector integrals

$$
\begin{aligned}
t_a^j &\equiv \int_a \rho^* \rho^{*\prime} \frac{\bar{v}'^2 (x - x')^j}{|\boldsymbol{x} - \boldsymbol{x}'|^3} \, d^3x' d^3x , \\
\mathcal{T}_a^j &\equiv \int_a \rho^* \rho^{*\prime} \frac{\bar{v}'^j \bar{\boldsymbol{v}}' \cdot (\boldsymbol{x} - \boldsymbol{x}')}{|\boldsymbol{x} - \boldsymbol{x}'|^3} \, d^3x' d^3x , \\
\mathcal{T}_a^{*j} &\equiv \int_a \rho^* \rho^{*\prime} \frac{\bar{v}^j \bar{\boldsymbol{v}}' \cdot (\boldsymbol{x} - \boldsymbol{x}')}{|\boldsymbol{x} - \boldsymbol{x}'|^3} \, d^3x' d^3x , \\
\mathcal{T}_a^{**j} &\equiv \int_a \rho^* \rho^{*\prime} \frac{[\bar{\boldsymbol{v}}' \cdot (\boldsymbol{x} - \boldsymbol{x}')]^2 (x - x')^j}{|\boldsymbol{x} - \boldsymbol{x}'|^5} \, d^3x' d^3x , \\
\Omega_a^j &\equiv \int_a \rho^* \rho^{*\prime} \rho^{*\prime\prime} \frac{(x - x')^j}{|\boldsymbol{x}' - \boldsymbol{x}''| |\boldsymbol{x} - \boldsymbol{x}'|^3} \, d^3x'' d^3x' d^3x ,
\end{aligned}
$$

$$\Omega_a^{*j} \equiv \int_a \rho^* \rho^{*\prime} \rho^{*\prime\prime} \frac{(\boldsymbol{x}' - \boldsymbol{x}'') \cdot (\boldsymbol{x} - \boldsymbol{x}')(x - x')^j}{|\boldsymbol{x}' - \boldsymbol{x}''|^3 |\boldsymbol{x} - \boldsymbol{x}'|^3} d^3x'' d^3x' d^3x,$$

$$\mathcal{P}_a^j \equiv \int_a \rho^* p^{*\prime} \frac{(x - x')^j}{|\boldsymbol{x} - \boldsymbol{x}'|^3} d^3x' d^3x,$$

$$\mathcal{E}_a^j \equiv \int_a \rho^* \rho^{*\prime} \Pi' \frac{(x - x')^j}{|\boldsymbol{x} - \boldsymbol{x}'|^3} d^3x' d^3x. \tag{6.42}$$

Considerable simplification of the equations results if we make use of virial identities that connect a number of integrals in Box 6.2. These identities are given, to Newtonian order, by

$$\frac{1}{2}\frac{d}{dt} I_a^{jk} = \frac{1}{2} S_a^{jk} + \int_a \rho^* \bar{v}^j \bar{x}^k \, d^3x,$$

$$\frac{1}{2}\frac{d^2}{dt^2} I_a^{jk} = 2\mathcal{T}_a^{jk} + \Omega_a^{jk} + \delta^{jk} P_a,$$

$$\frac{1}{2}\frac{d^3}{dt^3} I_a^{jk} = 4H_a^{(jk)} - 3K_a^{jk} + \delta^{jk} \dot{P}_a - 2L_a^{(jk)}. \tag{6.43}$$

Two vectorial identities are given by

$$\frac{d}{dt}\int_a \rho^* \bar{V}_a^j d^3x = \mathcal{T}_a^j + \mathcal{T}_a^{*j} + \Omega_a^j + \mathcal{P}_a^j,$$

$$\frac{d}{dt}\int_a \rho^* \bar{W}_a^j d^3x = -t_a^j - \mathcal{T}_a^j + \mathcal{T}_a^{*j} + 3\mathcal{T}_a^{**j} - \Omega_a^{*j} - \mathcal{P}_a^j, \tag{6.44}$$

where we have discarded post-Newtonian corrections as well as terms that arise from external gravitational potentials, that can be shown to vary as $O(R_a^2/r_{ab}^2)$, where R_a is the characteristic size of the ath body.

We now assume that each body is in dynamical equilibrium. By this we mean that each body has had time, under its own internal dynamics, to relax to a steady state in which its internal structure does not depend on time. This means, in particular, that the structure integrals listed in Box 6.2 can all be taken to be time-independent. Then we can set equal to zero any total time derivatives of purely internal quantities, such as the left-hand-sides of Eqs. (6.43) and (6.44). This is a reasonable approximation for systems with well separated bodies, since any secular changes in the structure of the bodies that would prevent the vanishing of such averaged time derivatives occur over timescales much longer than an orbital timescale.

The final form of the equations of motion is

$$\boldsymbol{a}_a = (\boldsymbol{a}_a)_{\text{self}} + (\boldsymbol{a}_a)_{\text{Newt}} + (\boldsymbol{a}_a)_{\text{Nbody}}, \tag{6.45}$$

where

$$(a_a^j)_{\text{self}} = -\frac{1}{m_a}\left[\frac{1}{2}(\alpha_3 + \zeta_1)t_a^j + \frac{1}{2}\zeta_1(2\mathcal{T}_a^j - 3\mathcal{T}_a^{**j}) + \zeta_2\Omega_a^j + \zeta_3\mathcal{E}_a^j + 3\zeta_4\mathcal{P}_a^j\right]$$
$$- \frac{1}{m_a}\alpha_3(w + v_a)^k H_a^{kj}, \tag{6.46a}$$

$$(a_a^j)_{\rm Newt} = -\frac{1}{m_a}(m_{\rm P})_a^{jk}\mathfrak{U}_{,k}\,, \tag{6.46b}$$

$$\begin{aligned}
(a_a^j)_{\rm Nbody} = &-\sum_{b\neq a}\frac{m_b n_{ab}^j}{r_{ab}^2}\Big\{\gamma v_a^2 - (2\gamma+2)(\boldsymbol{v}_a\cdot\boldsymbol{v}_b) + (\gamma+1)v_b^2 \\
&-\frac{3}{2}(\boldsymbol{n}_{ab}\cdot\boldsymbol{v}_b)^2 + \frac{1}{2}(\alpha_2+\alpha_3)|\boldsymbol{w}+\boldsymbol{v}_b|^2 \\
&-\frac{1}{2}\alpha_1(\boldsymbol{w}+\boldsymbol{v}_a)\cdot(\boldsymbol{w}+\boldsymbol{v}_b) - \frac{3}{2}\alpha_2[\boldsymbol{n}_{ab}\cdot(\boldsymbol{w}+\boldsymbol{v}_b)]^2 \\
&-\left(2\gamma+2\beta+1+\frac{1}{2}\alpha_1-\zeta_2\right)\frac{m_a}{r_{ab}} - (2\gamma+2\beta)\frac{m_b}{r_{ab}}\Big\} \\
&+\sum_{b\neq a}\frac{m_b}{r_{ab}^2}\boldsymbol{n}_{ab}\cdot\big[(2\gamma+2)\boldsymbol{v}_a - (2\gamma+1)\boldsymbol{v}_b\big](v_a^j - v_b^j) \\
&-\sum_{b\neq a}\frac{m_b}{r_{ab}^2}\boldsymbol{n}_{ab}\cdot\left[\alpha_2(\boldsymbol{w}+\boldsymbol{v}_b) + \frac{1}{2}\alpha_1(\boldsymbol{v}_a-\boldsymbol{v}_b)\right](w^j + v_b^j) \\
&+\sum_{b\neq a}\sum_{c\neq a,b}\frac{m_b m_c}{r_{ab}^2}n_{ab}^j\Big[(2\gamma+2\beta-2\xi)\frac{1}{r_{ac}} + (2\beta-1-2\xi-\zeta_2)\frac{1}{r_{bc}} \\
&-\frac{1}{2}(1+2\xi+\alpha_2-\zeta_1)\frac{r_{ab}}{r_{bc}^2}(\boldsymbol{n}_{ab}\cdot\boldsymbol{n}_{bc}) - \xi\frac{r_{bc}}{r_{ac}^2}(\boldsymbol{n}_{ac}\cdot\boldsymbol{n}_{bc})\Big] \\
&-\frac{1}{2}(4\gamma+3-2\xi+\alpha_1-\alpha_2+\zeta_1)\sum_{b\neq a}\sum_{c\neq a,b}\frac{m_b m_c}{r_{ab}r_{bc}^2}n_{bc}^j \\
&-\xi\sum_{b\neq a}\sum_{c\neq a,b}\frac{m_b m_c}{r_{ab}^3}(n_{ac}^k - n_{bc}^k)\left(\delta^{jk} - 3n_{ab}^j n_{ab}^k\right),
\end{aligned} \tag{6.46c}$$

where $\boldsymbol{n}_{ab} = \boldsymbol{x}_{ab}/r_{ab}$.

The terms within square brackets in $(\boldsymbol{a}_a)_{\rm self}$, Eq. (6.46a), depend only on the internal structure of the ath body, and thus represent "self-accelerations" of the body's center of mass. Such self-accelerations are associated with breakdowns in conservation of total momentum, since they depend on the PPN conservation-law parameters α_3, ζ_1, ζ_2, ζ_3, and ζ_4. In any semiconservative theory of gravity

$$\alpha_3 = \zeta_1 = \zeta_2 = \zeta_3 = \zeta_4 = 0\,, \tag{6.47}$$

and these self-accelerations are absent. Also note that spherically symmetric bodies, or bodies with reflection symmetry about the origin (as assumed in PW) suffer no acceleration regardless of the theory of gravity, since for them the terms t_a^j, $\mathcal{T}_a^j$, $\mathcal{T}_a^{**j}$, Ω_a^j, $\mathcal{E}_a^j$ and $\mathcal{P}_a^j$ are identically zero. The same is true for a composite massive body made up of two bodies in a nearly circular orbit, when the self-acceleration is averaged over an orbital period. Thus, there is little hope of testing the existence of these terms in the solar system because of the high degree of symmetry of the bodies and the high circularity of their orbits. However, in binary pulsars, where orbital eccentricities can be large, this effect has been tested, leading to a tight bound on ζ_2 (see Section 9.3).

The final term in Eq. (6.46a), $-m_a^{-1}\alpha_3(w+v_a)^k H_a^{kj}$, is a self-acceleration which involves the massive body's motion relative to the universe rest frame. It depends on the conservation-law/preferred-frame parameter α_3, which is zero in any semiconservative theory of gravity. For any static body, $\bar{\boldsymbol{v}} = 0$, thus H_a^{kj} is zero, but for a body that rotates uniformly with angular velocity $\boldsymbol{\omega}_a$, $\bar{\boldsymbol{v}} = \boldsymbol{\omega}_a \times \bar{\boldsymbol{x}}$, and thus

$$H_a^{kj} = \epsilon^{k\ell m}\omega_a^\ell \int_a \rho^* \rho^{*\prime} \frac{\bar{x}'^m (x - x')^j}{|\boldsymbol{x} - \boldsymbol{x}'|^3} d^3x' d^3x$$
$$= \epsilon^{k\ell m}\omega_a^\ell \Omega_a^{jm} . \tag{6.48}$$

For a nearly spherical body, the isotropic part of Ω_a^{jm} makes the dominant contribution to Eq. (6.48), that is,

$$\Omega_a^{jm} \simeq \frac{1}{3}\Omega_a \delta^{jm} , \quad H_a^{kj} \simeq \frac{1}{3}\epsilon^{k\ell j}\omega_a^\ell \Omega_a . \tag{6.49}$$

Then the final acceleration term in Eq. (6.46a) becomes

$$-\frac{1}{3}\alpha_3 \frac{\Omega_a}{m_a}(\boldsymbol{w} + \boldsymbol{v}_a) \times \boldsymbol{\omega}_a . \tag{6.50}$$

In Section 8.4 we will see that this term can produce strikingly large accelerations in spinning pulsars, leading to very strong bounds on α_3.

The next term, $(\boldsymbol{a}_a)_{\rm Newt}$ in Eq. (6.45), is the quasi-Newtonian acceleration of the body, but it contains post-Newtonian corrections that depend on the internal structure of both body a and the other bodies in the system. These corrections can be grouped together so as to appear as corrections to the "passive gravitational mass" of body a and to the "active gravitational mass" of the other bodies b. In fact, the passive mass turns out to be tensorial, given by

$$(m_{\rm P})_a^{jk} = m_a \Bigg\{ \delta^{jk} \left[1 + (4\beta - \gamma - 3 - 3\xi - \alpha_1 + \alpha_2 - \zeta_1)\frac{\Omega_a}{m_a} - 3\xi n_{ab}^\ell n_{ab}^m \frac{\Omega_a^{\ell m}}{m_a} \right]$$
$$+ (2\xi - \alpha_2 + \zeta_1 - \zeta_2)\frac{\Omega_a^{jk}}{m_a} \Bigg\} . \tag{6.51}$$

The quasi-Newtonian potential $\mathfrak{U}(\boldsymbol{x}_a)$ has the form

$$\mathfrak{U}(\boldsymbol{x}_a) = \sum_{b \neq a} \frac{[m_{\rm A}(\boldsymbol{n}_{ab})]_b}{r_{ab}} , \tag{6.52}$$

where the active gravitational mass of the bth body is given by

$$[m_{\rm A}(\boldsymbol{n}_{ab})]_b = m_b \Bigg\{ 1 + \left(4\beta - \gamma - 3 - 3\xi - \frac{1}{2}\alpha_3 - \frac{1}{2}\zeta_1 - 2\zeta_2 \right) \frac{\Omega_b}{m_b} + \zeta_3 \frac{E_b}{m_b}$$
$$+ \left(3\zeta_4 - \zeta_1 - \frac{3}{2}\alpha_3 \right) \frac{P_b}{m_b} + \frac{1}{2}(\zeta_1 - 2\xi) n_{ab}^j n_{ab}^k \frac{\Omega_b^{jk}}{m_b} \Bigg\} . \tag{6.53}$$

Note that the active and passive gravitational mass tensors may be functions of direction $\boldsymbol{n}_{ab}$. An alternative form of the quasi-Newtonian acceleration involves inertial, active and passive mass tensors that are *independent* of position, at the expense of having a complicated tensorial gravitational potential, as follows:

$$(\tilde{m}_{\rm I})_a^{jk}(a_a^k)_{\rm Newt} = (\tilde{m}_{\rm P})_a^{\ell m} \mathfrak{U}_{,j}^{\ell m} ,$$
$$\mathfrak{U}^{\ell m} \equiv \sum_{b \neq a} (\tilde{m}_{\rm A})_b^{mq} \frac{n_{ab}^q n_{ab}^\ell}{r_{ab}} , \tag{6.54}$$

where the inertial, passive, and active mass tensors are given by

$$(\tilde{m}_{\rm I})^{jk}_a \equiv m_a \left\{ \delta^{jk} \left[1 + (\alpha_1 - \alpha_2 + \zeta_1)\frac{\Omega_a}{m_a} \right] + (\alpha_2 - \zeta_1 + \zeta_2)\frac{\Omega^{jk}_a}{m_a} \right\},$$

$$(\tilde{m}_{\rm P})^{\ell m}_a \equiv m_a \left\{ \delta^{\ell m} \left[1 + (4\beta - \gamma - 3 - 3\xi)\frac{\Omega_a}{m_a} \right] - \xi \frac{\Omega^{\ell m}_a}{m_a} \right\},$$

$$(\tilde{m}_{\rm A})^{mq}_b \equiv m_b \left\{ \delta^{mq} \left[1 + \left(4\beta - \gamma - 3 - 3\xi - \frac{1}{2}\alpha_3 - \frac{1}{2}\zeta_1 - 2\zeta_2 \right) \frac{\Omega_b}{m_b} \right.\right.$$
$$\left.\left. + \zeta_3 \frac{E_b}{m_b} + \left(3\zeta_4 - \zeta_1 - \frac{3}{2}\alpha_3 \right) \frac{P_b}{m_b} \right] + \frac{1}{2}(\zeta_1 - 2\xi)\frac{\Omega^{mq}_b}{m_b} \right\}. \tag{6.55}$$

In Newtonian theory, the active gravitational mass, the passive gravitational mass, and the inertial mass are the same, hence each massive body's acceleration is independent of its mass or structure, in accord with the Newtonian equivalence principle. However, according to Eq. (6.55), passive gravitational mass need not be equal to inertial mass in a given metric theory of gravity, and in fact both may be anisotropic; their difference depends on several PPN parameters, and on the gravitational self-energy (Ω and Ω^{jk}) of the body. This is a breakdown in the gravitational Weak Equivalence Principle (GWEP) (see Section 3.3), also called the "Nordtvedt effect" after its discoverer (Nordtvedt, 1968a,b). The possibility of such an effect was first noticed by Dicke (1964) (see also Dicke (1969) and Will (1971c)). The observable consequences of the Nordtvedt effect will be discussed in Chapter 8. Its existence does not violate EEP or the Eötvös experiment (Chapter 2), because the laboratory-sized bodies considered in those situations have negligible self gravity, that is, $(\Omega/m)_{\rm lab\ bodies} < 10^{-39}$.

According to Eq. (6.55), the active gravitational mass for massive bodies may also differ from inertial mass and from passive gravitational mass. In Newtonian gravitation theory, the conservation of momentum of an isolated system is a result of the law "action equals reaction," that is, of the law "active gravitational mass equals passive gravitational mass." In the PPN formalism, one can still use such Newtonian language to describe the quasi-Newtonian acceleration $(\boldsymbol{a}_a)_{\rm Newt}$. From Section 4.4, we know that momentum conservation is a property of semiconservative theories of gravity, whose parameters satisfy $\alpha_3 = \zeta_k = 0$. By substituting these values into Eq. (6.55), we find that for semiconservative theories, the active and passive mass tensors are indeed equal, and are given by

$$(\tilde{m}_{\rm P})^{jk}_a = (\tilde{m}_{\rm A})^{jk}_a = m_a \left\{ \delta^{jk} \left[1 + (4\beta - \gamma - 3 - 3\xi)\frac{\Omega_a}{m_a} \right] - \xi \frac{\Omega^{jk}_a}{m_a} \right\}. \tag{6.56}$$

Equivalently, we can write $(\boldsymbol{a}_a)_{\rm Newt}$ to post-Newtonian order in the form

$$(\tilde{m}_{\rm I})^{jk}_a (a^k_a)_{\rm Newt} = \sum_{b \neq a} m_a m_b \left\{ \left[1 + (4\beta - \gamma - 3 - 3\xi)\left(\frac{\Omega_a}{m_a} + \frac{\Omega_b}{m_b} \right) \right] \frac{n^j_{ab}}{r^2_{ab}} \right.$$
$$\left. + \xi \left(\frac{\Omega^{k\ell}_a}{m_a} + \frac{\Omega^{k\ell}_b}{m_b} \right) \frac{n^\ell_{ab}}{r^2_{ab}} \left(2\delta^{jk} - 3 n^j_{ab} n^k_{ab} \right) \right\}. \tag{6.57}$$

The term in braces is manifestly antisymmetric under interchange of a and b, hence action equals reaction, and $\sum_a (\tilde{m}_{\rm I})^{jk}_a (a^k_a)_{\rm Newt} = 0$.

Note that in general relativity, the mass tensors of Eq. (6.55) are isotropic and equal to the inertial mass, that is, dropping the Kronecker deltas,

$$(\tilde{m}_{\rm I}) = (\tilde{m}_{\rm P}) = (\tilde{m}_{\rm A}) = m_a \quad [\text{general relativity}] \,. \tag{6.58}$$

There is no Nordtvedt effect in general relativity. However, in scalar-tensor theories, for example, the Nordtvedt effect is present; in those theories $\tilde{m}_{\rm I} = m_a$, but

$$\tilde{m}_{\rm P} = \tilde{m}_{\rm A} = m_a \left[1 + \frac{1+2\lambda}{2+\omega} \frac{\Omega_a}{m_a} \right] . \tag{6.59}$$

For most practical situations, we may assume that the bodies in question are spherically symmetric, then using the approximation $\Omega_a^{jk} \simeq \frac{1}{3}\delta^{jk}\Omega_a$ to simplify the mass tensors, we may write

$$(a_a^j)_{\rm Newt} = \frac{(m_{\rm P})_a}{m_a} \mathfrak{U}_{,j} \,, \qquad \mathfrak{U} \equiv \sum_{b \neq a} \frac{(m_{\rm A})_b}{r_{ab}} \,, \tag{6.60}$$

where we combine $(\tilde{m}_{\rm I}^{jk})^{-1}$ and $(\tilde{m}_{\rm P}^{lm})$ into a single quantity $m_{\rm P}$ to obtain

$$\frac{(m_{\rm P})_a}{m_a} = 1 + \left(4\beta - \gamma - 3 - \frac{10}{3}\xi - \alpha_1 + \frac{2}{3}\alpha_2 - \frac{2}{3}\zeta_1 - \frac{1}{3}\zeta_2 \right) \frac{\Omega_a}{m_a} \,, \tag{6.61a}$$

$$\begin{aligned} \frac{(m_{\rm A})_b}{m_b} &= 1 + \left(4\beta - \gamma - 3 - \frac{10}{3}\xi - \frac{1}{2}\alpha_3 - \frac{1}{3}\zeta_1 - 2\zeta_2 \right) \frac{\Omega_b}{m_b} \\ &\quad + \zeta_3 \frac{E_b}{m_b} + \left(3\zeta_4 - \zeta_1 - \frac{3}{2}\alpha_3 \right) \frac{P_b}{m_b} \,. \end{aligned} \tag{6.61b}$$

The remaining term $(a_a^j)_{\rm Nbody}$, in Eq. (6.45) is called the *N*-body term. It contains the post-Newtonian corrections to the Newtonian equations of motion which would result from treating each body as a "point mass" moving along a geodesic of the PPN metric produced by all the other bodies, assumed to be point masses, taking account of certain post-Newtonian terms generated by the gravitational field of the body itself. It is the *N*-body acceleration which produces the "classical" perihelion shift of the planets, as well as a host of other effects, to be examined in Chapters 8 and 9. For the case of general relativity, the *N*-body terms in Eq. (6.46c) are in agreement with the equations first obtained by Lorentz and Droste (1917), and later by de Sitter (1916) (once a crucial error in de Sitter's work has been corrected), Einstein, Infeld, and Hoffmann (1938), Levi-Civita (1965), and Fock (1964).

6.4 Two-Body Systems

6.4.1 Effective one-body equation of motion

We now truncate the PPN *N*-body equations of motion (6.45) and (6.46) to $N = 2$. We consider a system of two bodies of inertial masses m_1 and m_2 and self-gravitational energies Ω_1 and Ω_2. Both bodies will be assumed to be reflection symmetric about their centers of mass, so that the self-terms in Eq. (6.46a) can be dropped, but we will let the second body have a small quadrupole moment $I_2^{\langle ij\rangle}$ as defined in Eq. (6.30). We assume that the entire system is at rest with respect to the universe rest frame ($\boldsymbol{w} = 0$) and that there are no other gravitating bodies near the system. In Chapter 8 we will return to the effects of motion of the system and of distant bodies (preferred-frame and preferred-location effects) on the two-body problem. For the moment we ignore them. We work in a PPN coordinate system in which the center of mass of the system is at rest at the origin. Making use of the fact that each body is nearly spherical, $\Omega_a^{jk} \simeq \frac{1}{3}\delta^{jk}\Omega_a$, we obtain from Eqs. (6.45) and (6.46) the acceleration of each body

$$
\begin{aligned}
\boldsymbol{a}_1 = {} & \frac{(m_{\rm P})_1}{m_1}\boldsymbol{\nabla}\mathfrak{U}_1 + \frac{m_2\boldsymbol{n}}{r^2}\left[(2\gamma+2\beta)\frac{m_2}{r} + \left(2\gamma+2\beta+1+\frac{1}{2}\alpha_1-\zeta_2\right)\frac{m_1}{r} - \gamma v_1^2\right. \\
& \left. + \frac{1}{2}(4\gamma+4+\alpha_1)(\boldsymbol{v}_1\cdot\boldsymbol{v}_2) - \frac{1}{2}(2\gamma+2+\alpha_2+\alpha_3)v_2^2 + \frac{3}{2}(1+\alpha_2)(\boldsymbol{n}\cdot\boldsymbol{v}_2)^2\right] \\
& + \frac{m_2\boldsymbol{n}}{r^2}\cdot\left[(2\gamma+2)\boldsymbol{v}_1 - (2\gamma+1)\boldsymbol{v}_2\right]\boldsymbol{v}_1 \\
& - \frac{1}{2}\frac{m_2\boldsymbol{n}}{r^2}\cdot\left[(4\gamma+4+\alpha_1)\boldsymbol{v}_1 - (4\gamma+2+\alpha_1-2\alpha_2)\boldsymbol{v}_2\right]\boldsymbol{v}_2\,, \\
\boldsymbol{a}_2 = {} & \{1 \rightleftharpoons 2;\ \boldsymbol{n}\to-\boldsymbol{n}\}\,,
\end{aligned} \tag{6.62}
$$

where $\boldsymbol{x} \equiv \boldsymbol{x}_{12}$, $\boldsymbol{n} = \boldsymbol{x}/r$, and $m_{\rm P}$ is given by Eq. (6.61a). Including the Newtonian contribution of the quadrupole moment in the quasi-Newtonian potential produced by body 2 as shown in Eq. (6.29), we have

$$
\begin{aligned}
(\mathfrak{U}_{,j})_1 &= -(m_{\rm A})_2\frac{n^j}{r^2} - \frac{15}{2}I_2^{\langle k\ell\rangle}\frac{n^{\langle jk\ell\rangle}}{r^4}\,, \\
(\mathfrak{U}_{,j})_2 &= (m_{\rm A})_1\frac{n^j}{r^2} + \frac{15}{2}\frac{(m_{\rm A})_1}{(m_{\rm A})_2}I_2^{\langle k\ell\rangle}\frac{n^{\langle jk\ell\rangle}}{r^4}\,,
\end{aligned} \tag{6.63}
$$

where $m_{\rm A}$ is given by Eq. (6.61b), and $n^{\langle jk\ell\rangle}$ is the symmetric tracefree combination of unit vectors, given by

$$
n^{\langle jk\ell\rangle} \equiv n^j n^k n^\ell - \frac{1}{5}\left(n^j\delta^{k\ell} + n^k\delta^{j\ell} + n^\ell\delta^{jk}\right). \tag{6.64}
$$

In Eq. (6.63), we have ignored the difference between the inertial and active masses in the quadrupole contributions, and used active masses throughout. Even though the quadrupole terms are of Newtonian order, they are generally very small, and thus this is a reasonable approximation. For a body which is axially symmetric about an axis with direction $\boldsymbol{e}$, it can be shown that

$$
I_2^{\langle kn\rangle} = -(m_{\rm A})_2 R_2^2 J_{2(2)}\left(e^k e^n - \frac{1}{3}\delta^{kn}\right), \tag{6.65}
$$

where R_2 is the mean radius of body 2, and $J_{2(2)}$ is its dimensionless quadrupole moment ($\ell = 2$), given for general axisymmetric bodies by

$$
J_2 = \frac{C-A}{mR^2}\,, \tag{6.66}
$$

where C and A are the moments of inertia about the symmetry axis and about an axis in the equatorial plane, respectively.

Since the center of mass is at rest at the origin, we may replace $\boldsymbol{v}_1$ and $\boldsymbol{v}_2$ in the post-Newtonian terms in Eqs. (6.62) by

$$\boldsymbol{v}_1 = \frac{m_2}{m}\boldsymbol{v}, \quad \boldsymbol{v}_2 = -\frac{m_1}{m}\boldsymbol{v}, \tag{6.67}$$

where

$$\boldsymbol{v} \equiv \boldsymbol{v}_1 - \boldsymbol{v}_2, \quad m \equiv m_1 + m_2 . \tag{6.68}$$

We also define the dimensionless reduced mass parameter

$$\eta \equiv \frac{m_1 m_2}{(m_1+m_2)^2} . \tag{6.69}$$

Then the relative acceleration $\boldsymbol{a} \equiv \boldsymbol{a}_1 - \boldsymbol{a}_2$ takes the form of an effective one-body equation of motion,

$$\begin{aligned}
\boldsymbol{a} = &-\frac{m^*\boldsymbol{n}}{r^2} + \frac{3}{2}\frac{mR_2^2 J_{2(2)}}{r^4}\left[5\boldsymbol{n}(\boldsymbol{e}\cdot\boldsymbol{n})^2 - 2(\boldsymbol{e}\cdot\boldsymbol{n})\boldsymbol{e} - \boldsymbol{n}\right] \\
&+ \frac{m\boldsymbol{n}}{r^2}\left\{[2\gamma + 2\beta + \eta(2+\alpha_1 - 2\zeta_2)]\frac{m}{r}\right. \\
&\left. - \frac{1}{2}[2\gamma + \eta(6+\alpha_1+\alpha_2+\alpha_3)]v^2 + \frac{3}{2}\eta(1+\alpha_2)(\boldsymbol{v}\cdot\boldsymbol{n})^2\right\} \\
&+ \frac{m\boldsymbol{v}(\boldsymbol{n}\cdot\boldsymbol{v})}{r^2}[2\gamma + 2 - \eta(2-\alpha_1+\alpha_2)] ,
\end{aligned} \tag{6.70}$$

where

$$\begin{aligned}
m^* &\equiv \frac{(m_P)_2}{m_2}(m_A)_1 + \frac{(m_P)_1}{m_1}(m_A)_2 \\
&= m + [\text{self energy terms for bodies 1 and 2}] .
\end{aligned} \tag{6.71}$$

The self-energy terms from Eqs. (6.61b) and (6.61a) that appear in the above expression are constant if the bodies are in quasistationary equilibrium. Furthermore, observations of a binary system can only measure the mass m^*, via Kepler's third law, for example, and thus the self-energy corrections are unmeasurable. Henceforth, we will simply drop the (*) in Eq. (6.70).

6.4.2 Perturbed two-body orbits and the Lagrange planetary equations

Here we provide a brief review of standard orbital perturbation theory, used to compute deviations from Keplerian two-body motion induced by perturbing forces, described by the equation of motion

$$\boldsymbol{a} = -\frac{m\boldsymbol{n}}{r^2} + \delta\boldsymbol{a} , \tag{6.72}$$

where $\delta\boldsymbol{a}$ is a perturbing acceleration. For a general orbit described by $\boldsymbol{x}$ and $\boldsymbol{v} = d\boldsymbol{x}/dt$, we define the "osculating" Keplerian orbit using a set of orbit elements: the semilatus rectum p, eccentricity e, inclination ι, nodal angle Ω and pericenter angle ω, defined by the following set of equations (see Figure 6.1):

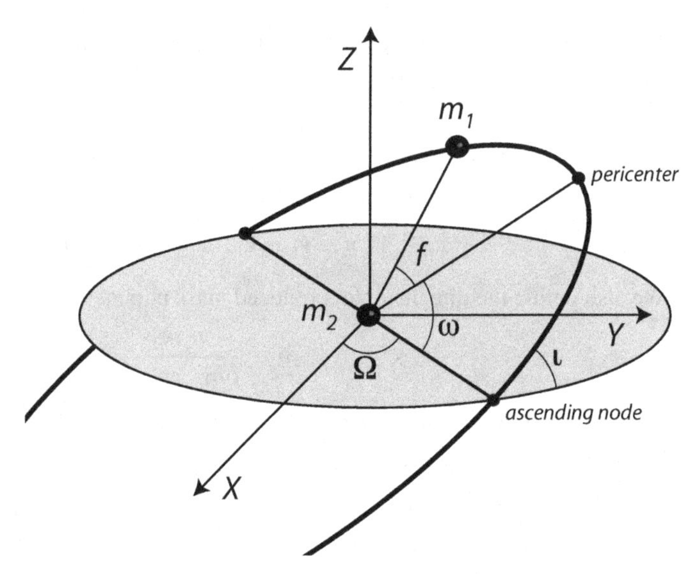

Fig. 6.1 Elements of a general relative two-body orbit.

$$\begin{aligned}
\boldsymbol{x} &\equiv r\boldsymbol{n}\,, \\
r &\equiv p/(1+e\cos f)\,, \\
\boldsymbol{n} &\equiv (\cos\Omega\cos\phi - \cos\iota\sin\Omega\sin\phi)\,\boldsymbol{e}_X \\
&\quad + (\sin\Omega\cos\phi + \cos\iota\cos\Omega\sin\phi)\,\boldsymbol{e}_Y \\
&\quad + \sin\iota\sin\phi\boldsymbol{e}_Z\,, \\
\boldsymbol{\lambda} &\equiv \partial\boldsymbol{n}/\partial\phi\,, \quad \hat{\boldsymbol{h}} = \boldsymbol{n}\times\boldsymbol{\lambda}\,, \\
\boldsymbol{h} &\equiv \boldsymbol{x}\times\boldsymbol{v} \equiv \sqrt{mp}\,\hat{\boldsymbol{h}}\,,
\end{aligned} \tag{6.73}$$

where $f \equiv \phi - \omega$ is the *true anomaly*, ϕ is the orbital phase measured from the ascending node, and $\boldsymbol{e}_A$ are chosen reference basis vectors. From the given definitions, we see that $\boldsymbol{v} = \dot{r}\boldsymbol{n} + (h/r)\boldsymbol{\lambda}$ and $\dot{r} = (he/p)\sin f$. The semilatus rectum p is related to the semimajor axis a by $p = a(1-e^2)$. One then introduces the radial $\mathcal{R}$, cross-track $\mathcal{S}$ and out-of-plane $\mathcal{W}$ components of the perturbing acceleration $\delta\boldsymbol{a}$, defined, respectively, by $\mathcal{R} \equiv \boldsymbol{n}\cdot\delta\boldsymbol{a}$, $\mathcal{S} \equiv \boldsymbol{\lambda}\cdot\delta\boldsymbol{a}$ and $\mathcal{W} \equiv \hat{\boldsymbol{h}}\cdot\delta\boldsymbol{a}$, and writes down the "Lagrange planetary equations" for the evolution of the orbit elements,

$$\begin{aligned}
\frac{dp}{dt} &= 2\sqrt{\frac{p^3}{m}}\frac{\mathcal{S}}{1+e\cos f}, \\
\frac{de}{dt} &= \sqrt{\frac{p}{m}}\left[\sin f\mathcal{R} + \frac{2\cos f + e + e\cos^2 f}{1+e\cos f}\mathcal{S}\right], \\
\frac{d\omega}{dt} &= \frac{1}{e}\sqrt{\frac{p}{m}}\left[-\cos f\mathcal{R} + \frac{2+e\cos f}{1+e\cos f}\sin f\mathcal{S} - e\cot\iota\frac{\sin(\omega+f)}{1+e\cos f}\mathcal{W}\right], \\
\frac{d\iota}{dt} &= \sqrt{\frac{p}{m}}\mathcal{W}\left(\frac{r}{p}\right)\cos\phi\,, \\
\frac{d\Omega}{dt} &= \sqrt{\frac{p}{m}}\mathcal{W}\left(\frac{r}{p}\right)\frac{\sin\phi}{\sin\iota}\,.
\end{aligned} \tag{6.74}$$

(see PW, section 3.3 for further discussion.) We have not listed a sixth equation for the variation of T, the "time of pericenter passage"; the six elements p, e, ω, ι, and Ω and T provide a complete characterization of the six orbital variables $\boldsymbol{x}(t)$ and $\boldsymbol{v}(t)$. The element T is relevant if the evolution of the orbit with time is the main focus. But for many applications, the evolution with orbital phase f or ϕ is the focus. In the latter case we use Eq. (6.74) and close the system by providing equations that connect f or ϕ with t, given by

$$\begin{aligned} \frac{df}{dt} &= \frac{h}{r^2} - \frac{d\omega}{dt} - \cos\iota\frac{d\Omega}{dt}, \\ \frac{d\phi}{dt} &= \frac{h}{r^2} - \cos\iota\frac{d\Omega}{dt}. \end{aligned} \tag{6.75}$$

These equations are *exact*: they are simply a reformulation of Eq. (6.72) in terms of new variables. However, they are particularly useful when the perturbations are small, so that solutions can be obtained by a process of iteration. At lowest order (no perturbations), the elements p, e, ω, Ω, and ι are constant, and $df/dt = \sqrt{mp}/r^2$; those solutions can then be plugged into the right-hand side and the equations integrated to find corrections, and so on. The corrections to the orbit elements tend to be of two classes: *periodic* corrections, which vary on an orbital timescale, and *secular* corrections, which vary on a longer timescale, depending upon the nature of the perturbations.

In many applications, such as planetary orbits in the solar system, the orbital inclinations are very small, and it is difficult to separate the variation of ω and of Ω, instead one measures variations in the pericenter direction within the orbital plane, relative to the X reference axis (the equinox direction, in the case of the solar system). The variable that describes this variation is

$$\begin{aligned} \frac{d\varpi}{dt} &\equiv \frac{d\omega}{dt} + \cos\iota\frac{d\Omega}{dt} \\ &= \frac{1}{e}\sqrt{\frac{p}{m}}\left[-\cos f\mathcal{R} + \frac{2+e\cos f}{1+e\cos f}\sin f\mathcal{S}\right]. \end{aligned} \tag{6.76}$$

We are mainly interested in "secular" variations of the orbital elements induced by the perturbations shown in Eq. (6.70). The components of the perturbations are given by

$$\begin{aligned} \mathcal{R} &= \frac{3}{2}\frac{mR^2J_2}{r^4}\left[3(\boldsymbol{e}\cdot\boldsymbol{n})^2 - 1\right] + \frac{m}{r^2}\left\{\left[2\gamma + 2\beta + \eta(2+\alpha_1 - 2\zeta_2)\right]\frac{m}{r}\right. \\ &\quad \left. - \frac{1}{2}\left[2\gamma + \eta(6+\alpha_1+\alpha_2+\alpha_3)\right]v^2 + \frac{1}{2}\left[4\gamma + 4 - \eta(1-2\alpha_1-\alpha_2)\right](\boldsymbol{v}\cdot\boldsymbol{n})^2\right\} \\ \mathcal{S} &= -3\frac{mR^2J_2}{r^4}(\boldsymbol{e}\cdot\boldsymbol{n})(\boldsymbol{e}\cdot\boldsymbol{\lambda}) + \frac{m}{r^2}(\boldsymbol{v}\cdot\boldsymbol{n})(\boldsymbol{v}\cdot\boldsymbol{\lambda})\left[2\gamma + 2 - \eta(2-\alpha_1+\alpha_2)\right], \\ \mathcal{W} &= -3\frac{mR^2J_2}{r^4}(\boldsymbol{e}\cdot\boldsymbol{n})(\boldsymbol{e}\cdot\hat{\boldsymbol{h}}), \end{aligned} \tag{6.77}$$

where we have dropped the subscript "2" from R and J_2. We will assume that the $X-Y-Z$ reference basis is chosen so that the symmetry axis of the body with the quadrupole moment is in the Z-direction. To calculate the changes in the orbit elements to first order in the perturbations, we substitute Eq. (6.73) into (6.77) and substitute $\mathcal{R}$, $\mathcal{S}$ and $\mathcal{W}$ into Eq. (6.74). We convert the equations from d/dt to d/df using $df/dt = h/r^2$, and integrate over f from 0 to 2π. The resulting secular changes in the elements per orbit are given by

$$
\begin{aligned}
\Delta p &= 0\,, \\
\Delta e &= 0\,, \\
\Delta \iota &= 0\,, \\
\Delta \omega &= \frac{6\pi m}{p}\left[\frac{1}{3}\left(2+2\gamma-\beta\right)+\frac{1}{6}\left(2\alpha_1-\alpha_2+\alpha_3+2\zeta_2\right)\right] \\
&\quad + 6\pi J_2\left(\frac{R}{p}\right)^2\left(1-\frac{5}{4}\sin^2\iota\right)\,, \\
\Delta \Omega &= -3\pi J_2\left(\frac{R}{p}\right)^2 \cos\iota\,.
\end{aligned} \tag{6.78}
$$

We will make use of the rate of pericenter advance when we turn to the orbit of Mercury in Chapter 7. We will employ the Lagrange planetary equations in Chapter 8 when we discuss orbital perturbations induced by the Nordtvedt effect and by preferred-frame and preferred-location effects, in Chapter 9 when we discuss orbital perturbations due to spin effects and in Chapters 10 and 12 when we discuss orbits involving compact bodies, such as binary pulsar systems and bodies orbiting supermassive black holes.

6.5 Semiconservative Theories and N-body Lagrangians

The overwhelming majority of modern alternative theories of gravity are based on an invariant action principle, and so they automatically generate a post-Newtonian limit that is semiconservative. It is therefore appropriate to re-express the post-Newtonian N-body equations of motion with the constraints $\alpha_3 = \zeta_k = 0$. In a PPN coordinate frame at rest with respect to the universe rest frame, the resulting equations consist of the standard Newtonian acceleration (here we ignore the GWEP violating effects on the inertial, passive and active masses) plus the post-Newtonian N-body acceleration $\boldsymbol{a}_{\rm Nbody}$, Eq. (6.46c) with $\boldsymbol{w} = 0$ and with semiconservative PPN parameters,

$$
\begin{aligned}
(\boldsymbol{a}_a)_{\rm Nbody} = &-\sum_{b\neq a}\frac{m_b \boldsymbol{n}_{ab}}{r_{ab}^2}\Bigg\{1+\gamma v_a^2-(2\gamma+2)(\boldsymbol{v}_a\cdot\boldsymbol{v}_b)+(\gamma+1)v_b^2-\frac{3}{2}(\boldsymbol{n}_{ab}\cdot\boldsymbol{v}_b)^2 \\
&+\frac{1}{2}\alpha_2|\boldsymbol{w}+\boldsymbol{v}_b|^2-\frac{1}{2}\alpha_1(\boldsymbol{w}+\boldsymbol{v}_a)\cdot(\boldsymbol{w}+\boldsymbol{v}_b)-\frac{3}{2}\alpha_2[\boldsymbol{n}_{ab}\cdot(\boldsymbol{w}+\boldsymbol{v}_b)]^2 \\
&-\left(2\gamma+2\beta+1+\frac{1}{2}\alpha_1\right)\frac{m_a}{r_{ab}}-(2\gamma+2\beta)\frac{m_b}{r_{ab}}\Bigg\} \\
&+\sum_{b\neq a}\frac{m_b}{r_{ab}^2}\boldsymbol{n}_{ab}\cdot\big[(2\gamma+2)\boldsymbol{v}_a-(2\gamma+1)\boldsymbol{v}_b\big](\boldsymbol{v}_a-\boldsymbol{v}_b) \\
&-\sum_{b\neq a}\frac{m_b}{r_{ab}^2}\boldsymbol{n}_{ab}\cdot\left[\alpha_2(\boldsymbol{w}+\boldsymbol{v}_b)+\frac{1}{2}\alpha_1(\boldsymbol{v}_a-\boldsymbol{v}_b)\right](\boldsymbol{w}+\boldsymbol{v}_b) \\
&+\sum_{b\neq a}\sum_{c\neq a,b}\frac{m_b m_c}{r_{ab}^2}\boldsymbol{n}_{ab}\Bigg[(2\gamma+2\beta-2\xi)\frac{1}{r_{ac}}+(2\beta-1-2\xi)\frac{1}{r_{bc}} \\
&-\frac{1}{2}(1+2\xi+\alpha_2)\frac{r_{ab}}{r_{bc}^2}(\boldsymbol{n}_{ab}\cdot\boldsymbol{n}_{bc})-\xi\frac{r_{bc}}{r_{ac}^2}(\boldsymbol{n}_{ac}\cdot\boldsymbol{n}_{bc})\Bigg]
\end{aligned}
$$

$$- \frac{1}{2}(4\gamma + 3 - 2\xi + \alpha_1 - \alpha_2) \sum_{b \neq a} \sum_{c \neq a,b} \frac{m_b m_c}{r_{ab} r_{bc}^2} \boldsymbol{n}_{bc}$$

$$- \xi \sum_{b \neq a} \sum_{c \neq a,b} \frac{m_b m_c}{r_{ab}^3} \left[(\boldsymbol{n}_{ac} - \boldsymbol{n}_{bc}) - 3\boldsymbol{n}_{ab} \cdot (\boldsymbol{n}_{ac} - \boldsymbol{n}_{bc}) \boldsymbol{n}_{ab} \right]. \tag{6.79}$$

Remarkably, these equations of motion can be derived from the Euler-Lagrange equations obtained by varying the trajectory $\boldsymbol{x}_q(t)$, $\boldsymbol{v}_q(t)$ of the qth particle in the action

$$I_{\text{Nbody}} = \int L dt, \tag{6.80}$$

where

$$L = -\sum_a m_a \left(1 - \frac{1}{2} v_a^2 - \frac{1}{8} v_a^4\right)$$

$$+ \frac{1}{2} \sum_a \sum_{b \neq a} \frac{m_a m_b}{r_{ab}} \left[1 - \frac{1}{2} v_a^2 - \frac{1}{4}(4\gamma + 3)|\boldsymbol{v}_a - \boldsymbol{v}_b|^2 - \frac{1}{2}(\boldsymbol{v}_a \cdot \boldsymbol{n}_{ab})(\boldsymbol{v}_b \cdot \boldsymbol{n}_{ab}) \right.$$

$$- \frac{1}{2}(\alpha_1 - \alpha_2)(\boldsymbol{w} + \boldsymbol{v}_a) \cdot (\boldsymbol{w} + \boldsymbol{v}_b) - \frac{1}{2}\alpha_2 (\boldsymbol{w} + \boldsymbol{v}_a) \cdot \boldsymbol{n}_{ab} (\boldsymbol{w} + \boldsymbol{v}_b) \cdot \boldsymbol{n}_{ab}$$

$$\left. - (2\beta - 1) \sum_{c \neq a} \frac{m_c}{r_{ac}} - \xi \frac{\boldsymbol{x}_{ab}}{r_{ab}^2} \cdot \sum_{c \neq ab} m_c \left(\frac{\boldsymbol{x}_{bc}}{r_{ac}} - \frac{\boldsymbol{x}_{ac}}{r_{bc}} \right) \right]. \tag{6.81}$$

A Hamiltonian H can then be constructed from L in the usual way by defining $p_a^j \equiv \partial L / \partial v_a^j$ and $H = \sum_a \boldsymbol{p}_a \cdot \boldsymbol{v}_a - L$, with the result

$$H = \sum_a \left(m_a + \frac{p_a^2}{2m_a} - \frac{p_a^4}{8m_a^3} \right)$$

$$- \frac{1}{2} \sum_a \sum_{b \neq a} \frac{m_a m_b}{r_{ab}} \left[1 + (2\gamma + 1)\frac{p_a^2}{m_a^2} - \frac{1}{2}(4\gamma + 3 + \alpha_1 - \alpha_2)\frac{\boldsymbol{p}_a \cdot \boldsymbol{p}_b}{m_a m_b} \right.$$

$$- \frac{1}{2}(1 + \alpha_2)\frac{(\boldsymbol{p}_a \cdot \boldsymbol{n}_{ab})(\boldsymbol{p}_b \cdot \boldsymbol{n}_{ab})}{m_a m_b} - (2\beta - 1) \sum_{c \neq a} \frac{m_c}{r_{ac}}$$

$$\left. - \xi \frac{\boldsymbol{x}_{ab}}{r_{ab}^2} \cdot \sum_{c \neq ab} m_c \left(\frac{\boldsymbol{x}_{bc}}{r_{ac}} - \frac{\boldsymbol{x}_{ac}}{r_{bc}} \right) \right], \tag{6.82}$$

where we have set $\boldsymbol{w} = 0$, for simplicity. A Hamiltonian formulation of the equations of motion is useful for analyzing the dynamics of many-body systems.

6.6 The Locally Measured Gravitational Constant

Here, we derive an equation which is not really an equation of motion, but is nevertheless a fundamental result within the PPN formalism. In the previous section, we found that some metric theories of gravity could predict a violation of GWEP (Nordtvedt effect). Such effects represent violations of the Strong Equivalence Principle (SEP). As discussed in Section 3.3, the existence of preferred-frame and preferred-location effects in local gravitational experiments would also represent violations of SEP. One such local gravitational experiment is a measurement of Newton's constant G, usually called the Cavendish experiment, in honor of the experiments carried out by Henry Cavendish in the late 1700s. In an idealized version of this experiment one measures the relative acceleration of two bodies as a function of their masses and of the distance between them. Distances and times are measured by means of physical rods and atomic clocks at rest in the laboratory. The gravitational constant G is then identified as that number with dimensions $\mathrm{cm}^3\mathrm{g}^{-1}\mathrm{s}^{-2}$ which appears in Newton's law of gravitation for the two bodies. This quantity is called the locally measured gravitational constant G_L.

The analysis of this experiment proceeds as follows: a body of mass m_1 (source) falls freely through spacetime. A test body with negligible mass moves through spacetime, maintained at a constant proper distance r_p from the source by a four-acceleration A provided by some nongravitational force. The line joining the pair of masses is nonrotating relative to asymptotically flat inertial space. An invariant "radial" unit four-vector E_r, points from the test mass toward the source. Then according to Newton's law of gravitation the radial component of the four-acceleration of the test mass is given by

$$\mathsf{A} \cdot \mathsf{E}_r \equiv -G_\mathrm{L} \frac{m_1}{r_p^2}, \tag{6.83}$$

for r_p small compared to the scale of inhomogeneities in the external gravitational fields. Since the quantity $\mathsf{A} \cdot \mathsf{E}_r$ is invariant, we can calculate it in a suitably chosen PPN coordinate system, then use Eq. (6.83) to read off the locally measured G_L.

Before carrying out the computation, however, it is instructive to ask what might be expected for the form of $\mathsf{A} \cdot \mathsf{E}_r$ to post-Newtonian order. We imagine that the source and the test body are moving with respect to the universe with velocity $\boldsymbol{w}_1$ and are in the presence of some external sources, idealized as point masses of mass m_a at location $\boldsymbol{x}_a$. It is simplest to do the calculation in a PPN coordinate system in which the source is momentarily at rest. Then we would expect $\mathsf{A} \cdot \mathsf{E}_r$ to contain post-Newtonian corrections to the equation $\mathsf{A} \cdot \mathsf{E}_r = m_1/r_p^2$ of the form

$$\mathsf{A} \cdot \mathsf{E}_r : \quad \frac{m_1}{r_p^2}\frac{m_1}{r_p}, \quad \frac{m_1}{r_p^2}\frac{m_a}{r_{1a}}, \quad \frac{m_1}{r_p}\frac{m_a}{r_{1a}^2}, \quad \frac{m_1}{r_p^2}(w_1^2), \tag{6.84}$$

where $r_{1a} = |\boldsymbol{x}_1 - \boldsymbol{x}_a|$. In obtaining these possible corrections, we have neglected the variation of the external gravitational potentials across the separation r_p. Such a variation will produce the standard Newtonian tidal gravitational force, which is of the form

$$(\mathsf{A} \cdot \mathsf{E}_r)_\mathrm{tidal} : \quad \frac{m_a}{r_{1a}^3} r_p, \tag{6.85}$$

and post-Newtonian corrections to this force. We will neglect such forces throughout. The first term in Eq. (6.84) represents post-Newtonian modifications in the two-body motion of the test body about the source; these can be understood and analyzed separately from a discussion of G_L. The third term represents effects due to the gradients of the external fields; however, if we fit $\mathsf{A} \cdot \mathsf{E}_r$ to an r_p^{-2} curve in order to determine G_L, these terms will have no effect (in most practical situations, they are smaller than the terms we are interested in by factors of r_p/r_{1a}). We will drop both types of terms throughout the analysis. Thus, we retain only terms of the form $(m_1/r_p^2)(m_a/r_{1a})$ or $(m_1/r_p^2)w_1^2$.

The form of the PPN metric that we will use is given by the expression in Box 4.1, where now the velocity $\boldsymbol{w}$ is the source's velocity relative to the mean rest frame of the universe, denoted $\boldsymbol{w}_1$. We label the test body by $a = 0$, the source by $a = 1$ and the remaining bodies by $a = 2, 3, \ldots$. Initially, both the source and test body are at rest, that is,

$$\boldsymbol{v}_1(t=0) = \boldsymbol{v}_0(t=0) = 0\,. \tag{6.86}$$

We separate the Newtonian gravitational potential U_1 due to the source from that due to the other bodies in the system:

$$U(\boldsymbol{x}) = U_1(r_1) + \sum_{a\neq 1} \frac{m_a}{r_a}\,, \tag{6.87}$$

where $r_1 = |\boldsymbol{x} - \boldsymbol{x}_1|$, $r_a = |\boldsymbol{x} - \boldsymbol{x}_a|$, and U_1 is assumed for simplicity to be spherically symmetric.

The proper distance between the test body and the source is given by applying Eq. (3.48), with the result

$$\begin{aligned} r_p &= \int_{\boldsymbol{x}_0}^{\boldsymbol{x}_1} \left[1 + \gamma U(\boldsymbol{x}) + O(\epsilon^2)\right] d|\boldsymbol{x}| \\ &= r_{01}\left(1 + \gamma \sum_{a\neq 1} \frac{m_a}{r_{1a}}\right)\,, \end{aligned} \tag{6.88}$$

where, in line with our starting assumptions, we have neglected the variation of the external gravitational potential across the separation r_{01} and we have dropped the term arising from the integral of U_1 from $\boldsymbol{x}_0$ to $\boldsymbol{x}_1$ because it leads to effects only in the two-body motion of the test body about the source. The proper distance r_p is to be kept constant by the four-acceleration A, thus

$$\frac{dr_p}{dt} = \frac{d^2 r_p}{dt^2} = 0\,. \tag{6.89}$$

Recalling that $\boldsymbol{v}_1 = \boldsymbol{v}_0 = 0$ at $t = 0$, it is simple to show that this implies that

$$\boldsymbol{n}_{01} \cdot \left(\frac{d\boldsymbol{v}_0}{dt} - \frac{d\boldsymbol{v}_1}{dt}\right) = 0\,, \tag{6.90}$$

where $\boldsymbol{n}_{01} \equiv \boldsymbol{x}_{01}/r_{01}$.

We now assume that the source follows a geodesic of spacetime, but that the test body undergoes a four-acceleration A, so that

$$\begin{aligned} u^\nu_{\rm source} u^\mu_{{\rm source};\nu} &= 0\,, \\ u^\nu_{\rm test} u^\mu_{{\rm test};\nu} &= A^\mu\,, \end{aligned} \tag{6.91}$$

with $u^\mu_{\rm test} A_\mu = 0$. In PPN coordinates, with $\boldsymbol{v}_1 = \boldsymbol{v}_0 = 0$, this may expressed at $t = 0$ as

$$\begin{aligned} \frac{dv_1^j}{dt} + \Gamma^j_{00}(\boldsymbol{x}_1) &= 0\,, \\ \frac{dv_0^j}{dt} + \Gamma^j_{00}(\boldsymbol{x}_0) &= \left(\frac{d\tau}{dt}\right)^2 A^j\,, \end{aligned} \tag{6.92}$$

with $A^0 = 0$, where, for the test body,

$$\left(\frac{d\tau}{dt}\right)^2 = 1 - 2\sum_{a\neq 1}\frac{m_a}{r_{1a}} + O(\epsilon^2), \tag{6.93}$$

where we again ignore spatial variations in external potentials and drop the contribution of the source potential. We use the PPN Christoffel symbol Γ^j_{00} from Box 6.1, and retain only the terms discussed earlier. Substituting Eq. (6.92) into Eq. (6.90), and including the Newtonian tidal force for illustration, we obtain

$$\begin{aligned} \mathsf{A}\cdot \boldsymbol{e} = & -3r_{10}e^j e^k \sum_{a\neq 1}\frac{m_a n_{1a}^{\langle jk\rangle}}{r_{1a}^3} \\ & - \boldsymbol{e}\cdot\boldsymbol{\nabla}U_1\left[1 - (4\beta + 2\gamma - 3 - \zeta_2 - 4\xi)\sum_{a\neq 1}\frac{m_a}{r_{1a}} - \frac{1}{2}(\alpha_1 - \alpha_3)w_1^2\right] \\ & - \boldsymbol{e}\cdot\boldsymbol{\nabla}X_{1,jk}\left[\frac{1}{2}\alpha_2 w_1^j w_1^k - \xi\sum_{a\neq 1}\frac{m_a n_{1a}^j n_{1a}^k}{r_{1a}}\right], \end{aligned} \tag{6.94}$$

where $\boldsymbol{e} \equiv \boldsymbol{n}_{01}$, and

$$n_{1a}^{\langle jk\rangle} = n_{1a}^j n_{1a}^k - \frac{1}{3}\delta^{jk}. \tag{6.95}$$

For a spherically symmetric body, the potential U_1 and superpotential X_1 are given by

$$U_1 = \frac{m_1}{r_{01}}, \qquad X_1 = m_1 r_{01} + \frac{1}{3}\frac{I_1}{r_{01}}, \tag{6.96}$$

where I_1 is the source's spherical moment of inertia, given by

$$I_1 = \int_1 \rho^* r^2 d^3x. \tag{6.97}$$

The radial unit four-vector E_r is proportional to the coordinate unit vector, $E_r^j = -\alpha n_{01}^j$, with $E_r^0 = 0$. Imposing the normalization $g_{\mu\nu}E_r^\mu E_r^\nu = 1$, and including in the metric only external potentials, we obtain

$$E_r^j = -n_{01}^j\left(1 - \gamma\sum_{a\neq 1}\frac{m_a}{r_{1a}}\right). \tag{6.98}$$

The final result for the invariant radial component of the four-acceleration A is

$$\begin{aligned} \mathsf{A}\cdot\mathsf{E}_r = & \sum_{a\neq 1}\frac{m_a r_p}{r_{1a}^3}\left[3(\boldsymbol{n}_{1a}\cdot\boldsymbol{e})^2 - 1\right] \\ & - \frac{m_1}{r_p^2}\left\{1 - (4\beta - \gamma - 3 - \zeta_2 - 3\xi)\sum_{a\neq 1}\frac{m_a}{r_{1a}} - \frac{1}{2}(\alpha_1 - \alpha_2 - \alpha_3)w_1^2\right. \\ & - \frac{1}{2}\alpha_2(\boldsymbol{w}_1\cdot\boldsymbol{e})^2 + \xi\sum_{a\neq 1}\frac{m_a}{r_{1a}}(\boldsymbol{n}_{1a}\cdot\boldsymbol{e})^2 \\ & \left. + \frac{I_1}{m_1 r_p^2}(3e^j e^k - \delta^{jk})\left[\frac{1}{2}\alpha_2 w_1^j w_1^k - \xi\sum_{a\neq 1}\frac{m_a}{r_{1a}}n_{1a}^j n_{1a}^k\right]\right\}. \end{aligned} \tag{6.99}$$

The first term in Eq. (6.99) is simply the Newtonian tidal acceleration; minimizing these accelerations is a significant challenge in real-life Cavendish experiments. From the second term, we can read off the locally measured gravitational constant,

$$G_L = 1 - [4\beta - \gamma - 3 - \zeta_2 - \xi(3+\chi)]\,U_{\rm ext} - \frac{1}{2}[\alpha_1 - \alpha_3 - \alpha_2(1-\chi)]\,w_1^2 - \frac{1}{2}\alpha_2(1-3\chi)(\boldsymbol{w}_1 \cdot \boldsymbol{e})^2 + \xi(1-3\chi)\,U^{jk}_{\rm ext} e^j e^k\,, \tag{6.100}$$

where

$$U_{\rm ext} \equiv \frac{m_a}{r_{1a}}\,, \qquad U^{jk}_{\rm ext} \equiv \frac{m_a}{r_{1a}} n^j_{1a} n^k_{1a}\,, \tag{6.101}$$

and

$$\chi \equiv \frac{I_1}{m_1 r_p^2}\,. \tag{6.102}$$

Here, we see a direct example of the possibility of violations of the Strong Equivalence Principle, via preferred-frame or preferred-location effects in local Cavendish experiments. The preferred-frame effects depend upon the velocity $\boldsymbol{w}_1$ of the source relative to the universe rest frame, and are present unless the PPN preferred-frame parameters α_1, α_2, and α_3 all vanish. The preferred-location effects depend upon the gravitational potentials $U_{\rm ext}$ and $U^{jk}_{\rm ext}$ of nearby bodies, and are present in general unless the PPN parameters satisfy $4\beta - \gamma - 3 - \zeta_2 = \xi = 0$. We note that general relativity predicts that $G_L = 1$.

For an alternative derivation of G_L that involves transforming the PPN metric to a locally comoving quasi-inertial frame of a chosen massive body and examining the effective "Newtonian" potential of that body in the presence of motion relative to the preferred frame and of external matter, see PW section 13.2.6.

6.7 Equations of Motion for Spinning Bodies

In Section 6.3 we ignored the finite extent of each body, except insofar as it would permit internal kinetic, gravitational and other forms of energy. We ignored any finite-size effects that would lead to tidal interactions and their post-Newtonian corrections, for example. But this meant that we also ignored the effects of rotation or spin of the bodies. But rotation is everywhere, and it is important to incorporate it in a description of the motion of an *N*-body system. In gravitational physics, it is becoming increasingly clear that spin effects play a central role in such phenomena as binary black-hole inspirals, gravitational collapse, accretion onto compact objects, and the emission of gravitational radiation. In addition, several key experimental tests of general relativity have involved the effects of spin. By spin, we mean the macroscopic rotation of an extended body, and not the quantum-mechanical spin of an elementary particle.

The motion of spinning bodies in curved spacetime has been a subject of considerable research for many years. This research has been aimed at discovering (i) how a body's spin alters its trajectory (deviations from geodesic motion) and (ii) how a body's motion in curved spacetime alters its spin. No rigorous solution is available for these problems because of the difficulties in defining a center of mass of a spinning body in curved spacetime. The extensive literature on this subject includes work by Mathisson (1937), Papapetrou (1951), Corinaldesi and Papapetrou (1951), Barker and O'Connell (1974), and Dixon (1979). For a review of this history, see Havas (1989).

However, in the post-Newtonian approximation, it is relatively straightforward to incorporate spin. In this section we sketch the method for obtaining both the equations of motion for the center-of-mass of each spinning body, and the evolution equations for the spin of each body. More details can be found in PW, chapters 9 and 13.

The starting point is a simple and natural definition of the spin tensor (see Box 6.2),

$$S_a^{jk} \equiv \int_a \rho^* \left(\bar{x}^j \bar{v}^k - \bar{x}^k \bar{v}^j\right) d^3x\,, \tag{6.103}$$

where $\bar{\boldsymbol{x}} \equiv \boldsymbol{x} - \boldsymbol{x}_{a(0)}$ and $\bar{\boldsymbol{v}} \equiv \boldsymbol{v} - \boldsymbol{v}_{a(0)}$ are the position and velocity of a fluid element relative to the body's baryonic center of mass, defined by Eq. (6.35). We also introduce a vectorial version of the spin angular momentum, defined by

$$\boldsymbol{S}_a \equiv \int_a \rho^* \bar{\boldsymbol{x}} \times \bar{\boldsymbol{v}}\, d^3x\,. \tag{6.104}$$

It is easy to show that the tensor and vector are related by

$$S_a^j = \frac{1}{2}\epsilon^{jpq} S_a^{pq}\,, \qquad S_a^{jk} = \epsilon^{jkp} S_a^p\,. \tag{6.105}$$

Box 6.3

World lines for spinning bodies

For spinning bodies, there is a basic ambiguity in the definition of the world line $\boldsymbol{x}_a$ of the body. To understand this, we define a four-tensor spin quantity for body a by the equation

$$J_a^{\mu\nu} \equiv 2 \int_a \rho^* (x^{[\mu} - \tilde{x}_a^{[\mu}) v^{\nu]} d^3x\,, \tag{6.106}$$

a form modeled after Eq. (4.69), with the center of the coordinate system there being replaced with the variable $\tilde{x}_a^\mu$ which denotes the "representative world line" of the body. Here $x^0 = \tilde{x}_a^0 = t$, and $v^0 = 1$. Then it is straightforward to show that

$$\begin{aligned} J_a^{0j} &= -m_a(x_a^j - \tilde{x}_a^j)\,, \\ J_a^{jk} &= S_a^{jk} + 2m_a(x_a^{[j} - \tilde{x}_a^{[j})v_a^{k]}\,. \end{aligned} \tag{6.107}$$

To relate the baryonic center of mass x_a^j to the world-line $\tilde{x}_a^j$ we must constrain J_a^{0j}. One possibility is to require $J_a^{0j} = 0$, so that $x_a^j = \tilde{x}_a^j$, and the spin tensor $J_a^{\mu\nu}$ is purely spatial in the PPN coordinate frame. But another possibility is to impose the "covariant" condition

$$J_a^{\mu\nu} u_{a\nu} = 0\,, \tag{6.108}$$

where u_a^μ is the body's four-velocity. This condition asserts that the spin tensor $J_a^{\mu\nu}$ is purely spatial in the body's rest-frame. To lowest PN order, it can be expressed in the form

$$J_a^{0j} = -J_a^{jk} v_a^k\,. \tag{6.109}$$

One can interpolate between the two conditions using a parameter λ, with $J_a^{0j} = -\lambda J_a^{jk} v_a^k$, leading to a relation between the center of mass and the world-line given to lowest PN order by

$$x_a^j = \tilde{x}_a^j + \lambda m_a^{-1} S_a^{jk} v_a^k . \tag{6.110}$$

Note that the difference between these world line definitions is of post-Newtonian order. The condition that fixes the representative world line of a spinning body is often called a "spin supplementary condition" (Barker and O'Connell, 1974). This is another example of the ambiguity inherent in defining the center of mass or the world line of an extended body. The important thing is to make a choice and to maintain that choice throughout all calculations.

It is then relatively straightforward to return to Section 6.3 and redo the calculation of the equations of motion for a system of bodies, but now including specific finite-size terms in which the combination $\bar{v}^j \bar{x}^k$ is integrated over a body, leading to a term proportional to S_a^{jk}. Only terms linear in the extent $\bar{x}^k$ for any given body are kept, with the result that only terms linear in spins, or involving the product of spins of two different bodies are kept. It turns out that one of the virial relations of Eq. (6.43) is modified by a term involving spin:

$$\frac{1}{2}\frac{d^3}{dt^3} I_a^{jk} = 4H_a^{(jk)} - 3K_a^{jk} + \delta^{jk}\dot{P}_a - 2L_a^{(jk)} + 3S_a^{p(j}\sum_b \frac{m_b}{r_{ab}^3}\left(n_{ab}^{k)} n_{ab}^p - \frac{1}{3}\delta^{k)p}\right) . \tag{6.111}$$

We also transform from the center of mass to the representative world line $\tilde{x}_a^j$ using Eq. (6.110) in both the acceleration da_a^j/dt and the Newtonian term $\sum_b m_b x_{ab}^j / r_{ab}^3$, to obtain the additional spin contributions to the equations of motion (dropping the tildes):

$$\delta \boldsymbol{a}_a = \boldsymbol{a}_a[\mathrm{SO}] + \boldsymbol{a}_a[\mathrm{SS}] , \tag{6.112}$$

where SO and SS denote spin-orbit and spin-spin terms, given by

$$\begin{aligned} a_a^j[\mathrm{SO}] = \frac{3}{2}\sum_b \frac{m_b}{r_{ab}^3}\Bigg\{ & n_{ab}^{\langle jk\rangle}\Big[v_a^p\Big((2\gamma+1+\lambda)\hat{S}_a^{kp} + (2\gamma+2)\hat{S}_b^{kp}\Big) \\ & - v_b^p\Big((2\gamma+1+\lambda)\hat{S}_b^{kp} + (2\gamma+2)\hat{S}_a^{kp}\Big)\Big] \\ & + n_{ab}^{\langle kp\rangle}(v_a - v_b)^p\Big((2\gamma+1-\lambda)\hat{S}_a^{jk} + (2\gamma+2)\hat{S}_b^{jk}\Big) \\ & + \frac{1}{2}\alpha_1 n_{ab}^{\langle jk\rangle}\Big[(w+v_a)^p \hat{S}_b^{kp} - (w+v_b)^p \hat{S}_a^{kp}\Big] \\ & + \frac{1}{2}\alpha_1 n_{ab}^{\langle kp\rangle}(v_a - v_b)^p \hat{S}_b^{jk} - \alpha_3 n_{ab}^{\langle jl\rangle}(w+v_b)^p \hat{S}_b^{lk}\Bigg\} , \end{aligned} \tag{6.113a}$$

$$a_a^j[\mathrm{SS}] = -\frac{15}{8}(4\gamma+4+\alpha_1)\sum_{b\neq a}\frac{m_b}{r_{ab}^4}\hat{S}_a^{kp}\hat{S}_b^{kq} n_{ab}^{\langle jpq\rangle} , \tag{6.113b}$$

where $\hat{S}_a^{jk} \equiv S_a^{jk}/m_a$ and where $n_{ab}^{\langle jk\rangle}$ and $n_{ab}^{\langle jpq\rangle}$ denote the symmetric tracefree combinations

$$\begin{aligned} n_{ab}^{\langle jk\rangle} &\equiv n_{ab}^j n_{ab}^k - \frac{1}{3}\delta^{jk} , \\ n_{ab}^{\langle jpq\rangle} &\equiv n_{ab}^j n_{ab}^p n_{ab}^q - \frac{1}{5}\left(n_{ab}^j \delta^{pq} + n_{ab}^p \delta^{jq} + n_{ab}^q \delta^{jp}\right) . \end{aligned} \tag{6.114}$$

It is useful to note that, when $\lambda = 1$, corresponding to the covariant spin condition, all reference to the velocity of the PPN coordinate system disappears. All terms depend on relative positions $\boldsymbol{x}_{ab}$, and on either relative velocities $\boldsymbol{v}_a - \boldsymbol{v}_b$ or the velocity of a given body relative to the universe rest frame, $\boldsymbol{w} + \boldsymbol{v}_a$ or $\boldsymbol{w} + \boldsymbol{v}_b$.

To obtain the equations of motion for the spin vector, we simply insert the PPN equations of hydrodynamics into the expression

$$\frac{dS_a^j}{dt} = \frac{1}{2}\epsilon^{jpq}\frac{dS_a^{pq}}{dt} = \epsilon^{jpq}\int_a \rho^* \bar{x}^p \frac{d\bar{v}^q}{dt} d^3x \,. \tag{6.115}$$

We again expand the potentials keeping only terms linear in the variable $\bar{x}^j$. We absorb back into the definition of S_a^j a set of PN terms that involve only the internal variables of body a and that can be expressed as a total time derivative. Finally, we define a PN-corrected spin called the "proper" spin, $\bar{\boldsymbol{S}}_a$ according to

$$\bar{\boldsymbol{S}}_a \equiv \boldsymbol{S}_a + \left[v_a^2 + (2\gamma + 1)\sum_b \frac{m_b}{r_{ab}} \right] \boldsymbol{S}_a - \frac{1}{2}(\boldsymbol{v}_a \cdot \boldsymbol{S}_a)\boldsymbol{v}_a \,. \tag{6.116}$$

This turns out to be the spin as measured in a quasi-Lorentz frame momentarily comoving with body a (see PW, section 9.5.7). The result is a spin evolution equation given by

$$\frac{d\bar{\boldsymbol{S}}_a}{dt} = \left(\frac{d\bar{\boldsymbol{S}}_a}{dt}\right)_{\text{self}} + \left(\frac{d\bar{\boldsymbol{S}}_a}{dt}\right)_{\Omega_a} + \left(\frac{d\bar{\boldsymbol{S}}_a}{dt}\right)_{\text{prec}} , \tag{6.117}$$

where the first two terms depend on the internal structure of the body:

$$\begin{aligned} \left(\frac{dS_a^j}{dt}\right)_{\text{self}} &= \epsilon^{jpq}\int_a \rho^* \bar{x}^p \left[\frac{1}{2}(\alpha_3 + \zeta_1)\bar{\Phi}_{1,q} + \zeta_2\bar{\Phi}_{2,q} + \zeta_3\bar{\Phi}_{3,q} \right. \\ &\qquad \left. + 3\zeta_4\bar{\Phi}_{4,q} - \frac{1}{2}\zeta_1\left(\bar{\Phi}_{6,q} + \bar{X}_{,00q}\right) \right] d^3x \,, \\ \left(\frac{dS_a^j}{dt}\right)_{\Omega_a} &= \epsilon^{jpq}\Omega_a^{pn} \left[\alpha_2 (w + v_a)^q (w + v_a)^n - 2\xi \sum_b \frac{m_b}{r_{ab}} n_{ab}^q n_{ab}^n \right] , \end{aligned} \tag{6.118}$$

where the barred potentials are generated only by internal variables of the body. The "self" terms are present only in non-conservative theories. For a body that is spherically symmetric to a good approximation, the integrals will all be proportional to δ^{pq}, which is killed by the contraction with ϵ^{jpq}. For a stationary axisymmetric body, the only quantities available to construct a two-index tensor are δ^{pq} and $e^p e^q$, where $\boldsymbol{e}$ is a unit vector in the direction of the symmetry axis, and these also are killed by a contraction with ϵ^{jpq}. The strange nonconservative precessions are therefore relevant only for rather oddly shaped bodies. The Ω_a term depends on the velocity of body a relative to the preferred frame, and on the distribution of distant matter via the Whitehead potential. For a spherically symmetric body, both terms are killed by the contraction with ϵ^{jpq}. In addition, for laboratory-sized spinning bodies, such as the gyroscopes of Gravity Probe B, these terms are utterly negligible.

The final term in Eq. (6.117) is the PPN equation of spin precession,

$$\left(\frac{d\bar{\boldsymbol{S}}_a}{dt}\right)_{\text{prec}} = \boldsymbol{\Omega}_a \times \bar{\boldsymbol{S}}_a \,, \tag{6.119}$$

where $\boldsymbol{\Omega}_a$ contains spin-orbit, spin-spin and preferred-frame contributions, given by

$$\boldsymbol{\Omega}_a = \boldsymbol{\Omega}_a[\text{SO}] + \boldsymbol{\Omega}_a[\text{SS}] + \boldsymbol{\Omega}_a[\text{PF}] \,, \tag{6.120}$$

where

$$\boldsymbol{\Omega}_a[\text{SO}] = \frac{1}{2}\sum_b \frac{m_b}{r_{ab}^2}\boldsymbol{n}_{ab} \times \left[(2\gamma+1)\boldsymbol{v}_a - (2\gamma+2)\boldsymbol{v}_b\right], \tag{6.121a}$$

$$\boldsymbol{\Omega}_a[\text{SS}] = \frac{1}{8}(4\gamma+4+\alpha_1)\sum_b \frac{1}{r_{ab}^3}\left[3\boldsymbol{n}_{ab}(\boldsymbol{n}_{ab}\cdot\bar{\boldsymbol{S}}_b) - \bar{\boldsymbol{S}}_b\right], \tag{6.121b}$$

$$\boldsymbol{\Omega}_a[\text{PF}] = -\frac{1}{4}\alpha_1\sum_b \frac{m_b}{r_{ab}^2}\boldsymbol{n}_{ab} \times (\boldsymbol{w}+\boldsymbol{v}_b) \,. \tag{6.121c}$$

The use of the proper spin ensures that this is a pure precession, in other words, the magnitude of the spin does not change as a result of its interaction with the other bodies in the system. The equations for spin evolution are independent of the parameter λ.

It is useful to note that $\boldsymbol{\Omega}_a$ can be expressed in the alternative form

$$\boldsymbol{\Omega}_a = \frac{1}{2}(2\gamma+1)\boldsymbol{v}_a \times \boldsymbol{\nabla}U - \frac{1}{2}\boldsymbol{\nabla}\times\boldsymbol{g} \,, \tag{6.122}$$

where $U = \sum_b (m_b/r_{ab})$ and $g^j = g_{0j}$, where g_{0j} may be read off from the PPN metric in Box 4.1, noting that, for a system of spinning bodies, the vector potential V_j has the form

$$\boldsymbol{V} = \sum_b \left[\frac{m_b \boldsymbol{v}_b}{r_{ab}} - \frac{1}{2}\frac{\boldsymbol{n}_{ab}\times\boldsymbol{S}_b}{r_{ab}^2} + O(R_b^2/r_{ab}^3)\right]. \tag{6.123}$$

Equation (6.119) gives the precession of the proper spin relative to the PPN coordinate frame, whose axes are non-rotating relative to distant matter. We will discuss the observable consequences of this precession in Chapter 9.

7 The Classical Tests

With the PPN formalism and its associated equations of motion in hand, we are now ready to confront the gravitation theories discussed in Chapter 5 with the results of experiments. In this chapter, we focus on the three "classical" tests of relativistic gravity, consisting of (i) the deflection of light, (ii) the time delay of light, and (iii) the perihelion advance of Mercury.

This usage of the term "classical" is a break with tradition. Traditionally, the term "classical tests" referred to the gravitational redshift experiment, the deflection of light, and the perihelion advance of Mercury. The reason is largely historical. These were the first observable effects of general relativity to be computed by Einstein, and he regarded these three as crucial tests of his theory. However, in Chapter 2 we saw that the gravitational redshift experiment is really not a test of general relativity, rather it is a test of the Einstein Equivalence Principle, upon which general relativity and every other metric theory of gravity are founded. Put differently, every metric theory of gravity automatically predicts the same redshift. For this reason, the the redshift effect has been dropped as a "classical" test (that is not to deny its importance, of course, as our discussion in Chapter 2 points out).

We can immediately replace it with an experiment that is as important as the other two, the Shapiro time delay of light. This effect is closely related to the deflection of light, as one might expect, since any physical mechanism in Maxwell's equations (refraction, dispersion, gravity) that bends light can also be expected to delay it. In fact, it is a mystery why Einstein did not discover this effect. It was discovered as a prediction of Einstein's theory in 1964, by radio astronomer Irwin Shapiro. The simplest explanation seems to be that Shapiro had the benefit of knowing that the space technology of the 1960s and 1970s would make feasible a measurement of a delay of the expected size (200 μs for a round trip signal to Mars). No such technology was known to Einstein. He was aware only of the known problem of Mercury's excess perihelion advance of 43 arcseconds per century, and of the potential ability to measure the deflection of starlight. But the lack of available technology may not be the whole story. After all, Einstein derived the gravitational redshift at a time when the hopes of measuring it were marginal at best (a reliable measurement was not performed until 1960), and other workers such as Lense and Thirring and de Sitter derived effects of general relativity, with little or no hope of seeing them measured using the technology of the day. Why then, did no one at the time take the step from deflection to time delay, if only as a matter of principle?

Nevertheless, despite its late arrival, the time delay deserves a place in the triumvirate of "classical" tests, not the least because it has given one of the most precise tests of general relativity to date!

We begin this chapter with measurements of the deflection of light, Section 7.1, then turn to the Shapiro time delay, Section 7.2, and in Section 7.3, we discuss the perihelion advance of Mercury.

7.1 Deflection of Light

Eq. (6.21) indicates that the path of a light ray is deflected by a distribution of mass, but it does not tells us how to measure the effect. We can measure the direction of an incoming photon from some source, but we have no information about the initial direction. In order to make this a testable effect we must use only quantities that can be observed directly. The key quantity is the angle *between* two sources. Then, if we can measure that angle at a time when the mass distribution is sufficiently far from the sources as seen on the sky that we can ignore any deflection, and then measure the angle when the mass is close by, we can measure the *differential* deflection.

We can give this angle a coordinate-invariant expression. Consider an observer at rest in the PPN coordinate system, with four velocity u^μ, who receives a signal from a target source and a reference source, whose trajectories are $x^\mu(t)$ and $x_r^\mu(t)$, respectively. The spatial directions of the incoming signals are given by projecting the tangent four-vectors $v^\mu \equiv dx^\mu/dt$ and $v_r^\mu \equiv dx_r^\mu/dt$ of the rays onto the hypersurface orthogonal to the observer's four-velocity, using the projection operator $P^\nu_\mu = \delta^\nu_\mu + u^\nu u_\mu$. The inner product between the resulting vectors is related to the cosine of the angle θ between the incoming directions by

$$\cos\theta \equiv \frac{v^\lambda P^\mu_\lambda v_r^\nu P_{\nu\mu}}{|v^\lambda P^\mu_\lambda|\,|v_r^\nu P_{\nu\mu}|} = 1 + \frac{v \cdot v_r}{(v\cdot u)(v_r \cdot u)} . \tag{7.1}$$

If we ignore the velocity of the observer, which only produces effects such as aberration, which can be readily accounted for if necessary, then Eq. (7.1) simplifies to

$$\cos\theta = 1 - (g_{00})^{-1} g_{\mu\nu} v^\mu v_r^\nu . \tag{7.2}$$

By substituting Eq. (6.17) into Eq. (7.2), we obtain, to post-Newtonian accuracy,

$$\cos\theta = \boldsymbol{k}\cdot\boldsymbol{k}_r + \boldsymbol{k}\cdot\boldsymbol{\alpha}_r + \boldsymbol{k}_r\cdot\boldsymbol{\alpha} . \tag{7.3}$$

We now consider the simple case where the source is a spherical body at the origin of the PPN coordinate system. Assuming that the ray is always outside the body, then in Eq. (6.13), $U_{,k} = -mx^k(t)/r(t)^3$, and Eq. (6.15) becomes $d\boldsymbol{\alpha}/dt = -(1+\gamma)m\boldsymbol{b}/r(t)^3$, where now $\boldsymbol{b} \equiv \boldsymbol{x_e} - (\boldsymbol{x_e}\cdot\boldsymbol{k})\boldsymbol{k}$. Employing the identity (6.18b) with $\boldsymbol{s} = \boldsymbol{x}(t)$ and integrating with respect to t, we obtain

$$\boldsymbol{\alpha}(t) = -(1+\gamma)m\frac{\boldsymbol{b}}{b^2}\left(\frac{\boldsymbol{x}\cdot\boldsymbol{k}}{r} - \frac{\boldsymbol{x_e}\cdot\boldsymbol{k}}{r_e}\right) . \tag{7.4}$$

Substituting Eq. (7.4) into Eq. (7.3) we obtain, to post-Newtonian order,

$$\cos\theta = \boldsymbol{k}\cdot\boldsymbol{k}_r - (1+\gamma)m\left\{\frac{\boldsymbol{b}\cdot\boldsymbol{k}_r}{b^2}\left(\frac{\boldsymbol{x_O}\cdot\boldsymbol{k}}{r_O} - \frac{\boldsymbol{x_e}\cdot\boldsymbol{k}}{r_e}\right) + \frac{\boldsymbol{b}_r\cdot\boldsymbol{k}}{b_r^2}\left(\frac{\boldsymbol{x_O}\cdot\boldsymbol{k}_r}{r_O} - \frac{\boldsymbol{x}_r\cdot\boldsymbol{k}_r}{r_r}\right)\right\} , \tag{7.5}$$

where $\boldsymbol{x}_O$ is the location of the observer, and the subscript r denotes the reference source. It is useful to note, to the post-Newtonian accuracy needed in Eq. (7.5), that

$$\boldsymbol{b} = \boldsymbol{x}_O - (\boldsymbol{x}_O \cdot \boldsymbol{k})\boldsymbol{k}\,, \quad \boldsymbol{b}_r = \boldsymbol{x}_O - (\boldsymbol{x}_O \cdot \boldsymbol{k}_r)\boldsymbol{k}_r\,. \tag{7.6}$$

We now define the angle θ_0 to be the angle between the unperturbed paths of the photons from the source and from the reference source, i.e.,

$$\cos\theta_0 \equiv \boldsymbol{k} \cdot \boldsymbol{k}_r\,, \tag{7.7}$$

and we define the "deflection" of the measured angle from the unperturbed angle to be

$$\delta\theta \equiv \theta - \theta_0\,. \tag{7.8}$$

There are two interesting cases to consider. This first is an idealized situation that leads to a simple formula. We suppose that the Sun itself is both the massive source *and* the reference source. Then the second term inside the braces never appears (since $\boldsymbol{b} = 0$ *a priori*), $\boldsymbol{b} \cdot \boldsymbol{k}_r/b = \sin\theta_0$, and

$$\delta\theta = \left(\frac{1+\gamma}{2}\right)\frac{2m}{b}\left(\frac{\boldsymbol{x}_\oplus \cdot \boldsymbol{k}}{r_\oplus} - \frac{\boldsymbol{x}_e \cdot \boldsymbol{k}}{r_e}\right). \tag{7.9}$$

For a photon emitted from a distant star or galaxy, $r_e \gg r_\oplus$, $\boldsymbol{x}_e \cdot \boldsymbol{k}/r_e \simeq -1$, and $\boldsymbol{x}_\oplus \cdot \boldsymbol{k}/r_\oplus \simeq \cos\theta_0$, and we obtain (Shapiro 1967; Ward 1970),

$$\delta\theta = \left(\frac{1+\gamma}{2}\right)\frac{4m}{b}\left(\frac{1+\cos\theta_0}{2}\right). \tag{7.10}$$

The deflection is a maximum for a ray which just grazes the Sun. For $\theta_0 \simeq 0$, $b \simeq R_\odot \simeq 6.96 \times 10^5$ km, $m = m_\odot = 1.476$ km, we obtain

$$\delta\theta_{\rm max} = 1.7504\left(\frac{1+\gamma}{2}\right)\text{arcsec}\,. \tag{7.11}$$

For rays coming from arbitrary directions, we use the fact that $b \simeq r_\oplus \sin\theta_0$ to express the deflection in the form

$$\begin{aligned}\delta\theta &= \left(\frac{1+\gamma}{2}\right)\frac{2m}{r_\oplus}\left(\frac{1+\cos\theta_0}{\sin\theta_0}\right)\\ &= 4.072\left(\frac{1+\gamma}{2}\right)\cot\frac{\theta_0}{2}\ \text{mas}\,.\end{aligned} \tag{7.12}$$

At 90° from the direction of the Sun, the deflection is about 4 mas.

The second case to consider is more closely related to the actual method of measuring the deflection of light using the techniques of radio or optical interferometry. In these techniques, one chooses a reference source near the observed source and monitors changes $\delta\theta$ in their angular separation. If we define Φ and Φ_r to be the angular separation between the Earth-Sun direction and the unperturbed direction of photons from the two sources, as in Figure 7.1, then

$$\cos\Phi = \frac{\boldsymbol{x}_\oplus \cdot \boldsymbol{k}}{r_\oplus} \qquad \cos\Phi_r = \frac{\boldsymbol{x}_\oplus \cdot \boldsymbol{k}_r}{r_\oplus}\,. \tag{7.13}$$

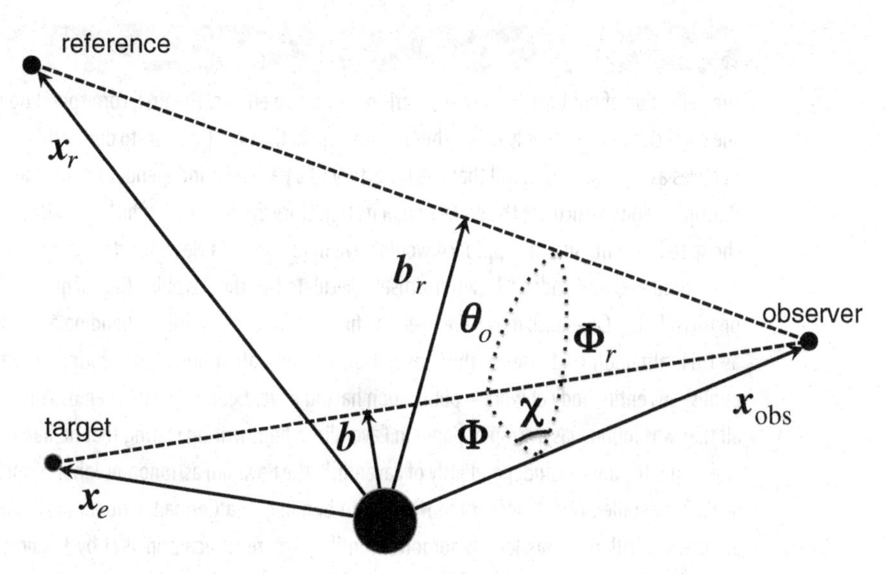

Fig. 7.1 Geometry of light deflection measurements.

Assuming again that the two sources are very distant and using the fact that $b = r_\oplus \sin\Phi$ and $b_r = r_\oplus \sin\Phi_r$, we obtain an expression purely in terms of measured angles on the sky:

$$\delta\theta = \left(\frac{1+\gamma}{2}\right)\frac{2m}{r_\oplus}\left[\frac{\cos\Phi_r - \cos\Phi\cos\theta_0}{(1-\cos\Phi)\sin\theta_0} + \frac{\cos\Phi - \cos\Phi_r\cos\theta_0}{(1-\cos\Phi_r)\sin\theta_0}\right]. \tag{7.14}$$

Note that the angle θ_0 is related to the angle χ between the Sun-source and Sun-reference directions projected on the plane of the sky (Figure 7.1) by the formula from spherical geometry

$$\cos\theta_0 = \cos\Phi_r\cos\Phi + \sin\Phi_r\sin\Phi\cos\chi\,. \tag{7.15}$$

If the observed source direction passes very near the Sun, while the reference source remains a decent angular distance away, we can approximate $\Phi \ll \Phi_r$, and thus

$$\theta_0 \simeq \Phi_r - \Phi\cos\chi + \frac{\cos\Phi_r\sin^2\chi}{2\sin\Phi_2}\Phi^2 + O(\Phi^3)\,. \tag{7.16}$$

The resulting deflection is

$$\delta\theta = \left(\frac{1+\gamma}{2}\right)\frac{4m}{b}\left[-\cos\chi + \frac{1+(1+2\sin^2\chi)\cos\theta_0}{2\sin\theta_0}\Phi + O(\Phi^2)\right]. \tag{7.17}$$

This result shows quite clearly how the relative angular separation between two distant sources may vary as the lines of sight of one of them passes near the Sun ($b \sim R_\odot$, χ varying).

7.1.1 Tests of the deflection of light

The first successful measurement of the bending of light by the Sun was carried out by Eddington and his colleagues during the total solar eclipse of May 29, 1919. The principle

Box 7.1 Why $\frac{1}{2}(1+\gamma)$?

The deflection of light can be viewed as arising from two effects. The first, corresponding to the "1/2" part of the coefficient, is commonly called the "Newtonian" deflection. One way to derive this is to assume that light behaves as a particle, to recall that the trajectory of a particle is independent of its mass (Weak Equivalence Principle), and to calculate the deflection of its trajectory in the limit in which the particle's speed approaches the speed of light. Such an approach would have made sense in Newton's day, when light was really viewed as a "corpuscle," and, indeed, Newton himself speculated on the possible effect of gravity on light. The English physicist Henry Cavendish may have been the first person to calculate the bending explicitly, possibly as early as 1784, although evidence for this was not discovered until around 1914, during an effort to compile and publish his entire body of work – publication having never been high on Cavendish's list of priorities. In fact, all that was found was a scrap of paper in Cavendish's handwriting stating that he had done the calculation, and giving the answer. Independently of Cavendish, the Bavarian astronomer Johann von Soldner did publish in 1803 a detailed calculation of the Newtonian bending in a German astronomical journal. Strangely, von Soldner's calculation was largely forgotten until it was resurrected in 1921 by Phillip Lenard as part of a campaign to discredit the "Jewish" relativity of Einstein by publicizing the earlier work of the "Aryan" von Soldner. Apparently, Lenard was not deterred by the fact that the 1919 measurements by Eddington actually favored general relativity over the Newtonian deflection.

Unaware of the earlier work, Einstein himself derived the "Newtonian" deflection in 1911. He argued that the principle of equivalence requires the replacement of the Minkowski metric of flat spacetime by the metric

$$ds^2 = -(1-2U)\,dt^2 + dx^2 + dy^2 + dz^2 .$$

Geodesic motion for a test particle in this spacetime reproduces Newtonian gravity, and geodesic motion for a photon gives the Newtonian deflection. Another derivation using only the equivalence principle imagines a sequence of freely falling frames through which a light ray passes as it travels near a gravitating body. Each frame is momentarily at rest at the moment the light ray enters it. Although the path of the ray is a straight line within each frame, the frame picks up a downward velocity during the ray's traversal, because of the body's gravitational attraction. When the adjacent, momentarily stationary frame receives the light ray, the ray is deflected toward the body because the downward motion of the previous frame induces aberration on the received ray. By adding up all the tiny aberrations over a sequence of frames, one arrives at the Newtonian deflection.

But a relativistic metric theory of gravity introduces an additional effect, because the spatial part of the metric now comes with the multiplying factor $(1+2\gamma U)$. This represents spatial curvature, which could not be taken into account either by Newtonian gravity or by the principle of equivalence. So the total deflection can be viewed as a sum of a Newtonian deflection relative to locally straight lines, plus the bending of locally straight lines relative to straight lines at infinity. Thus in the coefficient $(1+\gamma)/2$, the "1/2" holds in any metric theory, while the "$\gamma/2$" varies from theory to theory.

of the experiment was to make differential measurements. Photographs of the stars near the Sun taken during the eclipse were to be compared with photographs of the same stars taken at night some months earlier or later, and the displacement of each star compared to the reference photograph was to be carefully measured. But on the day of the eclipse at Eddington's site, a rainstorm started, and as the morning wore on, he began to lose all hope. But at the last moment, the weather began to change for the better, and when the partial eclipse was well advanced, the astronomers began to get a glimpse of the Sun. Of the sixteen photographs taken through the remaining cloud cover, only two had reliable images, totaling only about five stars. Nevertheless, comparison of the two eclipse plates with a comparison plate taken at the Oxford University telescope before the expedition yielded results in agreement with general relativity, corresponding to a deflection at the limb of the Sun (grazing ray) of 1.60 ± 0.31 arcseconds, corresponding to 0.91 ± 0.18 for $(1 + \gamma)/2$. The Sobral expedition, blessed with better weather, managed to obtain eight usable plates showing at least seven stars each. The nineteen plates taken on a second telescope were deemed worthless because the telescope apparently changed its focal length just before totality of the eclipse, possibly as a result of heating by the Sun. Analysis of the good plates yielded a grazing deflection of 1.98 ± 0.12 arcseconds, or 1.13 ± 0.07 for $(1 + \gamma)/2$ (Dyson et al., 1920). These results are indicated by the point and the arrow (off the chart) to the far left of Figure 7.2. Eddington's announcement in November 1919 that

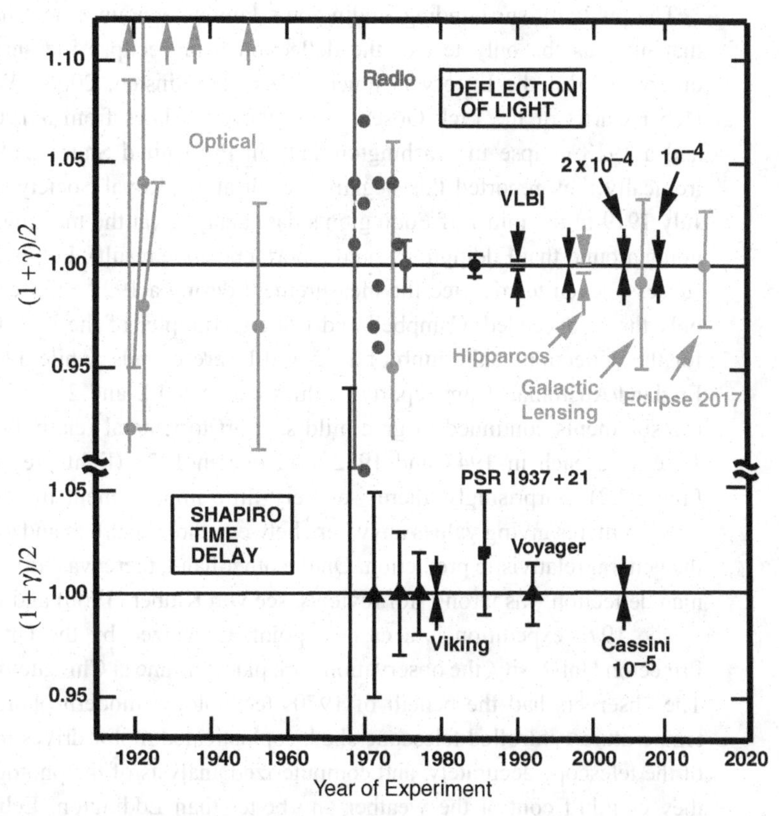

Fig. 7.2 Measurements of the coefficient $(1 + \gamma)/2$ from light deflection and time delay measurements. Its GR value is unity. The arrows at the top denote anomalously large values from early eclipse expeditions. The Shapiro time-delay measurements using the Cassini spacecraft yielded an agreement with GR to 10^{-3} percent, and VLBI light deflection measurements have reached 0.01 percent. Hipparcos denotes the optical astrometry satellite, which reached 0.2 percent.

the bending measurements were in agreement with general relativity helped make Einstein an international celebrity.

Questions were raised, however, about the results of Eddington's measurements. Given the poor quality of the data, did they really support Einstein or not? Was it proper for Eddington to discard the data from the second telescope at the Sobral site? More recently, some (Earman and Glymour, 1980) have wondered whether Eddington's enthusiasm for the theory of general relativity caused him to select or massage the data to get the desired result. Numerous reanalyses between 1923 and 1956 of the plates used by Eddington yielded the same results within 10 percent. In 1979, on the occasion of the centenary of Einstein's birth, astronomers at the Royal Greenwich Observatory reanalysed both sets of Sobral plates using a modern tool called the Zeiss Ascorecord and its data reduction software (Harvey, 1979). The plates from the first telescope yielded virtually the same deflection as that obtained by Eddington's team with the errors reduced by 40 percent. Despite the scale changes in the second set of Sobral plates, the analysis still gave a result 1.55 ± 0.34 arcseconds at the limb, consistent with general relativity, albeit with much larger errors. Looking back on Eddington's treatment of the data, Kennefick (2009) has argued that there is no credible evidence of bias on his part.

The publicity surrounding Eddington's famous announcement has left the impression that his was the only test of the deflection using eclipse measurements, successful or otherwise. But the history is much richer (Crelinsten, 2006). William Campbell and Heber Curtis of the Lick Observatory analyzed plates from a 1900 eclipse in Georgia and a 1918 eclipse in Washington state in the United States and found no deflection; ironically they reported this negative result at the Royal Society of London meeting in July 1919 in the midst of Eddington's data analysis (at the meeting, rumors were already going around that Eddington would report a positive result). Following up on Eddington's success, seven teams tried the measurement during a 1922 eclipse in Australia, although only three succeeded. Campbell and Robert Trumpler of the Lick team reported a result for the deflection at the limb of 1.72 ± 0.11 arcseconds, while a Canadian team and an England/Australian team reported values between 1.2 and 2.3 arcseconds. Later eclipse measurements continued to give mild support to general relativity: one in 1929, two in 1936, one each in 1947 and 1952, and one in 1973 (light grey points and arrows in Figure 7.2). Surprisingly, there was very little improvement in accuracy, with different measurements giving values anywhere between three-quarters and one and one-third times the general relativistic prediction. On the other hand, there was little doubt that the Newtonian deflection was wrong [for reviews, see von Klüber (1960) and Bertotti et al. (1962)].

The 1973 expedition is a case in point. Organized by the University of Texas and Princeton University, the observation took place in June at Chinguetti Oasis in Mauritania.[1] The observers had the benefit of 1970s technology: modern photographic emulsions, a temperature-controlled telescope shed, sophisticated motor drives to control the direction of the telescope accurately, and computerized analysis of the photographs. Unfortunately, they couldn't control the weather any better than Eddington. Eclipse morning brought

[1] The author vividly remembers Bryce DeWitt's slide show on scouting a site for the observations presented at the Les Houches Summer School on black holes in August 1972, including photos of DeWitt and Richard Matzner, precariously perched atop camels.

high winds, drifting sand, and dust too thick to see the Sun. But as totality of the eclipse approached, the winds died down, the dust began to settle, and the astronomers took a sequence of photographs during what they have described as the shortest six minutes of their lives. They had hoped to gather over 1000 star images, but the dust cut the visibility to less than 20 percent and only a disappointing 150 were obtained. After a follow-up expedition to the site in November to take comparison plates, the photographs were analyzed using the GALAXY Measuring Engine at the Greenwich Observatory, with a result 0.95 ± 0.11 times the general relativity prediction, essentially no improvement in accuracy over previous eclipse measurements (Brune et al., 1976; Jones, 1976).

During the 2017 solar eclipse that passed over the United States, a number of amateur and professional astronomers endeavored to improve upon these results by exploiting new technologies such as CCD cameras, together with precise reference locations of stars provided by space astrometry telescopes, such as GAIA. Bruns (2018) achieved a three percent result in agreement with general relativity.

The development of radio interferometry, and later of very-long-baseline radio interferometry (VLBI), produced greatly improved determinations of the deflection of light. These techniques now have the capability of measuring angular separations and changes in angles to accuracies better than 100 microarcseconds. Early measurements took advantage of the fact that groups of strong quasars annually pass very close to the Sun (as seen from the Earth), such as the group 3C273, 3C279 and 3C48. As the Earth moves in its orbit, changing the lines of sight of the quasars relative to the Sun, the angular separation between pairs of quasars varies according to Eq. (7.17). A number of measurements of this kind over the period 1969–1975 yielded determinations of the coefficient $\frac{1}{2}(1+\gamma)$, or equivalently of $\gamma - 1$, reaching levels of a percent. A 1995 VLBI measurement using 3C273 and 3C279 yielded $\gamma - 1 = (-8 \pm 34) \times 10^{-4}$ (Lebach et al., 1995), while a 2009 measurement using the VLBA targeting the same two quasars plus two other nearby radio sources yielded $\gamma - 1 = (-2 \pm 3) \times 10^{-4}$ (Fomalont et al., 2009).

In recent years, transcontinental and intercontinental VLBI observations of quasars and radio galaxies have been made partly to monitor the Earth's rotation and partly to establish a "reference frame" for precision astrometry ("VLBI" in Figure 7.2). These measurements are sensitive to the milliarcsecond level deflections of light over almost the entire celestial sphere. A 2004 analysis of almost 2 million VLBI observations of 541 radio sources, made by 87 VLBI sites yielded $\gamma - 1 = (-1.7 \pm 4.5) \times 10^{-4}$ (Shapiro et al., 2004). Analyses that incorporated data through 2010 yielded $\gamma - 1 = (-0.8 \pm 1.2) \times 10^{-4}$ (Lambert and Le Poncin-Lafitte, 2009, 2011).

To reach high precision at optical wavelengths requires observations from space. The Hipparcos optical astrometry satellite making global measurements of the deflection yielded a result for $(1+\gamma)/2$ at the level of 0.2 percent (Froeschlé et al., 1997). GAIA, a high-precision astrometric orbiting telescope launched by ESA in 2013 (Gaia Collaboration et al., 2016) possesses astrometric capability ranging from 10 to a few hundred microarcseconds, plus the ability to measure the locations of a billion stars down to the 20th magnitude; it could eventually measure the light-deflection and γ to the 10^{-6} level (Mignard and Klioner, 2010).

7.1.2 Gravitational lenses

In 1979, astronomers Dennis Walsh, Robert Carswell, and Ray Weymann discovered the "double quasar" Q0957+561, which consisted of two quasar images about 6 arcseconds apart, with almost the same redshift ($z = 1.41$) and very similar spectra (Walsh et al., 1979). It was immediately realized that there was just one quasar, but that intervening matter in the form of a galaxy or a cluster of galaxies was bending the light from the quasar and producing two separate images.

Since then, the "gravitational lens" has become an important tool for astronomers. The phenomenon has been exploited to map the distribution of mass around galaxies and clusters, and to search for dark matter, dark energy, compact objects, and extrasolar planets. Many subtopics of gravitational lensing have been developed to cover different astronomical realms: microlensing for the search for dim compact objects and extrasolar planets, the use of luminous arcs to map the distribution of mass and dark matter, and weak lensing to measure the properties of dark energy. Lensing has to be taken into account in interpreting certain aspects of the cosmic microwave background radiation, and in extracting information from gravitational waves emitted by sources at cosmological distances.

Gravitational lensing has also yielded a remarkable measurement of $(1+\gamma)/2$ on galactic scales (Bolton et al., 2006). It used data on gravitational lensing by 15 elliptical galaxies, collected by the Sloan Digital Sky Survey. The mass distribution of each lensing galaxy (including the contribution from dark matter) was derived from the observed velocity dispersion of stars within the galaxy. Comparing the observed lensing with the lensing predicted by the deflection formula (6.20) provided a five percent bound on $(1 + \gamma)/2$, in agreement with general relativity. Although the accuracy was only comparable to that of Eddington's 1919 measurements, this test of Einstein's light deflection was obtained on a galactic, rather than a solar-system scale.

7.2 The Shapiro Time Delay

Eqs. (6.17) or (6.21) reveal that, in addition to being deflected in the gravitational field of a massive body, a photon is also retarded in its motion. Its coordinate speed is less than unity (nevertheless, the speed measured by a local freely falling observer as the photon passes her is always unity). We consider the simple case of a light ray passing by a spherical body of mass m. Solving Eq. (6.21) for $t - t_e$, and recalling that $\boldsymbol{k} \cdot \delta \boldsymbol{x}_\perp = 0$, we obtain

$$t - t_e = |\boldsymbol{x}(t) - \boldsymbol{x}_e| + (1+\gamma)m \ln\left[\frac{(r(t) + \boldsymbol{x}(t)\cdot\boldsymbol{k})(r_e - \boldsymbol{x}_e\cdot\boldsymbol{k})}{b^2}\right], \tag{7.18}$$

where $\boldsymbol{b} \equiv \boldsymbol{x}_e - (\boldsymbol{x}_e \cdot \boldsymbol{k})\boldsymbol{k}$. For a signal emitted from the Earth, received at a planet or spacecraft at $\boldsymbol{x}_p$ and returned to Earth, the roundtrip travel time Δt is given by

$$\Delta t = 2|\boldsymbol{x}_\oplus - \boldsymbol{x}_p| + 2(1+\gamma)m \ln\left[\frac{(r_\oplus + \boldsymbol{x}_\oplus\cdot\boldsymbol{k})(r_p - \boldsymbol{x}_p\cdot\boldsymbol{k})}{b^2}\right], \tag{7.19}$$

where $\boldsymbol{k}$ is the direction of the photon on its return flight. Here we have ignored the motion of the Earth during the round trip of the signal. To be completely correct, the round trip travel time should be expressed in terms of the proper time elapsed during the round trip, as measured by an atomic clock on Earth; but this introduces no new effects so we will not do so here (in actual measurements of the delay, these effects are routinely taken into account).

The additional delay δt produced by the second term in Eq. (7.19) is a maximum when the planet is on the far side of the Sun from the Earth (superior conjunction), that is, when

$$\boldsymbol{x}_\oplus \cdot \boldsymbol{k} \simeq r_\oplus , \quad \boldsymbol{x}_p \cdot \boldsymbol{k} \simeq -r_p , \quad b \simeq R_\odot , \tag{7.20}$$

then

$$\begin{aligned} \delta t &= 2(1+\gamma) m \ln \left(\frac{4 r_\oplus r_p}{b^2} \right) \\ &= \left(\frac{1+\gamma}{2} \right) \left\{ 238.5\,\mu\text{s} - 19.7\,\mu\text{s} \ln \left[\left(\frac{b}{R_\odot} \right)^2 \frac{a}{r_p} \right] \right\} , \end{aligned} \tag{7.21}$$

where $R_\odot$ is the radius of the Sun, and $a = 1.496 \times 10^8$ km is the astronomical unit.

Soon after radio astronomer Irwin Shapiro discovered this effect as a theoretical prediction of general relativity, a program of measurements began, using radar ranging to targets passing through superior conjunction. The first of these was carried out by Shapiro's team (Shapiro et al., 1968, 1971) using radar signals bouncing off the surface of Mercury and Venus. Since one does not have access to a "Newtonian" signal against which to compare the round trip travel time of the observed signal, it is again necessary to do a differential measurement of the variations in round trip travel times as the target passes through superior conjunction, and to look for the characteristic logarithmic behavior of the delay. To achieve this accurately however, one must take into account the variations in $|\boldsymbol{x}_\oplus - \boldsymbol{x}_p|$ due to the orbital motion of the target relative to the Earth. This is done by using radar-ranging (and possibly other) data on the target taken when it is far from superior conjunction (i.e., when the time-delay term is negligible) to determine an accurate orbit for the target, using the orbit to predict the PPN coordinate trajectory $\boldsymbol{x}_p$ near superior conjunction, then combining that information with the trajectory of the Earth $\boldsymbol{x}_\oplus$ to determine all the relevant quantities in Eq. (7.19). The resulting predicted round trip travel times in terms of the unknown coefficient $\frac{1}{2}(1+\gamma)$ are then fit to the measured travel times using the method of least squares, and an estimate obtained $\frac{1}{2}(1+\gamma)$. This is an oversimplification, of course.

The targets employed included planets, such as Mercury or Venus, used as passive reflectors of the radar signals, and artificial satellites, such as Mariners 6 and 7, Voyager 2, the Viking Mars landers and orbiters, and the Cassini spacecraft to Saturn, used as active retransmitters of the radar signals.

The results for the coefficient $\frac{1}{2}(1+\gamma)$ of all radar time-delay measurements performed to date are shown in the bottom half of Figure 7.2. The 1976 Viking experiment resulted in a 0.1 percent measurement (Reasenberg et al., 1979).

A significant improvement was reported in 2003 from Doppler tracking of the Cassini spacecraft while it was on its way to Saturn (Bertotti et al., 2003), with a result $\gamma - 1 = (2.1 \pm 2.3) \times 10^{-5}$. This was made possible by the ability to do Doppler measurements using both X-band (7175 MHz) and Ka-band (34316 MHz) radar, thereby significantly reducing the dispersive effects of the solar corona. Note that with Doppler measurements, one is essentially measuring the time derivative of the Shapiro delay. In addition, the 2002 superior conjunction of Cassini was particularly favorable: with the spacecraft at 8.43 astronomical units from the Sun, the distance of closest approach of the radar signals to the Sun was only $1.6 R_\odot$. For scalar-tensor theories, the Cassini bound places a limit $\omega_0 > 40,000$ or $\alpha_0 < 3.5 \times 10^{-3}$. For the case of massive scalar fields, Alsing et al. (2012) displayed the bounds as a function of the mass. Improved measurements of the Shapiro delay, down to the level of parts per million, may be possible using data from the cruise phase of the joint European-Japanese BepiColombo project to place two orbiters around Mercury (Benkhoff et al., 2010), scheduled for launch in late 2018 (Imperi and Iess, 2017).

The Shapiro delay has been measured and plays an important role in analyzing binary pulsar data, most notably from the "double pulsar" J0737-3039 and even, recently, from the original Hulse-Taylor binary pulsar B1913+16 (see Chapter 12). In combination with the geometric delay caused by light deflection, the Shapiro delay also plays a role in efforts to measure the Hubble constant using time delays between images in gravitational lenses (see Meylan et al. (2006) for recent reviews).

7.3 The Perihelion Advance of Mercury

The explanation of the anomalous perihelion shift of Mercury's orbit was an early triumph of GR. Since Le Verrier's 1859 announcement of an unexplained advance in the perihelion of Mercury after the perturbing effects of the other planets had been accounted for, no credible explanation had been found, despite numerous attempts. The modern value for this discrepancy is about 43 arcseconds per century (see Table 7.1). General relativity

Table 7.1 Planetary contributions to Mercury's perihelion advance (in arcseconds per century).

Planet	Advance
Venus	277.8
Earth	90.0
Mars	2.5
Jupiter	153.6
Saturn	7.3
Total	531.2
Observed	574.1
Discrepancy	42.9

accounted for the anomalous shift in a natural way without disturbing the agreement with other planetary observations.

In Chapter 6, Eq. (6.78) we obtained the secular changes in the orbit elements of a two-body system in the PPN framework, including the effects of a Newtonian quadrupole moment for one of the bodies (but ignoring preferred-frame or preferred location effects for the moment). Since the orbit of Mercury is inclined by only 7° relative to the ecliptic plane of the solar system, and since changes in its perihelion are measured relative to the fixed reference system of the ecliptic plane, the relevant observable quantity is $\Delta\varpi = \Delta\omega + \cos\iota\Delta\Omega$. From Eq. (6.78), with $\iota \simeq 0$, we obtain

$$\Delta\varpi = \frac{6\pi m}{p}\left[\frac{1}{3}(2+2\gamma-\beta) + \frac{1}{6}\eta(2\alpha_1 - \alpha_2 + \alpha_3 + 2\zeta_2)\right] + \frac{3\pi}{2}J_2\left(\frac{R}{p}\right)^2, \quad (7.22)$$

where now J_2 and R are the quadrupole moment and radius of the Sun, and $m = m_\odot + m_☿$. We have ignored a small additional quadrupole contribution caused by the fact that the Sun's rotation axis is inclined relative to the ecliptic by $\theta \sim 7°$; this adds a correction to the quadrupole term that is too small to be measurable at present.

The first term in Eq. (7.22) is the "classical" perihelion shift, which depends upon the PPN parameters γ and β. The second term depends upon the dimensionless reduced mass of the two bodies; it is zero in any fully conservative theory of gravity ($\alpha_1 = \alpha_2 = \alpha_3 = \zeta_2 = 0$); it is also negligible for Mercury, since $\eta \simeq m_☿/m_\odot \simeq 2\times 10^{-7}$. We will drop this term henceforth. The third term depends upon the solar quadrupole moment J_2. For a Sun that rotates uniformly with its observed surface angular velocity, so that the quadrupole moment is produced by centrifugal flattening, one may estimate $J_2 \simeq 10^{-7}$. This actually agrees reasonably well with values inferred from rotating solar models that are in accord with observations of the normal modes of solar oscillations (helioseismology); the latest inversions of helioseismology data give $J_2 = (2.2 \pm 0.1)\times 10^{-7}$ (Mecheri et al., 2004; Antia et al., 2008); for a review of measurements of the solar quadrupole moment, see (Rozelot and Damiani, 2011). Substituting standard orbital elements and physical constants for Mercury and the Sun we obtain the rate of perihelion shift $d\varpi/dt$, in seconds of arc per century,

$$\frac{d\varpi}{dt} = 42.980\left[\frac{1}{3}(2+2\gamma-\beta) + 6.5\times 10^{-4}\left(\frac{J_2}{2.2\times 10^{-7}}\right)\right]. \quad (7.23)$$

(see Nobili and Will (1986) for a discussion of how the precise numerical value of the GR prediction is obtained).

In the current method of testing gravitational theories using the dynamics of Mercury, the actual value of the predicted advance is somewhat irrelevant. Instead, the equations of motion used in all ephemeris computer codes to predict the orbits of Mercury and of all the planets are the PPN *N*-body equations (6.46c). Those equations contain all the effects needed to determine the total advance of the perihelion of Mercury, including both the effects of the other planets and post-Newtonian effects. The modeling of the orbits depends on a host of parameters. These include masses and orbit elements at a chosen epoch of all the planets, of the Moon and many of the other moons of the planets, and of many asteroids. The effects of solar wind and radiation pressure on spacecraft must be modeled. Parameters associated

Box 7.2 The solar oblateness controversy

During the 1960s, Robert Dicke and Mark Goldenberg attempted to determine J_2 by measuring the Sun's visual shape. Because the surface of the Sun is an equipotential, its shape is affected by J_2 in a way that can be directly related to the deformation of the external gravitational field. The shape was measured by inserting a circular, opaque disk in front of a telescopic image of the Sun, leaving only a thin visible ring at the edge of the Sun, and measuring the difference in brightness of the visible ring between the equator and pole of the Sun. If the Sun were oblate, the ring at the equator should extend further beyond the occulting disk, and should therefore be brighter. But many factors had to be corrected for, including the effects of atmospheric distortion on the observed shape of the Sun, and the effects of possible temperature differences between the polar and equatorial regions of the Sun, which would lead to brightness differences not associated with the shape. Dicke and Goldenberg claimed to have measured a J_2 of the order of 2.5×10^{-5}, over 100 times larger than the currently accepted value. Dicke postulated that such a large oblateness would occur if the core of the Sun were rotating much faster than its outer layers, thereby generating more centrifugal flattening than would be expected on the basis of the observed surface rotation alone (see Dicke and Goldenberg (1974) for a review).

A value of J_2 this large would mean that solar oblateness contributes as much as 4 arcseconds per century to Mercury's perihelion advance, which would destroy the agreement of the measured advance with the prediction of general relativity. But it would have supported the Brans-Dicke theory of gravity if $\omega_{\rm BD}$ were chosen to be around 5, a value favored by Dicke for other reasons. Later observations of the visible shape of the Sun by Henry Hill and others, along with observations to try to better understand the temperature differences, did not fully resolve this controversy.

The resolution came with the advance of helioseismology. This was the discovery that the Sun vibrates in a superposition of thousands of normal modes with an array of frequencies, as could be observed by measuring the frequency spectrum of Doppler-shifted solar spectral lines. The specific pattern of frequencies depends on the Sun's angular-velocity profile. Through a systematic program of ground-based and space-based observations of the Sun, it became possible to determine the Sun's rotational profile over much of its interior. The conclusion was that the core does not rotate much faster than the surface, and solar models consistent with this information produced the currently accepted value of $J_2 = 2.2 \times 10^{-7}$; this is approximately what one would infer from a Sun that rotates uniformly at its observed surface rate.

with the effects of the interplanetary plasma and the Earth's atmosphere on the propagation of the tracking signals, and with variations in locations of radar tracking stations caused by Earth rotation changes and tidal effects are included. Finally, the PPN parameters are included. The relevant data include centuries of optical observations of planetary motions on the sky, radar measurements of planets and spacecraft, and laser ranging to the Moon. Predicted orbits are compared with the data, and the parameters are adjusted to obtain a best fit in the least-squares sense. The results include an estimate for the PPN parameter combination $(2+2\gamma-\beta)/3$, along with an associated error. Some data sets are particularly sensitive to the Shapiro time delay, and these provide accurate estimates of $(1+\gamma)/2$. Lunar laser ranging data are sensitive to the Nordtvedt effect (to be discussed in Chapter 8), and improve the parameters of the Earth's orbit about the Earth–Moon barycenter. As the data improve, either from more accurate range or Doppler measurements, or from new satellite missions, particularly planetary flyby and orbiter missions, the estimates for many parameters may change and the errors may decrease.

By the end of the 1970s, measurements of the factor $(2 + 2\gamma - \beta)/3$ had attained precisions of roughly 0.5 percent, through analyses of three centuries of optical observations of Mercury's orbit, and of radar measurements since the middle 1960s both to Mercury's surface and to the Mariner 10 spacecraft, which had close encounters with Mercury in 1974 and 1975. These estimates continued to improve somewhat during the 1980s and 1990s through the steady accumulation of data. However, a major advance in measuring the perihelion advance was made by exploiting Mercury MESSENGER. In 2011, MESSENGER became the first spacecraft to orbit Mercury, and range and Doppler measurements of the orbiter were made until the spacecraft ended its mission in 2015 with a controlled crash on the surface of Mercury. By 2013, MESSENGER data had already led to dramatically improved knowledge of Mercury's orbit. One analysis of all the available data yielded bounds on γ and β given by $\gamma - 1 = (-0.3 \pm 2.5) \times 10^{-5}$ and $\beta - 1 = (0.2 \pm 2.5) \times 10^{-5}$ (Fienga et al., 2011; Verma et al., 2014; Fienga et al., 2015). The bound on γ is consistent with that obtained from the dedicated test of the Shapiro time delay using Cassini (Section 7.2). The analysis also yielded an estimate $J_2 = (2.4 \pm 0.2) \times 10^{-7}$, consistent with the results from helioseismology. Taking into account the errors in γ and β at the level of 2.5×10^{-5}, this corresponds to an effective measurement of the relativistic perihelion advance of Mercury at 42.980 ± 0.001 arcseconds per century. This precision could be improved using future data from BepiColombo (Milani et al., 2002; Ashby et al., 2007).

A slightly weaker bound $\beta - 1 = (0.4 \pm 2.4) \times 10^{-4}$ from the perihelion advance of Mars (adopting the Cassini bound on γ *a priori*) was obtained by exploiting data from the Mars Reconnaissance Orbiter (Konopliv et al., 2011). Laser tracking of the Earth-orbiting satellite LAGEOS II led to a measurement of its relativistic *perigee* precession (3.3 arcseconds per year) in agreement with general relativity to 2 percent (Lucchesi and Peron, 2010, 2014).

8 Tests of the Strong Equivalence Principle

The next class of experiments that test relativistic gravitational effects may be called tests of the Strong Equivalence Principle (SEP). That principle states that (i) WEP is valid for self-gravitating bodies as well as for test bodies (GWEP), (ii) the outcome of any local test experiment, gravitational or nongravitational, is independent of the velocity of the freely falling apparatus, and (iii) the outcome of any local test experiment is independent of where and when in the universe it is performed. In Section 3.3, we pointed out that many metric theories of gravity (perhaps all except general relativity) can be expected to violate one or more aspects of SEP. In Chapter 6, working within the PPN framework, we saw explicit evidence of some of these violations: violations of GWEP in the equations of motion for massive self-gravitating bodies, Eq. (6.46b); preferred-frame effects in the N-body equations of motion, Eq. (6.46c); and preferred-frame and preferred-location effects in the locally measured gravitational constant, Eq. (6.100).

This chapter is devoted to the study of some of the observable consequences of such violations of SEP, and to the experiments that test for them. In Section 8.1, we consider violations of GWEP (the Nordtvedt effect), and its primary experimental test via Lunar laser ranging. Section 8.2 focuses on preferred-frame and preferred-location effects on the orbital motions of planets and binary systems, while Section 8.3 analyzes such effects on the structure of self-gravitating bodies. Another violation of SEP would be the variation with time of the gravitational constant G as a result of cosmic evolution. Tests of such a variation are described in Section 8.5.

8.1 The Nordtvedt Effect

The breakdown in the Weak Equivalence Principle for massive, self-gravitating bodies (GWEP), which many metric theories predict, has a variety of observable consequences. In Chapter 6, we saw that this violation could be expressed in quasi-Newtonian language by attributing to each massive body inertial and passive gravitational mass tensors $\tilde{m}_{\rm I}^{jk}$ and $\tilde{m}_{\rm P}^{jk}$, which may differ from each other. The quasi-Newtonian part of the body's acceleration may then be written [see Eq. (6.54)]

$$(\tilde{m}_{\rm I})_a^{jk}(a_a^k)_{\rm Newt} = (\tilde{m}_{\rm P})_a^{lm}\mathfrak{U}^{lm}_{,j}\,, \tag{8.1}$$

where $\mathfrak{U}^{lm}$ is a quasi-Newtonian gravitational potential, and the inertial and passive mass tensors, $\tilde{m}_{\rm I}^{jk}$ and $\tilde{m}_{\rm P}^{jk}$, are given by

$$(\tilde{m}_I)_a^{jk} \equiv m_a \left\{ \delta^{jk} \left[1 + (\alpha_1 - \alpha_2 + \zeta_1)\frac{\Omega_a}{m_a} \right] + (\alpha_2 - \zeta_1 + \zeta_2)\frac{\Omega_a^{jk}}{m_a} \right\},$$

$$(\tilde{m}_P)_a^{lm} \equiv m_a \left\{ \delta^{lm} \left[1 + (4\beta - \gamma - 3 - 3\xi)\frac{\Omega_a}{m_a} \right] - \xi\frac{\Omega_a^{lm}}{m_a} \right\}, \tag{8.2}$$

where Ω_a and Ω_a^{jk} are the body's internal gravitational energy and gravitational energy tensor (see Box 6.2 for definitions), and m_a is the total mass energy of the body. Most self-gravitating bodies (planets, stars) are very nearly spherically symmetric, so we may approximate

$$\Omega_a^{jk} \simeq \frac{1}{3}\Omega_a \delta^{jk}. \tag{8.3}$$

With these approximations, we write the quasi-Newtonian equation (8.1) in the form

$$\boldsymbol{a}_a = \left(\frac{m_P}{m}\right)_a \nabla \mathfrak{U}, \tag{8.4}$$

where

$$\left(\frac{m_P}{m}\right)_a = 1 + \left(4\beta - \gamma - 3 - \frac{10}{3}\xi - \alpha_1 + \frac{2}{3}\alpha_2 - \frac{2}{3}\zeta_1 - \frac{1}{3}\zeta_2\right)\frac{\Omega_a}{m_a},$$

$$\mathfrak{U} = \sum_{b \neq a} \frac{(m_A)_b}{r_{ab}}. \tag{8.5}$$

The most important consequence of the Nordtvedt effect is a polarization of the orbit of a binary system in the presence of a third body. If the self-gravitational energy per unit mass of one member of the binary is larger than that of its companion, then the two bodies may fall toward the distant third body with slightly different accelerations, leading to the orbital polarization. The most notable example is that of the Earth–Moon orbit in the presence of the Sun (Nordtvedt, 1968c). Other examples include binary systems (especially binary pulsars) in the presence of the galaxy, and the recently discovered pulsar in a triple system, consisting of a neutron star and two white-dwarf companions.

Even in pure Newtonian gravity, the three-body problem is notoriously difficult. However, if the system is "hierarchical," meaning that the average separation of one pair of bodies is small compared to the distance between that pair and the third body, then the problem is amenable to an expansion in terms of the small parameter r/R where r and R are the two separations (in some situations, a better expansion parameter is the ratio of the period of the inner binary to that of the third body, P_b/P_3). In hierarchical triple systems, the inner binary can be treated as a Keplerian two-body orbit to lowest order, with the third body producing small perturbations.

Using our quasi-Newtonian equation of motion (8.4), we obtain, for a two-body system in the presence of a third body

$$\boldsymbol{a}_1 = -\left(\frac{m_P}{m}\right)_1 \left[(m_A)_2 \frac{\boldsymbol{x}_{12}}{r_{12}^3} + (m_A)_3 \frac{\boldsymbol{x}_{13}}{r_{13}^3} \right],$$

$$\boldsymbol{a}_2 = \left(\frac{m_P}{m}\right)_2 \left[(m_A)_1 \frac{\boldsymbol{x}_{12}}{r_{12}^3} - (m_A)_3 \frac{\boldsymbol{x}_{23}}{r_{23}^3} \right]. \tag{8.6}$$

We denote the barycenter position of the two-body system by

$$\boldsymbol{x}_c \equiv \frac{m_1}{m}\boldsymbol{x}_1 + \frac{m_2}{m}\boldsymbol{x}_2 \,, \tag{8.7}$$

where $m \equiv m_1 + m_2$. Assuming that $r_{12} \ll r_{23}$, we expand

$$\frac{x^j_{13}}{r^3_{13}} = \frac{x^j_{c3}}{r^3_{c3}} - x^k_{1c}\left(\frac{1}{r_{c3}}\right)_{,jk} - \frac{1}{2}x^k_{1c}x^m_{1c}\left(\frac{1}{r_{c3}}\right)_{,jkm} + O(r^3_{1c}/r^5_{c3}) \,, \tag{8.8}$$

where $\boldsymbol{x}_{c3} \equiv \boldsymbol{x}_c - \boldsymbol{x}_3$ and $\boldsymbol{x}_{1c} \equiv \boldsymbol{x}_1 - \boldsymbol{x}_c$; we expand x^j_{23}/r^3_{23} in a similar fashion. We define $R \equiv r_{c3} = |\boldsymbol{x}_{c3}|$, $\boldsymbol{N} \equiv \boldsymbol{x}_{c3}/r_{c3}$, $\boldsymbol{n}_{1c} \equiv \boldsymbol{x}_{1c}/r_{1c}$, $\boldsymbol{n}_{2c} \equiv \boldsymbol{x}_{2c}/r_{2c}$, and notice that

$$r_{1c} = (m_2/m)r \,, \qquad r_{2c} = (m_1/m)r \,, \qquad \boldsymbol{n}_{1c} = -\boldsymbol{n}_{2c} = \boldsymbol{n} \,, \tag{8.9}$$

where $r \equiv |\boldsymbol{r}_{12}|$ and $\boldsymbol{n} \equiv \boldsymbol{x}_{12}/r_{12}$. With this notation we find that the relative acceleration $\boldsymbol{a} \equiv \boldsymbol{a}_1 - \boldsymbol{a}_2$ is given by

$$\begin{aligned} \boldsymbol{a} = &-\frac{m^*}{r^2}\boldsymbol{n} - \delta\frac{m_3}{R^2}\boldsymbol{N} + \alpha^*\frac{m_3 r}{R^3}\big[3\boldsymbol{N}(\boldsymbol{N}\cdot\boldsymbol{n}) - \boldsymbol{n}\big] \\ &- \frac{3}{2}\Delta^*\frac{m_3 r^2}{R^4}\big[5\boldsymbol{N}(\boldsymbol{N}\cdot\boldsymbol{n})^2 - 2\boldsymbol{n}(\boldsymbol{N}\cdot\boldsymbol{n}) - \boldsymbol{N}\big] \,, \end{aligned} \tag{8.10}$$

where m^* is defined by Eq. (6.71), we have dropped the A subscript on m_3, and we have defined

$$\begin{aligned} \alpha^* &\equiv \alpha + \frac{1}{2}\delta\Delta \,, \\ \Delta^* &\equiv \Delta\alpha + \frac{1}{2}\delta(1 - 2\eta) \,, \end{aligned} \tag{8.11}$$

where

$$\alpha \equiv \frac{1}{2}\left[\left(\frac{m_P}{m}\right)_1 + \left(\frac{m_P}{m}\right)_2\right] , \tag{8.12a}$$

$$\delta \equiv \left(\frac{m_P}{m}\right)_1 - \left(\frac{m_P}{m}\right)_2 , \tag{8.12b}$$

$$\Delta \equiv \frac{m_2 - m_1}{m_1 + m_2} . \tag{8.12c}$$

The first term in Eq. (8.10) is the standard Newtonian acceleration; since m^* represents the Kepler-measured mass of the two-body system we will replace it by m. The second term is a relative acceleration that stretches or shrinks the orbit along a line directed toward the third body—this is the Nordtvedt effect. From Eq. (8.5) we have that

$$\delta = \left(4\beta - \gamma - 3 - \frac{10}{3}\xi - \alpha_1 + \frac{2}{3}\alpha_2 - \frac{2}{3}\zeta_1 - \frac{1}{3}\zeta_2\right)\left(\frac{\Omega_1}{m_1} - \frac{\Omega_2}{m_2}\right) . \tag{8.13}$$

The third and fourth terms originate from the Newtonian tidal interactions with the third body, and they typically give rise to orbital perturbations that are substantially larger than the nominal size of the Nordtvedt effect.

We now consider a simplified situation in which the inner and outer orbits lie in the same plane (for the Earth–Moon-Sun system, their actual relative inclination is about 5°), the inner orbit is approximately circular (the eccentricity of the Moon-Earth orbit is 0.055) and the outer orbit is circular (the eccentricity of the Earth-Sun orbit is 0.017). We calculate the first-order perturbations of the inner orbit created by the last three terms in Eq. (8.10). In the limit of small eccentricity, we can write for the inner binary $d\phi/dt = \omega_b \approx$ constant, and for the outer binary, $d\phi/dt = \omega_3 =$ constant, so that

$$\begin{aligned} r &= a(1 - e\cos f) + O(e^2)\,, \\ f &= \phi - \omega = \omega_b t - \omega + O(e)\,, \\ \boldsymbol{n} &= \boldsymbol{e}_X \cos(\omega_b t) + \boldsymbol{e}_Y \sin(\omega_b t) + O(e)\,, \\ \boldsymbol{\lambda} &= -\boldsymbol{e}_X \sin(\omega_b t) + \boldsymbol{e}_Y \cos(\omega_b t) + O(e)\,, \\ \boldsymbol{N} &= \boldsymbol{e}_X \cos(\omega_3 t - \Phi) + \boldsymbol{e}_Y \sin(\omega_3 t - \Phi)\,, \end{aligned} \tag{8.14}$$

where e is the eccentricity of the inner orbit, and Φ is the initial phase of the third body. Note that

$$\begin{aligned} \boldsymbol{n}\cdot\boldsymbol{N} &= \cos(\Lambda t + \Phi)\,, \\ \boldsymbol{\lambda}\cdot\boldsymbol{N} &= -\sin(\Lambda t + \Phi)\,, \end{aligned} \tag{8.15}$$

where $\Lambda \equiv \omega_b - \omega_3$

We now use the Lagrange planetary equations reviewed in Section 6.4.2. This must be done with some care, because the orbital eccentricity is small, and ω loses its meaning when $e \to 0$. In such cases it is best to use the alternative variables $A \equiv e\cos\omega$ and $B \equiv e\sin\omega$, and to reexpress the Lagrange planetary in terms of A and B. We therefore write

$$r = a\big[1 - A\cos(\omega_b t) - B\sin(\omega_b t)\big] + O(e^2)\,, \tag{8.16}$$

and convert Eqs. (6.74) to the form

$$\frac{da}{dt} = \frac{2}{\omega_b}\mathcal{S} + O(e)\,, \tag{8.17}$$

$$\frac{dA}{dt} = \frac{1}{\omega_b p}\Big[\mathcal{R}\sin(\omega_b t) + 2\mathcal{S}\cos(\omega_b t)\Big] + O(e)\,, \tag{8.18}$$

$$\frac{dB}{dt} = \frac{1}{\omega_b p}\Big[-\mathcal{R}\cos(\omega_b t) + 2\mathcal{S}\sin(\omega_b t)\Big] + O(e)\,, \tag{8.19}$$

where $\mathcal{R}$ and $\mathcal{S}$ are the radial and tangential components of the perturbing acceleration, and we have used the fact that $(p/m)^{1/2} = (\omega_b p)^{-1} + O(e^2)$. To lowest order in e, $\mathcal{R}$ and $\mathcal{S}$ are given by

$$\begin{aligned} \mathcal{R} = \omega_3^2 a\bigg\{&\alpha^*\big[3\cos^2(\Lambda t + \Phi) - 1\big] - \frac{3}{2}\frac{a}{R}\Delta^*\cos(\Lambda t + \Phi)\big[5\cos^2(\Lambda t + \Phi) - 3\big] \\ &- \frac{R}{a}\delta\cos(\Lambda t + \Phi)\bigg\}\,, \end{aligned} \tag{8.20}$$

$$\begin{aligned} \mathcal{S} = \omega_3^2 a\bigg\{&-3\alpha^*\sin(\Lambda t + \Phi)\cos(\Lambda t + \Phi) + \frac{3}{2}\frac{a}{R}\Delta^*\sin(\Lambda t + \Phi)\big[5\cos^2(\Lambda t + \Phi) - 1\big] \\ &+ \frac{R}{a}\delta\sin(\Lambda t + \Phi)\bigg\}\,, \end{aligned} \tag{8.21}$$

where $\omega_3^2 \equiv m_3/R^3$. Because the orbital planes are taken to coincide, there are no perturbations of the inclination ι or line of nodes Ω.

Substituting Eq. (8.21) into Eq. (8.19), integrating subject to the boundary conditions $a = a_0$, $A = A_0$, $B = B_0$ at $t = 0$, and inserting the results into Eq. (8.16), we obtain

$$\begin{aligned}\delta r(t) &= a' \left[1 - A' \cos(\omega_b t) - B' \sin(\omega_b t)\right] \\ &+ a_0 \left(\frac{\omega_3}{\omega_b}\right)^2 \left\{ \frac{3}{2}\alpha^* \frac{\omega_b^2(1+\omega_b/\Lambda)}{\omega_b^2 - 4\Lambda^2} \cos 2(\Lambda t + \Phi) \right. \\ &- \frac{1}{8}\frac{a_0}{R}\Delta^* \left[\frac{3\omega_b^2(3+2\omega_b/\Lambda)}{\omega_b^2 - \Lambda^2} \cos(\Lambda t + \Phi) \right. \\ &+ \left. \frac{5\omega_b^2(3+2\omega_b/\Lambda)}{\omega_b^2 - 9\Lambda^2} \cos 3(\Lambda t + \Phi) \right] \\ &- \left. \delta \frac{R}{a_0} \frac{\omega_b^2(1+2\omega_b/\Lambda)}{\omega_b^2 - \Lambda^2} \cos(\Lambda t + \Phi) \,. \right\} , \end{aligned} \tag{8.22}$$

where a', A' and B' differ from a_0, A_0 and B_0 by small corrections induced by the perturbations, and represent unobservable constant shifts in the orbit elements. The remaining terms depend on the difference $\Lambda = \omega_b - \omega_3$, and thus have a very different time dependence from that of the eccentric orbit. Notice that, because we have a sinusoidal driving force acting on a sinusoidal oscillator, the solutions take the form of a resonant response, with denominators of the form $1/(\omega_b^2 - N^2\Lambda^2)$, with N representing the harmonic degree. If $N = 1$, then, if $\omega_3 \ll \omega_b$, the resonant factor becomes $\omega_b^2/(\omega_b^2 - \Lambda^2) \approx \omega_b/2\omega_3 \gg 1$, leading to an enhanced amplitude.

We first apply these results to the Earth–Moon system, with the Sun making up the third body. We set $m_2 = m_\oplus$, $m_1 = m_{☾} = m_\oplus/81.3$, $m_3 = m_\odot$, $\omega_b = \omega_{☾}$, the angular velocity of the Moon around the Earth, and $\omega_3 = \omega_\oplus$, the angular velocity of the Earth around the Sun, with $\omega_{☾} \approx 13.37\omega_\oplus = 2.66 \times 10^{-6}\,\mathrm{s}^{-1}$. In the Earth–Moon system, Λ is known as the *synodic frequency*, the angular frequency of the lunar orbit relative to the Sun. This is to be distinguished from the *sidereal frequency* $\omega_{☾}$, which refers to the global coordinate frame. We also have

$$\frac{\Omega_\oplus}{m_\oplus} = -4.6 \times 10^{-10}, \quad \frac{\Omega_{☾}}{m_{☾}} = -0.2 \times 10^{-10}, \tag{8.23}$$

and $R/a_0 \approx 390$. Inserting the relevant numbers for the Earth–Moon system, setting $\delta \simeq 4.4 \times 10^{-10}$, and taking into account the fact that $a_0 \approx 3.84 \times 10^5$ km, we see that the three tidal perturbations and the Nordtvedt effect have approximate amplitudes of 2770 km (2Λ), 3 km (3Λ), 73 km (Λ), and 8 meters (Λ), respectively. Two of the tidal perturbations occur at twice and three times the synodic frequency, whereas the Nordtvedt effect occurs at the synodic frequency; the tidal perturbations are much larger, but they can be cleanly separated from the Nordtvedt effect by observing many lunar orbits. The third tidal perturbation contributes at frequency Λ with an amplitude of 73 km. However, because this perturbation depends on parameters ($\omega_\oplus$, $\omega_{☾}$, R, a_0, Δ) that are very accurately measured by other means, it can be predicted in advance to an accuracy well below the amplitude of the Nordtvedt effect.

In fact, an accurate calculation of the relevant amplitudes would require us to go beyond the first-order perturbation analysis carried out here. The perturbations induced by each term affect the behavior of the other terms, and so it is necessary to go to higher order in the orbital perturbation equations, and to include higher values of the multipole index ℓ. The problem is that the solar perturbation of the lunar orbit is so large, as seen by the 2770 km leading tidal amplitude, that one must employ more sophisticated techniques, such as the Hill-Brown lunar theory, in order to find a sequence of perturbations that converges in a reasonable way. The final conclusion of such calculations is that the effective amplitude of the Nordtvedt term is increased from the amplitude shown in Eq. (8.22) by a factor approximately equal to $1 + 2\omega_\oplus/\omega_☾ \simeq 1.15$, leading to an amplitude of 9.2 meters. Similarly, the amplitude of the competing synodic term is increased from 73 km to 110 km, but it can still be predicted accurately enough to be subtracted from the data.

The resulting prediction for the Nordtvedt effect is

$$\delta r(t) \simeq 9.2\, \eta_N \cos(\Lambda t + \Phi)\ \mathrm{m}, \tag{8.24}$$

where η_N is the Nordtvedt parameter

$$\eta_N \equiv 4\beta - \gamma - 3 - \frac{10}{3}\xi - \alpha_1 + \frac{2}{3}\alpha_2 - \frac{2}{3}\zeta_1 - \frac{1}{3}\zeta_2 . \tag{8.25}$$

As we explain in Box 8.1, long-term monitoring of the lunar orbit has revealed no sign of a Nordtvedt effect, and η_N is currently constrained to be smaller than about 4×10^{-4}.

Tests of the Nordtvedt effect for neutron stars have also been carried out using a class of systems known as "wide-orbit binary millisecond pulsars" (WBMSP). These are pulsar-white-dwarf binaries with orbital periods ranging from 5 to 800 days, and eccentricities ranging from 10^{-3} to 10^{-5}. In the gravitational field of the galaxy, a non-zero Nordtvedt effect can induce an anomalous eccentricity pointed toward the galactic center (Damour and Schaefer, 1991). This can be bounded using statistical methods, given a sufficient number of WBMSPs. The current bound is $|\eta_N(\Omega/m)_{\rm NS}| < 4.6 \times 10^{-3}$, from data on more than 20 WBMSPs, including some highly circular systems (Stairs et al., 2005; Gonzalez et al., 2011). Because $(\Omega/m)_{\rm NS} \sim 0.1$ for typical neutron stars, this bound on η_N does not compete with the bound from LLR; on the other hand, the presence of neutron stars implies that these systems test the Strong Equivalence Principle in the strong-field regime. Freire et al. (2012) discuss ways to carry out this test in WBMSP systems without relying on statistical averages over many systems.

The remarkable pulsar in a triple system J0337+1715, reported by Ransom et al. (2014), may provide a test of the Nordtvedt effect that supercedes that from LLR. The system consists of a millisecond pulsar in a 1.6 day orbit with a white-dwarf companion of $0.2M_\odot$, with a more distant $0.4M_\odot$ white dwarf in a 327 day orbit around the pair. The two orbits are almost perfectly coplanar and highly circular. The key parameters of the system are shown in Table 8.1. Inserting those parameters into Eq. (8.22), we find that the three tidal perturbations have amplitudes 2.2×10^5 km (2Λ), 2×10^3 km (3Λ) and 1.1×10^4 km (Λ), while the Nordtvedt effect (Λ) has the amplitude

$$\delta r_N \sim 354\,\mathrm{km} \left(\frac{\eta_N}{10^{-4}}\right)\left(\frac{\Omega_1/m_1 - \Omega_2/m_2}{0.1}\right), \tag{8.26}$$

Box 8.1 Lunar laser ranging and the Nordtvedt effect

In the late 1950s, Dicke wondered whether one could measure a variation with time of Newton's constant G, a feature that he would soon incorporate into Brans-Dicke theory. By the early 1960s, the development of pulsed ruby lasers and the rapid build-up of the lunar space program led him and others to propose making very accurate measurements of the Earth–Moon distance by bouncing laser pulses off specially designed reflectors, to be placed on the lunar surface by either unmanned or manned landers. Such measurements would provide tests of general relativity, but they would also have other important scientific benefits, such as improving our understanding of the Earth–Moon orbit, the librations of the Moon, and even the motions of the Earth-bound laser sources resulting from tidal motions and continental drift.

The first retroreflector was deployed on the Moon by US Apollo 11 astronaut Neil Armstrong on July 21, 1969, and within a month, the first successful acquisition was made of a reflected laser signal. Two other US and two French-built reflectors were subsequently placed on the Moon by US astronauts and Soviet unmanned landers. Strangely, the French reflectors were never detected via laser bounces, until just recently.

Since that time a worldwide network of observatories has made regular measurements of the round-trip travel time to the three US lunar retroreflectors, with accuracies that are routinely at the 50 ps (1 cm) level, and that are approaching 5 ps (1 mm). These measurements are fit using the method of least-squares to a theoretical model for the lunar motion that takes into account perturbations created by the Sun and the other planets, tidal interactions, and post-Newtonian gravitational effects. The predicted round-trip travel times between retroreflector and telescope also take into account the librations of the Moon, the orientation of the Earth, the location of the observatories, and atmospheric effects on the signal propagation. The Nordtvedt parameter is then estimated in the least-squares fit, along with several other important parameters of the model.

From the first published analyses of lunar laser ranging (LLR) data in 1976 to the present, there has been absolutely no evidence, within experimental uncertainty, for the Nordtvedt effect. The residual orbital perturbation is $\delta r(t) = (2.8 \pm 4.1)\ \mathrm{mm}\ \cos(\Lambda t + \Phi)$.

At this level of precision, however, we can't regard the results of LLR as a completely "clean" test of the Nordtvedt effect until we consider the possibility of a compensating violation of the Weak Equivalence Principle for the Earth and Moon. This is because the chemical compositions of the Earth and Moon differ: the Earth is richer in the iron group elements, while the Moon is richer in silicates. To address this issue, the Eöt-Wash group at the University of Washington in Seattle carried out a novel torsion-balance test of WEP by fabricating laboratory bodies whose chemical compositions mimic that of the Earth and Moon. They found that the mini-Earth and mini-Moon fell with the same acceleration to 1.4 parts in 10^{13} (Baeßler et al., 1999; Adelberger, 2001). Combining this limit with the results for $\delta r(t)$ led to a bound on the Nordtvedt parameter

$$\eta_N = (4.4 \pm 4.5) \times 10^{-4}.$$

This represents a limit on a possible violation of the Strong Equivalence Principle of about 2 parts in 10^{13} (Williams et al., 2004b, 2009; Hofmann et al., 2010).

The Apache Point Observatory for Lunar Laser ranging Operation (APOLLO) project, a joint effort by researchers from the University of Washington, Seattle, and the University of California, San Diego, is using enhanced laser and telescope technology, together with a good, high-altitude site in New Mexico, hoping to improve the LLR bound by as much as an order of magnitude (Murphy et al., 2012). Strangely, the current limitation is not the laser ranging itself (which routinely reaches millimeter levels of precision) but the fact that many of the large ephemeris codes used to analyze the data have not kept pace with the improved data on solar-system dynamics. This effort will be aided by the fortuitous 2010 discovery by the Lunar Reconnaissance Orbiter of the precise landing site of the Soviet Lunokhod I rover, which deployed a French retroreflector in 1970. Its uncertain location made it effectively lost to lunar laser ranging for almost 40 years. Its location on the lunar surface will make it useful in improving models of the lunar libration (Murphy et al., 2011).

Lunar laser ranging also yielded important bounds on the PPN parameters α_1 and α_2 and on a time variation of G (see Sections 8.4 and 8.5).

Table 8.1 Parameters of the pulsar J0337+1715 triple system. Numbers in parentheses denote errors in the last digit.

Parameter	Value
Pulsar mass	$1.438(1)M_\odot$
Inner companion mass	$0.1975(2)M_\odot$
Outer companion mass	$0.4101(3)M_\odot$
Inner orbit	
Semimajor axis	1.9242(4) ls
Orbital period	1.629401788(5) d
Eccentricity	$6.9178(2) \times 10^{-4}$
Inclination	$39.25(1)^\circ$
Outer orbit	
Semimajor axis	118.04(3) ls
Orbital period	327.257541(7) d
Eccentricity	$3.5356196(2) \times 10^{-2}$
Inclination	$39.24(1)^\circ$

where we have scaled η_N by the approximate current bound from LLR, and the self-gravitational energy difference by 0.1, a typical value for neutron stars in general relativity (compare with the parts in 10^{10} for the Earth–Moon system and parts in 10^4 for the companion white dwarf). With timing residuals for the pulsar at the level of a microsecond, corresponding to errors in δr of order 0.3 km, the bound on η_N could potentially be improved by three orders of magnitude. This estimate is consistent with results of a simulation by Shao (2016).[1]

[1] In spring 2018, Archibald et al. (Nature, in press) reported an upper limit on the difference in acceleration between the neutron star and white dwarf of 2.6 parts in 10^6.

However, because neutron stars involve strong gravity, post-Newtonian theory does not necessarily give a good approximation to the consequences of SEP violations, so this discussion of WBMSPs and the pulsar triple system should be regarded as only qualitative. In Chapter 10, we will introduce a framework for treating the post-Newtonian motion of systems containing compact bodies (neutron stars and and black holes).

8.2 Preferred Frames and Locations: Orbits

There are a number of observable effects of a preferred-frame and preferred-location type in the orbital motions of bodies governed by the PPN N-body equation of motion (6.45). To determine these effects, we consider a two-body system whose barycenter moves relative to the universe rest frame with velocity $\boldsymbol{w}$, and that resides in the gravitational potential $U_{\rm ext}$ of a distant body (the galaxy will prove to be the dominant such body). In the N-body equations of motion (6.45), we will ignore all the self-acceleration terms except the term in Eq. (6.46a) that depends on α_3 and $\boldsymbol{w}$. We will also ignore the Newtonian acceleration, the Nordtvedt terms, and all the post-Newtonian terms that were included in the classical perihelion-shift calculation. Thus, from Eqs. (6.46a), (6.46c), and (6.50) we have the additional accelerations

$$\begin{aligned}
\delta\boldsymbol{a}_1 &= -\frac{1}{3}\alpha_3\frac{\Omega_1}{m_1}(\boldsymbol{w}+\boldsymbol{v}_1)\times\boldsymbol{\omega}_1 + \frac{m_2\boldsymbol{n}}{r^2}\left[(4\beta+2\gamma-1-\zeta_2-3\xi)\frac{m_G}{r_G}\right. \\
&\quad +\frac{1}{2}(\alpha_1-\alpha_2-\alpha_3)w^2+\frac{1}{2}\alpha_1\boldsymbol{w}\cdot\boldsymbol{v}_1+\frac{1}{2}(\alpha_1-2\alpha_2-2\alpha_3)\boldsymbol{w}\cdot\boldsymbol{v}_2 \\
&\quad \left.+\frac{3}{2}\alpha_2(\boldsymbol{w}\cdot\boldsymbol{n})^2+3\alpha_2(\boldsymbol{w}\cdot\boldsymbol{n})(\boldsymbol{v}_2\cdot\boldsymbol{n})\right] \\
&\quad +\xi\frac{m_2}{r^2}\frac{m_G}{r_G}\left[2\boldsymbol{n}_G(\boldsymbol{n}_G\cdot\boldsymbol{n})-3\boldsymbol{n}(\boldsymbol{n}_G\cdot\boldsymbol{n})^2\right]-\alpha_2\frac{m_2}{r^2}(\boldsymbol{n}\cdot\boldsymbol{w})\boldsymbol{v}_2 \\
&\quad -\frac{1}{2}\frac{m_2}{r^2}\boldsymbol{n}\cdot\left[\alpha_1\boldsymbol{v}_1-(\alpha_1-2\alpha_2)\boldsymbol{v}_2+2\alpha_2\boldsymbol{w}\right]\boldsymbol{w}, \\
\boldsymbol{a}_2 &= \{1\rightleftharpoons 2;\ \boldsymbol{n}\to-\boldsymbol{n}\},
\end{aligned} \tag{8.27}$$

where $\boldsymbol{n}\equiv\boldsymbol{x}_{12}/r$, $r_G\equiv|\boldsymbol{x}_{1G}|$, and $\boldsymbol{n}_G\equiv\boldsymbol{x}_{1G}/r_{1G}$. In obtaining Eq. (8.27), we have ignored terms of order $m_G r/r_G^2$, $m_G r^2/r_G^3$, and so on. The first two terms inside the braces in Eq. (8.27) are constant, therefore they can simply be absorbed into the Newtonian acceleration by redefining the gravitational constant (they are related to the constant corrections to $G_{\rm L}$ in Eq. (6.100)).

We will defer a discussion of the term proportional to $\alpha_3\Omega_1/m_1(\boldsymbol{w}+\boldsymbol{v}_1)\times\boldsymbol{\omega}_1$ to later in this section.

The relative acceleration $\boldsymbol{a}\equiv\boldsymbol{a}_1-\boldsymbol{a}_2$ is given by

$$\begin{aligned}
\boldsymbol{a} &= \frac{m\boldsymbol{x}}{r^3}\left[\frac{1}{2}\alpha_1\Delta\boldsymbol{w}\cdot\boldsymbol{v}+\frac{3}{2}\alpha_2(\boldsymbol{w}\cdot\boldsymbol{n})^2\right]-\frac{m\boldsymbol{x}}{r^3}\cdot\left[\frac{1}{2}\alpha_1\Delta\boldsymbol{v}+\alpha_2\boldsymbol{w}\right]\boldsymbol{w} \\
&\quad +\xi\frac{m}{r^2}\frac{m_G}{r_G}\left[2\boldsymbol{n}_G(\boldsymbol{n}_G\cdot\boldsymbol{n})-3\boldsymbol{n}(\boldsymbol{n}_G\cdot\boldsymbol{n})^2\right],
\end{aligned} \tag{8.28}$$

where we have made use of Eqs. (6.67) and (6.68).

Following the method outlined in Section 6.4.2, we calculate the secular changes in the orbit elements resulting from these perturbations. Using Eq. (6.73), we define

$$\begin{aligned}
\boldsymbol{e}_{\rm P} &\equiv \boldsymbol{n}|_{\phi=\omega} = \boldsymbol{e}_{\Omega} \cos\omega + \boldsymbol{e}_{\perp} \sin\omega\,, \\
\boldsymbol{e}_{\rm Q} &\equiv \boldsymbol{\lambda}|_{\phi=\omega} = -\boldsymbol{e}_{\Omega} \sin\omega + \boldsymbol{e}_{\perp} \cos\omega\,, \\
\hat{\boldsymbol{h}} &\equiv \boldsymbol{e}_{\rm P} \times \boldsymbol{e}_{\rm Q} = \boldsymbol{e}_{\Omega} \times \boldsymbol{e}_{\perp}\,,
\end{aligned} \tag{8.29}$$

where $\boldsymbol{e}_{\rm P}$ is a unit vector pointing toward the pericenter (it is the Runge-Lenz vector divided by the orbital eccentricity) and $\boldsymbol{e}_{\rm Q} = \hat{\boldsymbol{h}} \times \boldsymbol{e}_{\rm P}$; $\boldsymbol{e}_{\Omega}$ is a unit vector pointing along the ascending node (see Figure 6.1), and $\boldsymbol{e}_{\perp} = \hat{\boldsymbol{h}} \times \boldsymbol{e}_{\Omega}$. For any vector $\boldsymbol{A}$, we then define components $A_{\rm P}$, $A_{\rm Q}$, $A_{\rm h}$, A_{Ω}, and $A_{\perp}$ accordingly.

The secular changes in the orbit elements are then given by

$$\begin{aligned}
\Delta a &= 0\,, \\
\Delta e &= -\pi\alpha_1 \Delta \left(\frac{m}{p}\right)^{1/2} w_{\rm P}(1-e^2)F(e) + 2\pi\alpha_2 w_{\rm P} w_{\rm Q} e\sqrt{1-e^2}F(e)^2 \\
&\quad - 4\pi\xi\frac{m_G}{r_G} n_{GP} n_{GQ} e\sqrt{1-e^2}F(e)^2\,, \\
\Delta\varpi &= -\pi\alpha_1 \Delta \left(\frac{m}{p}\right)^{1/2} w_{\rm Q}\frac{\sqrt{1-e^2}F(e)}{e} - \pi\alpha_2\left(w_{\rm P}^2 - w_{\rm Q}^2\right)F(e)^2 \\
&\quad + 2\pi\xi\frac{m_G}{r_G}\left(n_{GP}^2 - n_{GQ}^2\right)F(e)^2\,, \\
\Delta\iota &= \pi\alpha_1 \Delta \left(\frac{m}{p}\right)^{1/2} w_{\rm h} \sin(\omega) e F(e) - 2\pi\alpha_2 w_{\rm h} w_{\rm R}\frac{F(e)}{\sqrt{1-e^2}} \\
&\quad + 4\pi\xi\frac{m_G}{r_G} n_{Gh} n_{GR}\frac{F(e)}{\sqrt{1-e^2}}\,, \\
\sin\iota\Delta\Omega &= -\pi\alpha_1 \Delta \left(\frac{m}{p}\right)^{1/2} w_{\rm h} \cos(\omega) e F(e) - 2\pi\alpha_2 w_{\rm h} w_{\rm S}\frac{F(e)}{\sqrt{1-e^2}} \\
&\quad + 4\pi\xi\frac{m_G}{r_G} n_{Gh} n_{GS}\frac{F(e)}{\sqrt{1-e^2}}\,,
\end{aligned} \tag{8.30}$$

where

$$F(e) \equiv \frac{1}{1+\sqrt{1-e^2}}\,, \tag{8.31}$$

and for any vector $\boldsymbol{A}$,

$$\begin{aligned}
A_{\rm R} &\equiv A_{\rm P}\cos(\omega) - A_{\rm Q}\sin(\omega)\sqrt{1-e^2}\,, \\
A_{\rm S} &\equiv A_{\rm P}\sin(\omega) + A_{\rm Q}\cos(\omega)\sqrt{1-e^2}\,.
\end{aligned} \tag{8.32}$$

In the small-eccentricity limit, to first order in e, these results simplify to

$$\Delta a = 0\,, \tag{8.33a}$$

$$\Delta e = -\frac{\pi}{2}\alpha_1\Delta\left(\frac{m}{a}\right)^{1/2} w_{\rm P} + \frac{\pi}{2}\alpha_2 w_{\rm P} w_{\rm Q} e - \pi\xi\frac{m_G}{r_G} n_{GP} n_{GQ} e\,, \tag{8.33b}$$

$$\Delta\varpi = -\frac{\pi}{2}\alpha_1\Delta\left(\frac{m}{a}\right)^{1/2}\frac{w_{\rm Q}}{e} - \frac{\pi}{4}\alpha_2\left(w_{\rm P}^2 - w_{\rm Q}^2\right) + \frac{\pi}{2}\xi\frac{m_G}{r_G}\left(n_{GP}^2 - n_{GQ}^2\right)\,, \tag{8.33c}$$

$$\Delta\iota = \frac{\pi}{2}\alpha_1\Delta\left(\frac{m}{a}\right)^{1/2} w_h e \sin(\omega) - \pi\alpha_2 w_h w_\Omega + 2\pi\xi\frac{m_G}{r_G} n_{Gh} n_{G\Omega}, \tag{8.33d}$$

$$\sin\iota\Delta\Omega = -\frac{\pi}{2}\alpha_1\Delta\left(\frac{m}{a}\right)^{1/2} w_h e \cos(\omega) - \pi\alpha_2 w_h w_\perp + 2\pi\xi\frac{m_G}{r_G} n_{Gh} n_{G\perp}, \tag{8.33e}$$

where, for $e = 0$, Eq. (8.32) implies that $A_R = A_\Omega$, and $A_S = A_\perp$.

It is tempting to interpret these secular changes as implying linearly growing values of the orbital elements. However, the expressions in Eq. (8.33) depend on ω, both from explicit ω dependence, and via the P and Q components of $\boldsymbol{w}$ and $\boldsymbol{n}_G$. The pericenter angle is actually advancing at an average rate $d\omega/d\phi \equiv \omega' \simeq 3m/p$ (see Eq. (6.78)), which we anticipate is much larger than the preferred-frame effects shown in Eq. (8.33c) – the goal is to set strong upper bounds on such effects. Thus the variations in the orbit elements will be modulated on a pericenter precession timescale and could even change sign. So in order to find the proper long-term evolution of the elements, we define, for a given element X_a, $dX_a/d\phi \equiv \Delta X_a/2\pi$, insert $\omega = \omega_0 + \omega'\phi$ in the right-hand sides (including in the P and Q components of various vectors) and integrate with respect to ϕ. Inserting the resulting integrals for e and ϖ, into the expression

$$r = a[1 - e\cos(\phi - \omega) + O(e^2)], \tag{8.34}$$

and expanding to first order in the preferred-frame perturbations, we obtain

$$\begin{aligned}\frac{r}{a} &= 1 - e_0\cos(\phi - \omega_0 - \omega'\phi) - \frac{1}{4}\alpha_1\Delta\left(\frac{m}{a}\right)^{1/2}\frac{w}{\omega'}(\hat{w}_\perp\cos\phi - \hat{w}_\Omega\sin\phi) \\ &+ \frac{e_0}{4}\alpha_2 w^2\frac{\sin\omega'\phi}{\omega'}\left[2\hat{w}_\Omega\hat{w}_\perp\cos(\phi+\omega_0) + \left(\hat{w}_\perp^2 - \hat{w}_\Omega^2\right)\sin(\phi+\omega_0)\right] \\ &+ \frac{e_0}{4}\xi\frac{m_G}{r_G}\frac{\sin\omega'\phi}{\omega'}\left[2n_{G\Omega}n_{G\perp}\cos(\phi+\omega_0) + \left(n_{G\perp}^2 - n_{G\Omega}^2\right)\sin(\phi+\omega_0)\right].\end{aligned} \tag{8.35}$$

The first term in Eq. (8.35) is the normal contribution to r/a resulting from the small eccentricity, with the pericenter advancing at a rate ω'. The second term is a forced eccentricity of the orbit, with an amplitude proportional to $(\hat{w}_\perp^2 + \hat{w}_\Omega^2)^{1/2} = (1 - \hat{w}_h^2)^{1/2} \equiv \sin\psi$, where ψ is the angle between the orbital angular momentum $\hat{\boldsymbol{h}}$ and the velocity $\boldsymbol{w}$ relative to the preferred frame, and a phase given by $\tan^{-1}(-\hat{w}_\Omega/\hat{w}_\perp)$ (Damour and Esposito-Farèse, 1992b). This effect is present even in the limit $e_0 \to 0$. The final terms are also polarizations of the orbit, proportional to e_0, with an amplitude modulated by the factor $\sin\omega'\phi/\omega'$. However, they vanish in the limit $e_0 \to 0$.

The other important effect of the preferred-frame perturbations is to cause the orbital angular momentum to precess. Since $\hat{\boldsymbol{h}} = \sin\iota(\sin\Omega\boldsymbol{e}_X - \cos\Omega\boldsymbol{e}_Y) + \cos\iota\boldsymbol{e}_Z$, variations in $\hat{\boldsymbol{h}}$ are given by

$$\Delta\hat{\boldsymbol{h}} = \sin\iota\Delta\Omega\boldsymbol{e}_\Omega - \Delta\iota\boldsymbol{e}_\perp. \tag{8.36}$$

Inserting Eqs. (8.33d) and (8.33e), and noting that $\boldsymbol{e}_\Omega\cos\omega + \boldsymbol{e}_\perp\sin\omega = \boldsymbol{e}_P$ and that $A_\perp\boldsymbol{e}_\Omega - A_\Omega\boldsymbol{e}_\perp = \boldsymbol{A}\times\hat{\boldsymbol{h}}$, we obtain

$$\Delta\hat{\boldsymbol{h}} = -\frac{\pi}{2}\alpha_1\Delta\left(\frac{m}{a}\right)^{1/2} w_h e\boldsymbol{e}_P - \pi\alpha_2 w_h(\boldsymbol{w}\times\hat{\boldsymbol{h}}) + 2\pi\xi\frac{m_G}{r_G}n_{Gh}(\boldsymbol{n}_G\times\hat{\boldsymbol{h}}), \tag{8.37}$$

leading to a precession of the angular momentum vector $\boldsymbol{h}$. Because of the variation of e in Eq. (8.33b), the magnitude of the angular momentum $h = \sqrt{ma(1-e^2)}$ will also change. The first two terms, dependent on the preferred-frame parameters α_1 and α_2 do not come as a surprise. We saw in Section 4.4 that, in semi-conservative theories of gravity, a system's total angular momentum is conserved only if the system is at rest in the preferred frame; for moving systems, angular momentum need not be conserved. The third term may seem surprising, since ξ can be non-zero even in a fully conservative theory, where total angular momentum is conserved. In this case, because of the distant mass m_G, the binary system is not a truly isolated system, and presumably there is a compensating change in the angular momentum of the orbit of the binary around the distant mass that conserves the *total* angular momentum. Changes in the angular momentum of a binary system induced by tidal interactions with a third body are well-known (Kozai-Lidov effects); here the effect is induced by a post-Newtonian violation of SEP.

We turn finally to the self-acceleration term in Eq. (8.27) proportional to α_3. We assume that only body 1 possesses enough self-gravitational binding energy and spin to generate such a term, so that the relative acceleration is given by

$$\boldsymbol{a} = -\frac{1}{3}\alpha_3 \frac{\Omega_1}{m_1} \boldsymbol{w} \times \boldsymbol{\omega}_1 , \tag{8.38}$$

where we drop the contribution from the orbital velocity $\boldsymbol{v}_1$. Calculating the secular changes in the orbit elements, we obtain

$$\Delta a = 0 , \tag{8.39a}$$

$$\Delta e = -\pi \frac{a^2\sqrt{1-e^2}}{m} B_{\mathrm{Q}} , \tag{8.39b}$$

$$\Delta \varpi = \pi \frac{a^2\sqrt{1-e^2}}{em} B_{\mathrm{P}} , \tag{8.39c}$$

$$\Delta \iota = \pi \frac{a^2 e}{m\sqrt{1-e^2}} B_{\mathrm{h}} \cos\omega , \tag{8.39d}$$

$$\sin\iota \Delta\Omega = \pi \frac{a^2 e}{m\sqrt{1-e^2}} B_{\mathrm{h}} \sin\omega , \tag{8.39e}$$

where $\boldsymbol{B} \equiv \alpha_3(\Omega_1/m_1)(\boldsymbol{w} \times \boldsymbol{\omega}_1)$. As before, we integrate with respect to ϕ including the dominant pericenter precession proportional to ω', assume that e is small, and substitute the results for e and ϖ into Eq. (8.34) to obtain

$$\begin{aligned}\frac{r}{a} &= 1 - e_0 \cos(\phi - \omega_0 - \omega'\phi) \\ &+ \frac{1}{2}\alpha_3 \frac{a^2}{m\omega'} \frac{\Omega_1}{m_1} \left[(\boldsymbol{w} \times \boldsymbol{\omega}_1) \cdot \boldsymbol{e}_\perp \sin\phi + (\boldsymbol{w} \times \boldsymbol{\omega}_1) \cdot \boldsymbol{e}_\Omega \cos\phi) \right] .\end{aligned} \tag{8.40}$$

Again we recognize a forced eccentricity, with amplitude proportional to $|\boldsymbol{w} \times \boldsymbol{\omega}_1| \sin\psi$, where ψ is the angle between $\hat{\boldsymbol{h}}$ and $\boldsymbol{w} \times \boldsymbol{\omega}_1$. If the body's spin vector $\boldsymbol{\omega}_1$ is perpendicular to the orbital plane, then $\psi = 90°$.

8.3 Preferred Frames and Locations: Structure of Massive Bodies

8.3.1 G_L and distortions of massive bodies

In Section 6.6, we found that some metric theories of gravity predict preferred-frame and preferred-location effects in the locally measured gravitational constant G_L as measured by means of a Cavendish experiment:

$$G_\mathrm{L} = 1 - [4\beta - \gamma - 3 - \zeta_2 - \xi(3+\chi)]\frac{m_G}{r_G} - \frac{1}{2}[\alpha_1 - \alpha_3 - \alpha_2(1-\chi)]w^2 - \frac{1}{2}\alpha_2(1-3\chi)(\boldsymbol{w}\cdot\boldsymbol{n})^2 + \xi(1-3\chi)\frac{m_G}{r_G}(\boldsymbol{n}_G\cdot\boldsymbol{n})^2, \quad (8.41)$$

where $\boldsymbol{w}$ is the velocity of the body relative to the preferred frame, m_G, r_G and $\boldsymbol{n}_G$ are the mass, distance, and unit vector for the external body (summed over all external masses, if necessary), and

$$\chi \equiv \frac{I}{mr^2}, \quad (8.42)$$

where m and I are the mass and spherical moment of inertia of the source body, and r and $\boldsymbol{n}$ are the distance and unit vector between the two masses being used to measure G_L.

These effects represent violations of SEP. Unfortunately, present-day laboratory measurements of G are only accurate to parts in 10^4 (Mohr and Taylor, 2005), and so are unlikely to shed light on the post-Newtonian corrections to G_L shown in Eq. (8.41), which are at levels of 10^{-6} and smaller in the solar system. However, such variations and anisotropies in G_L should affect the structure of self-gravitating bodies such as the Earth, in a manner that is analogous to the tidal effects of the Sun and Moon.

The tidal forces caused an external body induce variations in the local acceleration of gravity g on the surface of a chosen body. The measured variations are affected not only by the direct tidal accelerations, but also by the displacement of the surface relative to the center of the body and by the redistribution of mass inside the body. One can show that the variations in g are then related to the external driving variations by a numerical factor, a combination of so-called "Love numbers," which depend on the detailed structure of the body (see section 2.4 of PW for a pedagogical discussion). For the Earth, the numerical factor turns out to be approximately 1.16. These solid Earth tides are to be distinguished from ocean tides, which are much more complex.

On the Earth, variations in g are measured routinely and to very high precision by global arrays of gravimeters, as a means to learn about the structure of the Earth via measurement of the Love numbers. We wish to use such measurements to set limits on the PPN parameters that appear in Eq. (8.41). We will assume that, for the Earth, we can approximate

$$\frac{\Delta g}{g} = 1.16\frac{\Delta G_\mathrm{L}}{G_\mathrm{L}}. \quad (8.43)$$

A more accurate calculation of $\Delta g/g$ would consider the PPN equation of hydrostatic equilibrium obtained from the static limit of the PPN equation of hydrodynamics (6.25), with perturbations induced by the various preferred-frame and preferred-location terms, and then would proceed with the same kind of perturbation calculation that is done for tidal perturbations, as detailed in PW, section 2.4. In the Earth's interior the perturbing force generated by the variations in G_L is proportional to $\rho^* \nabla U$, whereas the tidal perturbing force is proportional to the distance from the center of the Earth. If the Earth's density were uniform, then $\rho^* \nabla U$ would be proportional to r, and the Love numbers would be the same as in the Newtonian tidal case. However, in Newtonian tidal theory, the Love number factor for gravimeter measurements (1.16), is not very sensitive (±5 percent) to variations in the model for the Earth, thus we do not expect it to be sensitive to a different disturbing force law.

Consider the first post-Newtonian term in Eq. (8.41). Because of the Earth's eccentric orbital motion, the external potential produced by the Sun varies yearly on Earth by only a part in 10^{10}, too small to be detected with confidence by Earth-bound gravimeters. The time-varying effects of other bodies (planets, the galaxy) are even smaller. The second term is proportional to $w^2 = w_0^2 + 2\boldsymbol{w}_0 \cdot \boldsymbol{v} + v^2$, where $\boldsymbol{w}_0 \simeq 10^{-3}$ is the velocity of the solar-system barycenter relative to the preferred frame and $\boldsymbol{v} \simeq 10^{-4}$ is the Earth's orbital velocity. The first term is constant and the third varies by only a part in 10^{10}, but the middle term varies with the full amplitude $\propto w_0 v$ because of the orbital motion of the Earth. Because of the rotation of the Earth (varying $\boldsymbol{n}$) the two anisotropic terms in Eq. (8.41) will generate variations in g at various harmonics of the Earth's rotation rate. In addition, $(\boldsymbol{w} \cdot \boldsymbol{n})^2$ will vary because of the orbital contribution to $\boldsymbol{w}$. Finally, the preferred-location term should include all gravitating matter that is not part of the cosmological background used to establish the asymptotically flat PPN coordinate system. Therefore, it should include the Sun, planets, stars, the Galaxy, and so on, at least out to scales where the matter distribution is sufficiently isotropic so as to wash out the variation with changing $\boldsymbol{n}$. In this case, the factor m_G/r_G is dominated by our galaxy ($\sim 5 \times 10^{-7}$) followed by the Sun ($\sim 10^{-8}$).

In order to illustrate how the variations in G_L would appear in practice in a time series of gravimeter readings, we will select the variations with largest amplitude in Eq. (8.41), and write

$$\frac{\Delta G_L}{G_L} = -\left(\alpha_1 - \alpha_3 - \frac{1}{2}\alpha_1\right) \boldsymbol{w}_0 \cdot \boldsymbol{v} + \frac{1}{4}\alpha_2 (\boldsymbol{w}_0 \cdot \boldsymbol{n})^2 - \frac{1}{2}\xi \frac{m_G}{r_G} (\boldsymbol{n}_G \cdot \boldsymbol{n})^2 , \tag{8.44}$$

where we have used the fact that $\chi \simeq 1/2$ for the Earth, and where here the subscript G refers to the galaxy. We work in geocentric ecliptic coordinates, and assume a circular Earth orbit, with the Earth at vernal equinox at $t = 0$. Then

$$\begin{aligned} \boldsymbol{v} &= v\,(\sin \omega t\, \boldsymbol{e}_x - \cos \omega t\, \boldsymbol{e}_y) , \\ \hat{\boldsymbol{w}}_0 &= \cos \beta_w \,(\cos \lambda_w\, \boldsymbol{e}_x + \sin \lambda_w \boldsymbol{e}_y) + \sin \beta_w \boldsymbol{e}_z , \\ \boldsymbol{n}_G &= \cos \beta_G \,(\cos \lambda_G\, \boldsymbol{e}_x + \sin \lambda_G \boldsymbol{e}_y) + \sin \beta_G \boldsymbol{e}_z . \end{aligned} \tag{8.45}$$

The latter two equations define the ecliptic coordinates (λ_w, β_w) and (λ_G, β_G) of the solar system's velocity and of the galactic center, respectively. For a gravimeter stationed at Earth latitude L,

$$
\begin{aligned}
\boldsymbol{n} = & \cos L \cos (\Omega t - \epsilon) \boldsymbol{e}_x + \left[\cos L \sin (\Omega t - \epsilon) \cos \theta + \sin L \sin \theta \right] \boldsymbol{e}_y \\
& - \left[\cos L \sin (\Omega t - \epsilon) \sin \theta - \sin L \cos \theta \right] \boldsymbol{e}_z \,,
\end{aligned} \tag{8.46}
$$

where ϵ is related to the longitude of the gravimeter on Earth, and θ is the "tilt" (23.5°) of the Earth relative to the ecliptic plane. Then the three amplitudes in Eq. (8.44) can be expressed as

$$
\begin{aligned}
\boldsymbol{w}_0 \cdot \boldsymbol{v} &= w_0 v \cos \beta_w \sin (\omega t - \lambda_w) \,, \\
(\boldsymbol{w}_0 \cdot \boldsymbol{n})^2 &= \frac{1}{2} w_0^2 \big[\cos^2 \delta_w \cos^2 L + 2 \sin^2 \delta_w \sin^2 L + \sin 2\delta_w \sin 2L \cos (\Omega t - \epsilon - \alpha_w) \\
& \qquad + \cos^2 \delta_w \cos^2 L \cos 2(\Omega t - \epsilon - \alpha_w) \big] \,, \\
(\boldsymbol{n}_G \cdot \boldsymbol{n})^2 &= \frac{1}{2} \big[\cos^2 \delta_G \cos^2 L + 2 \sin^2 \delta_G \sin^2 L + \sin 2\delta_G \sin 2L \cos (\Omega t - \epsilon - \alpha_G) \\
& \qquad + \cos^2 \delta_G \cos^2 L \cos 2(\Omega t - \epsilon - \alpha_G) \big] \,,
\end{aligned} \tag{8.47}
$$

where (α_w, δ_w) and (α_G, δ_G) are the Earth-oriented equatorial coordinates. The ecliptic and equatorial coordinates are related by

$$
\begin{aligned}
\sin \delta &= \sin \beta \cos \theta + \cos \beta \sin \theta \sin \lambda \,, \\
\cos \delta \cos \alpha &= \cos \beta \cos \lambda \,, \\
\cos \delta \sin \alpha &= - \sin \beta \sin \theta + \cos \beta \cos \theta \sin \lambda \,.
\end{aligned} \tag{8.48}
$$

The two anisotropic terms produce variations in $\Delta g/g$ at frequencies 2Ω and Ω, corresponding to periods of half a sidereal day ("semidiurnal") and a sidereal day ("diurnal"), respectively. The dominant Newtonian tides occur at frequencies $2(\Omega - \omega_☾)$ and $2(\Omega - \omega_\oplus)$, corresponding to half a lunar and solar day respectively. In fact, because of factors such as the tilt of the lunar orbit relative to the ecliptic, there are numerous sidebands of the diurnal and semidiurnal tides at $\pm\omega_\oplus$, $\pm 2\omega_\oplus$, $\pm\omega_☾$ and $\pm 2\omega_☾$. Similarly, if we had included the orbital contributions to $(\boldsymbol{w} \cdot \boldsymbol{n})^2$ and the solar contribution to the preferred-location term, there would be a corresponding array of sidebands in the PPN signal. Thus by analyzing gravimeter data taken over long periods of time, and incorporating Earth models that predict the Newtonian tidal amplitude to relatively high accuracy, one can endeavor to place bounds on α_2 and ξ. We will discuss specific experiments and the bounds they provide in Section 8.4.

The $\boldsymbol{w}_0 \cdot \boldsymbol{v}$ term produces an annual variation in the size of $G_{\rm L}$. Such variations would cause the Earth to expand and shrink slightly, changing its moment of inertia, and thereby changing its rotation rate on an annual basis. In Section 8.4, we will discuss the bounds placed by comparing such variations with the observations.

8.3.2 Precession of the spin axis of massive bodies

Another effect induced by preferred-frame and preferred-location violations of SEP is an apparent failure of conservation of angular momentum, leading to a precession of the axis of a rotating massive body. We have already seen an example of this in the precession of the angular momentum of a binary system in Section 8.2, Eq. (8.37). Here we will derive the analogous effect for a rotating, self-gravitating body at rest in the PPN coordinate frame. In the PPN equation of hydrodynamics (6.25), we split the various potentials U, Φ_2, Φ_W, and so on into an internal part and a part due to an external body, keeping only terms proportional to m_G/r_G. We also include the preferred-frame terms. Keeping only such terms along with the Newtonian Euler equation, we obtain

$$\begin{aligned}\rho^* \frac{dv^j}{dt} &= \rho^* U_{,j}\left[1 - \frac{m_G}{r_G}(4\beta + 2\gamma - 1 - \zeta_2 - 4\xi) + \frac{1}{2}(\alpha_3 - \alpha_1)w^2\right] \\ &\quad - \left[1 + (2-\gamma)\frac{m_G}{r_G}\right]p_{,j} - \rho^* X_{,jkm}\left[\xi\frac{m_G}{r_G}n_G^k n_G^m - \frac{1}{2}\alpha_2 w^k w^m\right] \\ &\quad + \frac{1}{2}(2\alpha_3 - \alpha_1)w^k V_{k,j} + \frac{1}{2}\alpha_1\left[w^j U_{,0} + 2\bar{v}^k w^{[j}U^{,k]}\right] - \alpha_2 w^k X_{,0jk}\,, \end{aligned} \tag{8.49}$$

where U, X, and V_k are potentials generated by the rotating body, with V_k dependent on $\bar{v}^k$, the velocity of the fluid relative to the center of mass. The coefficient of m_G/r_G in the first line of this expression differs slightly from the corresponding term in Eq. (8.41) because here we work in PPN coordinates, whereas G_L in Eq. (8.41) was defined using proper distances and times. We now wish to calculate the rate of change of the angular momentum tensor, $dJ^{ij}/dt \equiv 2\int \rho^* x^{[i}(dv^{j]}/dt)d^3x$. We use the fact that terms such as $\int \rho^* x^{[i}U^{,j]}d^3x$ or $\int x^{[i}p^{,j]}d^3x$ are already known to vanish in establishing the Newtonian conservation of angular momentum, and that quantities such as $\int \rho^* x^i V^{,j}_k d^3x$, $\int \rho^* x^i v_k U^{,j} d^3x$ or $\int \rho^* x^i \dot{U} d^3x$ vanish because they contain odd numbers of spatial indices, and we make our standard assumption of reflection symmetric bodies. The only terms that survive are

$$\frac{dJ^{ij}}{dt} = -\int \rho^* x^{[i}X^{,j]km}d^3x\left[2\xi\frac{m_G}{r_G}n_G^k n_G^m - \alpha_2 w^k w^m\right]. \tag{8.50}$$

Recalling that $X = \int \rho^* |\mathbf{x} - \mathbf{x}'| d^3x'$, it is straightforward to show that

$$\int \rho^* x^{[i}X^{,j]km}d^3x = \Omega^{k[i}\delta^{j]m} + \Omega^{m[i}\delta^{j]k}\,, \tag{8.51}$$

where Ω^{ij} is the self-gravitational energy tensor for the body. It satisfies the virial relation $2\mathcal{T}^{ij} + \Omega^{ij} + P\delta^{ij} = 0$, where P is the integral of the pressure, and $\mathcal{T}^{ij}$ is the kinetic energy tensor, given for a uniformly rotating body by

$$\mathcal{T}^{ij} = \frac{1}{2}\int \rho^* v^i v^j d^3x = \frac{1}{2}\epsilon^{ikm}\epsilon^{jpq}\omega^k\omega^p \int \rho^* x^m x^q d^3x\,. \tag{8.52}$$

For a nearly spherical body, $\int \rho^* x^m x^q d^3x \simeq \frac{1}{2}\delta^{pq} I_e$, where $I_e = \int \rho^* r^2 \sin^2\theta d^3x$ is the moment of inertia about the rotation axis $\boldsymbol{e}$, and thus

$$\mathcal{T}^{ij} = \frac{1}{4}I_e\omega^2\left(\delta^{ij} - e^i e^j\right). \tag{8.53}$$

Pulling everything together and recalling that the angular momentum vector is given by $J^k = \frac{1}{2}\epsilon^{kij}J^{ij}$ we find that the torque on the body is given by

$$\frac{d\boldsymbol{J}}{dt} = -\frac{1}{2}I_e\omega^2\left[2\xi\frac{m_G}{r_G}\boldsymbol{e}\cdot\boldsymbol{n}_G(\boldsymbol{e}\times\boldsymbol{n}_G) - \alpha_2\boldsymbol{e}\cdot\boldsymbol{w}(\boldsymbol{e}\times\boldsymbol{w})\right]. \tag{8.54}$$

This is completely equivalent to the change in orbital angular momentum shown in Eq. (8.37). There is no α_1 term in Eq. (8.54) because of our assumption of reflection symmetry for the body; this is equivalent to setting $e = 0$ (circular orbit) in Eq. (8.37).

8.4 Preferred Frames and Locations: Bounds on the PPN Parameters

Here we review the bounds that have been placed on the preferred-frame parameters α_1, α_2 and α_3 and on the preferred-location or Whitehead parameter ξ by observational searches for the effects described in Sections 8.2 and 8.3.

To be concrete, we must employ values for the velocity of the preferred frame and the galactic potential and their associated directions. In any theory of gravity with a preferred frame, it is natural to assume that the preferred frame is the rest frame of the overall mass distribution of the universe, as represented by the frame in which the cosmic microwave background (CMB) is isotropic. From the measured dipole anisotropy of the CMB, we know that the solar-system is moving with a speed $w_0 \simeq 370\,\mathrm{km\,s^{-1}} \simeq 1.2 \times 10^{-3}$. Table 8.2 shows the precise value along with the direction in galactic, ecliptic, and equatorial coordinates.

The galactic potential m_G/r_G is less precise, but we can estimate it within a factor of a few. The solar system's velocity within the galaxy is 230 km/s; assuming that $v^2 \sim m_G/r_G$ we can estimate $m_G/r_G \sim 5 \times 10^{-7}$. This assumes that the mass of the galaxy is more or less concentrated at the center, whereas we know that the bulk of the mass of the galaxy is in a roughly spherical halo of stars and dark matter, substantially larger in size than the visible Milky Way.

However it can be shown using a simple density model for the galaxy that the rough estimate we have adopted holds up within a factor of a few. First, we note that the important quantity $(m_G/r_G)n_G^i n_G^j$ is a single-mass version of the potential tensor $U^{ij} = \delta^{ij}U - X^{,ij}$. For a spherically symmetric distribution of matter, X is given by

$$X = rm(r) + \frac{4\pi}{3r}\int_0^r \rho' r'^4 dr' + \frac{4\pi}{3}\int_r^\infty \rho' r'(r^2 + 3r'^2)dr' , \tag{8.55}$$

where ρ is the mass density and $m(r)$ is the mass inside radius r. Then, for spherical symmetry,

$$\begin{aligned} X^{,ij} &= n^i n^j \frac{d^2X}{dr^2} - \frac{1}{r}\left(\delta^{ij} - n^i n^j\right)\frac{dX}{dr} \\ &= -n^i n^j \left[\frac{m(r)}{r} - \frac{I(r)}{r^3}\right] + \delta^{ij}\left[\frac{m(r)}{r} - \frac{I(r)}{3r^3} + \frac{8\pi r}{3}\int_r^\infty \rho' r' dr'\right] , \end{aligned} \tag{8.56}$$

where $I(r)$ is the spherical moment of inertia inside radius r. Since we are only interested in anisotropic effects, we will drop all terms proportional to δ^{ij}.

To compare with our rough estimate we consider a specific density distribution given by $4\pi\rho = \alpha/r_c^2$, for $r < r_c$, and $4\pi\rho = \alpha/r^2$, for $r > r_c$, where r_c is a core radius meant to

Table 8.2 Parameters of the preferred frame velocity $\boldsymbol{w}_0$ and the galactic potential at the solar system.

	Galactic (ℓ, b)	Ecliptic (λ, β)	Equatorial (α, δ)
Velocity, $368 \pm 2\,\text{km s}^{-1}$ (1.23×10^{-3})			
Longitude	263°.85	171°.55	167°.85
Latitude	48°.25	−11°.13	−6°.88
Galactic potential $\sim 5 \times 10^{-7}$			
Longitude	0°	266°.84	266°.41
Latitude	0°	−5°.54	−28°.94

represent the mass of the inner part of the galaxy, and α is a parameter. The $1/r^2$ density distribution is meant to model the dark matter halo, and to yield a flat rotation curve for the outer reaches of the Milky Way, in rough agreement with observations. By noting that a circular orbit in a spherical potential satisfies, $v^2/r = a_r = m(r)/r^2$, and considering the case $r > r_c$, we can fit $\alpha = v^2/(1 - 2q/3)$, where $q = r_c/r$, and find that

$$U_G^{ij} = \frac{2}{3} v^2 n_G^i n_G^j \frac{1 - q + q^3/5}{1 - 2q/3} \,. \tag{8.57}$$

For the case $r < r_c$, a similar calculation gives

$$U_G^{ij} = \frac{2}{5} v^2 n_G^i n_G^j \,, \tag{8.58}$$

independent of q. Thus for $v \sim 230$ km/s, we find an amplitude $2 - 4 \times 10^{-7}$. Note from Eq. (8.56) that only the matter inside our radius has an effect on the anisotropy. Even though the galaxy and its halo are not strictly spherically symmetric, this is unlikely to alter the estimate significantly. Table 8.2 shows our adopted value for the galactic potential along with the directions to the galactic center in various coordinates.

The anisotropies in G_L in Eq. (8.44) provided the first bounds on PPN preferred-frame and preferred-location parameters. Surveying the geophysical literature of the time, Will (1971a) pointed out that data on the semidurnal tides of the Sun and Moon agreed with Newtonian predictions to the level corresponding to amplitudes of 10^{-9}, resulting in the bounds $|\alpha_2| < 3 \times 10^{-2}$ and $|\xi| < 5 \times 10^{-3}$. An explicit experimental test was performed by Warburton and Goodkind (1976), who used an array of superconducting gravimeters at a site in southern California to take an 18-month record of tidal amplitudes and phases. They included the full array of PPN variations, not just the ones shown in Eq. (8.44), and carefully accounted for the disturbing effects of tides of the nearby Pacific Ocean. The result was the pair of bounds $|\alpha_2| < 4 \times 10^{-4}$ and $|\xi| < 10^{-3}$. Nordtvedt and Will (1972) placed modest bounds on various combinations of the parameters using perhelion advance information and limits on any anomalous annual variation in the rotation rate of the Earth, the latter induced by the $\boldsymbol{w}_0 \cdot \boldsymbol{v}$ term in Eq. (8.44).

A tighter bound on α_2 was set by Nordtvedt (1987), by considering the spin precession effect of Eq. (8.54) acting on the Sun. Focusing on the preferred-frame term, the precession takes the form $d\boldsymbol{J}/dt = \boldsymbol{\Omega} \times \boldsymbol{J}$, where $\boldsymbol{J} = I_e\omega\boldsymbol{e}$ and $\boldsymbol{\Omega} = -\frac{1}{2}\alpha_2\omega(\boldsymbol{e}\cdot\boldsymbol{w})\boldsymbol{w}$, where ω is the rotation angular velocity of the Sun. Over a time T, the change in direction of the spin axis would be of order

$$\frac{\Delta J}{J} \sim \frac{\omega T}{2}\alpha_2 w^2 \sin\beta_w \cos\beta_w \,. \tag{8.59}$$

Inserting $\omega \simeq 2.8 \times 10^{-6}\,\mathrm{s}^{-1}$, $T = 4 \times 10^9$ years, and requiring that the change be no larger than the known 7° tilt of the Sun's axis relative to the ecliptic, we obtain $|\alpha_2| < 2 \times 10^{-6}$.

Lunar laser ranging yielded the bounds $\alpha_1 = (-0.7 \pm 1.8) \times 10^{-4}$ and $\alpha_2 = (1.8 \pm 5.0) \times 10^{-5}$ from the absence of any induced eccentricity (Eq. (8.35)) in the lunar orbit (Müller et al., 1996, 2008).

Substantial improvements in bounds on these parameters came from analyzing timing data from a variety of isolated and binary pulsars.

In TEGP, we pointed out that, for an isolated spinning body moving relative to the preferred frame, the term proportional to α_3 in Eq. (8.27) produces a self-acceleration (recall that α_3 is also a conservation law parameter). If the body were a pulsar, this would induce a change in the observed pulse period P_p at a rate $dP_p/dt \simeq 2 \times 10^{-4}\alpha_3$, for $\Omega/m \simeq 0.1$ and $w \simeq 300$ km/s. Given that half of the observed pulsars have dP_p/dt lying between 10^{-14} and 10^{-15}, we inferred the upper bound $|\alpha_3| < 10^{-10}$. This bound was improved dramatically using data on ~ 21 WBMSPs. Searches for the induced eccentricity of Eq. (8.40), combined with statistical arguments to mitigate the effects of unknown orbital orientations of the observed population of WBMSPs led to the impressive bound $|\alpha_3| < 4 \times 10^{-20}$ (Bell and Damour, 1996; Stairs et al., 2005).

Similar searches for eccentricities induced by the α_1 term in Eq. (8.35) resulted in bounds on α_1 as small as a few parts in 10^5 (Wex, 2000; Shao and Wex, 2012). The tightest bound used a specific binary pulsar J1738+0333, whose orbit around its white-dwarf companion has an eccentricity 3.4×10^{-7}. The analysis was helped by the fact that the white dwarf is bright enough to be observed spectroscopically, leading to accurate determinations of the key orbital parameters. Furthermore, because the pericenter advances at a rate of about 1.6 deg yr^{-1}, the decade-long data span made it possible to partially separate any induced eccentricity, whose direction is fixed by the direction of $\boldsymbol{w}$, from the natural eccentricity, which rotates with the pericenter. For this system the result was $|\alpha_1| < 3.4 \times 10^{-5}$ (Shao and Wex, 2012).

Limits on α_2 and ξ were obtained by looking for the precession of the orbital plane of a binary system [see Eq. (8.37)]. Such a precession would lead to a variation in the "projected semimajor axis" of the pulsar, $a_p \sin\iota$, a quantity that is measured very accurately in binary pulsar timing (see Section 12.1). Combining data from the two WBMSP systems J1738+0333 and J1012+5307, Shao and Wex (2012, 2013) obtained the bounds $|\alpha_2| < 1.8 \times 10^{-4}$ and $|\xi| < 3.1 \times 10^{-4}$.

These bounds were improved substantially by searching for the precession of the spin of *isolated* pulsars (see Eq. (8.54)), the analogue of the precession effect studied by Nordtvedt (1987) in the solar context. If the spin of the pulsar precesses, then the pulse profile should change as the line of site to the pulsar slices different parts of the emission region. Using 15

Box 8.2 Neutron stars and PPN parameters

In this chapter, we have used a number of observations of binary and isolated pulsars to place bounds on preferred-frame and preferred-location parameters. And observations of the pulsar in a triple system may soon place a bound on the Nordtvedt parameter.

However, the PPN framework does not strictly apply to such systems, because the neutron stars are compact, strong-field objects, in the sense that the post-Newtonian expansion parameter ϵ can be of order $0.1 - 0.3$ near and inside these objects. And in theories of gravity with spontaneous scalarization effects, the effective internal values of ϵ can be much larger than unity.

Accordingly, one should treat bounds on the PPN parameters inferred from such systems with some caution. In some alternative theories, the effect of the compact body could be to rescale the relevant PPN parameter by a factor $1+\epsilon$, where $\epsilon \sim m/r$, where m and r are the mass and radius of the neutron star. Since the bounds on many of these parameters are very small, such corrections of order tens of percent will be insignificant. On the other hand, in some theories, additional parameters may be present for systems containing compact objects, and the relation between these paramaters and the true PPN parameters may be more complicated than a simple rescaling of the PPN parameter. This will be addressed in Section 10.3.1, where we will introduce a "modified EIH formalism," a parametrization of the Lagrangian for systems of compact bodies, rather than of the spacetime post-Newtonian metric. These parameters will depend on the theory of gravity and on the internal structure of the compact bodies in the system. In the limit where there are no compact bodies, these parameters will have specific limits in terms of the PPN parameters for semiconservative theories. For example, using the modified EIH formalism, we will derive an equation for the preferred-frame perturbations of a binary system of compact bodies that generalizes the preferred-frame terms in Eq. (8.28).

In the literature on tests of SEP using neutron stars, it is conventional to designate the relevant parameters using "hats," for example $\hat{\alpha}_1$, $\hat{\alpha}_2$, or $\hat{\xi}$, to emphasize that the parameters may not be quite the same as the corresponding PPN parameters.

years of observations of B1937+21 and J1744-1134, two isolated millisecond pulsars with very stable pulse profiles, Shao et al. (2013) and Shao and Wex (2013) found no evidence of pulse profile changes, and thereby established the strong bounds $|\alpha_2| < 1.6 \times 10^{-9}$ and $|\xi| < 3.9 \times 10^{-9}$. The use of isolated pulsars rather than binary pulsars was crucial, in order to avoid any complication from the well-known geodetic precession of spins that occurs in binary pulsar systems (see Section 12.1.2).

8.5 Constancy of Newton's Gravitational Constant

Most theories of gravity that violate SEP predict that the locally measured Newtonian gravitational constant may vary with time as the universe evolves. For the theories discussed in Chapter 5, the predictions for $\dot{G}/G$ can generally be written in terms of time derivatives of the asymptotic dynamical fields or of the asymptotic matching parameters.

Other, more heuristic proposals for a changing gravitational constant, such as those due to Dirac, cannot be written this way. Dyson (1972) gives a detailed discussion of these earlier proposals. Where G does change with cosmic evolution, its rate of variation should be of the order of the expansion rate of the universe, that is

$$\frac{\dot{G}}{G} \simeq \sigma H_0 , \tag{8.60}$$

where H_0 is the Hubble expansion parameter whose value is

$$H_0 = 70 \pm 2 \,\text{km s}^{-1}\,\text{Mpc}^{-1} \simeq 7 \times 10^{-11}\,\text{yr}^{-1} , \tag{8.61}$$

(Bennett et al., 2013; Planck Collaboration et al., 2014; Grieb et al., 2017), and σ is a dimensionless parameter whose value depends upon the theory of gravity under study and upon the detailed cosmological model assumed.

For general relativity, of course, G is precisely constant ($\sigma = 0$). For Brans-Dicke theory in a spatially flat Friedmann-Robertson-Walker (FRW) cosmology, $\sigma \simeq -(2+\omega)^{-1}$ [see, e.g., Section 16.4 of Weinberg (1972)]. In Einstein-Æther and Khronometric theory, G is constant, basically because the auxiliary vector field is constrained to have unit norm.

However, several observational constraints can be placed on $\dot{G}/G$, one kind coming from bounding the present rate of variation, another from bounding a difference between the present value and a past value. The first type of bound typically comes from LLR measurements, planetary radar-ranging measurements, and pulsar timing data. The second type comes from studies of the evolution of the Sun, stars and the Earth, Big-Bang nucleosynthesis, and analyses of ancient eclipse data. Given the present uncertainties in laboratory measurements of G (together with significant disagreements among measurements that are larger than the individual errors), it is unlikely that laboratory tests of $\dot{G}$ will contribute interesting bounds for the foreseeable future. For a time during the 1970s, van Flandern (1975) and others argued for a nonzero $\dot{G}$, based on observations of the recession of the Moon from the Earth. Today those claims are given little credence, particularly in the face of much more precise, independent bounds on $\dot{G}/G$.

The best limits on a current $\dot{G}/G$ come from improvements in the ephemeris of Mars using range and Doppler data from the Mars Global Surveyor (1998 – 2006), Mars Odyssey (2002 – 2008), and Mars Reconnaissance Orbiter (2006 – 2008), together with improved data and modeling of the effects of the asteroid belt (Pitjeva, 2005; Konopliv et al., 2011; Pitjeva and Pitjev, 2013). Since the bound is actually on variations of $GM_\odot$, any future improvements in $\dot{G}/G$ beyond a part in 10^{13} per year will have to take into account models of the actual mass loss from the Sun, due to radiation of photons and neutrinos ($\sim 0.7 \times 10^{-13}\,\text{yr}^{-1}$) and due to the solar wind ($\sim 0.6 \times 10^{-13}\,\text{yr}^{-1}$). Another bound comes from LLR measurements (Williams et al., 2004b). For earlier results see Dickey et al. (1994),Williams et al. (1996) and Müller et al. (1999). This precision could be improved by the future BepiColombo mission to Mercury, scheduled for launch in late 2018 (Milani et al., 2002; Ashby et al., 2007).

Although bounds on $\dot{G}/G$ from solar-system measurements can be correctly obtained in a phenomenological manner through the simple expedient of replacing G by $G_0 + \dot{G}(t - t_0)$ in Newton's equations of motion, the same does not hold true for pulsar and binary pulsar

Table 8.3 Tests of the constancy of G.

Method	$\dot{G}/G$ (10^{-13} yr^{-1})	Reference
Mars ephemeris	0.1 ± 1.6	Konopliv et al. (2011)
	-0.6 ± 0.4	Pitjeva and Pitjev (2013)
Lunar laser ranging	4 ± 9	Williams et al. (2004b)
	0.7 ± 3.8	Hofmann et al. (2010)
Binary and millisecond pulsars	-7 ± 33	Deller et al. (2008)
		Lazaridis et al. (2009)
Helioseismology	0 ± 16	Guenther et al. (1998)
Big-bang nucleosynthesis	0 ± 4	Copi et al. (2004)
		Bambi et al. (2005)

timing measurements. The reason is that, in theories of gravity that violate SEP, such as scalar-tensor theories, the "mass" and moment of inertia of a gravitationally bound body may vary with G. Because neutron stars are highly relativistic, the fractional variation in these quantities can be comparable to $\dot{G}/G$, the precise variation depending both on the equation of state of neutron star matter and on the theory of gravity in the strong-field regime. The variation in the moment of inertia affects the spin rate of the pulsar, while the variation in the mass can affect the orbital period of a binary in a manner that can subtract from the direct effect of a variation in G, given by $\dot{P}_b/P_b = -2\dot{G}/G$ (Nordtvedt, 1990). Thus, the bounds quoted in Table 8.3 for binary and millisecond pulsars are theory-dependent and must be treated as merely suggestive.

Bounds from helioseismology and Big-Bang nucleosynthesis (BBN) assume a model for the evolution of G over the multi-billion-year time spans involved. For example, the concordance of predictions for light elements produced around 3 minutes after the Big Bang with the abundances observed in the present universe indicate that G then was within 20 percent of G today. Assuming a power-law variation of $G \sim t^{-\alpha}$ then yields the bound on $\dot{G}/G$ today shown in Table 8.3.

9 Other Tests of Post-Newtonian Gravity

A number of tests of post-Newtonian gravitational effects do not fit into either of the two categories, classical tests or tests of SEP. These include the search for relativistic effects involving spinning bodies (Section 9.1), the de Sitter precession (Section 9.2), and tests of post-Newtonian conservation laws (Section 9.3). Some of these experiments provide limits on PPN parameters, in particular the conservation-law parameters ζ_1, ζ_2, ζ_3, and ζ_4, that were not constrained (or that were constrained only indirectly) by the classical tests and by tests of SEP. Such experiments provide new information about the nature of post-Newtonian gravity. Others, such as measurements of spin effects, constrain values for PPN parameters that are already better constrained by the experiments discussed in Chapters 7 and 8. Nevertheless, such experiments are important, for the following reasons:

(i) They may provide independent checks of the values of the PPN parameters, and thereby independent tests of gravitation theory. They are independent in the sense that the physical mechanism responsible for the effect being measured may be rather different than the mechanism that led to the prior limit on the PPN parameters. An example is the Lense-Thirring effect, the dragging of inertial frames produced by the rotation of a body. It is not a preferred-frame effect, yet it depends upon the parameter α_1. So while a bound on α_1 achieved by measuring the Lense-Thirring effect may not beat the strong bounds we saw in Chapter 8, it comes from rather different physical effects.

(ii) The structure of the PPN formalism is an assumption about the nature of gravity, one that, while seemingly compelling, could be incorrect or incomplete. According to this viewpoint, one should not prejudice the design, performance, and interpretation of an experiment by viewing it within any single theoretical framework. Thus, the γ parameters measured by light-deflection and time-delay experiments could in principle be different, while according to the PPN formalism they must be identical. We agree with this viewpoint because, although theoretical frameworks such as the PPN formalism have proven to be very powerful tools for analyzing both theory and experiment, they should not be used in a prejudicial way to reduce the importance of experiments that have independent, compelling justifications for their performance.

(iii) Any result in disagreement with general relativity would be of extreme interest.

9.1 Testing the Effects of Spin

Post-Newtonian gravity introduces an entirely new class of phenomena that are not present in Newtonian gravity – the effects of spin or rotation. These effects occur because, at post-Newtonian order, the velocity of matter becomes important, and consequently the internal velocity of a rotating body, even if it is at rest, can affect the gravitational field it generates, and can influence how it responds to an external field. These effects, known in various contexts as spin-orbit and spin-spin coupling, frame-dragging, Lense-Thirring effects, gravitomagnetism, and so on, are fundamentally relativistic. This is also true in quantum mechanics, where spin effects arise only when special relativity is merged with quantum theory to produce the Dirac equation. In post-Newtonian theory, the term "spin" refers only to the rotation of a finite-size, massive body. It does not refer to quantum mechanical spin, although there are numerous parallels between effects in the two realms.

There *are* precessions of spins in Newtonian gravity, but the nature of the effects is completely different. For example, the "precession of the equinoxes" is caused by the precession of the Earth's rotation axis with a period of about 26,000 years. However this is the result of the interaction between the Earth's quadrupole moment (which is caused by the centrifugal forces associated with its rotation) and the gravitational fields of the Sun and Moon. It is a nonrelativistic, quadrupole interaction involving nonspherical mass distributions. By contrast, the relativistic effects of spin are dipole in nature.

9.1.1 Effects of spin in two-body systems

In Section 6.7 we derived the post-Newtonian effects of the spin of a body on its motion, and the effects of its motion in a gravitational field on its spin. We found that a spinning body experiences spin-orbit and spin-spin accelerations [Eq. (6.113)] produced by both its own spin and the spins of other bodies in the system. At the same time, its own spin experiences a precession, caused by spin-orbit, spin-spin and preferred-frame couplings. The key equations, (6.113), and (6.119)–(6.121), were derived for a general N-body system of spinning masses. Here we specialize to two bodies and derive results that will allow us to discuss several important tests of the effects of spin, including Gravity Probe B, LAGEOS tracking, and precessions in binary pulsars.

We consider a binary system of bodies with masses m_1 and m_2 and spins $\boldsymbol{S}_1$ and $\boldsymbol{S}_2$, in an elliptical orbit with elements p, e, ι, Ω, and ω. We will first work out the precession of the spin of body 1, averaged over an orbit. Using Eqs. (6.119) and (6.121) restricted to two bodies, along with the fact that $\boldsymbol{v}_1 = (m_2/m)\boldsymbol{v}$ and $\boldsymbol{v}_2 = -(m_1/m)\boldsymbol{v}$, we obtain

$$\frac{d\boldsymbol{S}_1}{dt} = \frac{1}{2}\left(2\gamma + 1 + \frac{m_1}{m}\right)\frac{m_2}{r^3}\boldsymbol{h} \times \boldsymbol{S}_1 + \frac{1}{8}\left(4\gamma + 4 + \alpha_1\right)\frac{1}{r^3}\left[3\boldsymbol{n}(\boldsymbol{n}\cdot\boldsymbol{S}_2) - \boldsymbol{S}_2\right] \times \boldsymbol{S}_1 , \tag{9.1}$$

where $r \equiv |\boldsymbol{x}_{12}|$, $\boldsymbol{n} \equiv \boldsymbol{x}_{12}/r$, and $\boldsymbol{h} \equiv \boldsymbol{x} \times \boldsymbol{v}$. By suitably interchanging labels $1 \rightleftharpoons 2$, we can write down the corresponding equation for spin 2. Using our Keplerian orbit formulae of Eq. (6.73), and setting both spins to be constant on the right-hand-side, we can integrate over one orbit to find the change $\Delta\boldsymbol{S}_1$. We make use of the fact that, when integrated over one period,

$$\int \frac{1}{r^3}dt = \frac{2\pi}{hp} , \quad \int \frac{n^i n^j}{r^3}dt = \frac{\pi}{hp}\left(\delta^{ij} - \hat{h}^i\hat{h}^j\right) , \tag{9.2}$$

where $h = \sqrt{mp}$ and $p = a(1 - e^2)$. The change in $\boldsymbol{S}_1$ over one orbit is then given by

$$\Delta \boldsymbol{S}_1 = \pi \left(2\gamma + 1 + \frac{m_1}{m}\right) \frac{m_2}{p} \hat{\boldsymbol{h}} \times \boldsymbol{S}_1 + \frac{\pi}{8hp}(4\gamma + 4 + \alpha_1)\left[\boldsymbol{S}_2 - 3\hat{\boldsymbol{h}}(\hat{\boldsymbol{h}} \cdot \boldsymbol{S}_2)\right] \times \boldsymbol{S}_1 . \tag{9.3}$$

The first term is frequently called the geodetic precession; it is a spin-orbit effect, as it involves the interaction between the spin $\boldsymbol{S}_1$ and the orbital angular momentum $\boldsymbol{h}$. For the geodetic effect, the change in $\boldsymbol{S}_1$ is always confined to the orbital plane (orthogonal to $\boldsymbol{h}$); it vanishes if the spin is parallel to $\boldsymbol{h}$ and is maximized if the spin lies in the orbital plane. Geometrically, the effect is partially linked to the fact that the parallel transport of a vector around a closed path in a curved spacetime causes a change in the direction of the vector related to the spatial curvature and the area enclosed by the path; the "2γ" part of the coefficient is related to this spatial curvature effect. The remaining piece of the coefficient is related to the special relativistic Thomas precession of a spin moving in a curved orbit.

The second term, a spin-spin interaction, is referred to in many ways: Lense-Thirring precession, dragging of inertial frames and "Schiff" precession. If the companion's spin is fixed, $\boldsymbol{S}_1$ undergoes a precession about a fixed vector that is a combination of the spin of the companion and the orbital angular momentum.

To find the spin-orbit and spin-spin contributions to the relative equation of motion for a binary system, $\boldsymbol{a} \equiv \boldsymbol{a}_1 - \boldsymbol{a}_2$, we take Eq. (6.113), set $\boldsymbol{w} = 0$ and $\alpha_3 = 0$, and obtain

$$\begin{aligned} \boldsymbol{a}[\mathrm{SO}] = \frac{3}{2r^3}\Bigg\{ & \boldsymbol{n}(\boldsymbol{n} \times \boldsymbol{v}) \cdot \left[(2\gamma + 1 + \lambda)\boldsymbol{\sigma} + \frac{1}{2}(4\gamma + 4 + \alpha_1)\boldsymbol{S}\right] \\ & + (\boldsymbol{n} \cdot \boldsymbol{v})\boldsymbol{n} \times \left[(2\gamma + 1 - \lambda)\boldsymbol{\sigma} + \frac{1}{2}(4\gamma + 4 + \alpha_1)\boldsymbol{S}\right] \\ & - \frac{2}{3}\boldsymbol{v} \times \left[(2\gamma + 1)\boldsymbol{\sigma} + \frac{1}{2}(4\gamma + 4 + \alpha_1)\boldsymbol{S}\right]\Bigg\}, \end{aligned} \tag{9.4a}$$

$$\boldsymbol{a}[\mathrm{SS}] = \frac{3}{8}(4\gamma + 4 + \alpha_1)\frac{m}{r^4}\left[5\boldsymbol{n}(\boldsymbol{n} \cdot \hat{\boldsymbol{S}}_1)(\boldsymbol{n} \cdot \hat{\boldsymbol{S}}_2) - \boldsymbol{n}\hat{\boldsymbol{S}}_1 \cdot \hat{\boldsymbol{S}}_2 - 2\hat{\boldsymbol{S}}_{(1}(\boldsymbol{n} \cdot \hat{\boldsymbol{S}}_{2)})\right], \tag{9.4b}$$

where

$$\begin{aligned} \boldsymbol{\sigma} &\equiv \frac{m_2}{m_1}\boldsymbol{S}_1 + \frac{m_1}{m_2}\boldsymbol{S}_2 , \\ \boldsymbol{S} &\equiv \boldsymbol{S}_1 + \boldsymbol{S}_2 , \\ \hat{\boldsymbol{S}}_i &\equiv \frac{\boldsymbol{S}_i}{m_i} . \end{aligned} \tag{9.5}$$

Calculating the radial, cross-track and out-of-plane components of these perturbing accelerations, substituting into Eq. (6.74), and integrating over one orbit assuming that the spins are constant over an orbital timescale, we obtain the secular changes in the orbit elements. For the spin-orbit terms we find

$$\begin{aligned} \Delta a &= 0 , \\ \Delta e &= 0 , \\ \Delta i &= \frac{\pi}{hp}(C_{\mathrm{P}} \cos\omega - C_{\mathrm{Q}} \sin\omega) = \frac{\pi}{hp}C_{\Omega} , \end{aligned}$$

$$\sin\iota\Delta\Omega = \frac{\pi}{hp}(C_{\rm P}\sin\omega + C_{\rm Q}\cos\omega) = \frac{\pi}{hp}C_{\perp},$$

$$\Delta\varpi = -\frac{2\pi}{hp}C_{\rm h}, \tag{9.6}$$

where

$$\boldsymbol{C} = (2\gamma+1)\boldsymbol{\sigma} + \frac{1}{2}(4\gamma+4+\alpha_1)\boldsymbol{S}, \tag{9.7}$$

and the subscripts P, Q and h denote components along the pericenter direction, the direction in the orbital plane given by $\boldsymbol{e}_{\rm Q} = \hat{\boldsymbol{h}} \times \boldsymbol{e}_{\rm P}$, and the direction $\hat{\boldsymbol{h}}$ normal to the orbital plane, respectively, while the subscripts Ω and $\perp$ denote components along the line of nodes and along the direction given by $\boldsymbol{e}_{\perp} = \hat{\boldsymbol{h}} \times \boldsymbol{e}_{\Omega}$. Note that the arbitrary parameter λ, related to the choice of world line of the spinning body (Box 6.3) drops out.

Although the inclination of the orbit relative to the reference system changes, the inclination relative to the vector $\boldsymbol{C}$ does not. The rate of change of $\hat{\boldsymbol{h}}$ is given by $\Delta\hat{\boldsymbol{h}} = \Delta\Omega\sin\iota\,\boldsymbol{e}_{\Omega} - \Delta\iota\,\boldsymbol{e}_{\perp} = (\pi/2hp)(C_{\perp}\boldsymbol{e}_{\Omega} - C_{\Omega}\boldsymbol{e}_{\perp})$, which is perpendicular to $\boldsymbol{C}$. Thus the orbital angular momentum vector precesses around $\boldsymbol{C}$ with $\hat{\boldsymbol{h}}\cdot\boldsymbol{C} = $ constant.

For the spin-spin contributions we obtain

$$\Delta a = 0,$$

$$\Delta e = 0,$$

$$\Delta i = -\frac{3\pi}{8p^2}(4\gamma+4+\alpha_1)\left[(S_1)_{\rm h}(S_2)_{\Omega} + (S_2)_{\rm h}(S_1)_{\Omega}\right],$$

$$\sin\iota\Delta\Omega = -\frac{3\pi}{8p^2}(4\gamma+4+\alpha_1)\left[(S_1)_{\rm h}(S_2)_{\perp} + (S_2)_{\rm h}(S_1)_{\perp}\right],$$

$$\Delta\varpi = -\frac{3\pi}{8p^2}(4\gamma+4+\alpha_1)\left[\boldsymbol{S}_1\cdot\boldsymbol{S}_2 - 3(S_1)_{\rm h}(S_2)_{\rm h}\right]. \tag{9.8}$$

9.1.2 Tests of spin precession

In 2011 the Relativity Gyroscope Experiment (Gravity Probe B or GPB) carried out by Stanford University, NASA, and Lockheed Martin Corporation, finally brought to a conclusion a space mission to detect the frame-dragging and geodetic precessions of an array of gyroscopes orbiting the Earth (Everitt et al., 2011; Everitt et al., 2015). The story of GPB is told in Box 9.1.

To analyze the experiment, we use Eq. (9.3), with the mass of the gyroscope m_1 set equal to zero; m_2 and $\boldsymbol{S}_2$ are the mass and spin of the Earth, and p is the orbital semilatus rectum. To maximize the geodetic precession, it is useful to orient the spin of the gyroscope to be perpendicular to $\hat{\boldsymbol{h}}$, in other words to lie in the orbital plane. Then the spin will precess within the orbital plane, in the same sense as the orbital motion. To maximize the frame-dragging precession, one should maximize $|\boldsymbol{S}_2 - 3\hat{\boldsymbol{h}}(\hat{\boldsymbol{h}}\cdot\boldsymbol{S}_2)| = (S_2^2 + 3(\hat{\boldsymbol{h}}\cdot\boldsymbol{S}_2)^2)^{1/2}$, which means putting the orbit on the equatorial plane so that $\hat{\boldsymbol{h}}$ and $\boldsymbol{S}_2$ are parallel, and orienting the spin to be in the orbital plane so that $\boldsymbol{S}_1$ is orthogonal to both $\hat{\boldsymbol{h}}$ and $\boldsymbol{S}_2$. However the resulting precession of the spin will also be in the orbital plane, adding to or subtracting from the geodetic term. This makes it impossible to separate the two effects. An alternative

Box 9.1

Gravity Probe B

Gravity Probe B will very likely go down in the history of science as one of the most ambitious, difficult, expensive, and controversial relativity experiments ever performed. It was almost 50 years from inception to completion and US$ 750 million in the making, although only about half of the total time was spent as a full-fledged, approved space flight program.

The origin of GPB dates back to around 1960, when Leonard Schiff at Stanford calculated the geodetic and Lense-Thirring precessions of a gyroscope and suggested the possibility of measuring them. Independently of Schiff, George Pugh at the Weapons Systems Evaluation Group of the US military had performed the same calculations a few months earlier. Pugh was assessing the use of high-performance gyroscopes in missile and aircraft guidance. He wondered how large the relativistic effects would be, and what it would take for a space experiment to measure them. Pugh's classified work could not be published in the open literature, and so Schiff was initially given credit for the idea; only later was Pugh's work discovered and recognized. It is worth noting that all this occurred in the very early days of space exploration, only two years after the launch by the Soviet Union of *Sputnik*, the world's first artificial satellite.

GPB started officially in late 1963 when NASA funded the initial R&D work at Stanford to identify the new technologies needed to make such a difficult measurement possible. Francis Everitt became Principal Investigator of GPB in 1981, and the project moved to the mission design phase in 1984. At that time, plans called for a preliminary flight on board the Space Shuttle to test key technologies to be used in GPB (similar in intent to the LISA Pathfinder mission of 2015), followed by full launch from the Shuttle a few years later. Unfortunately the 1986 Space Shuttle Challenger catastrophe forced a cancellation of the technology test, and a complete redesign of the spacecraft for a launch from a Delta rocket. The additional costs and delays arising from this engendered increasing criticism of the project, and NASA asked the National Academy of Sciences to conduct a thorough review of the project in 1994. Following the endorsement of that committee, GPB was approved for flight development, and began to collaborate with Lockheed-Martin and NASA's Marshall Space Flight Center. The satellite finally was launched on April 20, 2004 for a 16-month mission, but another five years of data analysis were needed to tease out the effects of relativity from a background of other disturbances of the gyros.

There were four gyroscopes aboard GPB. Each gyroscope was a fused silica rotor, about four centimeters in diameter, machined to be spherical and homogeneous to better than a part per million, and coated with a thin film of niobium. The gyroscope assembly, which sat in a dewar of 2440 liters of liquid helium, was held at 1.8 degrees Kelvin. At this temperature, niobium is a superconductor, and the supercurrents in the niobium of each spinning rotor produce a "London" magnetic moment parallel to its spin axis. Extremely sensitive magnetometers (superconducting quantum interference detectors, or "SQUIDs") attached to the wall of the spherical chamber housing each rotor were capable of detecting minute changes in the orientation of the rotor's magnetic moment and hence the precession of its rotation axis.

After the spacecraft reached its orbit, the four gyros were aligned to spin along the symmetry axis of the spacecraft. This axis was also the optical axis of a telescope directly mounted on the end of the structure housing the rotors. Spacecraft thrusters oriented the telescope to point precisely toward IM Pegasi (except when the Earth intervened, once per orbit). In order to average out numerous unwanted nonrelativistic torques on the gyros, the spacecraft rotated about its axis once every 78 seconds. The satellite was placed in an almost perfectly circular polar orbit with an altitude of 642 kilometers above the Earth's surface.

Almost every aspect of the spacecraft, its subsystems, and the science instrumentation performed extremely well, some far better than expected. Still, the success of such a complex and delicate experiment boils down to figuring out the sources of error. In particular, having an accurate calibration of the electronic readout from the SQUID magnetometers with respect to the tilt of the gyros was essential. The plan for calibrating the SQUIDs was to exploit the aberration of starlight, which causes a precisely calculable misalignment between the rotors and the telescope as the latter shifts its pointing toward the guide star by up to 20 arcseconds to compensate for the orbital motion of the spacecraft and of the Earth. However, three important, but unexpected, phenomena were discovered during the experiment that affected the accuracy of the results.

First, because each rotor is not exactly spherical, its principal axis rotates around its spin axis with a period of several hours, with a fixed angle between the two axes. This is the familiar "polhode" motion of a spinning top. In fact this polhoding was essential in the calibration process because it led to modulations of the SQUID output via the residual trapped magnetic flux on each rotor (about 1 percent of the London moment). The polhode period and angle of each rotor were expected to remain constant throughout the mission, leading to accurate calibrations, but instead these quantities actually decreased monotonically with time, implying the presence of some damping mechanism, and this significantly complicated the calibration analysis. In addition, each rotor was found to make occasional, seemingly random "jumps" in its orientation—some as large as 100 milliarcseconds. Some rotors displayed more frequent jumps than others. Without being able to continuously monitor the rotors' orientation, the GPB team could not fully exploit the calibrating effect of the stellar aberration in their analysis. Finally, during a planned 40-day, end-of-mission calibration phase, the team discovered that when the spacecraft was deliberately pointed away from the guide star by a large angle, the misalignment induced much larger torques on the rotors than expected. From this, they inferred that even the very small misalignments that occurred during the science phase of the mission had induced torques that were probably several hundred times larger than the designers had estimated.

What ensued during the data analysis phase following the mission was worthy of a detective novel. The critical clue came from the calibration tests. Here, they took advantage of the residual trapped magnetic flux on the gyroscope. Superconducting lead shielding was used to suppress stray magnetic fields before the niobium coated gyroscopes were cooled, but no shielding is ever perfect. This flux adds a periodic modulation to the SQUID output, which the team used to plot the phase and polhode angle of each rotor throughout the mission. This helped them to discover that interactions between random patches of *electrostatic* potential fixed to the surface of each rotor, and similar patches on the inner surface of its spherical housing, were causing the extraneous torques. In principle, the rolling spacecraft should have suppressed these effects via averaging, but they were larger than expected.

Fortunately, the patches are fixed on the various surfaces, and so it was possible to build a parametrized model of the patches on both surfaces using multipole expansions, and to calculate the torques induced by those interactions when the spin and spacecraft axes are misaligned, as a function of the parameters. One prediction of the model was that the induced torque should be perpendicular to the plane formed by the two axes, and this was clearly seen in the data. Another prediction was that, when the slowly decreasing polhode period crosses an integer multiple of the spacecraft roll period, the torques fail to average over the roll period, whereupon the spin axis precesses about its initial direction in an opening Cornu spiral, then migrates to a new direction along a closing Cornu spiral. This is known as a loxodromic path, familiar to navigators as a path of fixed bearing on the Earth's surface. Detailed observation of the orientation of the rotors during such "resonant jumps" showed just such loxodromic behavior. In the end, every jump of every rotor could be identified by its "mode number," the integer relating its polhode period to the spacecraft roll period.

The original goal of GPB was to measure the frame-dragging precession with an accuracy of one percent, but the problems discovered over the course of the mission dashed the initial optimism that this would be possible. Although the GPB team were able to model the effects of the patches, they had to pay the price of the increase in error that comes from using a model with so many parameters. The experiment uncertainty quoted in the final result—roughly 20 percent for frame dragging—is almost totally dominated by those errors (Table 9.1). Nevertheless, after the model was applied to each rotor, all four gyros showed consistent Lense-Thirring precessions. Gyro 2 was particularly "unlucky"—it had the largest uncertainties because it suffered the most resonant jumps. Numerous cross-checks were carried out, including estimating the relativity effect during different segments of the 12-month science phase (various events, including computer reboots and a massive solar storm in January 2005, caused brief interruptions in data taking), increasing and decreasing the number of parameters in the torque model, and so on.

When GPB was first conceived in the early 1960s, tests of general relativity were few and far between, and most were of limited precision. But during the ensuing decades, there was enormous progress in experimental gravity in the solar system and in binary pulsars. By the middle 1970s, some argued that the PPN parameters γ and α_1 were already known to better accuracy than GPB could ever achieve. Given its projected high cost, critics argued for the cancellation of the GPB mission. The counterargument was that all such assertions involved theoretical assumptions about the class of theories encompassed by the PPN approach, and that all existing bounds on the post-Newtonian parameters involved phenomena entirely different from the precession of a gyroscope. All these issues were debated, for example, in the 1994 National Academy review of GPB that recommended its continuation.

Disclosure: The author was a member of the 1994 NAS panel that recommended completion of GPB, and served as Chair of an external NASA Science Advisory Committee for Gravity Probe B from 1998 to 2011.

strategy, and the one adopted by GPB, is to put the gyros into a polar orbit, so that $\hat{\boldsymbol{h}}\cdot\boldsymbol{S}_2 = 0$. This reduces the amplitude by a half, but with $\boldsymbol{S}_1$ lying in the orbital plane to maximize the geodetic effect, the frame-dragging precession will now be perpendicular to the orbital plane. By separately measuring the north-south (geodetic) precession and the east-west (frame-dragging) precession, one can make a clean measurement of both effects, at least in principle. Dividing Eq. (9.3) by the orbital period $P = 2\pi(a^3/m)^{1/2}$, we obtain the rate of angular precession in the two directions given by

$$\frac{d\hat{\boldsymbol{S}}_1}{dt} = \Omega_{[\mathrm{Geo}]}\boldsymbol{e}_{\mathrm{NS}} + \Omega_{[\mathrm{FD}]}\boldsymbol{e}_{\mathrm{EW}}\,, \tag{9.9}$$

where

$$\begin{aligned}\Omega_{[\mathrm{Geo}]} &= \frac{1}{2}(2\gamma+1)\frac{m_{\oplus}^{3/2}}{a^{5/2}}(1-e^2)^{-1} \\ &= \frac{1}{3}(2\gamma+1)\,6638\ \mathrm{mas}\ \ \mathrm{yr}^{-1}\,,\end{aligned}$$

$$\Omega_{[FD]} = \frac{1}{16}(4\gamma + 4 + \alpha_1)\frac{J_\oplus}{a^3}(1 - e^2)^{-3/2}\sin\psi$$
$$= \frac{1}{8}(4\gamma + 4 + \alpha_1)\ 39.2\,\text{mas yr}^{-1}, \tag{9.10}$$

where a and e are the semimajor axis and eccentricity of the orbit, and ψ is the angle between the spin axes of the gyroscope and of the Earth. In obtaining the numerical values for GPB, we used the fact that the satellite was in a circular orbit at 642 km altitude ($a = 7013$ km). The angular momentum of the Earth is known from combining data from the luni-solar precession of the Earth's axis and from the multipole moments of the Earth, and is given by $J_\oplus = 0.3307 m_\oplus R_\oplus^2 \omega_\oplus$, where $R_\oplus = 6371$ km is the Earth's mean radius, and $\omega_\oplus = 7.2921 \times 10^{-5}\,\text{s}^{-1}$ is the Earth's rotational angular velocity. In a perfect world, the angle ψ would be chosen to be 90°, but in fact it was about 73°. This is because the precessions of the gyroscopes had to be compared with directions in inertial space, as established to high precision, for example, by the distant quasars. To make that comparison, an on-board telescope was oriented toward a "guide" star, and the spin directions were aligned to be parallel to the optical axis of the telescope. This alignment was operationally necessary because the spacecraft itself rotated about that same axis in order to average out numerous environmentally induced torques on the gyros. Many candidate guide stars were considered, but in the end the star IM Pegasi (HR 8703) was selected, in part because it was optically bright and relatively isolated in the sky, and in part because it was bright in the radio, so that its own proper motion relative to the quasars could be measured using VLBI to an accuracy better than the goal for measuring the gyro precessions. IM Peg lies at a distance of 96 pc, and at a declination of 16.83°, thus $\psi = 73.2°$. In fact, VLBI measurements of IM Peg made before, during and after the GPB mission gave a proper motion $34.3 \pm 0.1\,\text{mas yr}^{-1}$, well within the error requirements for GPB (Bartel et al., 2015).

Two additional relativistic geodetic precessions had to be accounted for, one induced by the quadrupole moment of the Earth ($\sim 7\,\text{mas yr}^{-1}$), and the other induced by the gyroscopes' orbital motion around the Sun ($\sim 20\,\text{mas yr}^{-1}$). The final results for the GPB experiment are shown in Table 9.1; the GPB saga is told in Box 9.1. The complete technical details of GPB can be found in a special issue of *Classical and Quantum Gravity*, Vol. 32, 2015.

Another example of relativistic spin precession is in binary pulsars. In this case, the masses of the two bodies are comparable, and using the fact that each mass is given by $m_i = m(1 \pm \sqrt{1 - 4\eta})/2$, we can express the two precession rates as

Table 9.1 Final results of Gravity Probe B.

	Measured	Predicted
Geodetic Precession (mas)	6602 ± 18	6606
Frame-dragging (mas)	37.2 ± 7.2	39.2

$$\Omega_{[\mathrm{Geo}]} = 5.8 \times 10^{-2}\,\mathrm{deg}\ \ \mathrm{yr}^{-1}\left(\frac{m}{m_\odot}\right)^{2/3}\left(\frac{1\,\mathrm{day}}{P_\mathrm{b}}\right)^{5/3}\frac{\alpha}{1-e^2}\,, \tag{9.11a}$$

$$\Omega_{[\mathrm{FD}]} = 3.3 \times 10^{-5}\,\mathrm{deg}\ \ \mathrm{yr}^{-1}\frac{m_2}{m}\left(\frac{R_2}{10\,\mathrm{km}}\right)^2\left(\frac{1\,\mathrm{day}}{P_\mathrm{b}}\right)^2\left(\frac{\nu_p}{1\,\mathrm{mHz}}\right)\frac{\beta}{(1-e^2)^{3/2}}\,, \tag{9.11b}$$

where ν_p is the spin frequency of the spinning companion, P_b is the orbital period, $\beta = I_2/m_2R_2^2$ is the moment-of-inertia factor, and

$$\alpha \equiv \frac{6}{7}\left(1+\frac{2}{3}\eta \pm \sqrt{1-4\eta}\right) = \begin{cases} 12/7 & : \ m_2 \gg m_1 \\ 1 & : \ m_2 = m_1 \\ 16\eta/7 & : \ m_2 \ll m_1 \end{cases} \tag{9.12}$$

For simplicity we have set $\gamma = 1$ and $\alpha_1 = 0$.

It is clear that the frame dragging precession is negligible in binary pulsar systems, whereas the geodetic effect could be measurable. Indeed, the first binary pulsar B1913+16 provided the first evidence of this effect (see Section 12.1 for details about binary pulsars). Observations (Kramer, 1998; Weisberg and Taylor, 2002) indicated that the pulse profile is varying with time, which suggested that the pulsar is undergoing geodetic precession on a 300-year timescale. The amount was consistent with GR, assuming that the pulsar's spin is suitably misaligned with the orbital angular momentum. Unfortunately, the evidence also suggested that the pulsar beam may precess out of our line of sight by 2025.

A better test was provided by the "double pulsar system," J0737–3039, discovered in 2003 (see Section 12.1.2). Because the orbit is seen to be almost edge-on, the radio signal from the primary pulsar A is partially eclipsed by the magnetosphere of pulsar B. By monitoring the long-term evolution of the amplitude variations of pulsar A's signal during eclipses and building a simple model of the magnetosphere of pulsar B, Breton et al. (2008) were able to measure the orientation of both the spin and magnetic axes of pulsar B relative to the orbital plane, as well as the rate of precession of the spin axis. Since both masses are known along with the orbital period and eccentricity (see Table 12.2), we can use Eq. (9.11a) to predict a value $\Omega_{[\mathrm{Geo}]} = 5.07\,\mathrm{deg}\ \ \mathrm{yr}^{-1}$. The value inferred from the model was $4.7 \pm 0.7\,\mathrm{deg}\ \ \mathrm{yr}^{-1}$, a 13% test in agreement with general relativity. In fact the precession is so significant that in 2008 the radio beam of pulsar B went out of sight.

9.1.3 Tests of spin effects on orbits

In Section 9.1.1, we saw that spin-orbit and spin-spin interactions can induce precessions of orbital planes and of orbital pericenters. In the solar system, the effect of the spin of the Sun on most orbits is too small to be detectable; the best hope might be Mercury, where the additional contribution to the perihelion advance due to the Sun's spin is $\sim 3 \times 10^{-3}$ arcseconds per century, of the same order as the current measurement uncertainties.

On the other hand, the effect *has* been measured using an array of Earth-orbiting satellites, tracked to high precision using laser ranging. For a satellite with negligible mass and spin, $\boldsymbol{\sigma} = 0$, so the vector $\boldsymbol{C}$ that appears in Eq. (9.7) is given by

$$\boldsymbol{C} = \frac{1}{2}(4\gamma + 4 + \alpha_1)\boldsymbol{S}_\oplus\,. \tag{9.13}$$

In a reference system in which $S_\oplus$ points in the z-direction, $C_\Omega = 0$, $C_\perp = C \sin\iota$ and $C_h = C \cos\iota$. Thus from Eq. (9.6), the inclination of the orbit relative to the Earth's spin axis stays unchanged, while the line of nodes and the pericenter advance at the rates

$$\begin{aligned}
\left(\frac{d\Omega}{dt}\right)_{\rm FD} &= \frac{1}{4}(4\gamma + 4 + \alpha_1)\frac{S_\oplus}{a^3(1-e^2)^{3/2}} \\
&= \frac{1}{8}(4\gamma + 4 + \alpha_1)\left(\frac{R_\oplus}{a}\right)^3 \frac{0.218}{(1-e^2)^{3/2}} \text{ as yr}^{-1}, \\
\left(\frac{d\omega}{dt}\right)_{\rm FD} &= -\frac{1}{8}(4\gamma + 4 + \alpha_1)\frac{S_\oplus}{a^3(1-e^2)^{3/2}}\cos\iota \\
&= -\frac{1}{8}(4\gamma + 4 + \alpha_1)\left(\frac{R_\oplus}{a}\right)^3 \frac{0.655}{(1-e^2)^{3/2}} \text{ as yr}^{-1},
\end{aligned} \tag{9.14}$$

where we use the fact that $\dot\omega = \dot\varpi - \dot\Omega\cos\iota$. Unfortunately, these effects are tiny compared to the effects of the Newtonian multipole moments of the Earth, given by

$$\begin{aligned}
\left(\frac{d\Omega}{dt}\right)_{\rm N} &= -\frac{3}{2}J_2 \frac{m_\oplus^{1/2} R_\oplus^2}{a^{7/2}(1-e^2)^{3/2}}\cos\iota \\
&= -3645\left(\frac{R_\oplus}{a}\right)^{7/2}\frac{\cos\iota}{(1-e^2)^{3/2}} \text{ deg yr}^{-1}, \\
\left(\frac{d\omega}{dt}\right)_{\rm N} &= 3J_2 \frac{m_\oplus^{1/2} R_\oplus^2}{a^{7/2}(1-e^2)^{3/2}}\left(1 - \frac{5}{4}\sin^2\iota\right) \\
&= 7291\left(\frac{R_\oplus}{a}\right)^{7/2}\frac{1}{(1-e^2)^{3/2}}\left(1 - \frac{5}{4}\sin^2\iota\right) \text{ deg yr}^{-1},
\end{aligned} \tag{9.15}$$

where J_2 is the quadrupole moment coefficient of the Earth, given by $J_2 = 1.0864 \times 10^{-3}$. Notice that the Newtonian nodal advance rate is proportional to $\cos\iota$, hence it vanishes for a polar orbit; for two orbits with the same orbital elements but with supplementary inclinations ($\iota_2 = \pi - \iota_1$), the nodal advances are equal and opposite. This dependence is true for *all* the even multipole moments J_4, J_6 and so on. Thus in order to measure the relativistic precession, one has to measure or suppress the Newtonian effect to extremely high accuracy.

Van Patten and Everitt (1976) proposed measuring the frame dragging effect using two satellites in counter-rotating polar orbits. The Newtonian nodal precession would be suppressed, at least up to the effects of the errors $\Delta\iota_1$ and $\Delta\iota_2$ involved in achieving true polar orbits, and those errors could be measured to sufficient precision by having the satellites track each other as they passed one another over the poles. Ground tracking would measure the nodal precession, and the contribution of the Newtonian precession could be subtracted. This proposal never came to fruition.

The 1976 launch of the "Laser Geodynamics Satellite" (LAGEOS) and of a second LAGEOS in 1992 changed the situation. These satellites are massive spheres, 60 cm in diameter and weighing about 400 kg, placed in nearly circular orbits with semimajor axes approximately equal to $2R_\oplus$ (the precise orbital elements are listed in Table 9.2). The spheres are covered with laser retroreflectors, similar to those used for lunar laser ranging.

Table 9.2 Orbital elements of laser-ranged satellites.

Satellite	Semimajor axis (km)	Orbital period (min)	Eccentricity	Inclination to equator (°)
LAGEOS I	12,257	225	0.0045	109.84
LAGEOS II	12,168	223	0.0135	52.64
LARES	7,821	115	0.0007	69.5

Because of their large mass-to-area ratio, the spheres are less affected by atmospheric drag than other satellites at similar altitudes. This, combined with the high precision of laser ranging (which routinely achieves millimeter-level precision), means that their orbits can be determined extremely precisely. The LAGEOS satellites were launched primarily to carry out studies in geodesy and geodynamics, but it was soon recognized that they were potentially capable of measuring the relativistic nodal advance, which for the values for LAGEOS I shown in Table 9.2 amounts to approximately 30.7 mas/yr.

Indeed, Ciufolini (1986) pointed out that if a second LAGEOS satellite were to have an inclination supplementary to that of LAGEOS I, then the Newtonian advance would be equal and opposite for the two satellites, so that if one measured the nodal advance of both satellites simultaneously and added them together, the large Newtonian effect would cancel out, up to the contributions of errors in achieving the precise orbital inclinations.

Ciufolini and other relativists campaigned vigorously to have LAGEOS II launched with $\iota_2 = 70.16°$, but other considerations prevailed in the end. LAGEOS II was launched with $\iota_2 = 52.64°$, mainly to optimize coverage by the world's network of laser tracking stations, which was important for geophysics and geodynamics research. The fall-back option was then to combine the data from the two satellites as they were. One could still eliminate the largest Newtonian contribution coming from J_2 with a suitable linear combination of the two measured nodal advances, thereby revealing the relativistic contribution and those coming from higher-order multipole moments. The uncertainties in the values of those moments would contribute to the error made in measuring the relativistic effect. For a time, Ciufolini and collaborators tried to include a third piece of data, the perigee advance of LAGEOS II (which has a small eccentricity), as a way to also eliminate J_4, but this turned out to be plagued with systematic errors that were large and hard to control (Ciufolini et al., 1997, 1998; Ciufolini, 2000).

The turning point came with CHAMP and GRACE. Europe's CHAMP (Challenging Minisatellite Payload) and NASA's GRACE (Gravity Recovery and Climate Experiment) missions, launched in 2000 and 2002, respectively, use precision tracking of spacecraft to measure variations in the Earth's gravity on scales as small as several hundred kilometers, with unprecedented accuracies. GRACE consists of a pair of satellites flying in close formation (200 kilometers apart) on polar orbits. Each satellite carries an on-board accelerometer to measure nongravitational perturbations, a satellite to satellite K-band radar link to measure variations in the Earth's gravity gradient on short scales, and a GPS tracking unit to measure larger scale variations. With the dramatic improvements on J_ℓ obtained by CHAMP and GRACE, Ciufolini and his colleagues could now treat J_4

and higher multipole moments as known, and use the two LAGEOS nodal advances to determine J_2 and the relativistic contribution. The final outcome was a successful test of the relativistic prediction at the level of 10 percent (Ciufolini et al., 2011).

On February 13, 2012, a third laser-ranged satellite, known as LARES (Laser Relativity Satellite) was launched by the Italian Space Agency. Its inclination was 69.5°, very close to the required supplementary angle relative to LAGEOS I, and its eccentricity was very nearly zero (see Table 9.2). However, the launch failed to achieve the same semimajor axis as that of the two LAGEOS satellites, again preventing a perfect cancellation of the Newtonian effect. Nevertheless, combining data from all three satellites with continually improving Earth data from GRACE, the LARES team reported a test of frame-dragging at the 5 percent level (Ciufolini et al., 2016).

9.2 De Sitter Precession

This effect, an advance of the line of nodes of the Earth–Moon orbit, was first calculated by de Sitter (1916) as a consequence of the post-Newtonian perturbations of the lunar orbit induced by the gravitational potential of the Sun and by the orbital motion of the Earth–Moon system around the Sun (for details of the derivation in general relativity in modern language, see section 10.1.6 of PW). However, it can equally well be regarded as a *geodetic precession* of the Earth–Moon system's angular momentum vector (its "spin") about the angular momentum vector of the Earth-Sun orbit. From the geodetic term in Eq. (9.3), the change in $\boldsymbol{S}_{\rm EM}$, the Earth–Moon's spin, over one orbit around the Sun is given by

$$\Delta \boldsymbol{S}_{\rm EM} = \pi(2\gamma + 1)\frac{m_\odot}{p}\hat{\boldsymbol{h}} \times \boldsymbol{S}_{\rm EM}\,, \tag{9.16}$$

where we have approximated $m_1 = m_\oplus + m_☾ \ll m$, and $m_2 = m_\odot \simeq m$. This precession leads directly to an advance of the lunar node, given by

$$\frac{d\Omega}{dt} = \frac{1}{2}(2\gamma + 1)\frac{m_\odot^{3/2}}{a^{5/2}}\frac{1}{1 - e^2} = 1.917\ \text{as/century}\,, \tag{9.17}$$

where a and e are the semimajor axis and eccentricity of the Earth–Moon barycenter orbit around the Sun. This effect has been measured to about 0.6 percent using lunar laser ranging data (Dickey et al., 1994; Williams et al., 2004a, 2004b).

9.3 Tests of Post-Newtonian Conservation Laws

Here we discuss tests of the "conservation law" PPN parameters ζ_1, ζ_2, ζ_3, ζ_4, and α_3. We have already seen in Section 8.4 that, by virtue of its additional role as a preferred-frame parameter, α_3 has been constrained to vanish to a few parts in 10^{20} using data from wide-binary millisecond pulsars.

In addition, there is strong theoretical evidence that ζ_4, which is related to the gravity generated by fluid pressure, is not really an independent parameter. In any reasonable theory of gravity there should be a connection between the gravity produced by kinetic energy (ρv^2), internal energy ($\rho\Pi$), and pressure (p). The idea is to consider an alternative PPN metric generated by a system of charged point masses, with gravitational potentials generated by masses, microscopic velocities, charges, and so on. Under suitable coarse graining as used in thermodynamics, this metric should transform smoothly to the hydrodynamic PPN metric. From such considerations, there follows (Will, 1976b) the additional theoretical constraint

$$6\zeta_4 = 3\alpha_3 + 2\zeta_1 - 3\zeta_3 . \tag{9.18}$$

In addition the ratio of active to passive mass for such a bound system of point charges should take the form

$$\frac{m_A}{m_P} = 1 + \frac{1}{2}\zeta_3 \frac{E_e}{m_P} , \tag{9.19}$$

where E_e is the electrostatic energy of the system of point charges.

A nonzero value for ζ_3 would result in a violation of conservation of momentum, or of Newton's third law in gravitating systems, based on differences in chemical composition.

A classic test of Newton's third law for gravitating systems was carried out by Kreuzer (1968), in which the gravitational attraction of fluorine and bromine were compared. Kreuzer's experiment used a Cavendish balance to compare the Newtonian gravitational force generated by a cylinder of Teflon (76 percent fluorine by weight) with the force generated by that amount of a liquid mixture of trichloroethylene and dibromomethane (74 percent bromine by weight) that had the same passive gravitational mass as the cylinder, namely the amount of liquid displaced by the cylinder at neutral buoyancy. In the actual experiment, the Teflon cylinder was moved back and forth in a container of the liquid, with the Cavendish balance placed near the container. Had the active masses of Teflon and displaced liquid differed at neutral buoyancy, a periodic torque would have been experienced by the balance. The absence of such a torque led to the conclusion that the ratios of active to passive mass for fluorine and bromine are the same to 5 parts in 10^5. Using the semi-empirical mass formula for nuclear electrostatic energies (which are the dominant contribution in this case), $E_e/m \simeq 7.6 \times 10^{-4} Z(Z-1)A^{-4/3}$, where Z and A are the atomic number and mass number, respectively, we obtain the bound $|\zeta_3| < 0.06$.

An improved bound was reported by Bartlett and van Buren (1986). They noted that current understanding of the structure of the Moon involves an iron-rich, aluminum-poor mantle whose center of mass is offset about 10 km from the center of mass of an aluminum-rich, iron-poor crust. The direction of offset is toward the Earth, about 14° to the east of the Earth–Moon line. Such a model accounts for the basaltic maria that face the Earth, and the aluminum-rich highlands on the Moon's far side, and for a 2 km offset between the observed center of mass and center of figure for the Moon. Because of this asymmetry, a violation of Newton's third law for aluminum and iron would result in a momentum nonconserving self-force on the Moon, whose component along the orbital direction would contribute to the secular acceleration of the lunar orbit. Improved knowledge of the lunar orbit through lunar laser ranging, and a better understanding of tidal effects in the Earth–Moon system (which also contribute to the secular acceleration) through satellite data, severely limit any anomalous secular acceleration, with the resulting limit of 4×10^{-12} on the difference between the active-to-passive mass ratios for aluminum and iron. This leads to the bound $|\zeta_3| < 10^{-8}$. Nordtvedt (2001) examined whether this bound could be improved by considering the asymmetric distribution of ocean water on Earth; the self-acceleration of the Earth in the polar direction would then lead to an effect in the Earth–Moon orbit that could be tested by lunar laser ranging.

Another test of PPN conservation laws involves an effect first pointed out, incorrectly, by Levi-Civita (1937). The effect is the secular acceleration of the center of mass of a binary system. Levi-Civita pointed out that general relativity predicted a secular acceleration in the direction of the periastron of the orbit, and found a binary system candidate in which he felt the effect might one day be observable. Eddington and Clark (1938) repeated the calculation using the N-body equations of motion of de Sitter (1916). After first finding a secular acceleration of opposite sign to that of Levi-Civita, they then discovered an error in de Sitter's equations of motion, and concluded finally that the secular acceleration was zero. Robertson (1938) independently reached the same conclusion using the Einstein-Infeld-Hoffmann equations of motion, and Levi-Civita later verified that result.

In fact, the center-of-mass acceleration does exist, but only in nonconservative theories of gravity (Will, 1976a). The simplest way to derive this result is to treat the two-body system as a single composite "body" in otherwise empty space, and to focus on the self acceleration terms in the equation of motion (6.46a). Setting $\alpha_3 = 0$, dropping the fluid terms $\mathcal{E}^j$ and $\mathcal{P}^j$, specializing the remaining terms to two bodies and then averaging over one orbit, we obtain

$$\boldsymbol{a}_{\mathrm{CM}} = -\frac{1}{2}\zeta_2 \frac{m^2}{a^3} \eta \Delta \frac{e}{(1-e^2)^{3/2}} \boldsymbol{e}_{\mathrm{P}}, \tag{9.20}$$

where $\boldsymbol{e}_{\mathrm{P}}$ is a unit vector in the direction of the pericenter of body 1, $\eta = m_1 m_2/(m_1+m_2)^2$ and $\Delta = (m_2-m_1)/(m_1+m_2)$. A consequence of this acceleration would be non-vanishing values for d^2P/dt^2, where P denotes the period of any intrinsic process in the system (orbital period, spectral frequencies, pulsar periods). The observed upper limit on d^2P_{p}/dt^2 of the binary pulsar B1913+16 places a strong constraint on such an effect, resulting in the bound $|\zeta_2| < 4 \times 10^{-5}$ (Will, 1992b).

No feasible experiment or observation has ever been proposed that would set a direct limit on the PPN parameters ζ_1 or ζ_4. Note, however, that these parameters do appear in combination with other PPN parameters in observable effects, for example, ζ_1 in the Nordtvedt effect (see Section 8.1).

10 Structure and Motion of Compact Objects

The landscape for testing gravitational theory underwent a dramatic upheaval in September 1974, with the discovery by Hulse and Taylor of the Binary Pulsar (see Section 12.1 for an account of the discovery). Until then, all tests of relativistic gravitation were carried out under conditions of weak gravitational fields and slow motions, and were well described by the PPN framework of Chapter 4. Tests involving the orbits of bodies, such as those of Mercury and other planets around the Sun, could be safely studied by treating the planets as test bodies and calculating their geodesic motion; the notable exception to this was the Nordtvedt effect in lunar motion. And no tests prior to 1974 even remotely involved gravitational radiation.

The discovery of the binary pulsar B1913+16 changed all that. Here was a system consisting of a neutron star and a compact companion star of almost equal mass; each star moves in a curved spacetime generated both by the companion star and by itself, and both stars are moving at substantial speeds. Neutron stars are *not* weak-field objects; the self-gravitational binding energy of a neutron star can be as much as 20 percent of its rest mass. Thus the PPN framework no longer applies, or at least must be applied with great caution. And within a few weeks of the announcement of the discovery, it was realized that, with orbital velocities of order 200 km s^{-1}, the system was so relativistic that the dissipative effects on its orbit resulting from the emission of gravitational radiation might eventually be detectable. And indeed, by 1979, the inspiral of the orbit had been measured in an amount that agreed with the prediction of general relativity, providing the first evidence of the existence of gravitational radiation.

In this chapter and the two chapters that follow, we will focus on this new landscape for testing relativistic gravity. This chapter will deal with the structure and motion of bodies with strong internal gravity, such as neutron stars and black holes; "compact body" has become the standard term for such astronomical objects. This is known as the "strong-field" regime of gravity since the spacetime geometry inside and near such objects is characterized by $m/r \sim 1$, as compared with 10^{-6} near the solar surface. Chapter 11 will treat gravitational radiation in general relativity and alternative theories of gravity. This is frequently termed the "dynamical" regime of gravity, where motions are sufficiently rapid that the gravitational waves generated can be significant, both for direct detection, and for the back-reaction on the evolution of the system. Using the basic frameworks developed in these two chapters, we will describe in Chapter 12 specific strong-field and dynamical tests of relativistic gravity, including tests involving binary pulsars, tests using gravitational waves from inspiralling binaries of compact bodies, such as the waves detected by the LIGO/Virgo observatories beginning in 2015, and tests involving orbits of stars and matter outside black holes.

We will begin by discussing the structure of neutron stars and black holes in general relativity and in a range of alternative theories in Sections 10.1 and 10.2, and then will turn to their motion in Section 10.3.

10.1 Structure of Neutron Stars

Neutron stars were first suggested as theoretical possibilities in the 1930s (Baade and Zwicky, 1934). They are highly condensed stars where gravitational forces are sufficiently strong to crush atomic electrons together with the nuclear protons to form neutrons, to raise the density of matter above nuclear density ($\rho \geq 3 \times 10^{14}$ g cm^{-3}), and to cause the neutrons to be quantum-mechanically degenerate. At such high densities, exotic phenomena can occur, including superfluidity, strange quark matter, and quark-gluon plasmas. A typical neutron-star model has a mass between 0.7 and 2.5 $M_\odot$, and a radius of order 10 km. However, they remained just theoretical possibilities until the discovery of pulsars in 1967 (Hewish et al., 1968) and their subsequent interpretation as rotating neutron stars. There followed significant progress in calculating the structure of neutron stars using different nuclear equations of state, in obtaining firm estimates for the maximum mass of neutron stars, and in developing robust methods for modelling rapidly rotating neutron stars.

Today, with over 2300 pulsars known (Lorimer and Kramer, 2012), neutron stars are fully accepted as members of the astronomical zoo, and indeed comprise a number of different classes. There is the class of mostly isolated pulsars of periods between 0.1 and 10 s, representing the bulk of the population. The class of "recycled" pulsars with periods between one and 100 milliseconds, found mostly in binary systems, includes objects whose spin periods are so stable that they rival the best Earth-bound atomic clocks, yielding methods both for testing general relativity and for detecting gravitational waves. The class of accreting, X-ray emitting neutron stars has provided laboratories for studing the behavior of matter and magnetic fields under extreme conditions. The statistics of neutron stars also provide constraints on models of supernovae and on the evolution of binary systems containing massive stellar progenitors. Pulsars are regarded as important laboratories for testing theories of the equation of state of matter at ultrahigh densities. They could also be useful for testing alternative theories of gravity, and this section is devoted to discussing the structure of these objects in a range of theories.

In Newtonian gravitation theory, the equations of stellar structure for a static, spherically symmetric star composed of matter at zero temperature are given by

$$\begin{aligned} \frac{dp}{dr} &= \rho \frac{dU}{dr}, \quad \text{[Hydrostatic equilibrium]}, \\ p &= p(\rho), \quad \text{[Equation of state]}, \\ \frac{d}{dr}\left(r^2 \frac{dU}{dr}\right) &= -4\pi r^2 \rho, \quad \text{[Field equation]}, \end{aligned} \tag{10.1}$$

where $p(r)$ is the pressure, $\rho(r)$ is the density and $U(r)$ is the Newtonian gravitational potential. Note that $T = 0$ (more precisely, $T \ll$ the Fermi temperature of the degenerate matter in the star) is an adequate approximation for neutron-star matter for all but newly formed hot neutron stars; this results in the simple equation of state $p(\rho)$.

In any metric theory of gravity, it is simple to write down the equations corresponding to the first two of these three equations because they follow from the Einstein Equivalence Principle (Chapter 2), which states that in local freely falling frames the nongravitational laws of physics are those of special relativity, $T^{\mu\nu}{}_{,\nu} = 0, p = p(\rho)$. Thus, we have in any basis,

$$T^{\mu\nu}{}_{;\nu} = 0\,, \quad p = p(\rho)\,. \tag{10.2}$$

For a perfect fluid, $T^{\mu\nu} = (\rho + p)u^\mu u^\nu + pg^{\mu\nu}$, where we have lumped the internal energy $\rho\Pi$ into ρ (compare Eq. (3.77)).

It is useful to rewrite these equations in a form that parallels the first two parts of Eq. (10.1). For a static, spherically symmetric spacetime, there exists a coordinate system in which the metric has the form

$$ds^2 = -e^{2\Phi(r)}dt^2 + e^{2\Lambda(r)}dr^2 + e^{2\mu(r)}r^2 d\Omega^2\,, \tag{10.3}$$

where $d\Omega^2 = d\theta^2 + \sin^2\theta d\phi^2$ is the element of the two-sphere. For theories of gravity with a preferred frame, established by an auxiliary vector or tensor field, for example, the assumption of spherical symmetry is only valid for stars at rest with respect to the preferred frame. There exists the freedom to change the coordinate r by $r = g(\tilde{r})$. If the radial coordinate is chosen so that $\mu(r) = 0$, the coordinates are called "curvature coordinates." In such coordinates, $2\pi r$ measures the proper circumference of circles of constant r. In general relativity, they are known as Schwarzschild coordinates. If the radial coordinate is chosen so that $\mu(r) = \Lambda(r)$, the coordinates are called "isotropic coordinates."

Now, for hydrostatic equilibrium, the equations of motion $T^{\mu\nu}{}_{;\nu} = 0$ may be written in the form

$$p_{,j} = -(\rho + p)u^\nu u_{j;\nu}\,, \tag{10.4}$$

where j runs over r, θ, and ϕ. For spherical symmetry only the $j = r$ component is nontrivial, and using the fact that $\boldsymbol{u} = (e^{-\Phi(r)}, 0, 0, 0)$, we obtain

$$\frac{dp}{dr} = -(\rho + p)\frac{d\Phi}{dr}\,, \qquad p = p(\rho)\,. \tag{10.5}$$

Notice that in the Newtonian limit, $p \ll \rho$, $\Phi \simeq -U$ and we recover the first two of Eqs. (10.1). Eq. (10.5) are valid independently of the theory of gravity, assuming static spherical symmetry. The field equations for Φ, Λ, and μ, along with field equations for any auxiliary fields, will depend upon the theory.

In constructing a stellar model, boundary conditions must be imposed. One set imposes a continuity condition for the matter, and defines the stellar surface at radius R:

$$\left.\frac{dp}{dr}\right|_{r=0} = 0\,, \qquad p(R) = 0\,. \tag{10.6}$$

Assuming that the spacetime far from the neutron star can be approximated as asymptotically flat, the metric functions are constrained to behave as $r \to \infty$ according to [see Eq. (5.6)]

$$
\begin{aligned}
e^{2\Phi(r)} &\to c_0 - 2c_0c_1\frac{m}{r}\,, \\
e^{2\Lambda(r)} &\to c_1\,, \\
e^{2\mu(r)} &\to c_1\,.
\end{aligned}
\tag{10.7}
$$

These conditions guarantee that after transforming to asymptotically Lorentz coordinates, and converting to geometrized units ($G_{\rm today} = 1$), the metric will have the form

$$
g_{00} \to -1 + \frac{2m}{r}\,, \quad g_{0j} \to 0\,, \quad g_{jk} \to \delta_{jk}\,, \tag{10.8}
$$

and thus that the Kepler-measured mass of the star will be m. The asymptotic boundary conditions on the auxiliary fields of the theory are assumed to take the form, as $r \to \infty$, and in the preferred rest frame (if any):

$$
\begin{aligned}
\phi(r) &\to \phi_0\,, \\
K^\mu(r) &\to (K, 0, 0, 0)\,, \\
B^{\mu\nu}(r) &\to \mathrm{diag}(b_0, b_1, b_2r^2, b_2r^2\sin^2\theta)\,.
\end{aligned}
\tag{10.9}
$$

These boundary values ϕ_0, K, b_i could depend on time through cosmological evolution, or they could depend on the presence of nearby matter, such as a companion star in a binary system. The presence of nearby matter could affect the internal structure of the neutron star because of violations of the strong equivalence principle (see Section 3.3 for discussion). If the timescale over which these parameters change is long compared to the internal dynamical timescale of the neutron star, one can still approximate the star's structure as quasistatic, with its structure evolving adiabatically as the external parameters vary. If the timescales are comparable, such as near the end of an inspiral of two compact bodies, then one must resort to a fully dynamical calculation.

Another problem that must be confronted in building neutron star models in alternative theories of gravity is that real neutron stars are in motion because of the motion of the host galaxy relative to the cosmic background radiation, and in some cases because of orbital motion around a companion. This means that in theories of gravity with auxiliary vector or tensor fields, one *cannot* automatically assume spherical symmetry, since these fields will single out a preferred direction in space from the point of view of the neutron star's rest frame. Depending on the theory, it may be possible to treat these nonspherical effects as small perturbations of an underlying spherical model.

For a range of nuclear equations of state, one then can determine such important quantities as the mass-radius relation, the maximum mass, the surface gravitational redshift, the innermost stable circular orbit, and, for rotating neutron stars, the moment of inertia, the quadrupole moment and the Love numbers. These astrophysically relevant parameters could then be compared with available data in order to place constraints on the alternative theory (as well as on the equation of state).

Another important set of parameters, known as "sensitivities," encodes the degree to which such quantities as the mass, radius and moment of inertia vary when the parameters ϕ_0, K, and b_i vary. These quantities will play an important role in the equations of motion for compact bodies in alternative theories of gravity (see Section 10.3).

Unfortunately, this exhausts the generic features of the equations of relativistic stellar structure, so we must now turn to specific theories. Here we describe neutron stars in only a sampling of theories of gravity. Berti et al. (2015) give a more thorough review, covering a wide range of alternative theories.

General relativity

In curvature coordinates ($\mu(r) \equiv 0$), and in units where $G = 1$, the field equations (5.13) take the form

$$\frac{d}{dr}\left[r(1 - e^{-2\Lambda}\right] = 8\pi r^2 \rho ,$$

$$\frac{d\Phi}{dr} = \frac{m(r) + 4\pi r^3 p}{r\,(r - 2m(r))} . \tag{10.10}$$

The first equation has the solution

$$e^{2\Lambda} = \left(1 - \frac{2m(r)}{r}\right)^{-1} , \tag{10.11}$$

where

$$m(r) = 4\pi \int_0^r r'^2 \rho(r')dr' . \tag{10.12}$$

These equations, together with Eq. (10.5) and the boundary conditions (10.7) (with $c_0 = c_1 = 1$), are sufficient to calculate neutron-star models. Called the TOV (Tolman, Oppenheimer, Volkoff) equations for hydrostatic equilibrium, they are the standard tool for the study of nonrotating neutron stars within general relativity. For reviews, including models of rotating neutron stars, see Arnett and Bowers (1977), Baym and Pethick (1979), Salgado et al. (1994), Cook et al. (1994), Stergioulas (2003) and Lattimer and Prakash (2007). The maximum masses of neutron star models range from around $1.4\,M_\odot$ for soft equations of state to $2.7\,M_\odot$ for stiff equations of state. The discovery of pulsars with masses in excess of $2\,M_\odot$, coupled with the assumption that general relativity is correct, has made it possible to rule out a number of soft equations of state.

Scalar-tensor theories

Using curvature coordinates ($\mu(r) \equiv 0$), and *defining*

$$e^{2\Lambda} \equiv \left[1 - \frac{2m(r)}{r}\right]^{-1} , \tag{10.13}$$

we can put the field equations (5.35) for scalar-tensor theories into the form

$$\frac{dm}{dr} = \frac{4\pi G r^2}{\phi}\left[\rho - \frac{\rho - 3p}{3 + 2\omega} - \frac{1}{8\pi G}\left(1 - \frac{2m}{r}\right)\left\{\Phi_{,r}\phi_{,r} - \frac{{\phi_{,r}}^2}{2\phi}\left(\omega - \frac{2\omega'\phi}{3 + 2\omega}\right)\right\}\right] ,$$

$$\frac{d\Phi}{dr} = \frac{1}{(1 + r\phi_{,r}/2\phi)}\left[\frac{m + 4\pi G r^3 p/\phi}{r\,(r - 2m)} + \frac{\omega r {\phi_{,r}}^2}{4\phi^2} - \frac{\phi_{,r}}{\phi}\right] ,$$

$$\frac{d}{dr}\left(r^2 e^{\Phi - \Lambda}\frac{d\phi}{dr}\right) = -\frac{r^2 e^{\Phi - \Lambda}}{3 + 2\omega}\left[\frac{8\pi G r(\rho - 3p)}{r - 2m} + \omega' {\phi_{,r}}^2\right] , \tag{10.14}$$

where G is the gravitational coupling constant, and $\omega' \equiv d\omega/d\phi$. The present value of G is related to the asymptotic value of ϕ by (see Eq. (5.54)) by

$$G_{\text{today}} \equiv \frac{G}{\phi_0}\left(\frac{4+2\omega(\phi_0)}{3+2\omega(\phi_0)}\right) = 1. \tag{10.15}$$

For the special case of Brans-Dicke theory (ω = constant), neutron star models differ only slightly from those of general relativity for almost all values of $\omega > 0$ (Salmona, 1967; Hillebrandt and Heintzmann, 1974). For general scalar-tensor theories with ω' positive and with relatively large values of ω_0, the differences are typically of order $1/\omega_0$ (Zaglauer, 1992). However, within a class of scalar-tensor theories for which $A(\varphi) = e^{\beta_0\varphi^2/2}$ [see Eq. (5.38)], and for negative values $\beta_0 = 2\omega_0'\phi_0/(3+2\omega_0)^2 < -4.85$, a phenomenon known as "spontaneous scalarization" can occur, whereby nonlinear interactions between gravity and the scalar field can lead to large deviations of the scalar field inside the star from its values outside, resulting in dramatically different neutron-star structure (Damour and Esposito-Farèse, 1993, 1996, 1998). For a semi-analytic analysis of neutron-star structure in scalar-tensor theories, along with an extensive bibliography, see Horbatsch and Burgess (2011).

Neutron star models have also been studied in various versions of $f(R)$ and chameleon scalar-tensor theories (Kobayashi and Maeda, 2008; Upadhye and Hu, 2009; Babichev and Langlois, 2009, 2010; Cooney et al., 2010; Jaime et al., 2011).

Vector-Tensor Theories

Neutron-star models in Einstein-Æther theory were studied by Eling et al. (2007), extending earlier work by Eling and Jacobson (2006). The star was assumed to be isolated in an asymptotically flat spacetime and at rest relative to the preferred frame induced by the vector field. They first imposed the constraints of Eq. (5.65), so that $\alpha_1 = \alpha_2 = 0$. They restricted the parameters further by requiring that propagating modes of gravitational waves be stable, carry positive energy and have speeds ≥ 1. They assumed that the vector field had only a time component ($K_r = 0$). This is not strictly required by the assumption of a static system, so it may be overly restrictive. Since the vector field in Einstein-Æther theory has unit norm, this implies that $K^0(r) = e^{-\Phi(r)}$. In static spherical symmetry with K_r forced to vanish, it turns out that the constants c_2 and c_3 play no role, and that the only relevant parameter is c_4, or equivalently the combination c_{14}. Recall from Chapter 5 that $0 < c_{14} < 2$. Eling et al. (2007) found, for example that the maximum mass was in general a monotonically decreasing function of c_{14}, ranging from about $2.4\,M_\odot$ for $c_{14} = 0$ to about $1.8\,M_\odot$ for $c_{14} = 1.9$ for a stiff hadronic equation of state, and from $2.2\,M_\odot$ to about $1.7\,M_\odot$ for a medium hadronic equation of state. For any MIT bag model equation of state with strange quarks, and for a soft hadronic equation of state, the maximum mass was less than $2\,M_\odot$ for all values of c_{14}. The observation of neutron stars with masses $\geq 2\,M_\odot$ thus constrains both the equation of state and the parameter c_{14}. Yagi et al. (2014a) extended this to a broad range of equations of state compatible with $2M_\odot$ neutron stars, and also studied models in Khronometric theory.

Tensor-Vector-Scalar Theories

Neutron stars in TeVeS were first studied by Lasky et al. (2008). While that analysis used the original flawed version of TeVes (Skordis, 2008), Lasky (2009) showed that, for static spherical symmetry, with the constraint $K_r = 0$, the equations used are equivalent to those obtained from the improved TeVeS theory, as described in Section 5.5. The models depend on the asymptotic value ϕ_0 of the scalar field ($0 \leq \phi_0 \ll 1$), on the vector coupling constant c_{14}, and on the scalar coupling constant $k \sim 0.03$. They used a polytropic equation of state with $\Gamma = 2.46$, which leads to a maximum mass in general relativity of only $1.65\,M_\odot$ (this was before the discovery of a $2\,M_\odot$ neutron star). As in Einstein-Æther theory, the maximum mass was found to be a decreasing function of c_{14}, holding k and ϕ_0 fixed. For $c_{14} = 0.5$ and $\phi_0 = 0.003$, the maximum mass decreased slightly with increasing k. In the limit of small values of the three parameters c_{14}, k and ϕ_0, the models approached their general relativistic counterparts.

Quadratic gravity theories

Pani et al. (2011) studied neutron star models in a class of quadratic theories described by the action

$$I = \int \Big[f_0(|\phi|)R + f_1(|\phi|)R^2_{\rm GB} + f_4(|\phi|)\, {}^*RR - \gamma(|\phi|)g^{\mu\nu}\partial_\mu\phi^*\partial_\nu\phi - V(\phi)\Big](-g)^{1/2}d^4x + I_{\rm NG}(q_A, A^2(|\phi|)g_{\mu\nu}) , \tag{10.16}$$

where $R^2_{\rm GB}$ is the Gauss-Bonnet invariant, given by

$$R^2_{\rm GB} = R^2 - 4R_{\alpha\beta}R^{\alpha\beta} + R_{\alpha\beta\gamma\delta}R^{\alpha\beta\gamma\delta} , \tag{10.17}$$

and ϕ is potentially a complex scalar field. The use of $R^2_{\rm GB}$ in the action (10.16) guarantees that the field equations contain no more than two derivatives of the metric (compare with Eq. (5.77)). They then specialized to Einstein-Dilaton-Gauss-Bonnet gravity, with ϕ real, $f_0 = (16\pi G)^{-1}$, and $V(\phi) = 0$. In spherical symmetry, the Chern-Simons term ${}^*RR$ makes no contribution, so f_4 can be set equal to zero. Finally, they used a parametrization $f_1(\phi) = (\alpha/16\pi)e^{\beta\phi}$ where α and β are constants, with α having dimensions of (length)^2. They chose boundary conditions in which the metric was asymptotically flat, and in which $\phi \to 0$ as $r \to \infty$. They found that $\phi \ll 1$ throughout the neutron star, and that the maximum mass was a decreasing function of the product $\alpha\beta$. For a reasonably stiff equation of state, the maximum mass ranged from $2.4\,M_\odot$ for small values of $\alpha\beta$, corresponding to the general relativistic limit, to $0.6\,M_\odot$ for $\alpha\beta = 100M^2_\odot$. Recall that the "natural" scale for α is $M^2_{\rm Planck} = 10^{-76}M^2_\odot$.

While spherical neutron stars are not affected by the Chern-Simons term, rotating neutron stars are. Slowly rotating neutron star models in dynamical Chern-Simons theory were constructed by Yunes et al. (2010), Ali-Haïmoud and Chen (2011), and Yagi et al. (2013). Unfortunately, modifications to the maximum mass were found to be degenerate with the equation of state. The stars develop a "dipole scalar charge" that can affect orbital motions in binary pulsars, but the effects were found to be too small to provide meaningful constraints on the theory (Yagi et al., 2013).

10.2 Structure of Black Holes

In a certain sense, black hole theory has a longer history than neutron-star theory, as it dates back to a suggestion by the Reverend John Michell (1784) that, according to Newtonian gravity, objects might exist in which the escape velocity from the surface could exceed the speed of light. Laplace made the same suggestion independently in 1796 (Laplace, 1808, 1799). Remarkably, Michell went on argue that such a body, though invisible, could be detected by measuring the motion of a companion star that might be in orbit around it. Michell had already used a sophisticated statistical analysis of William Herschel's data on close associations of stellar pairs and triples to conclude that some of them *must* be self-gravitating systems. It would take 20 more years for Herschel to confirm the existence of binary stars. For a history of the ideas of Michell and Laplace concerning black holes, see Montgomery et al. (2009).

Within general relativity, two key events in the history of black holes were the discovery of the Schwarzschild metric (Schwarzschild, 1916) and the analysis of gravitational collapse across the Schwarzschild horizon (Oppenheimer and Snyder, 1939). However, theoretical black hole physics really came into its own with the discovery of the true nature of the "event horizon" of Schwarzschild's solution (Finkelstein, 1958; Kruskal, 1960) and the discovery in 1963 of the Kerr metric (Kerr, 1963), now known to be the unique solution for a stationary, vacuum, rotating black hole (with the Schwarzschild metric being the limiting case corresponding to no rotation). Other important developments in the theory of black holes include the full understanding of event horizons as nonsingular, causal boundaries between an exterior spacetime and an interior from which there is no escape, the establishment of black-hole uniqueness and stability, the formulation of singularity theorems, the development of black-hole accretion theory, and the discovery of black-hole thermodynamics and Hawking radiation. For a history, see Israel (1987).

It was the discovery in 1971 of the rapid variations of the X-ray source Cygnus X-1 by telescopes aboard the UHURU satellite that took black holes out of the realm of pure theory (Bolton, 1972; Webster and Murdin, 1972). The source of X-rays was observed to be in a binary system with the companion star HDE 226868; analysis of the nature of the companion and of its orbit around the X-ray source, and detailed study of the X-rays, led to the conclusion that the unseen body was a compact object with a mass exceeding $9\,M_\odot$. Since the maximum masses of white dwarfs and neutron stars are approximately $1.4\,M_\odot$ and $\sim 3\,M_\odot$, respectively, the simplest conclusion was that the object was a black hole. The source of the X-rays was believed to be the hot, inner regions of an accretion disk around the black hole, formed by gas stripped from the atmosphere of the companion star. Since 1971, many black hole candidates in X-ray binary systems have been found (Celotti et al., 1999).

Studies of the central regions of quasars, galaxies and globular clusters have indicated that supermassive black holes may be ubiquitous (see Kormendy and Richstone (1995) for an early review); indeed, every massive spiral galaxy appears to contain a central massive black hole. There is even a $4.3 \times 10^6\, M_\odot$ black hole in the center of the Milky Way, the precise measurement of its mass being made possible by observations of stars in orbit around it (Eckart and Genzel, 1996; Ghez et al., 1998). Although the masses of such black holes are trivial compared to the total mass of their host galaxies, their apparent ability to generate extraordinary mass and energy outflow from accreting matter can have important feedback effects on the structure of the galaxy, including such phenomena as star formation.

The capstone to the remarkable history of black holes is the September 14, 2015 detection by LIGO of gravitational waves from the inspiral and merger of two black holes of masses 29 and $36\, M_\odot$ (Abbott et al., 2016c).

General relativity predicts the existence of black holes. Black holes are the end products of catastrophic gravitational collapse in which the collapsing matter crosses an event horizon, a surface whose radius depends upon the mass and angular momentum of matter that has fallen across it, and which is a one-way membrane for timelike or null world lines. Such world lines can cross the horizon moving inward but not outward. The interior of the black hole is causally disconnected from the exterior spacetime. There is considerable evidence to support the claim that any gravitational collapse involving sufficient mass, whether spherically symmetric or not, with zero net charge and zero net angular momentum, results in a black hole whose metric (at late times after the black hole has become stationary) is the Schwarzschild metric, given by

$$ds^2 = -\left(1 - \frac{2M}{r}\right) dt^2 + \left(1 - \frac{2M}{r}\right)^{-1} dr^2 + r^2 \left(d\theta^2 + \sin^2\theta d\phi^2\right) , \qquad (10.18)$$

where M is the mass of the black hole. If the collapsing body has net rotation, the black hole is described by the Kerr metric.

While curved spacetime is essential to the existence of event horizons as one-way membranes for the physical interactions, the existence of black holes is not an automatic byproduct of curved spacetime, but depends on the specific theory of gravity.

The general approach for finding black hole solutions in a given theory is the same as for neutron stars, except that one assumes vacuum everywhere. In general relativity, solving Eqs. (10.10) with $\rho = p = 0$ leads immediately to Eq. (10.18). To see if a particular vacuum solution corresponds to a black hole, one must check whether a non-singular event horizon exists, and whether any true physical singularities in the spacetime are safely hidden behind the event horizon. In some theories, different fundamental speeds or "null cones" may be associated with different fields (see Section 11.3), and thus there may be more than one "event" horizon.

Once again, we provide a mere sampling of black hole structure in alternative theories of gravity; see Berti et al. (2015) for a more thorough review, including various attempts to "parametrize" black holes in a manner similar to the PPN formalism.

Scalar-tensor theories

As one might expect, scalar-tensor theories, being in some sense the least violent modification of general relativity, predict black holes. However, what is unexpected is that they predict black holes whose geometry is *identical* to the Schwarzschild geometry of general relativity. Roger Penrose was probably the first to conjecture this, in a 1970 talk at the Fifth Texas Symposium on Relativistic Astrophysics. Motivated by this remark, Thorne and Dykla (1971) showed that during gravitational collapse to form a black hole, the Brans-Dicke scalar field is radiated away, leaving only its constant asymptotic value and a Schwarzschild black hole. Hawking (1972) argued on general grounds that the scalar field ϕ must be constant throughout the exterior of the horizon, given by its asymptotic cosmological value ϕ_0. Thus, the vacuum field equation (5.35a) for the metric is Einstein's vacuum field equation, and the solution is the Schwarzschild solution. The scalar field has no effect other than to determine the value of the gravitational constant $G_{\rm local}$. (This result also holds for rotating and charged black holes.)

In Brans-Dicke theory for instance, the most direct way to verify this (Hawking, 1972) is to use the vacuum field equation for $\varphi \equiv \phi - \phi_0$, $\Box\varphi = 0$, and to integrate the quantity $\varphi\Box\varphi$ over the exterior of the horizon between two spacelike hypersurfaces at different values of coordinate time. After an integration by parts, we obtain

$$\int \varphi_{,\alpha}\varphi^{,\alpha}\sqrt{-g}d^4x - \frac{1}{2}\int \left(\varphi^2\right)^{,\alpha} d\Sigma_\alpha = 0\,. \tag{10.19}$$

Now the surface integrals over the spacelike hypersurfaces cancel because the situation is stationary. The integral over the hypersurface at infinity vanishes because $\varphi \sim r^{-1}$ asymptotically. The integral over the horizon vanishes because $d\Sigma_\alpha$ is parallel to the generators of the horizon and is thus in a direction generated by the symmetry transformations of the black hole (Killing direction) whereas $(\varphi^2)_{,\alpha}$ vanishes along that symmetry direction. Thus

$$\int \varphi_{,\alpha}\varphi^{,\alpha}\sqrt{-g}d^4x = 0\,. \tag{10.20}$$

But $\varphi_{,\alpha}$ is spacelike since φ is stationary, so $\varphi_{,\alpha}\varphi^{,\alpha} \geq 0$ everywhere. Equation (10.20) thus implies that $\varphi_{,\alpha} = 0$. Sotiriou and Faraoni (2012) extended Hawking's theorem to the class of $f(R)$ theories that can be transformed into generalized scalar-tensor theories.

These theorems can be evaded, for example, if the scalar field is time dependent (e.g., because of cosmic evolution) or complex, or if it is massive and the black hole is rotating (see Berti et al. (2015) for a review).

Vector-tensor theories

Black-hole solutions have been studied extensively in Einstein-Æther theory and Hořava gravity (Eling and Jacobson, 2006; Tamaki and Miyamoto, 2008; Barausse et al., 2011; Barausse and Sotiriou, 2013a). Barausse et al. (2011) obtained numerical black hole solutions for a range of values of the two parameters c_+ and c_- between zero and unity (the equations are so complicated that no analytic solutions were found). They evaluated such observable or invariant quantities as the angular velocity of a particle at the innermost stable circular orbit (ISCO), the angular velocity of the circular photon orbit, the maximum redshift of a photon emitted by a particle on the ISCO, and the proper circumference of the outer event horizon in units of the black hole mass. For $c_+ = c_- = 0$, the solutions were numerically equivalent to the Schwarzschild solution, as expected. These invariant quantities were insensitive to the value of c_-, and depended fairly weakly on c_+, deviating from the general relativistic values by between five and 15 percent for $c_+ \sim 0.6$, reaching deviations of 50 percent for $c_+ \sim 0.9$. In Hořava gravity, the deviations from general relativity were consistently smaller in most of the parameter space. Slowly rotating black hole solutions have also been analyzed (Barausse and Sotiriou, 2013b; Barausse et al., 2016).

Tensor-vector-scalar theories

Black-hole solutions in TeVeS were studied by Giannios (2005) and Lasky (2009). For a black hole at rest in the preferred frame, and for the restricted case in which the vector field has only a time component, the physical metric turns out to be identical to the Schwarzschild metric; in the unrestricted case ($K_r \neq 0$) only the asymptotic properties of the metric and fields were studied.

Quadratic gravity theories

Non-rotating black hole solutions in a range of quadratic gravity theories were studied with a view toward possible observational signatures by Pani and Cardoso (2009), Yunes and Stein (2011) and Kleihaus et al. (2011); for a brief review, see Blázquez-Salcedo et al. (2017). In Chern-Simons theory, spherically symmetric black holes are identical to their general relativistic counterparts, but rotating black holes are not. Slowly rotating black hole models were analysed by Yunes and Pretorius (2009), Konno et al. (2009) and Yagi et al. (2012).

10.3 The Motion of Compact Objects

In Chapter 6, we derived the N-body equations of motion for massive, self-gravitating bodies within the PPN framework [see Eqs. (6.45) and (6.46)]. A key assumption that went into that analysis was that the weak-field, slow-motion limit of gravitational theory applied everywhere, in the interiors of the bodies as well as between them. This assumption restricted the applicability of the equations of motion to systems such as the solar system. However, when dealing with systems containing a neutron star or a black hole with highly relativistic spacetimes near or inside them, one can no longer apply the assumptions of the post-Newtonian limit everywhere, except possibly in the interbody region between the relativistic bodies. Instead, one must employ a method for deriving equations of motion for compact objects that, within a chosen theory of gravity, involves (a) solving the full, relativistic equations for the regions inside and near each body, (b) solving the post-Newtonian equations in the interbody region, and (c) matching these solutions in an appropriate way in a "matching region" surrounding each body. This matching presumably leads to constraints on the motions of the bodies (as characterized by suitably defined centers of mass); these constraints would be the sought-after equations of motion. Such a procedure amounts to a generalization of the Einstein-Infeld-Hoffmann (EIH) approach (Einstein et al., 1938).

Let us first ask what would be expected from such an approach within general relativity. In the fully weak-field post-Newtonian limit, we found that the motion of post-Newtonian bodies is independent of their internal structure, that is, there is no Nordtvedt effect. Each body moves on a geodesic of the post-Newtonian interbody metric generated by the other bodies, with proper allowance for post-Newtonian terms contributed by its own interbody field. This is the EIH result.

It turns out however, that this structure independence seems to be valid even when the bodies are highly compact (neutron stars or black holes). The only restriction is that they be quasistatic, nearly spherical, and sufficiently small compared to their separations that tidal interactions may be neglected. This would be a bad approximation for a neutron star about to spiral into a stellar-mass black hole, for example, but is a good approximation for binary pulsars or for the inspiral of a binary system of compact bodies before the last few orbits.

Although this conclusion has not been proven rigorously, a strong argument for its plausibility can be presented by considering in more detail the matching procedure discussed earlier. We first note that the solution for the relativistic structure and gravitational field of each body is independent of the interbody gravitational field, since we can always choose a coordinate system for each body that is freely falling and approximately Minkowskian in the matching region and in which the body is at rest. Thus, there is no way for the external fields to influence the body or its field, provided that we can neglect tidal effects due to inhomogeneities of the interbody field across the interior of the matching region. Only the velocity and acceleration of the body are affected. Now, if the body is static and spherically symmetric to sufficient accuracy, its external gravitational field is characterized only by its Kepler-measured mass m and is independent of its internal structure; the field is given essentially by the Schwarzschild metric. Thus, the matching procedure described above must yield the same result, whether the body is a black hole of mass m or a post-Newtonian body of mass m. In the latter case, the result is the EIH equations of motion (see Section 6.2), so those equations must be valid in all cases. A slightly different way to see this is to note that because the local field of the body in the freely falling frame is spherically symmetric, depends only on the constant mass m, and is unaffected by the external geometry, the acceleration of the body in the freely falling frame must vanish, so its trajectory must be a geodesic of some metric. The metric to be used is a post-Newtonian interbody metric that includes post-Newtonian terms contributed by the body itself, but that excludes self-fields. This was shown for nonrotating black holes in a seminal paper by D'Eath (1975), and was extended to a variety of contexts by Rudolph and Börner (1978a,b), Kates (1980), Thorne and Hartle (1985), Anderson (1987), Itoh et al. (2000), and Taylor and Poisson (2008). This structure independence was also verified explicitly to 2PN order for self-gravitating fluid bodies by Mitchell and Will (2007). For a pedagogical treatment of this approach at 1PN order, see PW, section 9.4.

A key element of these derivations is the validity of the Strong Equivalence Principle within general relativity (see Chapter 3 for discussion), which guarantees that the structure of each body is independent of the surrounding gravitational environment. By contrast, most alternative theories of gravity possess additional gravitational fields, whose values in the matching region can influence the structure of each body, and thereby can affect its motion. Consider as a simple example a theory with an additional scalar field (scalar-tensor theory). In the local freely falling coordinates, although the interbody metric is Minkowskian up to tidal terms, the scalar field has a value $\phi_0(t)$. In a solution for the structure of the body, this boundary value of $\phi_0(t)$ could influence the resulting mass, for example, by controlling the locally measured value of G, thus turning the mass into a function $m(\phi_0)$ of the scalar field. Thus, the asymptotic metric of the body in the matching region may depend upon its internal structure via the dependence of m on ϕ_0 (essentially, the matching conditions will depend upon m, $dm/d\phi, \ldots$). Furthermore, the acceleration of the body in the freely falling frame need not be zero, as we saw in Sections 2.4 and 3.3. If the mass-energy of a body varies as a result of a variation in an external parameter, we found, using cyclic gedanken experiments that assumed only conservation of energy, that (see Eqs. (3.97) and (3.99))

$$\boldsymbol{a} - \boldsymbol{a}_{\text{geodesic}} \sim \frac{1}{m}\boldsymbol{\nabla} E_{\text{B}}(\boldsymbol{X}, \boldsymbol{V}) \sim \frac{\partial \ln m}{\partial \phi}\boldsymbol{\nabla}\phi . \tag{10.21}$$

Thus, the bodies need not follow geodesics of any metric, rather their motion may depend on their internal structure.

In practice, the matching procedure described above is very cumbersome (D'Eath, 1975). Within general relativity, a simpler method for obtaining the EIH equations of motion is to treat each body as a "point" mass of inertial mass m_a and to solve Einstein's equations using a point-mass matter action or energy-momentum tensor, with proper care to neglect or regularize infinite "self" fields. In the action for general relativity, we thus write

$$I = \frac{1}{16\pi G}\int R\sqrt{-g}\, d^4x - \sum_a m_a \int d\tau_a , \tag{10.22}$$

where τ_a is proper time along the world line of the ath body. By solving the field equations to 1PN order, it is then possible to derive straightforwardly from the matter action an N-body EIH action in the form

$$I_{\text{EIH}} = \int L(\boldsymbol{x}_1, \ldots \boldsymbol{x}_N, \boldsymbol{v}_1, \ldots \boldsymbol{v}_N)dt , \tag{10.23}$$

with a Lagrangian L written purely in terms of the variables $(\boldsymbol{x}_a, \boldsymbol{v}_a)$ of the bodies. The result is Eq. (6.81), with the PPN parameters set to their general relativistic values. The N-body EIH equations of motion are then given by

$$\frac{d}{dt}\frac{\partial L}{\partial v_a^j} - \frac{\partial L}{\partial x_a^j} = 0, \quad a = 1, \ldots N . \tag{10.24}$$

In alternative theories of gravity, we assume that the only difference is the possible dependence of the mass on the boundary values of the auxiliary fields. In the quasi-Newtonian limit (Sections 2.4 and 3.3), this was sufficient to yield the complete quasi-Newtonian acceleration of composite bodies including modifications (Nordtvedt effect) due to their internal structure. Thus, following the suggestion of Eardley (1975), we merely replace the constant inertial mass m_a in the matter action with the variable inertial mass $m_a(\psi_A)$, where ψ_A represents the values of the external auxiliary fields, evaluated at the body (we neglect their variation across the interior of the matching region), with infinite self-field contributions excluded. The functional dependence of m_a upon the variable ψ_A will depend on the nature and structure of the body. Thus, we write the action of the alternative theory in the form

$$I = I_G - \sum_a \int m_a\left(\psi_A[\mathbf{x}_a(\tau_a)]\right) d\tau_a\,, \tag{10.25}$$

where I_G is the action for the metric and auxiliary fields ψ_A. In varying the action with respect to the fields $g_{\mu\nu}$ and ψ_A, the variation of m_a must now be taken into account. In the post-Newtonian limit, where the fields ψ_A are expanded about asymptotic values $\psi_A^{(0)}$ according to $\psi_A = \psi_A^{(0)} + \delta\psi_A$, it is generally sufficient to expand $m_a(\psi_A)$ in the form

$$m_a(\psi_A) = m_a(\psi_A^{(0)}) + \sum_A \frac{\partial m_a}{\partial \psi_A^{(0)}}\delta\psi_A + \frac{1}{2}\sum_{A,B} \frac{\partial^2 m_a}{\partial \psi_A^{(0)}\partial \psi_B^{(0)}}\delta\psi_A\delta\psi_B + \dots . \tag{10.26}$$

Thus, the final form of the metric and of the N-body Lagrangian will depend on $m_a \equiv m_a(\psi_A^{(0)})$ and on the parameters $\partial m_a/\partial\psi_A^{(0)}$, and so on. We will use the term "sensitivities" to describe these parameters, since they measure the sensitivity of the inertial mass to changes in the fields ψ_A. Thus, we define

$$\begin{aligned} s_a^{(A)} &\equiv \frac{\partial \ln m_a}{\partial \ln \psi_A^{(0)}}\,, \\ s_a'^{(AB)} &\equiv \frac{\partial^2 \ln m_a}{\partial \ln \psi_A^{(0)}\partial \ln \psi_B^{(0)}}\,, \end{aligned} \tag{10.27}$$

and so on, where the derivatives are typically taken holding the total baryon number of the body fixed, for bodies made of matter. For black holes, some other method must be used to identify the sensitivities. Similar sensitivities can be defined for the radius and moment of inertia of the compact body. The final result is a "modified EIH formalism."

Gralla (2010, 2013) developed a more general theory of the motion of "small" bodies characterized by such parameters as mass, spin and charge in an environment of external fields, and argued that Eardley's ansatz is a special case of that general framework.

10.3.1 A modified EIH formalism

By analogy with the PPN formalism, a general EIH formalism can be constructed using arbitrary parameters whose values depend both on the theory under study and on the nature of the bodies in the system. However, to keep the resulting formalism simple, we will make some restrictions. First, we restrict attention to semiconservative theories of gravity; every Lagrangian-based metric theory of gravity falls into this class. We will also restrict attention to theories of gravity that have no Whitehead term in the post-Newtonian limit; all of the currently viable theories described in Chapter 5 satisfy this constraint. This generalizes the formalism presented in the first edition of this book, which made a restriction to fully conservative theories. Nordtvedt (1985) discussed a similar formalism for compact-body dynamics.

Each body is characterized by an inertial mass m_a, defined to be the quantity that appears in the conservation laws for energy and momentum that emerge from the EIH Lagrangian. We then write for the metric, valid in the interbody region and far from the system,

$$\begin{aligned} g_{00} &= -1 + 2\sum_a \alpha_a^* \frac{m_a}{|\boldsymbol{x} - \boldsymbol{x}_a|} + O(\epsilon^2) , \\ g_{0j} &= O(\epsilon^{3/2}) , \\ g_{jk} &= \left(1 + 2\sum_a \gamma_a^* \frac{m_a}{|\boldsymbol{x} - \boldsymbol{x}_a|}\right) \delta_{jk} + O(\epsilon^2) , \end{aligned} \tag{10.28}$$

where α_a^* and γ_a^* are functions of the parameters of the theory and of the structure of the ath body. For test-body geodesics in this metric, the quantities $\alpha_a^* m_a$, and $\sum_a \alpha_a^* m_a$ are the Kepler-measured active gravitational masses of the individual bodies and of the system as a whole. In general relativity, $\alpha_a^* \equiv \gamma_a^* \equiv 1$.

To obtain the modified EIH Lagrangian, we first generalize the post-Newtonian semi-conservative N-body Lagrangian, Eq. (6.81), in a natural way, to obtain

$$\begin{aligned} L_{\text{EIH}} &= -\sum_a m_a \left[1 - \frac{1}{2} v_a^2 - \frac{1}{8} (1 + \mathcal{A}_a) v_a^4\right] \\ &+ \frac{1}{2} \sum_a \sum_{b \neq a} \frac{m_a m_b}{r_{ab}} \left[\mathcal{G}_{ab} + 3\mathcal{B}_{ab} v_a^2 - \frac{1}{2} \left(\mathcal{G}_{ab} + 6\mathcal{B}_{(ab)} + \mathcal{C}_{ab}\right) \boldsymbol{v}_a \cdot \boldsymbol{v}_b \right. \\ &\left. - \frac{1}{2} (\mathcal{G}_{ab} + \mathcal{E}_{ab}) (\boldsymbol{v}_a \cdot \boldsymbol{n}_{ab})(\boldsymbol{v}_b \cdot \boldsymbol{n}_{ab})\right] \\ &- \frac{1}{2} \sum_a \sum_{b \neq a} \sum_{c \neq a} \mathcal{D}_{abc} \frac{m_a m_b}{r_{ab}} \frac{m_c}{r_{ac}} , \end{aligned} \tag{10.29}$$

where $\boldsymbol{n}_{ab} = \boldsymbol{x}_{ab}/r_{ab}$. Notice that we did not introduce a parameter in front of the v_a^2 term in Eq. (10.29). Any such parameter can always be absorbed into a new definition of the inertial mass m_a' of body a. We are then free to change the constant term $-\sum_a m_a$ to be the sum of the new inertial masses $-\sum_a m_a'$. This has no effect on the equations of motion, but does allow the Hamiltonian derived from L_{EIH} to be the sum of the new inertial masses at lowest order. We also did not include a term of the form $(m_a m_b / r_{ab})(\boldsymbol{v}_b \cdot \boldsymbol{n}_{ab})^2$; such a term can be associated (via a total time derivative in the Lagrangian) with the Whitehead term in Eq. (6.81), which we have chosen to reject.

The quantities $\mathcal{A}_a$, $\mathcal{G}_{ab}$, $\mathcal{B}_{ab}$, $\mathcal{C}_{ab}$, $\mathcal{E}_{ab}$ and $\mathcal{D}_{abc}$ are functions of the parameters of the theory and of the structure of each body, and satisfy

$$\mathcal{G}_{ab} = \mathcal{G}_{(ab)}, \quad \mathcal{C}_{ab} = \mathcal{C}_{(ab)}, \quad \mathcal{E}_{ab} = \mathcal{E}_{(ab)}, \quad \mathcal{D}_{abc} = \mathcal{D}_{a(bc)} . \tag{10.30}$$

Note that $\mathcal{B}_{ab}$ has no special symmetry, in general.

In general relativity, $\mathcal{G}_{ab} = \mathcal{B}_{ab} = \mathcal{D}_{abc} = 1$, while $\mathcal{A}_a = \mathcal{C}_{ab} = \mathcal{E}_{ab} = 0$. In the post-Newtonian limit of semiconservative theories (with $\xi = 0$), for structureless masses (no self-gravity), the parameters have the values [compare Eq. (6.81)]

$$\begin{aligned} &\mathcal{G}_{ab} = 1 , \quad \mathcal{B}_{ab} = \frac{1}{3}(2\gamma + 1) , \quad \mathcal{D}_{abc} = 2\beta - 1 , \\ &\mathcal{A}_a = 0 , \quad \mathcal{C}_{ab} = \alpha_1 - \alpha_2 , \quad \mathcal{E}_{ab} = \alpha_2 . \end{aligned} \tag{10.31}$$

In the fully conservative case, including contributions of the self-gravitational binding energies of the bodies, the parameters have the values

$$\mathcal{G}_{ab} = 1 + (4\beta - \gamma - 3)\left(\frac{\Omega_a}{m_a} + \frac{\Omega_b}{m_b}\right),$$
$$\mathcal{B}_{ab} = \frac{1}{3}(2\gamma + 1), \quad \mathcal{D}_{abc} = 2\beta - 1,$$
$$\mathcal{A}_a = \mathcal{C}_{ab} = \mathcal{E}_{ab} = 0, \tag{10.32}$$

where Ω_a is the self-gravitational energy of the ath body.

In a theory of gravity that singles out a preferred frame, such as the Einstein-Æther theory, the Lagrangian (10.29) must be understood to be defined in that frame. To obtain the Lagrangian in a moving frame, we must carry out a post-Galilean transformation of the Lagrangian (see Section 4.3 for discussion). We make a low-velocity Lorentz transformation (post-Galilean transformation) from $(t, \boldsymbol{x})$ to $(\tau, \boldsymbol{\xi})$ coordinates, given by

$$\boldsymbol{x} = \boldsymbol{\xi} + \left(1 + \frac{1}{2}w^2\right)\boldsymbol{w}\tau + \frac{1}{2}(\boldsymbol{\xi}\cdot\boldsymbol{w})\boldsymbol{w} + \xi \times O(\epsilon^2),$$
$$t = \tau\left(1 + \frac{1}{2}w^2 + \frac{3}{8}w^4\right) + \left(1 + \frac{1}{2}w^2\right)\boldsymbol{\xi}\cdot\boldsymbol{w} + \tau \times O(\epsilon^3). \tag{10.33}$$

From the transformation (10.33), we have

$$\boldsymbol{v}_a = \boldsymbol{\nu}_a + \boldsymbol{w} - \frac{d\boldsymbol{\nu}_a}{d\tau}(\boldsymbol{\xi}_a\cdot\boldsymbol{w}) - \frac{1}{2}w^2\boldsymbol{\nu}_a - (\boldsymbol{\nu}_a\cdot\boldsymbol{w})\left(\boldsymbol{\nu}_a + \frac{1}{2}\boldsymbol{w}\right),$$
$$\frac{1}{r_{ab}} = \frac{1}{\xi_{ab}}\left[1 + \frac{1}{2}(\boldsymbol{w}\cdot\boldsymbol{n}'_{ab})^2 + \frac{1}{\xi_{ab}}\left((\boldsymbol{w}\cdot\boldsymbol{\xi}_a)(\boldsymbol{\nu}_a\cdot\boldsymbol{n}'_{ab}) - (\boldsymbol{w}\cdot\boldsymbol{\xi}_b)(\boldsymbol{\nu}_b\cdot\boldsymbol{n}'_{ab})\right)\right], \tag{10.34}$$

where $\boldsymbol{\xi}_a$, $\boldsymbol{\xi}_b$ and $\boldsymbol{\nu}_a \equiv d\boldsymbol{\xi}_a/d\tau$ are to be evaluated at the same time, given by a clock at the spatial origin ($\boldsymbol{\xi} = 0$) of the moving coordinate system, and $\boldsymbol{n}'_{ab} \equiv \boldsymbol{\xi}_{ab}/\xi_{ab}$. The new Lagrangian is given by

$$L(\boldsymbol{\xi}, \tau) = L(\boldsymbol{x}, t)\frac{dt}{d\tau} - \frac{df}{d\tau}, \tag{10.35}$$

where $dt/d\tau$ is evaluated at $\boldsymbol{\xi} = 0$, and where we are free to subtract a total time derivative of a function f to simplify the new Lagrangian. Substituting these results into Eqs. (10.29) and (10.35), dropping constants and total time derivatives, and replacing ξ_{ab} and $\boldsymbol{\nu}_a$ with r_{ab} and $\boldsymbol{v}_a$, we obtain the Lagrangian in the moving frame

$$L = L_{\rm EIH} + \frac{1}{4}\sum_a m_a\mathcal{A}_a\left[v_a^2 w^2 + 2v_a^2(\boldsymbol{v}_a\cdot\boldsymbol{w}) + 2(\boldsymbol{v}_a\cdot\boldsymbol{w})^2\right]$$
$$-\frac{1}{4}\sum_{a\neq b}\frac{m_a m_b}{r_{ab}}\left\{\mathcal{C}_{ab}w^2 - 2\left(6\mathcal{B}_{[ab]} - \mathcal{C}_{ab}\right)(\boldsymbol{v}_a\cdot\boldsymbol{w})\right.$$
$$\left. + \mathcal{E}_{ab}\left[(\boldsymbol{w}\cdot\boldsymbol{n}_{ab})^2 + 2(\boldsymbol{w}\cdot\boldsymbol{n}_{ab})(\boldsymbol{v}_a\cdot\boldsymbol{n}_{ab})\right]\right\}. \tag{10.36}$$

Notice that the Lagrangian is post-Galilean invariant if and only if

$$\mathcal{A}_a \equiv \mathcal{B}_{[ab]} \equiv \mathcal{C}_{ab} \equiv \mathcal{E}_{ab} \equiv 0. \tag{10.37}$$

These quantities are the preferred-frame parameters of our modified EIH formalism.

Since our ultimate goal is to apply this formalism to binary systems containing compact objects, let us now restrict attention to two-body systems. We obtain from L the two-body equations of motion

$$
\begin{aligned}
\boldsymbol{a}_1 = & -\frac{m_2 \boldsymbol{n}}{r^2}\Big\{\mathcal{G}_{12} - (3\mathcal{G}_{12}\mathcal{B}_{12} + \mathcal{D}_{122})\frac{m_2}{r} \\
& -\frac{1}{2}\left[2\mathcal{G}_{12}^2 + 6\mathcal{G}_{12}\mathcal{B}_{(12)} + 2\mathcal{D}_{211} + \mathcal{G}_{12}(\mathcal{C}_{12} + \mathcal{E}_{12})\right]\frac{m_1}{r} \\
& +\frac{1}{2}\left[3\mathcal{B}_{12} - \mathcal{G}_{12}(1 + \mathcal{A}_1)\right] v_1^2 + \frac{1}{2}(3\mathcal{B}_{21} + \mathcal{G}_{12} + \mathcal{E}_{12}) v_2^2 \\
& -\frac{1}{2}\left(6\mathcal{B}_{(12)} + 2\mathcal{G}_{12} + \mathcal{C}_{12} + \mathcal{E}_{12}\right)\boldsymbol{v}_1 \cdot \boldsymbol{v}_2 - \frac{3}{2}\left(\mathcal{G}_{12} + \mathcal{E}_{12}\right)(\boldsymbol{n}\cdot\boldsymbol{v}_2)^2 \\
& +\frac{1}{2}\left(\mathcal{C}_{12} + \mathcal{G}_{12}\mathcal{A}_1\right) w^2 + \frac{1}{2}\left(\mathcal{C}_{12} - 6\mathcal{B}_{[12]} + \mathcal{E}_{12} + 2\mathcal{G}_{12}\mathcal{A}_1\right)\boldsymbol{v}_1 \cdot \boldsymbol{w} \\
& +\frac{1}{2}\left(\mathcal{C}_{12} + 6\mathcal{B}_{[12]} - \mathcal{E}_{12}\right)\boldsymbol{v}_2 \cdot \boldsymbol{w} + \frac{3}{2}\mathcal{E}_{12}\left[(\boldsymbol{w}\cdot\boldsymbol{n})^2 + 2(\boldsymbol{w}\cdot\boldsymbol{n})(\boldsymbol{v}_2\cdot\boldsymbol{n})\right]\Big\} \\
& + \frac{m_2 \boldsymbol{v}_1}{r^2}\boldsymbol{n}\cdot\left\{\left[3\mathcal{B}_{12} + \mathcal{G}_{12}(1 + \mathcal{A}_1)\right]\boldsymbol{v}_1 - 3\mathcal{B}_{12}\boldsymbol{v}_2 + \mathcal{G}_{12}\mathcal{A}_1 \boldsymbol{w}\right\} \\
& - \frac{1}{2}\frac{m_2 \boldsymbol{v}_2}{r^2}\boldsymbol{n}\cdot\Big\{\left(6\mathcal{B}_{(12)} + 2\mathcal{G}_{12} + \mathcal{C}_{12} + \mathcal{E}_{12}\right)\boldsymbol{v}_1 \\
& \quad - \left(6\mathcal{B}_{(12)} + \mathcal{C}_{12} - \mathcal{E}_{12}\right)\boldsymbol{v}_2 + 2\mathcal{E}_{12}\boldsymbol{w}\Big\} \\
& - \frac{1}{2}\frac{m_2 \boldsymbol{w}}{r^2}\boldsymbol{n}\cdot\Big\{\left(\mathcal{C}_{12} - 6\mathcal{B}_{[12]} + \mathcal{E}_{12} - 2\mathcal{G}_{12}\mathcal{A}_1\right)\boldsymbol{v}_1 - \left(\mathcal{C}_{12} - 6\mathcal{B}_{[12]} - \mathcal{E}_{12}\right)\boldsymbol{v}_2 \\
& \quad -2\left(\mathcal{G}_{12}\mathcal{A}_1 - \mathcal{E}_{12}\right)\boldsymbol{w}\Big\}\,, \\
\boldsymbol{a}_2 = & \{1 \rightleftharpoons 2;\, \boldsymbol{n} \to -\boldsymbol{n}\}\,,
\end{aligned} \tag{10.38}
$$

where $\boldsymbol{a}_a \equiv d\boldsymbol{v}_a/dt$, $\boldsymbol{x} \equiv \boldsymbol{x}_1 - \boldsymbol{x}_2$, $r \equiv |\boldsymbol{x}|$, and $\boldsymbol{n} \equiv \boldsymbol{x}/r$. If the conditions of Eq. (10.37) hold, then the preferred-frame terms vanish.

To Newtonian order, the center of mass $\boldsymbol{X} = (m_1\boldsymbol{x}_1 + m_2\boldsymbol{x}_2)/m$ of the system is unaccelerated, thus, to sufficient accuracy in the post-Newtonian terms, we can set $\boldsymbol{X} = \dot{\boldsymbol{X}} = 0$ and write

$$
\begin{aligned}
\boldsymbol{x}_1 &= \left[\frac{m_2}{m} + O(\epsilon)\right]\boldsymbol{x}\,, \\
\boldsymbol{x}_2 &= -\left[\frac{m_1}{m} + O(\epsilon)\right]\boldsymbol{x}\,.
\end{aligned} \tag{10.39}
$$

As in Section 6.4.1, we define

$$
\begin{aligned}
&\boldsymbol{v} \equiv \boldsymbol{v}_1 - \boldsymbol{v}_2\,, \quad \boldsymbol{a} \equiv \boldsymbol{a}_1 - \boldsymbol{a}_2\,, \\
&m \equiv m_1 + m_2\,, \quad \eta \equiv \frac{m_1 m_2}{m^2}\,, \quad \Delta = \frac{m_2 - m_1}{m}\,.
\end{aligned} \tag{10.40}
$$

We also define the coefficients

$$
\begin{aligned}
&\mathcal{G} \equiv \mathcal{G}_{12}\,, \quad \mathcal{B}_+ \equiv \mathcal{B}_{(12)}\,, \quad \mathcal{B}_- \equiv \mathcal{B}_{[12]}\,, \quad \mathcal{D} \equiv \frac{m_2}{m}\mathcal{D}_{122} + \frac{m_1}{m}\mathcal{D}_{211}\,, \\
&\mathcal{C} \equiv \mathcal{C}_{12}\,, \quad \mathcal{E} \equiv \mathcal{E}_{12}\,, \quad \mathcal{A}^{(n)} \equiv \left(\frac{m_2}{m}\right)^n \mathcal{A}_1 - \left(-\frac{m_1}{m}\right)^n \mathcal{A}_2\,.
\end{aligned} \tag{10.41}
$$

Then the equation of motion for the relative orbit takes the form

$$\boldsymbol{a} = \boldsymbol{a}_{\rm L} + \boldsymbol{a}_{\rm PF}\,, \tag{10.42}$$

where the purely two-body, or "local" contributions have the form (we use "hats" to denote parameters associated with compact bodies)

$$\boldsymbol{a}_{\rm L} = -\frac{\mathcal{G}m\boldsymbol{n}}{r^2} + \frac{m}{r^2}\left[\boldsymbol{n}\left(\hat{A}_1 v^2 + \hat{A}_2 \dot{r}^2 + \hat{A}_3\frac{m}{r}\right) + \dot{r}\boldsymbol{v}\hat{B}_1\right], \tag{10.43}$$

where

$$\begin{aligned}
\hat{A}_1 &= \frac{1}{2}\left\{\mathcal{G}(1-6\eta) - 3\mathcal{B}_+ - 3\Delta\mathcal{B}_- - \eta(\mathcal{C}+2\mathcal{E}) + \mathcal{G}\mathcal{A}^{(3)}\right\},\\
\hat{A}_2 &= \frac{3}{2}\eta(\mathcal{G}+\mathcal{E}),\\
\hat{A}_3 &= \mathcal{G}\left[2\eta\mathcal{G} + 3\mathcal{B}_+ + \eta\,(\mathcal{C}+\mathcal{E}) + 3\Delta\mathcal{B}_-\right] + \mathcal{D},\\
\hat{B}_1 &= \mathcal{G}(1-2\eta) + 3\mathcal{B}_+ + 3\Delta\mathcal{B}_- + \eta\mathcal{C} + \mathcal{G}\mathcal{A}^{(3)},
\end{aligned} \tag{10.44}$$

and the preferred-frame contributions have the form

$$\begin{aligned}
\boldsymbol{a}_{\rm PF} = +\frac{m}{r^2}\Bigg\{&\boldsymbol{n}\left[\left(\frac{1}{2}\hat{\alpha}_1 + 2\mathcal{G}\mathcal{A}^{(2)}\right)(\boldsymbol{w}\cdot\boldsymbol{v}) + \frac{3}{2}\left(\hat{\alpha}_2 + \mathcal{G}\mathcal{A}^{(1)}\right)(\boldsymbol{w}\cdot\boldsymbol{n})^2\right]\\
&-\boldsymbol{w}\left[\frac{1}{2}\hat{\alpha}_1(\boldsymbol{n}\cdot\boldsymbol{v}) + \hat{\alpha}_2(\boldsymbol{n}\cdot\boldsymbol{w})\right] + \mathcal{G}\mathcal{A}^{(2)}\boldsymbol{v}(\boldsymbol{n}\cdot\boldsymbol{w})\Bigg\},
\end{aligned} \tag{10.45}$$

where

$$\begin{aligned}
\hat{\alpha}_1 &= \Delta\,(\mathcal{C}+\mathcal{E}) - 6\mathcal{B}_- - 2\mathcal{G}\mathcal{A}^{(2)},\\
\hat{\alpha}_2 &= \mathcal{E} - \mathcal{G}\mathcal{A}^{(1)}.
\end{aligned} \tag{10.46}$$

These two parameters play the role of compact-body analogues of the PPN parameters α_1 and α_2 in the context of binary systems.

In the Newtonian limit of the orbital motion, we have $\boldsymbol{a} = \mathcal{G}m\boldsymbol{x}/r^3$, with Keplerian orbit solutions given by Eq. (6.73), but with m replaced by $\mathcal{G}m$. The post-Newtonian terms in Eq. (10.43) can be inserted into the Lagrange planetary equations (6.74). Using the methods outlined in Section 6.4.2, we find that the pericenter advance per orbit is given by

$$\Delta\omega = \frac{6\pi m}{p}\mathcal{P}\mathcal{G}^{-1}, \tag{10.47}$$

where $p = a(1-e^2)$, and where

$$\mathcal{P} = \mathcal{G}\mathcal{B}_+ + \frac{1}{6}\left(\mathcal{G}^2 - \mathcal{D}\right) + \frac{1}{6}\mathcal{G}\left[6\Delta\mathcal{B}_- + \eta\mathcal{G}(2\mathcal{C}+\mathcal{E}) + \mathcal{G}\mathcal{A}^{(3)}\right]. \tag{10.48}$$

This is the only secular perturbation produced by the post-Newtonian terms in Eq. (10.43). In the PPN limit, this result agrees with Eq. (6.78) for semiconservative theories.

We can convert $\Delta\omega$ to a rate by dividing by the orbital period and can use $P_b = 2\pi(a^3/\mathcal{G}m)^{1/2}$, to eliminate a to obtain

$$\langle\dot{\omega}\rangle = \frac{6\pi}{P_b(1-e^2)}\left(\frac{2\pi m}{P_b}\right)^{2/3}\mathcal{P}\mathcal{G}^{-4/3}. \tag{10.49}$$

This expression will be useful for discussing binary pulsar measurements.

We can also use the preferred-frame contributions to the equations of motion $\boldsymbol{a}_{\rm PF}$ to work out the secular perturbations of the orbit elements of a binary system, extending the analysis of Section 8.2 to systems with compact bodies. The results are identical to Eqs. (8.2), (8.33), (8.35), and (8.37), with $\xi = 0$, and with $\hat{\alpha}_1$ and $\hat{\alpha}_2$ replacing α_1 and α_2, respectively.

Returning to the full N-body Lagrangian (10.29), and working at quasi-Newtonian order, we can derive the leading effects in a hierarchical three-body system, such as the pulsar triple J0337+1715, including the Nordtvedt effect. The resulting equation of motion for the inner binary is given by Eqs. (8.10)–(8.12), with

$$\begin{aligned} m^* &= \mathcal{G}_{12} m , \\ \alpha &= \frac{1}{2} \left(\mathcal{G}_{13} + \mathcal{G}_{23} \right) , \\ \delta &= \mathcal{G}_{13} - \mathcal{G}_{23} . \end{aligned} \tag{10.50}$$

As an illustration of this modified EIH framework for compact bodies, we will again focus on specific theories where calculations have been carried out. As we have already discussed, a variety of approaches have shown that the EIH equations of motion for compact objects within general relativity are identical to those of the post-Newtonian limit with weak fields everywhere. In other words, in general relativity, $\mathcal{G}_{ab} \equiv \mathcal{B}_{ab} \equiv \mathcal{D}_{abc} \equiv 1$, and the remaining coefficients vanish, independently of the nature of the bodies.

10.3.2 Scalar-tensor theories

The modified EIH formalism was first developed by Eardley (1975) for application to Brans-Dicke theory. It makes use of the fact that only the scalar fleld ϕ produces an external influence on the structure of each compact body via its boundary values in the matching region. This boundary value of ϕ is related to the local value of the gravitational constant as felt by the compact body by

$$G_{\rm local} = \frac{G}{\phi} \left(\frac{4 + 2\omega}{3 + 2\omega} \right) , \tag{10.51}$$

where G is the fundamental gravitational coupling constant. Thus we will treat the inertial mass m_a of each body as a being a function of ϕ. Then, by defining the deviation of ϕ from its asymptotic value ϕ_0 as in Section 5.3.2, $\phi \equiv \phi_0(1 + \Psi)$, we can write down the expansion

$$m_a(\phi) = m_a \left[1 + s_a \Psi + \frac{1}{2} (s_a^2 + s_a' - s_a) \Psi^2 + O(\Psi^3) \right] , \tag{10.52}$$

where $m_a \equiv m_a(\phi_0)$, and we define the dimensionless sensitivities

$$\begin{aligned} s_a &\equiv \left(\frac{d \ln m_a(\phi)}{d \ln \phi} \right)_0 , \\ s_a' &\equiv \left(\frac{d^2 \ln m_a(\phi)}{d (\ln \phi)^2} \right)_0 . \end{aligned} \tag{10.53}$$

The action for the theory is then written

$$I = \frac{1}{16\pi G}\int\left[\phi R - \frac{\omega(\phi)}{\phi}g^{\mu\nu}\phi_{,\mu}\phi_{,\nu}\right]\sqrt{-g}d^4x - \sum_a \int m_a(\phi)d\tau_a\,, \tag{10.54}$$

where the integrals over proper time τ_a are to be taken along the world line of each body a. We have dropped the scalar self-interaction potential $U(\phi)$. It is straightforward to vary the action with respect to $g_{\mu\nu}$ and ϕ to obtain the field equations,

$$G_{\mu\nu} = \frac{8\pi G}{\phi}T_{\mu\nu} + \frac{\omega(\phi)}{\phi^2}\left(\phi_{,\mu}\phi_{,\nu} - \frac{1}{2}g_{\mu\nu}\phi_{,\lambda}\phi^{,\lambda}\right)$$
$$+ \frac{1}{\phi}\left(\phi_{;\mu\nu} - g_{\mu\nu}\Box_g\phi\right)\,, \tag{10.55a}$$

$$\Box_g\phi = \frac{1}{3+2\omega(\phi)}\left(8\pi GT - 16\pi G\phi\frac{\partial T}{\partial\phi} - \frac{d\omega}{d\phi}\phi_{,\lambda}\phi^{,\lambda}\right)\,, \tag{10.55b}$$

where

$$T^{\mu\nu} = (-g)^{-1/2}\sum_a m_a(\phi)u_a^\mu u_a^\nu (u_a^0)^{-1}\delta^3(\mathbf{x}-\mathbf{x}_a)\,, \tag{10.56}$$

where u_a^μ is the four-velocity of body a. The equations of motion take the form

$$T^{\mu\nu}{}_{;\nu} - \frac{\partial T}{\partial\phi}\phi^{,\nu} = 0\,. \tag{10.57}$$

In harmonic gauge, the relaxed field equations for $\tilde{h}^{\alpha\beta}$ are given by Eqs. (5.43), while for the scalar field, the relaxed field equation is given by Eqs. (5.45) and (5.46), but with the replacement

$$\tilde{T} \to \frac{\phi_0^2}{\phi^2}\left(T - 2\phi\frac{\partial T}{\partial\phi}\right)\,. \tag{10.58}$$

where $T = T^{\mu\nu}g_{\mu\nu}$.

Repeating the post-Newtonian calculation using the relaxed field equations, as described in Section 5.3.2, we obtain, to lowest order

$$\Psi = 2\zeta\sum_b \frac{m_b}{r_b}(1-2s_b) + O(\epsilon^2)\,,$$

$$g_{00} = -1 + 2\sum_b \frac{m_b}{r_b}(1-2\zeta s_b) + O(\epsilon^2)\,,$$

$$g_{0j} = -4(1-\zeta)\sum_b \frac{m_b v_b^j}{r_b} + O(\epsilon^{5/2})\,,$$

$$g_{jk} = \delta_{jk}\left[1 + 2\sum_b \frac{m_b}{r_b}(1-2\zeta+2\zeta s_b)\right]\,, \tag{10.59}$$

where $\zeta = 1/(4+2\omega_0)$, $r_b = |\boldsymbol{x}-\boldsymbol{x}_b|$, and we have chosen units in which $G_{\rm local} = 1$. For the explicit $O(\epsilon^2)$ terms in g_{00} and Ψ, see Mirshekari and Will (2013). Notice that the active gravitational mass as measured by test-body Keplerian orbits far from each body is given by

$$(m_A)_a = m_a(1 - 2\zeta s_a) . \tag{10.60}$$

To compare with the post-Newtonian limit, we must connect the sensitivity s_a to the self-gravitational energy Ω_a, defined in Box 6.2. The gravitational energy is actually proportional to $G_{\rm local}$. Thus if the total inertial mass of a weak-field, gravitationally bound body is $m = m_0 + \mathcal{T} + G_{\rm local}\Omega + E_{\rm int}$, where $\mathcal{T}$ and $E_{\rm int}$ are the kinetic and internal energies, respectively, then $\Omega = dm/dG_{\rm local}$. We can then express the sensitivity s for a given body in the form

$$\begin{aligned} s_a &= \left(\frac{d\ln m_a(\phi)}{d\ln G_{\rm local}}\right)_0 \left(\frac{d\ln G_{\rm local}}{d\ln\phi}\right)_0 , \\ &= -(1+2\lambda)\frac{\Omega_a}{m_a} , \end{aligned} \tag{10.61}$$

where we used Eq. (10.51), including the fact that ω also depends on ϕ, and used the definition (see Eq. (5.55)),

$$\lambda \equiv \frac{\phi_0\omega_0'}{(3+2\omega_0)(4+2\omega_0)} . \tag{10.62}$$

The active mass in the weak field limit is then given by

$$(m_A)_a = m_a\left[1 + 2\zeta(1+2\lambda)\frac{\Omega_a}{m_a}\right] , \tag{10.63}$$

in agreement with Eq. (6.59).

From the complete post-Newtonian solution for $g_{\mu\nu}$ and Ψ, we can obtain the matter action for the ath body, given by

$$I_a = -\int m_a(\phi)\left(-g_{00} - 2g_{0j}v_a^j - g_{jk}v_a^j v_a^k\right)^{1/2} dt . \tag{10.64}$$

To obtain an N-body action in the form of Eq. (10.23), we first make the gravitational terms in I_a, manifestly symmetric under interchange of all pairs of particles, then take one of each such term generated in I_a, and sum the result over a. The resulting N-body Lagrangian then has the form of Eq. (10.29) with

$$\begin{aligned} \mathcal{G}_{ab} &= 1 - 2\zeta\left(s_a + s_b - 2s_a s_b\right) , \\ \mathcal{B}_{ab} &= \frac{1}{3}\left[\mathcal{G}_{ab} + 2(1-\zeta)\right] , \\ \mathcal{D}_{abc} &= \mathcal{G}_{ab}\mathcal{G}_{ac} + 2\zeta(1-2s_b)(1-2s_c)\left[\lambda(1-2s_a) + 2\zeta s_a'\right] , \\ \mathcal{A}_a &= \mathcal{B}_{[ab]} = \mathcal{C}_{ab} = \mathcal{E}_{ab} = 0 . \end{aligned} \tag{10.65}$$

The quasi-Newtonian equations of motion obtained from the EIH Lagrangian are

$$\boldsymbol{a}_a = -\sum_{b\neq a}\frac{m_b\boldsymbol{x}_{ab}}{r_{ab}^3}\left[1 - 2\zeta\left(s_a + s_b - 2s_a s_b\right)\right] . \tag{10.66}$$

For bodies with weak internal fields the product term $s_a s_b$ may be neglected, and the acceleration may be written

$$\boldsymbol{a}_a = -\frac{(m_{\rm P})_a}{m_a} \sum_{b \neq a} \frac{(m_{\rm A})_b \boldsymbol{x}_{ab}}{r_{ab}^3} , \tag{10.67}$$

where $(m_{\rm A})_b$ is given by Eq. (10.60) and where

$$(m_{\rm P})_a = m_a(1 - 2\zeta s_a) , \tag{10.68}$$

in agreement with our results of Section 6.3, Eq. (6.59). However, if the bodies are sufficiently compact that $s_a \sim s_b \sim 1$, then because of the product term $s_a s_b$, it is impossible to describe the quasi-Newtonian equations simply in terms of active and passive masses of individual bodies.

Mirshekari and Will (2013) extended this analysis to higher post-Newtonian orders, obtaining the equations of motion for compact binary systems through 2.5PN order. At 1PN order, the equations have the form

$$\frac{d^2\boldsymbol{x}}{dt^2} = -\frac{\mathcal{G}m}{r^2} \left[\boldsymbol{n}\,(1 - A_{\rm PN}) - \dot{r}\boldsymbol{v}B_{\rm PN}\right] , \tag{10.69}$$

where

$$\begin{aligned} A_{\rm PN} &= -(1 + 3\eta + \bar{\gamma})v^2 + \frac{3}{2}\eta\dot{r}^2 + 2(2 + \eta + \bar{\gamma} + \bar{\beta}_+ + \Delta\bar{\beta}_-)\frac{\mathcal{G}m}{r} , \\ B_{\rm PN} &= 2(2 - \eta + \bar{\gamma}) , \end{aligned} \tag{10.70}$$

where the coefficients $\bar{\gamma}$, $\bar{\beta}_+$, and $\bar{\beta}_-$ are defined by

$$\begin{aligned} \bar{\gamma} &\equiv -2\mathcal{G}^{-1}\zeta(1 - 2s_1)(1 - 2s_2) , \\ \bar{\beta}_\pm &\equiv \frac{1}{2}\left(\bar{\beta}_1 \pm \bar{\beta}_2\right) , \\ \bar{\beta}_1 &\equiv \mathcal{G}^{-2}\zeta(1 - 2s_2)^2\left[\lambda(1 - 2s_1) + 2\zeta s_1'\right] , \\ \bar{\beta}_2 &\equiv \mathcal{G}^{-2}\zeta(1 - 2s_1)^2\left[\lambda(1 - 2s_2) + 2\zeta s_2'\right] . \end{aligned} \tag{10.71}$$

Roughly speaking, the sensitivity $s \sim$ [self-gravitational binding energy]/[mass], so $s_\oplus \sim 10^{-10}$, $s_\odot \sim 10^{-6}$, $s_{\rm white\ dwarf} \sim 10^{-4}$. For neutron stars, whose equation of state is of the form $p = p(\rho)$, a model is uniquely determined (for a given function $\omega(\phi)$) by the local value of ϕ_0 (or $G_{\rm local}$) and by the central density ρ_c, in other words $m = m(G, \rho_c)$ and $N = N(G, \rho_c)$, where N is the total baryon number. The sensitivity can be expressed in terms of G as

$$s_{\rm NS} = -(1 + 2\lambda)\left(\frac{\partial \ln m}{\partial \ln G}\right)_N . \tag{10.72}$$

However, in any theory of gravity in which the local gravitational structure depends upon a single external parameter whose role is that of a gravitational "constant," it can be shown that

$$\left(\frac{\partial \ln m}{\partial \ln G}\right)_N = -\left(\frac{\partial \ln m}{\partial \ln G}\right)_{\rho_c} + \left(\frac{\partial \ln m}{\partial \ln N}\right)_G \left(\frac{\partial \ln N}{\partial \ln G}\right)_{\rho_c} . \tag{10.73}$$

For fixed central density and fixed equation of state, a simple scaling argument reveals that m and N scale as $G^{-3/2}$, so that

$$s_{NS} = \frac{3}{2}(1 + 2\lambda)\left[1 - \left(\frac{\partial \ln m}{\partial \ln N}\right)_G\right]. \tag{10.74}$$

Note that $(\partial m/\partial N)_G$ is the injection energy per baryon.

For the special case of Brans-Dicke theory, Salmona (1967) showed that neutron star models differed very little from their general relativistic counterparts, for values of ω as small as 2.5. Using general relativistic neutron star models Eardley (1975) obtained sensitivities ranging from 0.07 to 0.39 for masses ranging from $0.5M_\odot$ to $1.41M_\odot$ (the maximum mass for the 1975-era equation of state used). Zaglauer (1992) obtained sensitivities for neutron stars with masses between about $1.1M_\odot$ and $1.46M_\odot$ for the B, O and M equations of state tabulated by Arnett and Bowers (1977). The sensitivities ranged from 0.19 to 0.30 for the soft B equation of state, and from 0.08 to 0.12 for the stiff M equation of state.

Similar results hold for general scalar-tensor theories with ω_0' positive (corresponding to positive values of the Damour-Esposito-Farèse parameter β_0). However, when ω_0' is sufficiently negative (corresponding to $\beta_0 < -4$), spontaneous scalarization can occur, leading to neutron star models whose structure is very different from those in general relativity, with dramatically different values of the sensitivities (Damour and Esposito-Farèse, 1993, 1996). In the DEF approach (see Section 5.3), the relevant parameter is $\alpha^2 \sim (1 - 2s)/(4 + 2\omega_0)$. For example, for a theory corresponding to a value $\beta_0 = -6$, with asymptotic conditions on the theory chosen to agree with the 1993 solar-system bound of $\omega_0 > 1500$, they found $\alpha^2 \sim 0.4$, corresponding to $s \sim -600$! As we will see in Chapter 12, binary pulsar observations make it possible to place very tight bounds on this regime of scalar-tensor theories.

For black holes, we found in Section 10.2 that the scalar field is constant in the exterior of the black hole. Thus from Eq. (10.59) we must have $s_{BH} = 1/2$. Another way to see this is to note that, because all information about the matter that formed the black hole has vanished behind the event horizon, the only scale on which the mass of the hole can depend is the Planck scale, $m_{Planck} = (\hbar c/G)^{1/2}$, and thus $m_{BH} \propto m_{Planck} \propto G^{-1/2} \propto \phi^{1/2}$, and hence $s_{BH} = 1/2$. Substituting $s_a = s_b = s_c = 1/2$ in Eqs. (10.65), we obtain $\mathcal{G}_{ab} = \mathcal{B}_{ab} = 1 - \zeta$ and $\mathcal{D}_{abc} = (1 - \zeta)^2$. Then it is straightforward to see that if we replace each mass in the Lagrangian (10.29) by $m/(1 - \zeta) = m(4 + 2\omega_0)/(3 + 2\omega_0)$, then apart from a meaningless overall mutiplicative factor of $(1 - \zeta)^{-1}$, the Lagrangian is *identical* to that for general relativity. Since black-hole masses cannot be measured by any means other than via their orbital dynamics, this implies that, at 1PN order, the motion of a system of black holes in scalar-tensor theory is indistinguishable from the motion in general relativity. In fact, Mirshekari and Will (2013) showed that this is valid through 2.5PN order.

For binary systems containing one black hole, the equations are again identical to those of general relativity through 1PN order. However, at 1.5PN order (where dipole radiation reaction begins) and beyond, the equations differ from those of general relativity by terms that depend on the single parameter $Q = \zeta(1-\zeta)^{-1}(1-2s_1)^2$, where s_1 is the sensitivity of the companion to the black hole (see Mirshekari and Will (2013) for details).

10.3.3 Constrained vector-tensor theories

Because the norm of the vector field in Einstein-Æther and Khronometric theories is constrained to be -1, the structure of a spherically symmetric compact body at rest with rest to the preferred rest frame does not depend on it. However it could depend on the time component K^0, or more properly on the invariant quantity $\boldsymbol{K} \cdot \boldsymbol{u}$, where $\boldsymbol{u}$ is the body's four-velocity. (The structure of a rotating body could also depend on the projection of $\boldsymbol{K}$ along the body's spin axis, but here we will focus on non-rotating bodies.) We define for a body with four-velocity $\boldsymbol{u}$,

$$\gamma \equiv -K^\mu u_\mu \equiv 1 + \Psi \,, \tag{10.75}$$

where we assume that far from the system, for a test body at rest, $\boldsymbol{K} \cdot \boldsymbol{u} = -1$. We define the sensitivities

$$\begin{aligned} s_a &\equiv \left(\frac{d \ln m_a(\gamma)}{d \ln \gamma} \right)_{\gamma=1} , \\ s'_a &\equiv \left(\frac{d^2 \ln m_a(\gamma)}{d (\ln \gamma)^2} \right)_{\gamma=1} , \end{aligned} \tag{10.76}$$

where the derivatives are to be taken holding baryon number fixed. Then in terms of Ψ, we can expand

$$m_a(\gamma) = m_a \left[1 + s_a \Psi + \frac{1}{2} \left(s_a^2 + s'_a - s_a \right) \Psi^2 + O(\Psi^3) \right] . \tag{10.77}$$

With this assumption, Foster (2007) and Yagi et al. (2014a) derived the metric and equations of motion to post-Newtonian order for systems of compact bodies in Einstein-Æther theory. The metric is given by

$$\begin{aligned} g_{00} &= -1 + 2U - 2U^2 - 2\Phi_2 + 3\Phi_{1s} + O(\epsilon^3) , \\ g_{0j} &= g^j + O(\epsilon^{5/2}) , \\ g_{jk} &= (1 + 2U)\delta_{jk} + O(\epsilon^2) , \end{aligned} \tag{10.78}$$

and the vector field is given by

$$\begin{aligned} K^0 &= (-g_{00})^{-1/2} + O(\epsilon^3) , \\ K^j &= k^j + O(\epsilon^{5/2}) , \end{aligned} \tag{10.79}$$

where

$$U = \sum_b \frac{G_N m_b}{r_b} , \quad \Phi_2 = \sum_{b,c} \frac{G_N^2 m_b m_c}{r_b r_{bc}} , \quad \Phi_{1s} = \sum_b \frac{G_N m_b}{r_b} v_b^2 (1 - s_b) ,$$

$$g^j = \sum_b \frac{G_N m_b}{r_b} \left[B_b^- v_b^j + B_b^+ n_b^j (\boldsymbol{n}_b \cdot \boldsymbol{v}_b) \right] ,$$

$$k^j = \sum_b \frac{G_N m_b}{r_b} \left[C_b^- v_b^j + C_b^+ n_b^j (\boldsymbol{n}_b \cdot \boldsymbol{v}_b) \right] , \tag{10.80}$$

where $G_N = 2G/(2 - c_{14})$ and $c_{14} = c_1 + c_4$. The quantities $B_b^\pm$ and $C_b^\pm$ are complicated expressions involving the constants c_1, c_2, c_3, and c_4 of Einstein-Æther theory and the sensitivities s_b (see Eqs. (23) and (29) of Foster (2007)).

From Eqs. (10.78) and (10.79), we obtain

$$\Psi(\boldsymbol{x}_a) = \frac{1}{2} v_a^2 + \frac{3}{8} v_a^4 + 2 v_a^2 U(\boldsymbol{x}_a) - v_a^j k^j(\boldsymbol{x}_a) + O(\epsilon^3) . \tag{10.81}$$

Note that, when $v_a^j = 0$, $\Psi = 0$, resulting in no dependence of the inertial mass on the vector field. Writing the action for the ath body as

$$I_a = - \int m_a(\Psi) \left(-g_{00} - 2 g_{0j} v_a^j - g_{jk} v_a^j v_a^k \right)^{1/2} dt , \tag{10.82}$$

we expand to post-Newtonian order, make the action manifestly symmetric under interchange of all pairs of particles, select one of each term, and sum over a. After rescaling each mass by $m_a \to m_a/(1 - s_a)$ and replacing the constant term in the Lagrangian by the sum of the rescaled masses, we obtain the modified EIH Lagrangian in the form of Eq. (10.29), with

$$\begin{aligned}
\mathcal{G}_{ab} &= \frac{G_N}{(1 - s_a)(1 - s_b)} , \\
\mathcal{B}_{ab} &= \mathcal{G}_{ab}(1 - s_a) , \\
\mathcal{D}_{abc} &= \mathcal{G}_{ab}\mathcal{G}_{ac}(1 - s_a) , \\
\mathcal{A}_a &= s_a - \frac{s'_a}{1 - s_a} , \\
\mathcal{C}_{ab} &= \mathcal{G}_{ab} \left[\alpha_1 - \alpha_2 + 3 (s_a + s_b) - \mathcal{Q}_{ab} - \mathcal{R}_{ab} \right] , \\
\mathcal{E}_{ab} &= \mathcal{G}_{ab} \left[\alpha_2 + \mathcal{Q}_{ab} - \mathcal{R}_{ab} \right] ,
\end{aligned} \tag{10.83}$$

where

$$\begin{aligned}
\mathcal{Q}_{ab} &= \frac{1}{2} \left(\frac{2 - c_{14}}{2c_+ - c_{14}} \right) (\alpha_1 - 2\alpha_2)(s_a + s_b) + \frac{2 - c_{14}}{c_{123}} s_a s_b , \\
\mathcal{R}_{ab} &= \frac{8 + \alpha_1}{4c_1} \left[c_- (s_a + s_b) + (1 - c_-) s_a s_b \right] .
\end{aligned} \tag{10.84}$$

Here α_1 and α_2 are the PPN parameters of Einstein-Æther theory, given by Eqs. (5.63), $c_\pm = c_1 \pm c_3$, and $c_{123} = c_1 + c_2 + c_3$. The two-body equations of motion that follow from the Lagrangian with these coefficients agree with Eq. (33) of Foster (2007) (after correcting a sign and a parenthesis in Eqs. (34) and (35) of that paper).

Note that the Lagrangian is in general not Lorentz invariant, and therefore will exhibit preferred-frame effects. Even when the parameters c_i are constrained so as to enforce $\alpha_1 = \alpha_2 = 0$, making the dynamics Lorentz invariant in the fully weak field post-Newtonian limit, the dynamics of compact bodies will still be dependent on the overall motion of the system via the motion-induced sensitivities of the bodies.

The corresponding equations in Khronometric theory can be obtained from these by setting $c_1 = -\epsilon$, $c_2 = \lambda_K$, $c_3 = \beta_K + \epsilon$, and $c_4 = \alpha_K + \epsilon$, and taking the limit $\epsilon \to \infty$ (see Yagi et al. (2014a)). The parameters of the modified EIH Lagrangian are given by Eq. (10.83), but now with $G_N = 2G/(2 - \alpha_K)$, and

$$\mathcal{Q}_{ab} = \frac{1}{\beta_K + \lambda_K} \left[(\alpha_K + \beta_K + 3\lambda_K)(s_a + s_b) + (2 - \alpha_K) s_a s_b \right] ,$$
$$\mathcal{R}_{ab} = \frac{1}{2}(8 + \alpha_1) \left[s_a + s_b - s_a s_b \right] , \tag{10.85}$$

where α_1 is the PPN parameter of Khronometric theory, given by Eq. (5.68).

Yagi et al. (2014a) calculated neutron star sensitivities in both Einstein-Æther and Khronometric theories. In order to do so, it was necessary to construct models for neutron stars moving uniformly relative to the preferred frame. From Eqs. (10.77) and (10.81), assuming uniform motion with no external bodies ($U = k^j = 0$), the sensitivity is given by $s = v^{-1} d \ln m/dv$. They chose the coefficients c_2 and c_4 so that the PPN parameters α_1 and α_2 saturate the bounds from solar-system measurements, and obtained fitting formulae for the sensitivities as a function of c_+, c_- and the compactness M/R of the neutron star.

11 Gravitational Radiation

In this chapter, we develop the tools needed to treat gravitational radiation in general relativity and alternative theories of gravity. This is the "dynamical" regime of relativistic gravity, and since gravitational waves are intimately linked to the motion of masses, we begin (Section 11.1) with a brief history of the "problem of motion and radiation" in general relativity. This history has been at times muddled, at times controversial. We follow this (Section 11.2) with a short account of the status of today's gravitational-wave detectors. We then turn to the theory of gravitational waves in general relativity and alternative theories, treating their speed (Section 11.3), polarization (Section 11.4) and generation, mainly by binary systems (Section 11.5).

11.1 The Problem of Motion and Radiation

At the most primitive level, the problem of motion in general relativity is relatively straightforward, and was an integral part of the theory as proposed by Einstein.[1] This is geodesic motion, the motion of a test particle in a fixed spacetime provided by some distribution of matter, for example, that of a spherically symmetric star. The motion of a test particle does not generate gravitational waves, of course. The first attempts to treat the motions of *multiple* bodies, each with a finite mass and thus each contributing to the spacetime geometry, were made by de Sitter (1916) and Droste (1917) (see also Lorentz and Droste (1917)). They derived the metric and equations of motion for a system of N bodies, in what today would be called the first post-Newtonian approximation of general relativity (de Sitter's equations turned out to contain some important errors, however).

Einstein (1916, 1918) studied gravitational radiation, using the linearized approximation of his theory to derive the energy emitted by a body such as a rotating rod or dumbbell, held together by nongravitational forces. The first paper was marred by rather egregious errors; the second paper was correct, apart from a trivial numerical error, later spotted and corrected by Eddington (1922). The underlying conclusion that dynamical systems would radiate gravitational waves was correct.

However, there was ongoing confusion over whether gravitational waves are real or are artifacts of general covariance. Although Eddington (1922) clearly elucidated the difference between the physical, coordinate independent modes and modes that were purely coordinate artifacts, he made the unfortunate remark that some gravitational waves

[1] This history is adapted from Will (2011). See also Damour (1987).

propagate "with the speed of thought." He was referring to the coordinate modes, but the understanding of the role of coordinates at the time was sufficiently poor that the remark cast a pall of doubt over the entire subject.

For example, in 1936, in a paper submitted to *The Physical Review*, Einstein and Rosen claimed to prove that gravitational waves could not exist. However, the anonymous referee of their paper found that they had made an error. He pointed out that the singularity in their solution that caused them offer their negative conclusion was actually a harmless coordinate singularity. Upset that the journal had sent his paper to a referee (a newly instituted practice), Einstein withdrew the paper and refused to publish in *The Physical Review* again. Einstein became duly convinced that the referee was right, however, and a corrected paper showing that gravitational waves *do* exist – cylindrical waves in this case – was published elsewhere (Einstein and Rosen, 1937). Fifty years later it was revealed that the anonymous referee had been H. P. Robertson (Kennefick, 2005).

The next significant advance in the problem of motion came a few years later. In 1938, Einstein, Infeld, and Hoffman published the now legendary EIH paper, a calculation of the N-body equations of motion using only the vacuum field equations of general relativity (Einstein et al., 1938). They treated each body in the system as a spherically symmetric object whose nearby vacuum exterior geometry approximated that of the Schwarzschild metric of a static spherical star. They then solved the vacuum field equations for the metric between each body in the system in a weak field, slow-motion approximation, equivalent to the post-Newtonian approximation. Then, using a version of what today would be called "matched asymptotic expansions" they showed that, in order for the nearby metric of each body to match smoothly to the interbody metric at each order in the expansion, certain conditions on the motion of each body had to be met. This is essentially the matching procedure described in Chapter 10. Together, these conditions turned out to be equivalent to the Droste–Lorentz N-body equations of motion. The internal structure of each body was irrelevant, apart from the requirement that its nearby field be approximately spherically symmetric.[2]

During the 1960s, Fock in the Soviet Union and Chandrasekhar in the United States independently developed and systematized the post-Newtonian approximation in a form that laid the foundation for modern post-Newtonian theory (Fock, 1964; Chandrasekhar, 1965). They developed a full post-Newtonian hydrodynamics, with the ability to treat realistic, self-gravitating bodies of fluid, such as stars and planets. In a suitable limit of "point" particles, or bodies whose size is small enough compared to the interbody separations that finite-size effects such as spin and tidal interactions can be ignored, their equations of motion could be shown to be equivalent to the EIH and the Droste-Lorentz equations of motion.[3]

[2] At the end of their paper, EIH noted that the detailed calculations were so incredibly complex that they could not possibly be published, but instead had been deposited in the US Library of Congress. The remark seems somewhat quaint in light of today's dependence on algebraic software to handle the millions of terms involved in deriving equations of motion to high post-Newtonian orders.

[3] The author carried out this calculation as a term paper in Kip Thorne's general relativity course in 1969. Evidently the paper was a success, for Thorne invited the author to join his research group the following summer.

Meanwhile, Pirani, Bondi, and others had demonstrated rigorously that gravitational waves were physically real, that they propagate with the speed of light and that they carry energy. At the same time, Joseph Weber began to build detectors for gravitational waves, based on suspended massive cylinders of aluminum that would vibrate in a resonant manner if a gravitational wave of the appropriate frequency passed them. In 1969 and 1970, Weber made the stunning announcement that he had detected gravitational waves, most likely coming from the central regions of the Milky Way. The strength of the signals and rate of the detections were orders of magnitude higher than could be explained by reasonable astrophysical sources, and numerous follow-up investigations by physicists who built their own detectors failed to confirm Weber's claims. By 1980, a consensus had emerged that Weber had *not* detected gravitational waves. Despite this negative outcome, Weber's work had the positive effect of putting gravitational wave theory and detection on the scientific map, and of bringing into the field distinguished experimentalists from other branches of physics, such as Vladimir Braginsky, William Fairbank, Heinz Billing, Edoardo Amaldi, Ronald Drever, and Rainer Weiss. These scientists helped change the field from one dominated by theorists to one characterized by a healthy interplay between theory and experiment. Some of them began to work on a new type of gravitational-wave detection, based on laser interferometry.

The discovery of the binary pulsar in 1974 revealed an inconvenient truth about the "problem of motion," however. As Jürgen Ehlers and colleagues pointed out in an influential 1976 paper (Ehlers et al., 1976), the general relativistic theory of motion and radiation, as it was understood at the time, was full of holes large enough to drive trucks through. They pointed out that most treatments of the problem used "delta functions" as a way to approximate the bodies in the system as point masses. As a consequence, the "self-field", the gravitational field of the body evaluated at its own location, becomes infinite. While this is not a major issue in Newtonian gravity or classical electrodynamics, the non-linear nature of general relativity requires that this infinite self-field contribute to gravity. In the past, such infinities had been simply swept under the rug. Similarly, because gravitational energy itself produces gravity, it thus acts as a source throughout spacetime. This means that, when calculating radiative fields, integrals of the multipole moments of the source that are so useful in treating electromagnetic radiation begin to diverge in the case of gravity. These divergent integrals had also been routinely swept under the rug. Ehlers et al. further pointed out that the true boundary condition for any problem involving radiation by an isolated system should be one of "no incoming radiation" from the past. Connecting this boundary condition with the routine use of retarded solutions of wave equations was not a trivial matter in general relativity. The response of the source to the emitted flux of radiation was usually calculated by assuming that the flux of energy and angular momentum was balanced by an equal loss of those quantities in the motion of the source. While this balance assumption made sense, Ehlers et al. argued that there was no actual proof that it was valid. Finally, they pointed out that there was no evidence that the post-Newtonian approximation, so central to the problem of motion, was a convergent or even asymptotic sequence. Nor had the approximation been carried out to high enough order to make credible error estimates.

During this time, some authors even argued that the "quadrupole formula" for the gravitational energy emitted by a system (see Section 11.5.2), while correct for a rotating dumbbell as calculated by Einstein in the linearized approximation, was actually *wrong* for a binary system moving under its own gravity. The discovery in 1979 that the rate of decay of the orbit of the binary pulsar was in agreement with the standard quadrupole formula made some of these arguments moot. Yet the question raised by Ehlers et al. was still relevant: Is the quadrupole formula for binary systems an actual prediction of general relativity?

Motivated by the Ehlers et al. critique, numerous workers began to address the holes in the problem of motion, and by the late 1990s most of the criticisms had been answered, particularly those related to divergences. For a detailed history of the ups and downs of the subject of motion and gravitational waves, see Kennefick (2007).

Additional motivation for this theoretical work came in the early 1990s from the experimentalists. By this time, proposals to construct large-scale laser interferometric gravitational wave observatories were well advanced, eventually culminating in the LIGO interferometers in the United States, and Virgo and GEO600 in Europe. A leading candidate source of waves detectable by these devices is the inspiral, driven by gravitational radiation damping, of a binary system of compact objects (neutron stars or black holes) (for a review of sources of gravitational waves, see Sathyaprakash and Schutz (2009)). It was realized that the analysis of signals from such systems would require theoretical predictions from general relativity that are extremely accurate, well beyond the leading-order prediction of Newtonian or even post-Newtonian gravity for the orbits, and well beyond the leading-order quadrupole formula for gravitational waves (Cutler et al., 1993).

This presented a substantial theoretical challenge: to calculate the motion and radiation of systems of compact objects to very high post-Newtonian order, a formidable algebraic task, while addressing the issues of principle raised by Ehlers et al. sufficiently well to ensure that the results were physically meaningful. This challenge was largely met, and a post-Newtonian description of the inspiral orbit and gravitational wave signal is now in hand, approaching the fourth post-Newtonian (4PN) order beyond the approximation used to account for the binary pulsar (Blanchet, 2014).

Meanwhile another major approach to the problem of motion, known as *numerical relativity*, finally began to bear fruit. This is the method whereby one solves Einstein's equations for complex problems, such as the evolution of the orbit of two compact bodies and the gravitational waves emitted, by purely numerical means. The effort began with the pioneering work by Hahn and Lindquist (1964) and Smarr et al. (1976, 1977), focused on the head-on collision of two black holes. Subsequent progress was slow, however, as researchers struggled to find formulations of Einstein's equations that were better suited to stable numerical analysis, to learn how to treat the coordinate freedom inherent in the theory, and to find schemes to deal with the event horizons of the black holes. Progress was also limited by the available computer speed, power and memory. The first full numerical solution for the head on collision of two nonspinning black holes was achieved in 1993 (Anninos et al., 1993; Matzner et al., 1995). The more complicated problem of two orbiting and inspiralling black holes proved very challenging, until major breakthroughs occurred in 2005, paving the way for rapid progress (Pretorius, 2005; Baker et al., 2006; Campanelli et al., 2006). Baumgarte and Shapiro (2010) provide a pedagogical review of the field.

Today it is common to see published gravitational waveforms from black-hole inspirals and mergers that cover many orbits, that include a range of mass ratios, that incorporate black-hole spins together with the complex precessions of both the spins and the orbital angular momentum, and that give accurate results for the merger of the two black holes and the quasi-normal mode "ringdown" of the final black hole. Remarkably, there is a substantial region of overlap where post-Newtonian theory and numerical relativity are in "unreasonably" good agreement. This has permitted the development of phenomenological approaches that yield accurate analytic formulae that match post-Newtonian theory at early stages in the inspiral, and numerical relativity at the final stage. One of the most prominent of these is the "effective one-body" (EOB) approach, pioneered by Buonanno and Damour (1999). These formulae played a crucial role in the recent detections of gravitational waves by LIGO and Virgo (Abbott et al., 2016b, 2016c, 2017b, 2017c).

At the same time, there has been significant progress in applying numerical relativity to systems containing matter, such as neutron stars. Here one must deal with the uncertainties in the equation of state of nuclear matter, the effects of shock waves and of magnetic fields, and microphysical processes such as radiation and neutrino transport (see Shibata and Taniguchi (2011) and Faber and Rasio (2012) for reviews). These advances became particularly relevant with the 2017 detection of a binary neutron star inspiral and merger by LIGO and Virgo, together with observations of the accompanying electromagnetic signals by numerous observatories on the ground and in space (Abbott et al., 2017d, 2017e).

Needless to say, the "problem of motion" in general relativity looks quite different today than it did in 1974!

11.2 Gravitational Wave Detectors

The attempts to detect gravitational waves, from Weber's pioneering yet flawed experiments to the present international effort involving laser interferometry, pulsar timing, and cosmic microwave background observations, is a story rich in sociology, history, technological development, and big-science politics. That story, and the details of the methods of gravitational-wave detection are beyond the scope of this book. Readers should consult the books by Saulson (1994), Maggiore (2007), and Creighton and Anderson (2011) for those details.

The resonant bar detectors used by Weber and others have been almost completely phased out, in favor of kilometer-scale laser interferometer observatories. The US Laser Interferometer Gravitational-Wave Observatory (LIGO) was constructed between 1994 and 1999, operated between 2002 and 2010 without making any detections, and then after a major upgrade to "advanced LIGO" with higher sensitivity, resumed operations in September 2015. LIGO consists of two interferometers, one in Hanford, Washington, and the other 3000 km away in Livingston, Louisiana, each with arms four kilometers long. The Virgo observatory, operated by a European consortium is in Cascina, Italy, near Pisa, with three-kilometer long arms. It operated between 2003 and 2011, was upgraded, and resumed operation as advanced Virgo in August 2017. A smaller interferometer, GEO600, located in Hannover, Germany, is operated by a German-UK consortium; because its 600-meter arms mean lower sensitivity, GEO600 is today used mainly as a test-bed for advanced technologies for future detectors, for specialized searches for gravitational waves, and as a backup when the other observatories are off-line for maintenance or upgrading. The Kamioka Gravitational-wave detector (KAGRA), an underground three-kilometer interferometer in Japan, is nearing completion. Plans are well advanced for LIGO-India, a four-kilometer interferometer to be located in Maharashtra state, using interferometer and laser instrumentation that was built, but not used, in the final advanced LIGO observatory in Hanford. LIGO-India may join the other interferometers around 2024. These ground-based detectors are sensitive to gravitational waves in the frequency band between about 10 Hz, where seismic and gravity-gradient noise dominate the detector response, and about 500 Hz, where photon-counting or "shot" noise dominates.

In addition, plans are being developed for the Laser Interferometer Space Antenna (LISA). Consisting of three spacecraft orbiting the Sun in a triangular formation separated from each other by 2.5 million km, LISA will be sensitive primarily to waves in the low frequency band between 10^{-4} and 10^{-1} Hz. Following the success of LISA Pathfinder, a mission designed to test and demonstrate some of the key technologies needed for the ambitious mission (Armano et al., 2016), LISA was formally selected in June 2017 by the European Space Agency for a launch around 2034, with NASA participating as a junior partner.

In its most schematic realization, a laser interferometric gravitational-wave detector works just like the interferometer used by Michelson in the late 1800s to measure the speed of light and search for evidence of an "aether." The real-life interferometers are much more complex than this, but this simple model is adequate and captures the essential physics.

A laser interferometer consists of a laser source, a beam splitter, and two end mirrors mounted on test masses imagined to be freely moving in spacetime (although in reality they are suspended by thin wires). The arms of the interferometers are taken to be perpendicular to each other, although this is not essential (in LISA they make angles of 60°). The laser beam is divided in two at the beam splitter, and each beam travels along one arm of the interferometer, reflects off the test mass, and returns to the beam splitter to be recombined with the other beam. The relative phase of the beams determines whether they produce a bright or dark spot when the recombined beam is measured by a photon detector. Since the initial phases at the beam splitter are identical, the phase difference depends on the difference in travel time along the two arms. If a passing gravitational wave induces a displacement ΔL that changes the relative lengths of the two arms, it will induce a varying interference pattern. The ground-based advanced LIGO and Virgo are sensitive to strains $\Delta L/L$ at the level of 10^{-22} at 100 Hz.

In LISA, the multimillion-kilometer long arms make reflection of the laser signal impractical. Instead each spacecraft is equipped with an optical system that records the phase of each incoming and outgoing laser signal referenced to the location of a freely floating test mass inside the spacecraft. By combining the phase information from these six laser links (three in each direction) in a technique known as time-delay interferometry, one can measure the relative displacements of each pair of spacecraft. LISA's peak strain sensitivity is expected to be of order 10^{-21} at about a millihertz frequency (Amaro-Seoane et al., 2012).

The pulsar timing array (PTA) technique is another variant on the interferometry method. Here, an array of very stable millisecond pulsars distributed around the sky provides the analogue of the laser signal. The phases of the pulsed radio signals are measured as they arrive at Earth and monitored over multiyear timescales. Gravitational waves passing through the galaxy will induce phase variations in the pulsar signals that are sufficiently coherent that they can be distinguished from intrinsic pulsar noise, atomic clock noise on Earth and dispersion from the interstellar medium. This technique is sensitive to gravitational waves with frequencies of order the inverse of the multiyear observation time, or in the nanoHertz regime (for a review, see Manchester (2015)).

Finally, gravitational waves of frequencies of the order of the inverse Hubble time can be probed by looking for a specific class of fluctuations (so-called "B-modes") in the polarization of the cosmic background radiation.

We now turn to a study of gravitational waves in metric theories of gravity.

11.3 Speed of Gravitational Waves

The Einstein Equivalence Principle demands that in every local, freely falling frame, the speed of light must be the same – unity, if one works in geometrized units. The speed of propagation of all zero-rest-mass, nongravitational fields must also be the same as that of light. However, EEP demands nothing about the speed of gravitational waves. That speed is determined by the detailed structure of the field equations of each metric theory of gravity. Some theories of gravity predict that weak, short-wavelength gravitational waves propagate with exactly the same speed as light. By weak, we mean that the dimensionless amplitude $h_{\mu\nu}$ that characterizes the waves is in some sense small compared to the metric of the background spacetime through which the wave propagates. that is,

$$|h_{\mu\nu}|/|g^{(B)}_{\mu\nu}| \ll 1 , \tag{11.1}$$

and by short wavelength, we mean that the wavelength λ is small compared to the typical radius of curvature $\mathcal{R}$ of the background spacetime, that is,

$$|\lambda/\mathcal{R}| \ll 1 . \tag{11.2}$$

This is equivalent to the geometrical optics limit, discussed in Chapter 3 for electromagnetic radiation. In the case of general relativity, for example, one can show (see MTW, exercise 35.15) that the gravitational wave vector $\boldsymbol{\ell}$ is tangent to a null geodesic with respect to the "background" spacetime, that is,

$$\ell^\mu \ell^\nu g^{(B)}_{\mu\nu} = 0 , \quad \ell^\nu \ell^\mu{}_{|\nu} = 0 , \tag{11.3}$$

where "|" denotes a covariant derivative with respect to the background metric. In a local, freely falling frame, where $g^{(B)}_{\mu\nu} = \eta_{\mu\nu}$, the speed of the radiation is thus the same as that of light. Gravitational radiation propagates along the "light cones" of electromagnetic radiation.

General relativity

A simple method to derive this result in general relativity, which can then be applied to other metric theories, is to solve the vacuum field equations, linearized about a background metric chosen locally to be the Minkowski metric. Physically, this is tantamount to solving the propagation equations for the radiation in a local Lorentz frame. As long as the wavelength is short compared to the radius of curvature of the background spacetime, this method will yield the same results as a full geometrical-optics computation. Referring to Section 5.2.2, where we defined the potential $h^{\alpha\beta}$ in Eq. (5.18), we see that the linearized vacuum field equation (5.19) in harmonic coordinates takes the form

$$\Box h^{\alpha\beta} = 0 , \tag{11.4}$$

where $\Box$ denotes the flat-spacetime d'Alembertian. The plane-wave solutions of this equation are

$$h^{\alpha\beta} = \mathcal{A}^{\alpha\beta} e^{i\ell_\mu x^\mu} , \quad \ell_\mu \ell_\nu \eta^{\mu\nu} = 0 , \tag{11.5}$$

where $\mathcal{A}^{\alpha\beta}$ and ℓ_μ are the constant amplitude tensor and wave vector, respectively. Thus, the electromagnetic and gravitational light cones coincide, that is, the gravitational waves are null.

Scalar-tensor theories

Here the linearized vacuum field equations in harmonic gauge, Eqs. (5.43) and (5.45), take the form

$$\Box \tilde{h}^{\alpha\beta} = 0 , \quad \Box \phi = 0 . \tag{11.6}$$

where $\tilde{h}^{\alpha\beta}$ is the potential related to the auxiliary gothic inverse metric by $\tilde{h}^{\alpha\beta} \equiv \eta^{\alpha\beta} - \tilde{\mathfrak{g}}^{\alpha\beta}$. Writing $\phi = \phi_0(1 + \Psi)$, where ϕ_0 is the asymptotic value of ϕ far from the system, we obtain the plane wave solutions

$$\begin{aligned} \tilde{h}^{\alpha\beta} &= \mathcal{A}^{\alpha\beta} e^{i\ell_\mu x^\mu} , \quad \ell_\mu \ell_\nu \eta^{\mu\nu} = 0 , \\ \Psi &= \mathcal{B} e^{i\ell'_\mu x^\mu} , \quad \ell'_\mu \ell'_\nu \eta^{\mu\nu} = 0 , \end{aligned} \tag{11.7}$$

where $\mathcal{B}$ is a constant amplitude. Recalling that $\tilde{g}_{\mu\nu} \equiv (\phi/\phi_0) g_{\mu\nu}$, we see that $\tilde{h}^{\alpha\beta}$ is related to the gothic metric potential $h^{\alpha\beta}$ to linear order by

$$\tilde{h}^{\alpha\beta} = h^{\alpha\beta} - \Psi \eta^{\alpha\beta} . \tag{11.8}$$

Although the tensor and scalar waves do not necessarily have the same wave vector (depending on the source, ℓ_ν need not be the same as ℓ'_ν), both types of wave are null.

This will no longer be true if the scalar field has a potential $V(\phi)$. If, for example, $V(\phi) \sim m^2\phi^2$, then the scalar field will propagate like a massive field with a speed less than that of light, while the waves of the field $\tilde{h}^{\alpha\beta}$ will remain null.

Vector-tensor and TeVeS theories

In vector-tensor theories the linearized field equations are much more complex than in the scalar-tensor case, with the propagation of linearized metric perturbations being strongly influenced by the background cosmological value of the vector field $\boldsymbol{K}$. In general there are ten different wave modes (six for the metric, four for the vector field), each with its own characteristic speed and polarization, although various gauge symmetries within a specific theory may reduce this number. For example, in the Einstein-Æther theory (see Section 5.4.1), there are five modes: two pure metric modes (no vector waves), with speed $(1 - c_+)^{-1/2}$, where $c_+ = c_1 + c_3$ and three mixed metric-vector modes, with speeds that are complicated functions of the parameters $c_1 \dots c_4$ of the theory (Jacobson and Mattingly, 2004).

TeVeS theories have the same general vector-tensor structure as the general vector-tensor theories (see Section 5.5), and as a result, they predict the same modes with the same speeds. At linearized order, the scalar modes decouple from the tensor and vector modes, but because of the nature of the interpolating function $\mathcal{F}(k\ell^2 h^{\mu\nu}\phi_{,\mu}\phi_{,\nu})$ that appears in the TeVeS action (5.69), the speed of the scalar waves is generally much lower than the speed of light (Sagi, 2010).

Massive gravity

Many massive gravity theories (Section 5.7) are motivated by a desire to have the metric, linearized about a flat spacetime background, obey a massive vacuum wave equation, of the form

$$\Box h^{\alpha\beta} - \left(\frac{m_g}{\hbar}\right)^2 h^{\alpha\beta} = 0\,, \tag{11.9}$$

where m_g is the mass associated with the field. Substituting the ansatz $h^{\alpha\beta} = \mathcal{A}^{\alpha\beta}\exp(i\ell_\mu x^\mu)$, we obtain $\eta_{\mu\nu}\ell^\mu\ell^\nu = -(m_g/\hbar)^2$. With $\ell^0 \equiv \omega$ and $\ell^j \equiv k^j$, we obtain the dispersion relation $\omega^2 - |\boldsymbol{k}|^2 = (m_g/\hbar)^2$. With $E = \hbar\omega$ and $p = \hbar k$, this is equivalent to $E^2 - p^2 = m_g^2$. Then it is straightforward to show that the propagation speed of the wave, $v_g \equiv d\omega/dk$ is given by

$$v_g = \left(1 - \frac{m_g^2}{E^2}\right)^{1/2} = \left(1 - \frac{\lambda^2}{\lambda_g^2}\right)^{1/2}, \tag{11.10}$$

where $\lambda_g = h/m_g$ is the "graviton" Compton wavelength, and λ is the wavelength of the gravitational waves. This wavelength-dependent propagation speed is analogous to the dispersion experience by an electromagnetic wave propagating through a tenuous plasma, a phenomenon well-known to radio astronomers. In Section 12.2.3, we will see how the gravitational-wave signals detected by LIGO in 2015 and 2017 were used to place a bound on the Compton wavelength λ_g.

11.4 Polarization of Gravitational Waves

11.4.1 The $E(2)$ framework

General relativity predicts that weak gravitational radiation has two independent states of polarization, conventionally called the "+" and "×" modes, or the +2 and −2 helicity states, to use the language of quantum field theory. However, general relativity is probably unique in that prediction; every other known, viable metric theory of gravity predicts more than two polarizations for the generic gravitational wave. In fact, the most general weak gravitational wave that a theory may predict is composed of six modes of polarization, expressible in terms of the six "electric" components of the Riemann tensor R_{0j0k}, that govern the driving forces in a detector (Eardley et al., 1973a,b).

Consider an observer in a local freely falling frame. In the neighborhood of a chosen fiducial event $\mathcal{P}$ along a world line γ, construct a locally Lorentz orthonormal coordinate system (τ, ξ^j) with τ as proper time along the world line and $\mathcal{P}$ as spatial origin ("Fermi normal coordinates"). The metric has the form (PW, section 5.2.5)

$$
\begin{aligned}
g_{00} &= -1 - R_{0p0q}(\gamma)\xi^p\xi^q + O(\xi^3)\,, \\
g_{0j} &= \frac{2}{3}R_{jpq0}(\gamma)\xi^p\xi^q + O(\xi^3)\,, \\
g_{jk} &= \delta_{jk} - \frac{1}{3}R_{jpkq}(\gamma)\xi^p\xi^q + O(\xi^3)\,.
\end{aligned}
\tag{11.11}
$$

For a test particle with spatial coordinates ξ^j, momentarily at rest in the frame, the acceleration relative to the origin is

$$
\frac{d^2\xi_j}{dt^2} = \frac{1}{2}g_{00,j} = -R_{0j0q}(\gamma)\xi^q\,. \tag{11.12}
$$

Note that despite the possible presence of auxiliary gravitational fields in a given metric theory of gravity, the acceleration is sensitive only to the Riemann tensor. This is not necessarily true if the body has self-gravitational energy (Lee, 1974).

Thus, a gravitational wave may be completely described in terms of the Riemann tensor it produces. We characterize a weak, plane, nearly null gravitational wave in any metric theory (Eardley et al., 1973b), as observed in some local Lorentz frame, by a linearized Riemann tensor with components that depend only on a retarded time $\tilde{u}$, that is,

$$
R_{\alpha\beta\gamma\delta} \equiv R_{\alpha\beta\gamma\delta}(\tilde{u})\,. \tag{11.13}
$$

The "wave vector" $\tilde{\ell}_\mu$ defined by

$$
\tilde{\ell}_\mu \equiv -\tilde{u}_{,\mu}\,, \tag{11.14}
$$

is normal to surfaces of constant $\tilde{u}$ and is almost null with respect to the local Lorentz metric, that is,

$$
\eta^{\mu\nu}\tilde{\ell}_\mu\tilde{\ell}_\nu \equiv \epsilon\,, \quad |\epsilon| \ll 1\,, \tag{11.15}
$$

where ϵ is related to the difference in speed, as measured in a local Lorentz frame at rest in the universal rest frame, between light and the propagating gravitational wave, that is,

$$\epsilon = \left(\frac{c}{v_g}\right)^2 - 1\,. \tag{11.16}$$

We now wish to analyze the general properties of the Riemann tensor for a weak, plane, nearly null gravitational wave. To do this, it is useful to introduce a locally null basis instead of the local Lorentz frame's Fermi normal basis (τ, ξ^j). Consider a null plane wave (light, for instance) propagating in the $+Z$ direction in the local Lorentz frame. The wave is described by functions of retarded time u, where $u = t - Z$. (we use units in which the locally measured speed of light is unity). A similar wave propagating in the $-Z$ direction would be described by functions of advanced time v, where $v = t + Z$. We now define the four-vector fields $\boldsymbol{\ell} \equiv \ell^\mu \boldsymbol{e}_\mu$ and $\boldsymbol{n} \equiv n^\mu \boldsymbol{e}_\mu$, where

$$\ell_\mu = -u_{,\mu}\,, \quad n_\mu = -\frac{1}{2} v_{,\mu}\,. \tag{11.17}$$

These vectors are tangent to the propagation directions of the two null plane waves. In the (τ, ξ^j) basis they have the form

$$\ell^\mu = (1,\,0,\,0,\,1)\,, \quad n^\mu = \frac{1}{2}(1,\,0,\,0,\,-1)\,, \tag{11.18}$$

and are null with respect to $\boldsymbol{\eta}$, that is, $\ell^\mu \ell^\nu \eta_{\mu\nu} = n^\mu n^\nu \eta_{\mu\nu} = 0$; note that $\ell^\mu n^\nu \eta_{\mu\nu} = -1$. We also introduce the complex null vectors $\boldsymbol{m} \equiv m^\mu \boldsymbol{e}_\mu$ and $\bar{\boldsymbol{m}} \equiv \bar{m}^\mu \boldsymbol{e}_\mu$, where the bar denotes complex conjugation, and where

$$m^\mu = \frac{1}{\sqrt{2}}(0,\,1,\,i,\,0)\,, \quad \bar{m}^\mu = \frac{1}{\sqrt{2}}(0,\,1,\,-i,\,0)\,. \tag{11.19}$$

These vectors are also null, since $m^\mu m^\nu \eta_{\mu\nu} = 0$, and $\bar{m}^\mu \bar{m}^\nu \eta_{\mu\nu} = 0$; note that $m^\mu \bar{m}^\nu \eta_{\mu\nu} = 1$. These orthogonality relations imply that

$$\eta^{\mu\nu} = -2\ell^{(\mu} n^{\nu)} + 2m^{(\mu} \bar{m}^{\nu)}\,. \tag{11.20}$$

For the remainder of this section, we will use Roman subscripts (excluding the usual spatial indices i, j, k) to denote components of tensors with respect to the null tetrad basis $\boldsymbol{\ell}$, $\boldsymbol{n}$, $\boldsymbol{m}$, $\bar{\boldsymbol{m}}$, that is, $P_{apb\ldots} \equiv P_{\alpha\beta\gamma\ldots} a^\alpha p^\beta b^\gamma \ldots$, where a, b, c, $\ldots$ run over $\boldsymbol{\ell}$, $\boldsymbol{n}$, $\boldsymbol{m}$, and $\bar{\boldsymbol{m}}$, while p, q, r, $\ldots$ run over only $\boldsymbol{\ell}$, $\boldsymbol{m}$, and $\bar{\boldsymbol{m}}$.

Because the null tetrad $\boldsymbol{\ell}$, $\boldsymbol{n}$, $\boldsymbol{m}$, and $\bar{\boldsymbol{m}}$ is a complete set of basis vectors, we may expand the gravitational wave vector $\tilde{\boldsymbol{\ell}}$ terms of them. But since the gravitational wave is not exactly null, this expansion will depend in general upon the velocity of the observer's local frame relative to the universe rest frame. Choose a "preferred" observer, whose frame is at rest in the universe, and let $\tilde{\ell}^\mu$ in this frame have the form

$$\tilde{\ell}^\mu = \ell^\mu (1 + \epsilon_\ell) + \epsilon_n n^\mu + \epsilon_m m^\mu + \bar{\epsilon}_m \bar{m}^\mu\,, \tag{11.21}$$

where $\{\epsilon_\ell,\, \epsilon_n,\, \epsilon_m\} \sim O(\epsilon)$. However, this observer is free (i) to orient his spatial basis so that the gravitational wave and his null wave are parallel, so that $\tilde{\ell}^j \propto \ell^j$, and (ii) to choose the frequency of his positively propagating null wave to be equal to that of the gravitational wave, so that $\tilde{\ell}^0 = \ell^0$. This implies that $\epsilon_m = 0$, $\epsilon_\ell = -\epsilon_n/2$, and

$$\tilde{\ell}^\mu = \ell^\mu - \epsilon_n \left(\frac{1}{2}\ell^\mu - n^\mu \right) . \tag{11.22}$$

Now, because the Riemann tensor is a function of retarded time $\tilde{u}$ alone, we have that

$$R_{\alpha\beta\gamma\delta,\mu} \equiv -\tilde{\ell}_\mu \frac{\partial R_{\alpha\beta\gamma\delta}}{\partial \tilde{u}} \equiv -\tilde{\ell}_\mu \dot{R}_{\alpha\beta\gamma\delta} . \tag{11.23}$$

This, together with the orthogonality relations among the null tetrad vectors, implies that

$$\begin{aligned} R_{\alpha\beta\gamma\delta,\ell} &= \epsilon_n \dot{R}_{\alpha\beta\gamma\delta} , \\ R_{\alpha\beta\gamma\delta,n} &= \left(1 - \frac{1}{2}\epsilon_n\right) \dot{R}_{\alpha\beta\gamma\delta} , \\ R_{\alpha\beta\gamma\delta,m} &= 0 . \end{aligned} \tag{11.24}$$

The linearized Bianchi identities $R_{ab[cd,e]} = 0$ then yield

$$R_{abpq,n} = O(\epsilon_n \dot{R}) , \tag{11.25}$$

which, except for a trivial nonwavelike constant, implies that

$$R_{abpq} = R_{pqab} = O(\epsilon_n R) . \tag{11.26}$$

Thus the only components of the Riemann tensor that are not $O(\epsilon_n)$ are of the form R_{npnq}. There are only six such components, and all other components of the Riemann tensor can be expressed in terms of them. They can be related to particular tetrad components of the irreducible parts of the Riemann tensor known as Newman-Penrose quantities (Newman and Penrose, 1962), specifically the Weyl tensor Ψ_A and the tracefree Ricci tensor Φ_{AB} (see MTW section 13.5 for definitions). For our nearly null plane wave in the preferred tetrad, they have the form

$$\begin{aligned} \Psi_2 &= -\frac{1}{6} R_{n\ell n\ell} + O(\epsilon_n R) , \\ \Psi_3 &= -\frac{1}{2} R_{n\ell n\bar{m}} + O(\epsilon_n R) , \\ \Psi_4 &= -R_{n\bar{m}n\bar{m}} , \\ \Phi_{22} &= -R_{nmn\bar{m}} , \end{aligned} \tag{11.27}$$

(Ψ_3 and Ψ_4 are complex). These results are valid for a gravitational wave as detected by the preferred observer. Now, in order to discuss the polarization properties of the waves, we must consider the behavior of these components as observed in local Lorentz frames related to the preferred frame by boosts and rotations. However, we must restrict attention to observers who agree with the preferred observer on the frequency of the gravitational wave and on its direction; such "standard" observers can most readily analyze the intrinsic polarization properties of the waves. The Lorentz frames of these standard observers are related by a subgroup of the group of Lorentz transformations that leave $\tilde{\ell}$ unchanged. The most general such transformation is given by

$$
\begin{aligned}
\boldsymbol{\ell}' &= (1-\alpha\bar{\alpha}\epsilon_n)\,\boldsymbol{\ell} - \epsilon_n\,(\bar{\alpha}\boldsymbol{m} + \alpha\bar{\boldsymbol{m}}) + O(\epsilon_n^2)\,,\\
\boldsymbol{n}' &= (1-\alpha\bar{\alpha}\epsilon_n)\,(\boldsymbol{n} + \alpha\bar{\alpha}\boldsymbol{\ell} + \bar{\alpha}\boldsymbol{m} + \alpha\bar{\boldsymbol{m}}) + O(\epsilon_n^2)\,,\\
\boldsymbol{m}' &= (1-\alpha\bar{\alpha}\epsilon_n)\,e^{i\phi}\,(\boldsymbol{m} + \alpha\boldsymbol{\ell}) - \epsilon_n\alpha e^{i\phi}\,(\boldsymbol{n} + \alpha\bar{\boldsymbol{m}}) + O(\epsilon_n^2)\,,\\
\bar{\boldsymbol{m}}' &= (1-\alpha\bar{\alpha}\epsilon_n)\,e^{-i\phi}\,(\bar{\boldsymbol{m}} + \bar{\alpha}\boldsymbol{\ell}) - \epsilon_n\bar{\alpha} e^{-i\phi}\,(\boldsymbol{n} + \bar{\alpha}\boldsymbol{m}) + O(\epsilon_n^2)\,,
\end{aligned}
\tag{11.28}
$$

where α is a complex number that produces null rotations (combinations of boosts and rotations) and ϕ is an arbitrary real phase ($0 \le \phi \le 2\pi$) that produces a rotation about $\boldsymbol{e}_Z$. The parameter α is arbitrary except for the restriction $\alpha\bar{\alpha} \ll \epsilon_n^{-1}$. This expresses the fact that our results are valid as long as the velocity w of the frame is not too close to the speed either of light or of the gravitational wave, whichever is less; note that for nearly null waves $\epsilon_n^{-1} \gg 1$, and almost any velocity that is not infinitesimally close to unity is permitted, since $\alpha\bar{\alpha} \sim w^2/(1-w^2)$. For exactly null waves, $\epsilon_n = 0$ and arbitrary velocities $w < 1$ are permitted.

Under this set of transformations, the amplitudes of the gravitational wave change according to

$$
\begin{aligned}
\Psi_2' &= \Psi_2 + O(\epsilon_n R)\,,\\
\Psi_3' &= e^{-i\phi}\,(\Psi_3 + 3\bar{\alpha}\Psi_2) + O(\epsilon_n R)\,,\\
\Psi_4' &= e^{-2i\phi}\,(\Psi_4 + 4\bar{\alpha}\Psi_3 + 6\bar{\alpha}^2\Psi_2) + O(\epsilon_n R)\,,\\
\Phi_{22}' &= \Phi_{22} + 2\alpha\Psi_3 + 2\bar{\alpha}\bar{\Psi}_3 + 6\alpha\bar{\alpha}\Psi_2 + O(\epsilon_n R)\,.
\end{aligned}
\tag{11.29}
$$

Consider a set of observers related to each other by pure rotations about the direction of propagation of the wave ($\alpha = 0$). A quantity that transforms under rotations by a multiplicative factor $e^{is\phi}$ is said to have helicity s as seen by these observers. Thus, ignoring the correction terms of $O(\epsilon_n R)$, we see that the amplitudes Ψ_2, Ψ_3, Ψ_4 and Φ_{22} have helicities

$$
\begin{aligned}
\Psi_2 &: \quad s = 0\,, & \Phi_{22} &: \quad s = 0\,,\\
\Psi_3 &: \quad s = -1\,, & \bar{\Psi}_3 &: \quad s = +1\,,\\
\Psi_4 &: \quad s = -2\,, & \bar{\Psi}_4 &: \quad s = +2\,.
\end{aligned}
\tag{11.30}
$$

However, these amplitudes are not observer-independent quantities, as can be seen from Eq. (11.29). For example, if in one frame $\Psi_4 \neq 0$ but $\Psi_2 \neq 0$, then there exists a frame in which $\Psi_4' = 0$. Thus, the presence or absence of the components of various helicities depends upon the frame. Nevertheless, certain frame-invariant statements can be made about the amplitudes, within the small corrections of $O(\epsilon_n R)$. These statements comprise a set of quasi-Lorentz invariant classes of gravitational waves. Each class is labeled by the Petrov type of its nonvanishing Weyl tensor and the maximum number of nonvanishing amplitudes as seen by any observer. These labels are independent of observer. For exactly null waves ($\epsilon_n = 0$), the classes are:

- Class II_6: $\Psi_2 \neq 0$. All standard observers measure the same value for Ψ_2, but disagree on the presence or absence of all other modes.
- Class III_5: $\Psi_2 = 0$, $\Psi_3 \neq 0$. All standard observers agree on the absence of Ψ_2 and on the presence of Ψ_3, but disagree on the presence or absence of Ψ_4 and Φ_{22}.

- Class N_3: $\Psi_2 = 0$, $\Psi_3 = 0$, $\Phi_{22} \neq 0$, $\Psi_4 \neq 0$. Presence or absence of all modes is observer-independent.
- Class N_2: $\Psi_2 = 0$, $\Psi_3 = 0$, $\Phi_{22} = 0$, $\Psi_4 \neq 0$. Independent of observer.
- Class O_1: $\Psi_2 = 0$, $\Psi_3 = 0$, $\Phi_{22} \neq 0$, $\Psi_4 = 0$. Independent of observer.
- Class O_0: $\Psi_2 = \Psi_3 = \Phi_{22} = \Psi_4 = 0$. Independent of observer: No wave.

For nearly null waves, simply replace the vanishing of modes (= 0) with the nearly vanishing of modes [$\simeq O(\epsilon_n R)$]. This scheme, developed by Eardley et al. (1973b), is known as the $E(2)$ classification for gravitational waves, since in the case of exactly null plane waves ($\epsilon_n = 0$), the transformation equations (11.28), are the "little group" $E(2)$ of transformations, a subgroup of the Lorentz group. The $E(2)$ class of a particular metric theory is defined to be the class of its most general wave.

Although we have confined our attention to plane gravitational waves, one can show straightforwardly (Eardley et al., 1973b) that these results also apply to spherical waves far from an isolated source provided one considers the dominant $1/R$ part of the outgoing waves, where R is the distance from the source.

11.4.2 $E(2)$ class of specific metric theories

To determine the $E(2)$ class of a particular theory, it is sufficient to examine the linearized vacuum field equations of the theory in the limit of plane waves (observer far from source of waves). Here we present some examples. Some useful identities that can be obtained from Eqs. (11.20) and (11.26) are

$$
\begin{aligned}
R_{n\ell} &= R_{n\ell n\ell} + O(\epsilon_n R)\,, \\
R_{nn} &= 2R_{nmn\bar{m}} + O(\epsilon_n R)\,, \\
R_{nm} &= R_{n\ell nm} + O(\epsilon_n R)\,, \\
R &= -2R_{n\ell} + O(\epsilon_n R)\,.
\end{aligned}
\tag{11.31}
$$

If the Riemann tensor is computed from a linearized metric perturbation $p_{\mu\nu}(\tilde{u})$, then

$$
\begin{aligned}
R_{\mu\nu\alpha\beta} &= \frac{1}{2}\left(p_{\mu\beta,\alpha\nu} - p_{\mu\alpha,\beta\nu} + p_{\nu\alpha,\beta\mu} - p_{\nu\beta,\alpha\mu}\right) \\
&= \frac{1}{2}\left(\tilde{\ell}_\alpha \tilde{\ell}_\nu \ddot{p}_{\mu\beta} - \tilde{\ell}_\beta \tilde{\ell}_\nu \ddot{p}_{\mu\alpha} + \tilde{\ell}_\beta \tilde{\ell}_\mu \ddot{p}_{\nu\alpha} - \tilde{\ell}_\alpha \tilde{\ell}_\mu \ddot{p}_{\nu\beta}\right),
\end{aligned}
\tag{11.32}
$$

and

$$
\begin{aligned}
\Psi_2 &= \frac{1}{12}\ddot{p}_{\ell\ell} + O(\epsilon_n R)\,, \\
\Psi_3 &= \frac{1}{4}\ddot{p}_{\ell\bar{m}} + O(\epsilon_n R)\,, \\
\Psi_4 &= \frac{1}{2}\ddot{p}_{\bar{m}\bar{m}} + O(\epsilon_n R)\,, \\
\Phi_{22} &= \frac{1}{2}\ddot{p}_{m\bar{m}} + O(\epsilon_n R)\,.
\end{aligned}
\tag{11.33}
$$

General relativity

The vacuum field equations are $R_{\mu\nu} = 0$. The waves are null ($\epsilon_n = 0$). Eq. (11.31) then implies that

$$R_{n\ell n\ell} = R_{nmn\bar{m}} = R_{n\ell nm} = 0\,, \tag{11.34}$$

and thus that

$$\Psi_2 = \Psi_3 = \Phi_{22} = 0\,. \tag{11.35}$$

The only unconstrained mode is $\Psi_4 \neq 0$, so general relativity is of $E(2)$ class N_2. There are only two polarization modes, of helicity ± 2.

Scalar-tensor theories

In the massless scalar case, the linearized vacuum field equations (5.35a) and (5.35b) take the form

$$\begin{aligned} \Box_\eta \phi &= 0\,, \\ R_{\mu\nu} - \frac{1}{2}\eta_{\mu\nu}R &= \frac{1}{\phi_0}\left(\phi_{,\mu\nu} - \eta_{\mu\nu}\Box_\eta\phi\right)\,, \\ R &= O({\phi_{,\mu}}^2)\,, \end{aligned} \tag{11.36}$$

where ϕ_0 is the cosmological boundary value of the scalar field ϕ. The solution to the first of Eqs. (11.36) for a plane wave is

$$\phi = \phi_0 \mathcal{B} e^{i\ell_\mu x^\mu}\,, \tag{11.37}$$

where $\eta_{\mu\nu}\ell^\mu\ell^\nu = 0$. Then from the remaining two of Eq. (11.36),

$$R_{\mu\nu} = -\mathcal{B} e^{i\ell_\mu x^\mu}\ell_\mu\ell_\nu\,. \tag{11.38}$$

Thus $R_{n\ell} = R_{nm} = 0$, but $R_{nn} \neq 0$, thus

$$\Psi_2 = \Psi_3 = 0\,, \quad \Phi_{22} \neq 0\,, \Psi_4 \neq 0\,, \tag{11.39}$$

and scalar-tensor theories are of class N_3.

Einstein-Æther theory

Depending on how the source excites waves in the vector field, there are three different modes of gravitational wave in Einstein-Æther theory (Jacobson and Mattingly, 2004). In one mode, the vector field is unexcited, leaving a pure gravity mode of class N_2 ($\Psi_2 = \Psi_3 = \Phi_{22} = 0$) exactly as in general relativity. A second mode has tranverse excitations of the vector field, producing gravitational waves of class III_5 ($\Psi_2 = 0$). The third mode has longitudinal excitations of the vector field, producing waves of the most general class, II_6. Each mode has a different gravitational wave speed. The modes and their speeds are summarized in Table 11.1.

Table 11.1 Gravitational-wave polarizations and speeds in Einstein-Æther theory.

Mode	v_g^2 and small c_i limit	$E(2)$ class
Pure metric	$\frac{1}{1-c_{13}} \to 1$	N_2
Transverse vector	$\frac{2c_1-c_1^2+c_3^2}{2c_{14}(1-c_{13})} \to \frac{c_1}{c_{14}}$	III_5
Longitudinal vector	$\frac{c_{123}}{c_{14}} \frac{2-c_{14}}{2(1+c_2)^2-c_{123}(1+c_2+c_{123})} \to \frac{c_{123}}{c_{14}}$	II_6

11.4.3 Measurement of gravitational-wave polarizations

Consider an idealized experiment in which an observer uses an array of devices to determine via Eq. (11.12) the six components R_{0j0k} of the Riemann tensor for an incident wave. These are generally referred to as the "electric" components, by analogy with the electric field components F^{0j} of the electromagnetic field tensor. Let us suppose that the waves come from a single localized source with spatial wave vector $\boldsymbol{k}$ chosen to be in the Z direction. Basis vectors $\boldsymbol{e}_X$ and $\boldsymbol{e}_Y$ are chosen to be perpendicular to the Z direction, and therefore to lie in the plane of the sky, with the orientation chosen to reflect some characteristic of the source (for example, the line of the ascending node of an orbit inclined with respect to the plane of the sky). If the observer expresses her data on the displacements ξ^k as a 3×3 symmetric matrix, defined by

$$R_{0j0k} \equiv -\frac{1}{2R}\frac{d^2}{dt^2}A_{jk}(t)\,, \tag{11.40}$$

where R is the distance to the source, then, to first order in the small displacement ξ^k, Eq. (11.12) gives

$$\xi^j(t) = \xi^j(0) + \frac{1}{2R}A_{jk}(t)\xi^k(0)\,, \tag{11.41}$$

where A_{jk} can be written in the form

$$A_{jk} = \begin{pmatrix} A_+ + A_S & A_\times & A_{V1} \\ A_\times & -A_+ + A_S & A_{V2} \\ A_{V1} & A_{V2} & A_L \end{pmatrix}. \tag{11.42}$$

The standard XYZ orientation of the matrix elements is assumed. From Eqs. (11.18), (11.19), and (11.27), the amplitudes in A_{jk} are related to the Newman-Penrose quantities by

$$\begin{aligned} \mathrm{Re}\Psi_4 &= \frac{1}{2R}\ddot{A}_+\,, & \mathrm{Im}\Psi_4 &= -\frac{1}{2R}\ddot{A}_\times\,, \\ \mathrm{Re}\Psi_3 &= \frac{1}{4\sqrt{2}R}\ddot{A}_{V1}\,, & \mathrm{Im}\Psi_3 &= -\frac{1}{4\sqrt{2}R}\ddot{A}_{V2}\,, \\ \Phi_{22} &= \frac{1}{2R}\ddot{A}_S\,, & \Psi_2 &= \frac{1}{12R}\ddot{A}_L\,. \end{aligned} \tag{11.43}$$

For each polarization mode, Eq. (11.41) produces a specific distortion of, say, a sphere of freely moving particles, as displayed in Figure 11.1.

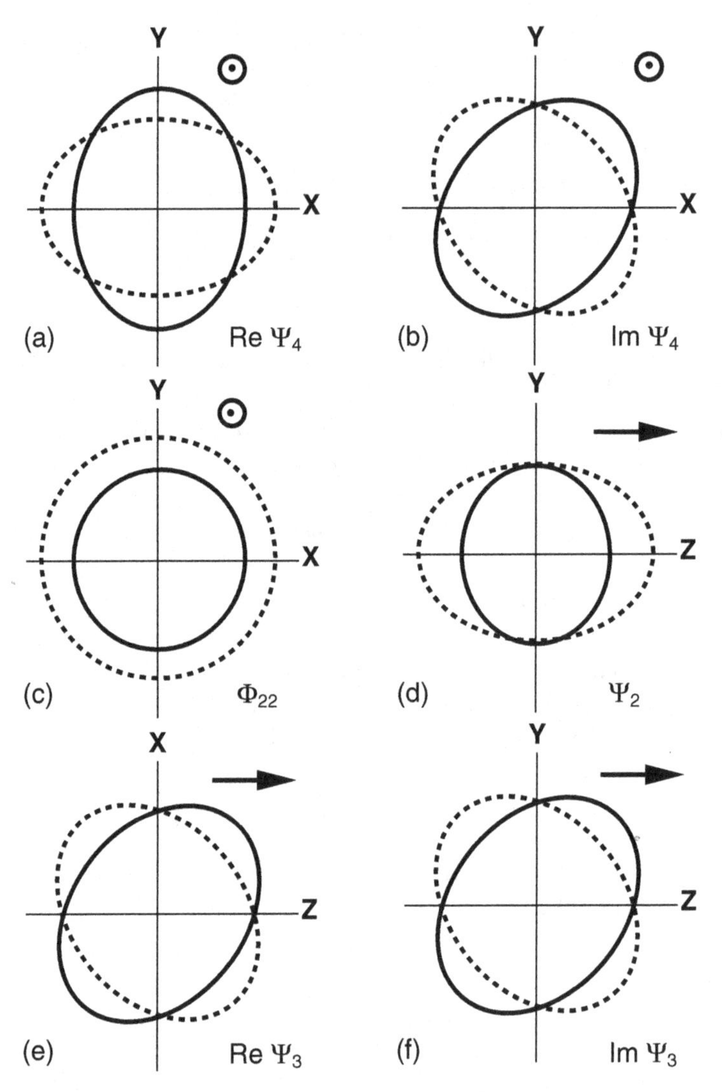

Fig. 11.1 Displacements induced on a sphere of free particles by each polarization mode of a plane gravitational wave permitted in any metric theory of gravity. The wave propagates in the $+Z$ direction and has a sinusoidal time dependence. The solid and dashed lines are snapshots at $\omega t = 0$ and π, respectively. There are no displacements perpendicular to the plane of the figure. In (a), (b) and (c), the wave propagates out of the page; in (d), (e), and (f), the wave propagates in the direction of the arrow.

Now, if the observer knows the direction $\boldsymbol{k}$ *a priori*, either by associating the wave with an electromagnetic counterpart, such as a gamma-ray burst, or by triangulating the direction using the arrival times of signals detected at widely spaced antennas, then by choosing the Z-axis parallel to $\boldsymbol{k}$, she can determine uniquely the amplitudes as given in Eq. (11.42), and thereby obtain the class of the incident wave. Because a specific source need not emit the most general wave possible, the $E(2)$ class determined by this method would be the least general class permitted by any metric theory of gravity.

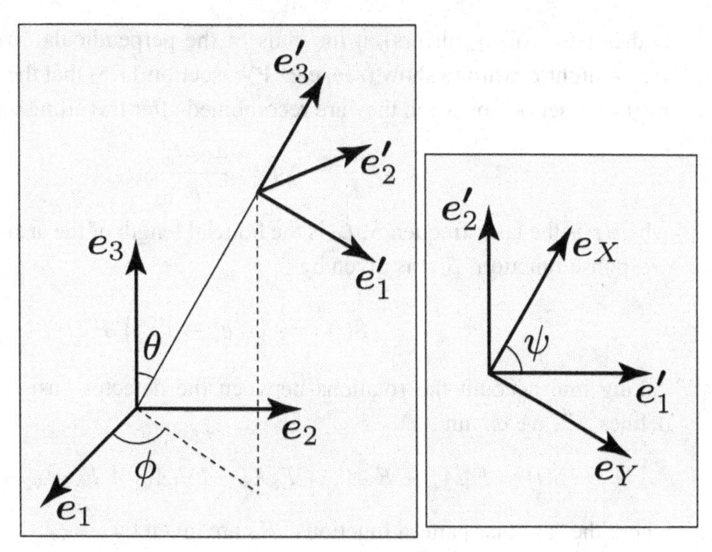

Fig. 11.2 Relation between the detector basis ($\boldsymbol{e}_1$, $\boldsymbol{e}_2$, $\boldsymbol{e}_3$) and the source basis ($\boldsymbol{e}_X$, $\boldsymbol{e}_Y$, $\boldsymbol{e}_Z = -\boldsymbol{e}'_3$).

However, if the observer does not know the direction *a priori*, it is not possible to determine the $E(2)$ class uniquely, since there are eight unknowns (six amplitudes and two direction angles) and only six observables A_{jk}. In particular, any observed A_{jk} can be fit by an appropriate wave of class II_6 and an appropriate direction. However, for certain observed A_{jk}, the $E(2)$ class may be limited in such a way as to provide a test of gravitational theory. For example, if the driving forces remain in a fixed plane and are pure quadrupole, that is, if there is a fixed coordinate system in which

$$A_{jk} = \begin{pmatrix} \lambda & \mu & 0 \\ \mu & -\lambda & 0 \\ 0 & 0 & 0 \end{pmatrix}, \tag{11.44}$$

then the wave may be either II_6 (unknown direction), or N_2, (direction parallel to the z-axis of new coordinate system). If this condition is *not* fulfilled, however, the class cannot be N_2. Such a result would be evidence against general relativity. Eardley et al. (1973b) provide a detailed enumeration of other possible outcomes of such idealized polarization measurements.

We must now link this idealized description of the polarization modes to measurements made by realistic gravitational wave detectors, for which the natural basis is defined by the detector, not the source.

For an Earth-bound laser interferometer, the natural basis is one where the directions of the two arms $\boldsymbol{e}_1$ and $\boldsymbol{e}_2$ define the x and y axes, and $\boldsymbol{e}_3$, perpendicular to the Earth's surface, defines the z axis (see Figure 11.2). The source direction is then characterized by the polar angles θ and ϕ, and the orientation of the source basis relative to the rotated detector basis (the primed basis in Figure 11.2) is characterized by a phase ψ (note that $\hat{\boldsymbol{k}} = \boldsymbol{e}_Z = -\boldsymbol{e}'_3$).

In addition, the detector response is somewhat more complicated than our ideal measurement, because it involves finding the *difference* in displacements of two free bodies (suspended mirrors) at the ends of the perpendicular arms of the interferometer. It is straightforward to show (see, e.g., PW, section 11.5) that the phase difference between the two laser beams when they are recombined after travelling along each arm is given by

$$\Delta\Phi = \frac{4\pi\nu L_0}{R} S(t) \,, \tag{11.45}$$

where ν is the laser frequency, L_0 is the fiducial length of the arms, and the interferometer's "response function" $S(t)$ is given by

$$S(t) = \frac{1}{2}\left(e_1^j e_1^k - e_2^j e_2^k\right) A^{jk}(t) \,. \tag{11.46}$$

Taking into account the rotations between the detector basis and the source basis that defines A^{jk}, we obtain

$$S(t) = F_+ A'_+ + F_\times A'_\times + F_S A_S + F_{V1} A'_{V1} + F_{V2} A'_{V2} + F_L A_L \,, \tag{11.47}$$

where the "angular pattern functions" F_A are given by

$$\begin{aligned}
F_+ &= \frac{1}{2}(1+\cos^2\theta)\cos 2\phi \,, \\
F_\times &= -\cos\theta \sin 2\phi \,, \\
F_S &= -\frac{1}{2}\sin^2\theta \cos 2\phi \,, \\
F_{V1} &= -\sin\theta\cos\theta\cos 2\phi \,, \\
F_{V2} &= \sin\theta \sin 2\phi \,, \\
F_L &= \frac{1}{2}\sin^2\theta\cos 2\phi \,,
\end{aligned} \tag{11.48}$$

and where

$$\begin{aligned}
A'_+ &\equiv A_+ \cos 2\psi + A_\times \sin 2\psi \,, \\
A'_\times &\equiv A_+ \sin 2\psi - A_\times \cos 2\psi \,, \\
A'_{V1} &\equiv A_{V1} \cos\psi + A_{V2} \sin\psi \,, \\
A'_{V2} &\equiv A_{V1} \sin\psi - A_{V2} \cos\psi \,.
\end{aligned} \tag{11.49}$$

Notice that the A'_A amplitudes are simply the original amplitudes in the source basis projected onto the basis vectors $\boldsymbol{e}'_1$ and $\boldsymbol{e}'_2$, with the factor 2ψ in A'_+ and $A'_\times$ arising because those modes have helicity 2. Also notice that, since $F_S = -F_L$, measurements using laser interferometers cannot measure the scalar and longitudinal modes separately, but can only measure the difference $A_S - A_L$. If the direction to the source is known, then a suitable array of five interferometers with distinct orientations, leading to five distinct values of θ, ϕ, can in principle measure the five amplitudes $A'_+, A'_\times, A'_{V1}, A'_{V2}$, and $A_S - A_L$. One can also show that, for an interferometer whose arms make an angle χ relative to each other, the angular pattern functions of Eqs. (11.48) are simply multiplied by a factor $\sin\chi$ (Cutler, 1998).

One might be tempted to imagine that, when the full network of kilometer-scale interferometers consisting of the two LIGO devices and Virgo, plus KAGRA and LIGO-India are online, such tests of polarization might become routine. However, the arms of the two LIGO interferometers in Hanford and Livingston are actually parallel to each other, apart from a small relative angle of about 27° induced by the curvature of the Earth's surface and a relative rotation of 90°. As a result, $\boldsymbol{e}_1^{\rm H} \approx \boldsymbol{e}_2^{\rm L}$, $\boldsymbol{e}_2^{\rm H} \approx -\boldsymbol{e}_1^{\rm L}$, and $S^{\rm H}(t) \approx -S^{\rm L}(t)$. This alignment is the result of a decision made early in the development of LIGO to forgo any sensitivity to multiple polarizations in favor of the redundancy provided by almost identical responses to any given signal. This decision was validated in the initial detection of GW150914, where one can see by eye in the almost raw data (mildly band-pass filtered to suppress low-frequency and high-frequency noise) that the chirp signals recorded in the Hanford and Livingston detectors are the precise negatives of each other.

Thus to carry out polarization tests, it will be necessary either to add an interferometer to the network, to combine data from sources of similar type (such as binary black hole mergers), or to focus on restricted classes of theories, such as N_3, N_4 or O_1. For specific analyses, see Wagoner and Kalligas (1997), Brunetti et al. (1999), Gasperini (1999), Magueijo (2003), Wen and Schutz (2005), and Nishizawa et al. (2009). By combining signals from various interferometers into a kind of "null channel" one can test for the existence of modes beyond the $+$ and $\times$ modes in a model-independent manner (Chatziioannou et al., 2012).

The space-based interferometer LISA can also measure polarizations. In this case, because the arm lengths ($\sim 2 \times 10^6$ km) may be comparable to or longer than the gravitational wavelength, our analysis based on local Fermi normal coordinates is no longer appropriate. Instead one uses an approach called "time delay interferometry" (TDI) (Armstrong et al., 1999), in which one determines the effect of the gravitational wave on each of the six laser beams propagating between the three spacecraft at the moment of emission and the moment of reception. The result is six "Doppler response functions," two for each spacecraft. Certain linear combinations of these functions completely suppress the contributions of laser noise, for example. Because the orientation of the spacecraft triangle changes as the constellation orbits the Sun, the response to the incoming gravitational wave varies continuously, permitting both localization of the source and measurement of the polarization modes (Tinto and Alves, 2010; Nishizawa et al., 2010). Gravitational-wave polarization can also be studied using pulsar timing arrays (Lee et al., 2008; Alves and Tinto, 2011; Chamberlin and Siemens, 2012).

11.5 Generation of Gravitational Waves

It is common knowledge that general relativity predicts that the lowest multipole emitted in gravitational radiation is quadrupole, in the sense that, if a multipole analysis of the gravitational field in the radiation zone far from an isolated system is performed in terms of tensor spherical harmonics, then only the harmonics with $\ell \geq 2$ are present (see Thorne (1980) for a thorough discussion of multipole-moment formalisms). For material sources, this statement can be reworded in terms of appropriately defined multipole moments of the matter and gravitational-field distribution within the near-zone surrounding the source: the lowest source multipole that generates radiation is quadrupole. For slow-motion, weak-field sources, such as binary star systems, quadrupole radiation is in fact the dominant multipole. The result is a gravitational waveform in the radiation zone given, to leading post-Newtonian order, by

$$h^{jk}(t) = \frac{2}{R}\ddot{I}^{jk}(\tau)\,, \tag{11.50}$$

where h^{jk} is defined by Eq. (5.18), R is the distance from the source, $\tau \equiv t - R$ is retarded time, dots denote time derivatives, and I^{jk} is the quadrupole moment tensor of the source, given by

$$I^{jk} \equiv \int \rho^* x^j x^k d^3x\,, \tag{11.51}$$

where ρ^* is the conserved density. The waveform h^{jk} can be related to the measured electric components of the Riemann tensor by Eq. (11.33),

$$R_{n\bar{m}n\bar{m}} = -\Psi_4 = -\frac{1}{2}\ddot{h}_{\bar{m}\bar{m}}\,. \tag{11.52}$$

The rate of change of energy of the system resulting from the flux of energy carried by the waves to infinity is given by

$$\frac{dE}{dt} = -\frac{1}{5}\dddot{I}^{\langle jk\rangle}\dddot{I}^{\langle jk\rangle}\,, \tag{11.53}$$

where $I^{\langle jk\rangle} \equiv I^{jk} - \frac{1}{3}\delta^{jk}I$ is the tracefree quadrupole moment tensor.

Eqs. (11.50) and (11.53) are collectively referred to as the "quadrupole formulae" for gravitational radiation. It has also become the accepted convention to refer to Eqs. (11.50) and (11.53) as the "Newtonian" waveform and energy flux. Even though Newtonian theory does not admit gravitational radiation because the interactions are considered to be instantaneous, the term "Newtonian" establishes these expressions as the lowest order terms in a post-Newtonian sequence of contributions, according to general relativity. Thus one can talk about 0.5PN, 1PN, 2PN corrections to these expressions.

For a binary system with total mass $m = m_1 + m_2$, reduced mass parameter $\eta = m_1 m_2/m^2$, orbital separation r and relative velocity $\boldsymbol{v}$, we obtain the waveform

$$h^{jk} = 4\eta\frac{m}{R}\left(v^j v^k - \frac{m}{r}n^j n^k\right)\,, \tag{11.54}$$

and the energy loss rate

$$\frac{dE}{dt} = -\frac{8}{15}\eta^2\left(\frac{m}{r}\right)^4\left(12v^2 - 11\dot{r}^2\right)\,, \tag{11.55}$$

where $\dot{r} = dr/dt$ (Peters and Mathews, 1963). This loss of energy results in a decrease of the orbital period P at a rate given, to lowest PN order, by $\dot{P}/P = -\frac{3}{2}\dot{E}/E$.

These comments apply to the asymptotic properties of the outgoing radiation field. However, if we are interested in the response of the source itself to the emission of the radiation we must deal with the field within the near zone at a level of approximation that incorporates the effects of gravitational radiation back reaction. A variety of computations in general relativity have yielded a local radiation-reaction force that, in a particular gauge, takes the form

$$F^j_{\rm React} = -\frac{2}{5}\overset{(5)}{I}{}^{\langle jk\rangle}x_k\,, \tag{11.56}$$

where the superscript (5) denotes five time derivatives. In the equations of motion, this represents a correction to Newtonian gravity at 2.5PN order. This leads to a loss of mechanical or orbital energy by the system that agrees with Eq. (11.53).

Quadrupole radiation also leads to a decrease of the orbital angular momentum and to a corresponding decrease of the orbital eccentricity.

Unlike general relativity, however, nearly every alternative metric theory of gravity predicts the presence of lower multipoles – monopole and dipole – in gravitational radiation, in addition to the quadrupole and higher multipole contributions (Eardley, 1975; Will and Eardley, 1977; Will, 1977). At zeroth PN order, the monopole moment is just the total rest mass of the system, which is conserved. Thus two time derivatives of the leading contribution to the monopole moment will vanish. At the first PN order, there are contributions to the monopole moment of order mv^2, which is of the same order as two time derivatives of the quadrupole moment, and so these contributions to the monopole moment serve mainly to modify the radiation at quadrupole order. At zeroth PN order, the dipole moment is proportional to the center of rest mass of the system, which at lowest order moves at constant velocity; thus two time derivatives of that quantity will vanish, by virtue of the Newtonian equations of motion. However, post-Newtonian corrections to the dipole moment need not share this property, and thus there could be dipole contributions to the radiation.

By contrast, in general relativity, the monopole moment turns out to be the *total* mass of the system, which is constant by virtue of the full equations of motion, at least up to the 2.5PN order at which the loss of mass to the flux of gravitational waves must be incorporated. Similarly, the dipole moment turns out to be the *full* dipole moment, or center of mass of the system, which is a linear function of time by virtue of the full equations of motion, at least up to the order at which the system experiences a recoil due to the emission of linear momentum in gravitational waves. These properties, which are a result of the remarkable adherence of general relativity to the Strong Equivalence Principle, account for the fact that quadrupole radiation leads the multipolar contributions to gravitational waves in general relativity.

In many alternative theories, the most important contribution to dipole radiation comes from a 1PN-order, residual dipole moment of gravitational energy, schematically of the form $D^j \sim \int \rho^* U x^j d^3x$. Focusing on the part of U that is the internal gravitational potential of each body, we can express the dipole moment in the form $D^j \sim \sum_a \Omega_a x_a^j$, where $\Omega_a = -\frac{1}{2}\int_a \rho^* U d^3x$. For a binary system with its center of mass placed at the origin of coordinates, we have to lowest PN order, $\boldsymbol{x}_1 = (m_2/m)\boldsymbol{x}$ and $\boldsymbol{x}_2 = -(m_1/m)\boldsymbol{x}$, and we have that

$$D^j \sim \eta m S x^j \,, \tag{11.57}$$

where

$$S \equiv \frac{\Omega_1}{m_1} - \frac{\Omega_2}{m_2} \tag{11.58}$$

is the difference in self-gravitational energy per unit mass between the two bodies. For compact bodies, S typically is related to the difference $s_1 - s_2$ between the sensitivities of the two bodies, as defined in Section 10.3. This varying dipole moment can thus contribute both to the gravitational waveform and to the fluxes of energy and angular momentum. Furthermore, if the bodies are compact bodies, such as neutron stars, where Ω/m could be as large as 0.3 or even black holes, where the analogue of Ω/m could be as large as unity, the contributions of dipole radiation could be larger than expected. In some theories, the flux of energy due to dipole radiation is similar to that in electromagnetism, namely $dE/dt \sim \ddot{\boldsymbol{D}} \cdot \ddot{\boldsymbol{D}} \sim \eta^2 (m/r)^4 \mathcal{S}^2$.

Generalizing the quadrupole energy flux of general relativity, Eq. (11.55) and adding the dipole energy flux, we can write down a parametrized formula for the leading energy loss rate from binary systems in a class of alternative theories, given by

$$\frac{dE}{dt} = -\eta^2 \mathcal{G}^{-1} \left(\frac{\mathcal{G} m}{r} \right)^4 \left[\frac{8}{15} \left(\kappa_1 v^2 - \kappa_2 \dot{r}^2 \right) + \frac{1}{3} \kappa_D \mathcal{S}^2 \right], \tag{11.59}$$

where $\mathcal{G} = \mathcal{G}_{12}$ is defined by Eq. (10.29), κ_1, κ_2 and κ_D are parameters that vary from theory to theory. In general relativity, $\mathcal{G} = 1$, $\kappa_1 = 12$, $\kappa_2 = 11$ and $\kappa_D = 0$.

The foregoing remarks are rather broad and qualitative. In the end, one must treat each theory in turn and calculate its gravitational waveform and fluxes of energy, angular momentum and linear momentum to the PN order needed. The next section attempts to provide a "recipe," largely based on experience with general relativity and scalar-tensor theories, for carrying out such calculations.

11.5.1 Calculating gravitational waves in alternative theories

In this section we will outline a general approach for calculating gravitational waves using a formalism developed for general relativity by Epstein and Wagoner (1975) and Wagoner and Will (1976), and extended by Will and Wiseman (1996) and Pati and Will (2000, 2002). This approach is closely related to the "post-Minkowskian" framework of Blanchet, Damour, and Iyer (for a thorough review, see Blanchet (2014)). Another approach, inspired by quantum field theory, is "Effective Field Theory" framework pioneered by Goldberger and Rothstein (2006) (for a review, see Foffa and Sturani (2014)).

We will focus our attention on gravitational waves from N-body systems (with $N=2$ being the most important case), assuming that the bodies are small compared to their separations. The two-body case is relevant for binary pulsar observations and for the first detections of gravitational radiation from compact binary inspiral.

Other potential sources of gravitational radiation, such as core collapse in supernovae, spinning deformed neutron stars, collisions of cosmic strings, and cosmological gravitational waves, involve very different assumptions, considerable input from nongravitational physics, and substantial numerical computation; these are beyond the scope of this book.

Because of the complexity of many alternative theories of gravitation beyond the post-Newtonian approximation, it has proven impossible to devise a formalism analogous to the PPN framework for describing gravitational radiation from dynamical sources. But we can provide a general description of the method used to arrive at the gravitational waveform and the energy and angular momentum fluxes within a chosen theory of gravity, emphasizing those features that are common to many currently viable theories, such as those described in Chapter 5. Later, we will treat specific theories. The method proceeds as follows:

Step 1: Select a theory, identify the variables and set the cosmological boundary conditions (follow Steps 1–3 in Section 5.1). Restrict the adjustable constants and cosmological matching parameters to values that give close agreement with solar-system tests.

Step 2: Working in the universal rest frame, expand the metric and any scalar, vector or tensor fields about their asymptotic values:

$$\begin{aligned} g_{\mu\nu} &= g^{(0)}_{\mu\nu} + p_{\mu\nu} , \\ \phi &= \phi_0(1+\Psi) , \\ K^\mu &= (K + k^0, k^1, k^2, k^3) , \\ B_{\mu\nu} &= B^{(0)}_{\mu\nu} + b_{\mu\nu} . \end{aligned} \tag{11.60}$$

We assume that the vector field is timelike, so that, in the universal rest frame far from the source, it has only a time component. All current theories with vector fields make this assumption. In theories with fields that single out a preferred rest frame, it may actually be advantageous to transform to the rest frame of the system under study at the beginning of the calculation, rather than at the end, as we did for the post-Newtonian limits of such theories.

Step 3: Derive the "relaxed field equations." Using any gauge freedom available, express the field equations in the form

$$\Box\psi^A = -16\pi\tau^A , \tag{11.61}$$

where ψ^A is one of the fields $p_{\mu\nu}$, Ψ, and so on, or possibly a linear combination of fields, and the "source" τ^A typically consists of matter and nongravitational field stress-energies, and of "gravitational" stress-energies consisting of terms quadratic and higher in the fields. The operator $\Box$ is given by

$$\Box \equiv -\frac{1}{v_{gA}^2}\frac{\partial^2}{\partial t^2} + \nabla^2 , \tag{11.62}$$

where the wave speed v_{gA} is a function of adjustible constants and matching parameters of the theory, and could be different for different ψ^A.

Because we will be dealing only with Lagrangian-based metric theories, we know that each theory admits a pseudotensorial object $\tau^{\mu\nu}$ that reduces to the matter energy-momentum tensor $T^{\mu\nu}$ in the absence of gravity, and that satisfies the equation

$$\tau^{\mu\nu}{}_{,\nu} = 0 , \tag{11.63}$$

(see Lee et al. (1974) and the discussion of Section 4.4). The system of Eqs. (11.61) and (11.63) is sometimes referred to as "relaxed" field equations, because one can solve Eq. (11.61) formally using the Green function for the operator $\Box$ as a *functional* of the matter variables, without specifying the behavior of the matter (this is the "relaxed" aspect of the procedure). Then one can use Eq. (11.63) to obtain the matter behavior as a function of time and thereby obtain the full solution for the fields of the theory (see PW section 6.2 for further discussion).

Box 11.1 Iterating the relaxed field equations

If one assumes that the fields are weak, so that $\psi^A \ll 1$, then it is natural to approximate solutions to Eq. (11.61) by a process of iteration. First substitute $\psi^A = 0$ in the right-hand-side of (11.61), so that τ^A consists of only matter contributions, and obtain the first iteration $\psi^A_{(1)}$. Then substitute those solutions back into the right-hand side of (11.61), and obtain the second iterated fields $\psi^A_{(2)}$. Continue the process until the desired accuracy is reached.

There is a persistent legend that, in order to obtain the gravitational wave signal and energy flux in general relativity at the leading quadrupole order, one only needs to obtain the first iterated field, which corresponds to linearized theory. While this is correct for the radiation from a source governed internally by nongravitational forces, such as an object in the laboratory, it is *incorrect* for a self-gravitating source, such as a binary star system. This is because the equations of motion that follow from the first iteration are given by $\tau^{\mu\nu}_{(0),\nu} = 0$, which do not contain gravitational interactions. A second iteration of the fields is required in order to arrive at a solution for gravitationally interacting systems that is compatible with the equations of motion $\tau^{\mu\nu}_{(1),\nu} = 0$, which now contain Newtonian gravity at leading order. The fact that the final answer, the so-called "quadrupole formula" (11.50), is formally the same for binary systems as for laboratory sources is both coincidental and misleading. The difference is in the meaning of the two time derivatives: for binary systems, one inserts the Newtonian equation of motion for the acceleration generated by two time derivatives, while for a laboratory system one inserts equations of motion appropriate for a system with only nongravitational interactions, such as a dumbell governed by internal atomic forces. The underlying reason for this formal coincidence is the Strong Equivalence Principle, which asserts that certain behaviors of systems, such as their motion or the waves they generate, are not sensitive to the internal structure of the bodies or systems, so that the leading-order waves generated by either a dumbell or a binary system depend only on the mass quadrupole moment.

This coincidence cannot be expected to hold in alternative theories of gravity, which generically violate SEP. Thus one cannot rely upon the linearized approximation to yield sensible results, but instead must carefully consider the structure of the relaxed field equations (11.61) and (11.63) in the theory and determine the number of iterations required to obtain solutions for the field and the equations of motion that are compatible with each other. For a detailed discussion within general relativity of this iteration process and of how to determine the number of iterations needed for a given situation, see PW, section 6.2.3.

Step 4: Obtain the formal solutions in the wave zone. The solution of Eq. (11.61) that has outward propagating disturbances at infinity is

$$\psi^A(t, \boldsymbol{x}) = 4 \int \frac{\tau^A(t - |\boldsymbol{x} - \boldsymbol{x}'|/v_{gA}, \boldsymbol{x}')}{|\boldsymbol{x} - \boldsymbol{x}'|} d^3x' \,. \tag{11.64}$$

The integration is to be carried out over the past "null cone" $\mathcal{C}(x)$ of the field point $(t, \boldsymbol{x})$, that is, over those points $(t', \boldsymbol{x}')$ such that $t' = t - |\boldsymbol{x} - \boldsymbol{x}'|/v_{gA}$ (see Figure 11.3). In dealing with slow-motion sources, it is conventional to divide the spatial domain into a near zone, with $r < \mathcal{R}$ and a far zone or wave zone, with $r > \mathcal{R}$, where $\mathcal{R} \sim \lambda_c \sim t_c$, where t_c and λ_c are the characteristic timescale of the source and the characteristic wavelength of the radiation, respectively. The vertical world tube $\mathcal{D}$ in Figure 11.3 is the interior region

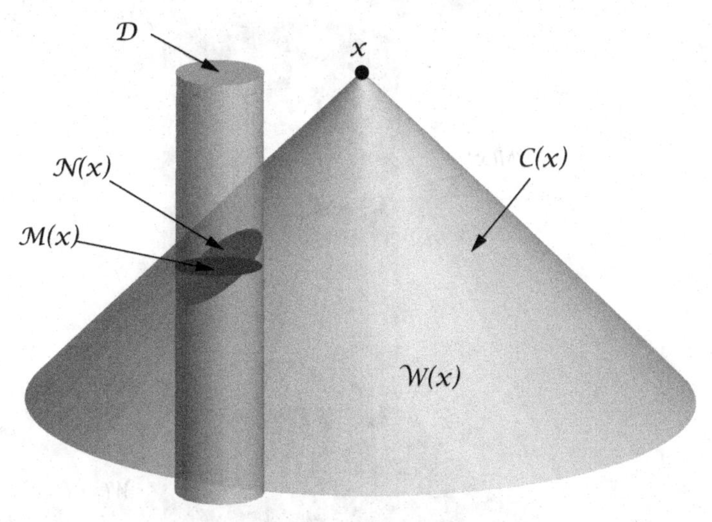

Fig. 11.3 Integration domains for the retarded solution (11.64) of the wave equation (11.61) with the field point in the wave zone: $\mathcal{C}(x)$ is the past "null" cone (with speed v_{gA}) of the field point $\boldsymbol{x}$; $\mathcal{D}$ is the world tube of the near zone of radius $\mathcal{R}$; $\mathcal{N}(x)$ is the intersection of $\mathcal{D}$ with $\mathcal{C}(x)$ and $\mathcal{W}(x)$ is the rest of the null cone; $\mathcal{M}(x)$ is a constant retarded-time hypersurface used for calculating multipole moments.

$r < \mathcal{R}$ projected in spacetime. The hypersurface $\mathcal{N}(x)$ is that piece of the null cone $\mathcal{C}$ within the near zone $\mathcal{D}$, and the hypersurface $\mathcal{W}(x)$ is the rest of the null cone. Thus the integral for the field ψ^A can be written as an integral over $\mathcal{N}(x)$ plus an integral over $\mathcal{W}(x)$, that is,

$$\psi^A(t, \boldsymbol{x}) = \psi^A_{\mathcal{N}}(t, \boldsymbol{x}) + \psi^A_{\mathcal{W}}(t, \boldsymbol{x}) . \tag{11.65}$$

To obtain the gravitational waveform and the energy flux, we want to evaluate the fields in the far wave zone, and retain only the leading contributions in powers of $1/R$, where R is the distance to the source. In the integral $\psi^A_{\mathcal{N}}$, $|\boldsymbol{x}'| < \mathcal{R}$, while $R \gg \mathcal{R}$, so we can expand $|\boldsymbol{x} - \boldsymbol{x}'|$ in powers of x'/R to obtain

$$\psi^A_{\mathcal{N}} = \frac{4}{R} \int_{\mathcal{N}} \tau^A \left(t - \frac{R}{v_{gA}} + \frac{\boldsymbol{N} \cdot \boldsymbol{x}'}{v_{gA}}, \boldsymbol{x}' \right) d^3x' + O(R^{-2}) , \tag{11.66}$$

where $\boldsymbol{N} \equiv \boldsymbol{x}/R$. For dynamical sources, $t_c \sim r_c/v_c$, where r_c is a characteristic scale of the material source and v_c is a characteristic velocity within the source, and thus for slow-motion sources ($v_c \ll 1$), $\mathcal{R} \sim t_c \gg r_c$, so the matter source is confined to a region deep within the near zone, and $\partial/\partial t \sim v_c/r_c$. Thus we can expand the retardation in Eq. (11.66) in the form

$$\psi^A_{\mathcal{N}} = \frac{4}{R} \sum_{m=0}^{\infty} \frac{1}{m!\,(v_{gA})^m} \left(\frac{\partial}{\partial t} \right)^m \int_{\mathcal{M}} \tau^A \left(t - \frac{R}{v_{gA}}, \boldsymbol{x}' \right) (\boldsymbol{N} \cdot \boldsymbol{x}')^m \, d^3x' + O(R^{-2}) , \tag{11.67}$$

where the integration is now over the constant retarded-time hypersurface $\mathcal{M}$ (see Figure 11.4). Note that

$$\psi^A_{\mathcal{N},j} = -(v_{gA})^{-1} N_j \psi^A_{\mathcal{N}} + O(R^{-2}) . \tag{11.68}$$

The source τ^A will contain contributions proportional to the matter variables ρ^*, p, Π, v, and so on, which have "compact support," that is, are nonzero only in finite spatial regions

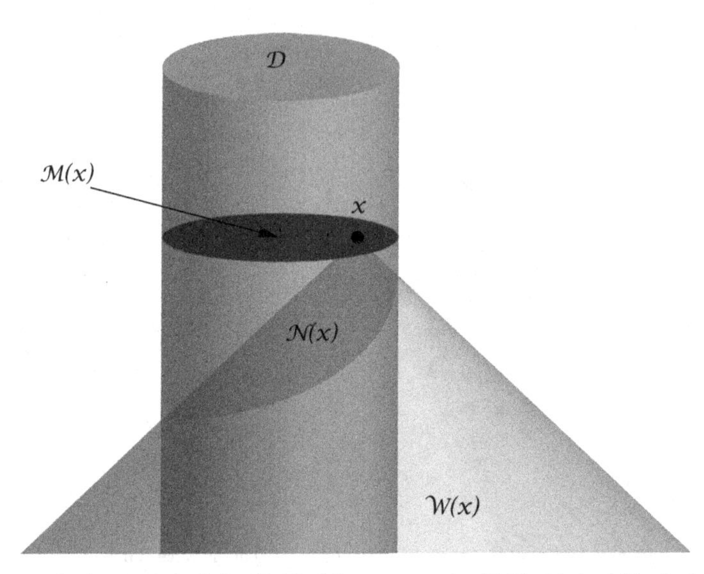

Fig. 11.4 Integration domains for the retarded solution (11.64) of the wave equation (11.61) with the field point in the near zone: here $\mathcal{M}(x)$ is a constant time hypersurface at time t within the near zone.

where the matter resides. However, at the second iteration and beyond, τ^A will also contain contributions proportional to products of fields and their derivatives. These contributions extend over all spacetime, although they generally decrease with distance from the source. Thus in integrating the multipole moments of τ^A over the hypersurface $\mathcal{M}$, the integrations must be extended all the way to the boundary of the near zone, at $\mathcal{R}$.

Because of these field contributions to τ^A, one must also include the integral $\psi^A_{\mathcal{W}}(t, \boldsymbol{x})$ over the rest of the past null cone, beyond the near zone. It turns out that this integration can best be done by changing variables from $d^3x' = r'^2 dr' d\Omega'$, where $d\Omega'$ is the element of solid angle, to a null-type variable $u' \equiv v_{gA}t - r' - |\boldsymbol{x} - \boldsymbol{x}'|$, and $d^3x' = r'(u', \theta', \phi')^2 du' d\Omega'$ (Wiseman and Will, 1991). This change is motivated by the fact that the fields that make up τ^A in the far zone are themselves retarded fields. Both the inner and the outer integrals will depend in general on the radius $\mathcal{R}$ where they terminate, but since that radius is chosen arbitrarily, the final result cannot depend on $\mathcal{R}$. This realization then provides a procedure for identifying and retaining in $\psi^A_{\mathcal{N}}(t, \boldsymbol{x})$ and in $\psi^A_{\mathcal{W}}(t, \boldsymbol{x})$ only contributions that do not depend on $\mathcal{R}$, and throwing away contributions that depend on $\mathcal{R}$. In fact it was shown explicitly that all $\mathcal{R}$-dependent contributions actually cancel each other in the sum $\psi^A_{\mathcal{N}} + \psi^A_{\mathcal{W}}$ (Pati and Will, 2000). In general relativity, it turns out that the contributions of $\psi^A_{\mathcal{W}}$ first occur 1.5PN orders beyond the leading quadrupole approximation; they are known as gravitational-wave "tails." While it is likely that, in an alternative theory of gravity, these outer integrals will contribute at a similar PN order, they must be carefully examined to determine the precise order at which they will contribute. In fact, in theories of gravity with dipole radiation, dipole "tails" from $\psi^A_{\mathcal{W}}$ can occur in the waveform at 1PN order, instead of 1.5PN order.

Step 5: Obtain a post-Newtonian expansion of the source τ^A. This will be needed both for evaluation of the multipole moments of τ^A in Eq. (11.67) to the required order, and for obtaining the equations of motion to the required order. For this purpose, we need the near-zone solution for the fields ψ^A. For $|\boldsymbol{x}-\boldsymbol{x}'| \ll \mathcal{R}$, we can carry out the expansion of the inner integral $\psi^A_{\mathcal{N}}$ in the form

$$\begin{aligned}
\psi^A_{\mathcal{N}} &= 4\sum_{m=0}^{\infty} \frac{(-1)^m}{m!\,(v_{gA})^m} \int_{\mathcal{M}} \tau^A(t,\boldsymbol{x}')|\boldsymbol{x}-\boldsymbol{x}'|^{m-1} d^3x' \,, \\
&= 4\int_{\mathcal{M}} \frac{\tau^A(t,\boldsymbol{x}')}{|\boldsymbol{x}-\boldsymbol{x}'|} d^3x' - \frac{4}{v_{gA}} \frac{\partial}{\partial t} \int_{\mathcal{M}} \tau^A(t,\boldsymbol{x}') d^3x' \\
&\quad + \frac{2}{v_{gA}^2} \frac{\partial^2}{\partial t^2} \int_{\mathcal{M}} \tau^A(t,\boldsymbol{x}')|\boldsymbol{x}-\boldsymbol{x}'| d^3x' + \dots \,.
\end{aligned} \tag{11.69}$$

The first term is an instantaneous solution of Poisson's equation with the source τ^A (called a Poisson potential), and, because $\partial/\partial t \sim v/r$, subsequent terms are of 0.5PN order, 1PN order, and so on. One must also evaluate the outer integral $\psi^A_{\mathcal{W}}$ for fields points in the near zone, although generally speaking, this integral begins to contribute only at very high post-Newtonian order (4PN in the case of general relativity).

Depending on the tensorial nature of τ^A, in the near zone it will typically be made up of terms of the form

$$\tau^A \sim \begin{cases} \rho^*,\ \rho^* v^2,\ p,\ \rho^* U,\ (\nabla U)^2,\ U_{,j}U_{,k},\ \dots & \text{scalar, tensor} \\ \rho^* v^j,\ \rho^* v^2 v^j,\ \rho^* U v^j,\ p v^j,\ \nabla V^j \cdot \nabla U,\ \dots & \text{vector}\,, \end{cases} \tag{11.70}$$

if the source is modeled as a fluid with weak internal gravitational fields, or

$$\tau^A \sim \begin{cases} m_a(\psi_A),\ m_a(\psi_A)v^2,\ m_a(\psi_A)U,\ (\nabla U)^2,\ U_{,j}U_{,k},\ \dots & \text{scalar, tensor} \\ m_a(\psi_A)v^j,\ m_a(\psi_A)v^2 v^j,\ m_a(\psi_A)U v^j,\ \nabla V^j \cdot \nabla U,\ \dots & \text{vector}\,, \end{cases} \tag{11.71}$$

if the source is modeled as a system of compact objects with masses dependent upon fields Ψ_A with appropriately defined "sensitivities" (see Section 10.3). In the compact-body case, each mass is accompanied by a delta function, augmented by a suitable prescription for regularizing infinite self-fields effects. Here "scalar" or "tensor" refers to quantities such as p_{00}, p_{jk}, Ψ, k^0, and so on, while "vector" refers to quantities such as p_{0j}, k^j, and so on. The wave-zone field is then determined by multipole moments of the source τ^A (Eq. (11.67)).

Various post-Newtonian conservation laws can often be used to simplify the multipole moments of τ^A in Eq. (11.67). For example, the monopole moment $\int_{\mathcal{M}} \rho^* d^3x$ is constant because of the conservation of baryon number of the system. If a collection of terms in a multipole moment can be combined to yield a component of the pseudotensorial quantity $\tau^{\mu\nu}$, then one can exploit a number of identities that follow from $\tau^{\mu\nu}{}_{,\nu} = 0$ to simplify or transform expressions. These identities include

$$\tau^{0j} = \left(\tau^{00} x^j\right)_{,0} + \left(\tau^{0k} x^j\right)_{,k} \,, \tag{11.72a}$$

$$\begin{aligned}
\tau^{(jk)} &= \frac{1}{2}\left(\tau^{00} x^j x^k\right)_{,00} + \left(\tau^{(pj)} x^k + \tau^{(pk)} x^j\right)_{,p} - \frac{1}{2}\left(\tau^{pq} x^j x^k\right)_{,pq} \\
&\quad + \left[\left(\tau^{[0p]} x^j x^k\right)_{,p} - \tau^{[0j]} x^k - \tau^{[0k]} x^j\right]_{,0} \,,
\end{aligned} \tag{11.72b}$$

$$\tau^{0j} x^k = \frac{1}{2}\left(\tau^{00} x^j x^k\right)_{,0} + \tau^{0[j} x^{k]} + \frac{1}{2}\left(\tau^{0p} x^j x^k\right)_{,p} \,. \tag{11.72c}$$

Note that, because we are dealing with semiconservative theories of gravity, $\tau^{\mu\nu}$ is not necessarily symmetric, so we have retained contributions of the antisymmetric parts of $\tau^{\mu\nu}$ where necessary.

Step 6: Determine the energy, momentum and angular momentum loss rate in terms of ψ. Generally speaking, this involves using the pseudotensorial quantity $\tau^{\mu\nu}$ and its associated conservation law $\tau^{\mu\nu}{}_{,\nu} = 0$. If one defines a quantity P^0 according to

$$P^\mu \equiv \int \tau^{\mu 0} d^3x \,, \tag{11.73}$$

then it is straightforward to show that

$$\frac{dP^\mu}{dt} = -\oint \tau^{\mu j} d^2S_j \,, \tag{11.74}$$

where the surface integral is to be carried out in the far wave zone. Since the matter energy-momentum tensor $T^{\mu\nu}$ has compact support, it will vanish in the far zone, and thus only the contributions from the fields ψ_A to $\tau^{\mu j}$ will be needed. And because $d^2S_j = R^2 n^j d\Omega$ and we wish to take the limit as $R \to \infty$, only the R^{-2} contributions to $\tau^{\mu j}$ are relevant. Making use of the far-zone behavior of Eq. (11.68), one can express the rate of change of P^0 and P^j in the generic form

$$\begin{aligned} \frac{dP^0}{dt} &= -\frac{R^2}{32\pi} \sum_{AB} \oint N^L \delta^M \dot\psi_A \dot\psi_B d\Omega \,, \\ \frac{dP^j}{dt} &= -\frac{R^2}{32\pi} \sum_{AB} \oint N^L \delta^M \dot\psi_A \dot\psi_B N^j d\Omega \,, \end{aligned} \tag{11.75}$$

where N^L denotes a product of L unit vectors, and δ^M denotes a product of M Kronecker deltas. The L and M indices are suitably contracted with indices on the ψ_As. In integrating over solid angles, one makes use of the identities

$$\oint N^L d\Omega = \begin{cases} \frac{4\pi}{(\ell+1)!!} \left[\delta^{L/2} + \text{sym}(q)\right] & L \text{ even}\,, \\ 0 & L \text{ odd}\,, \end{cases} \tag{11.76}$$

where sym(q) denotes all distinct terms obtained by permuting indices on the Kronecker deltas, with $q = (\ell - 1)!!$ denoting the total number of terms.

Similarly one can define an angular-momentum quantity, given by

$$J^{jk} \equiv 2 \int x^{[j} \tau^{k]0} d^3x \,, \tag{11.77}$$

whose rate of change is given by

$$\frac{dJ^{jk}}{dt} = -2 \int \tau^{[jk]} d^3x - 2 \oint x^{[j} \tau^{k]m} d^2S_m \,. \tag{11.78}$$

The first term in Eq. (11.78) is not related to gravitational radiation; it is the result of the fact that some theories of gravity are only semi conservative (see Section 4.4 and Eq. (4.70)). For the surface term, the presence of the factor $x^j \propto R$ means that one must calculate the subleading R^{-3} terms in $\tau^{k]m}$ in order to find a nontrivial result (the R^{-2} contributions give zero because of an odd number of unit vectors). This means obtaining the subleading R^{-2} terms in the ψ_A. Section 12.2.4 of PW gives a detailed derivation within general relativity.

In some theories of gravity, $\tau_{\mu\nu}$ is not unique; scalar-tensor theories are examples (Lee, 1974). Therefore one must determine the physical meaning of the quantity P^0. In general relativity, it is known to be the Kepler measured mass, or the Bondi mass of the system (when evaluated at spacelike infinity, it is also the Arnowitt-Deser-Misner (ADM) mass). In scalar-tensor theories, different choices for $\tau_{\mu\nu}$ lead to different masses: tensor mass, scalar mass or active (Kepler) mass. In the end, it is not so important which type of "mass" is represented by P^0, but rather what changes in observable parameters are induced by the change dP^0/dt. Therefore one must carefully calculate P^0 to an appropriate PN order in order to relate it to quantities such as orbital period, which are measurable.

Step 7: Apply to binary systems. We consider a system made of two bodies that are small compared to their separations ($d \ll r$); that is, we ignore all tidal interactions between them. We may thus treat each body's structure as static and spherical in its own rest frame. We then follow the procedure of Section 6.3: for a given element of matter in body a, we write $\boldsymbol{x} \equiv \boldsymbol{X}_a + \bar{x}$ and $\boldsymbol{v} \equiv \boldsymbol{v}_a$, where

$$\begin{aligned} \boldsymbol{X}_a &\equiv \frac{1}{m_a} \int_a \rho^* \left(1 + \Pi - \frac{1}{2}\bar{U}\right) \boldsymbol{x} d^3x \,, \\ m_a &\equiv \int_a \rho^* \left(1 + \Pi - \frac{1}{2}\bar{U}\right) d^3x \,, \\ \boldsymbol{v} &\equiv \frac{d\boldsymbol{X}}{dt} \,, \\ \bar{U} &\equiv \int_a \frac{\rho^*}{|\boldsymbol{x} - \boldsymbol{x}_a|} d^3x \,. \end{aligned} \tag{11.79}$$

We note that m_a is conserved to post-Newtonian order, as long as tidal forces are neglected. The full Newtonian potential U for spherical bodies is given by

$$\begin{aligned} U(\boldsymbol{x}, t) &= \bar{U} + \sum_{b \neq a} \frac{m_b}{|\boldsymbol{x} - \boldsymbol{X}_b|} \,, \quad \boldsymbol{x} \text{ inside body } a \\ &= \sum_b \frac{m_b}{|\boldsymbol{x} - \boldsymbol{X}_b|} \,, \quad \boldsymbol{x} \text{ outside body } a \,. \end{aligned} \tag{11.80}$$

In the compact-body approach, each body is located at the singularity of the delta function, internal gravitational fields are regularized appropriately and the masses are expanded in terms of sensitivities. Only the external potential is present, where now $m_a \equiv m_a(\psi_A^{(0)})$, where $\psi_A^{(0)}$ represents the asymptotic values of the relevant auxiliary fields.

Step 8: Apply balance equations to obtain the orbital evolution. Let us suppose that $\tau^{\mu\nu}$ has been chosen so that, to the appropriate orders, P^0 and $J^j = (1/2)\epsilon^{jpq} J^{pq}$, are given by

$$\begin{aligned} P^0 &= \sum_a \left[m_a + \frac{1}{2} m_a v_a^2 - \frac{1}{2} \sum_{b \neq a} \frac{\mathcal{G} m_a m_b}{r_{ab}} + O(m_a \epsilon^2) \right] , \\ \boldsymbol{J} &= \sum_a m_a \boldsymbol{x}_a \times \boldsymbol{v}_a [1 + O(\epsilon)] \,, \end{aligned} \tag{11.81}$$

where $r_{ab} \equiv |\boldsymbol{x}_a - \boldsymbol{x}_b|$ and where $\mathcal{G} \equiv \mathcal{G}_{12}$ is the parameter defined in the modified EIH Lagrangian, Eq. (10.29). It depends in general on the parameters of the theory and on the sensitivities.

For a binary system we may evaluate the orbital terms in Eq. (11.81) using the appropriate Keplerian equations, with the result

$$P^0 = m - \eta \frac{\mathcal{G}m^2}{2a} + O(m\epsilon^2)\,,$$
$$\boldsymbol{J} = \eta m \sqrt{\mathcal{G}mp}\,\hat{\boldsymbol{h}}[1 + O(\epsilon)]\,, \tag{11.82}$$

where $m = m_1 + m_2$, $\eta = m_1 m_2/m^2$, a is the semimajor axis, related to the binary orbital period by $(P_b/2\pi)^2 = a^3/\mathcal{G}m$, and $p = a(1-e^2)$.

In the emission of gravitational radiation whose source is the orbital motion, we will treat the quantities m_1 and m_2 as constant because of our neglect of tidal forces and internal motions. Then one can obtain a direct relation between dP^0/dt and $d\boldsymbol{J}/dt$ and variations in a, e, and P. Thus, for example, if the resulting energy loss rate for a binary system is given by Eq. (11.59), then one can average over an orbit to obtain

$$\frac{dP^0}{dt} = -\frac{32}{5}\eta^2 \mathcal{G}^{-1} \left(\frac{\mathcal{G}m}{a}\right)^5 F(e) - \frac{1}{3}\kappa_D \eta^2 \mathcal{S}^2 \mathcal{G}^{-1} \left(\frac{\mathcal{G}m}{a}\right)^4 G(e)\,, \tag{11.83}$$

where

$$F(e) = \frac{1}{12}(1-e^2)^{-7/2}\left[\kappa_1\left(1 + \frac{7}{2}e^2 + \frac{1}{4}e^4\right) - \frac{1}{2}\kappa_2 e^2 \left(1 + \frac{1}{2}e^2\right)\right],$$
$$G(e) = (1-e^2)^{-5/2}\left(1 + \frac{1}{2}e^2\right). \tag{11.84}$$

Combining Eqs. (11.82) and (11.83) with the relation between orbital period P_b and semimajor axis a, we can obtain an expression for the time-derivative $\dot{P}_b$,

$$\dot{P}_b = -\frac{192\pi}{5}\left(\frac{2\pi\mathcal{G}\mathcal{M}}{P_b}\right)^{5/3} F(e) - 2\pi\kappa_D \eta \mathcal{S}^2 \left(\frac{2\pi\mathcal{G}m}{P_b}\right) G(e)\,, \tag{11.85}$$

where $\mathcal{M}$ is called the "chirp mass," given by

$$\mathcal{M} \equiv \eta^{5/3} m\,. \tag{11.86}$$

If one has calculated dP^0/dt and $d\boldsymbol{J}/dt$ to higher PN orders, then it will be necessary to evaluate P^0 and $\boldsymbol{J}$ to a corresponding PN order.

Although the use of energy and angular momentum balance to infer the behavior of the source from flux formulae in the far zone was specifically criticized by Ehlers et al. (1976), it has actually been validated within general relativity and scalar-tensor theories to modest PN orders beyond the quadrupole formula via first-principles calculations of the radiation-reaction forces (Pati and Will, 2002; Will, 2005; Mirshekari and Will, 2013). Unfortunately, to reach a given PN order beyond the quadrupole formula, it is much easier to obtain flux results than it is to obtain radiation-reaction results. Therefore, in order to work at the high-PN orders required to analyze the inspiral of compact binaries, one is forced to adopt the balance assumption.

This completes our recipe for analyzing the generation of gravitational waves. To illustrate the method, we will now focus on two metric theories, general relativity and scalar-tensor theories where gravitational radiation from compact binaries has been analized in considerable detail. Partial results are also known for some of the alternative theories discussed in Chapter 5.

11.5.2 General relativity

In Section 5.2.2 we described the post-Minkowskian approach to solving Einstein's equations iteratively, using the Landau-Lifshitz formulation of the equations. We write the wave equation for the field $h^{\alpha\beta}$, Eq. (5.19) in the form

$$\Box h^{\alpha\beta} = -16\pi\tau^{\alpha\beta}\,, \tag{11.87}$$

where $\tau^{\alpha\beta} = (-g)(T^{\alpha\beta} + t^{\alpha\beta}_{\rm LL} + t^{\alpha\beta}_{\rm H})$, and where we have set $G = 1$. Note that, because of the harmonic gauge condition $h^{\alpha\beta}{}_{,\beta} = 0$, we have that $\tau^{\alpha\beta}{}_{,\beta} = 0$. Using Eq. (11.67), and recalling that $v_g = 1$ in general relativity, we obtain the multipole expansion in the far wave zone,

$$h^{\alpha\beta}_{\mathcal{N}} = \frac{4}{R}\sum_{m=0}^{\infty}\frac{1}{m!}\left(\frac{\partial}{\partial t}\right)^m \int_{\mathcal{M}} \tau^{\alpha\beta}\,(t-R,\boldsymbol{x}')\,(\boldsymbol{N}\cdot\boldsymbol{x}')^m\, d^3x' + O(R^{-2})\,. \tag{11.88}$$

Because of the harmonic gauge condition combined with Eq. (11.68), we can show that

$$\begin{aligned} h^{j0}_{\mathcal{N}} &= N_k h^{jk}_{\mathcal{N}} + O(R^{-2})\;[\text{modulo a constant}]\,,\\ h^{00}_{\mathcal{N}} &= N_j N_k h^{jk}_{\mathcal{N}} + O(R^{-2})\;[\text{modulo a constant}]\,, \end{aligned} \tag{11.89}$$

therefore, we need to determine only the jk components.

For the first iteration h^{jk}_1, the source τ^{jk}_0 is the special relativistic stress tensor for matter $T^{jk}_0 \sim \rho^* v^j v^k$. However, the equation of motion that is compatible with the gauge condition at the first iteration is $T^{\alpha\beta}_0{}_{,\beta} = 0$, which implies no gravitational interaction between the two bodies. Furthermore, an inspection of the Landau-Lifshitz pseudotensor $t^{\alpha\beta}_{\rm LL}$ reveals that it contributes terms to τ^{jk} of the form $U_{,j}U_{,k}$, where U is the near zone gravitational potential; these terms are of the *same* order as the purely kinetic terms in T^{jk}_0, and therefore they must be included. This is why *two* iterations of the relaxed equations are essential. However, it turns out that these details can be averted by exploiting the identity in Eq. (11.72b). At the lowest multipole order, we have that

$$\begin{aligned} h^{jk}_{\mathcal{N}} &= \frac{4}{R}\int_{\mathcal{M}} \tau^{jk}\,(t-R,\boldsymbol{x}')\,d^3x'\\ &= \frac{2}{R}\frac{d^2}{dt^2}\int_{\mathcal{M}} \tau^{00}\,(t-R,\boldsymbol{x}')\,x'^j x'^k d^3x'\,. \end{aligned} \tag{11.90}$$

In converting the divergence terms in Eq. (11.72b) to surface integrals at the boundary of $\mathcal{M}$, we must verify that there are no $\mathcal{R}$-independent contributions. Any $\mathcal{R}$-dependent contributions can be discarded, as we know that they will be cancelled by corresponding contributions from the outer integral $h^{kj}_{\mathcal{W}}$. To lowest PN order, $\tau^{00} = \rho^*$, thus we obtain

$$h^{jk}_{\mathcal{N}} = \frac{2}{R}\frac{d^2}{dt^2}\int_{\mathcal{M}} \rho^*(t-R,\boldsymbol{x}')x'^j x'^k d^3x' \,[1+O(\epsilon)]$$
$$= \frac{2}{R}\ddot{I}^{jk}(t-R)\,, \tag{11.91}$$

where $I^{jk} \equiv \sum_a m_a x_a^j x_a^k$ is the quadrupole moment tensor of the source. This is the well-known quadrupole formula for the gravitational waveform. To complete the solution we must add the term $h^{jk}_{\mathcal{W}}$ from the integration over the rest of the past null cone, however this is known not to contribute until 1.5PN order beyond the leading quadrupole term.

It can be demonstrated that the total mass-energy of an isolated system is given by

$$E \equiv P^0 = \int \tau^{00} d^3x\,. \tag{11.92}$$

Then, because of the condition $\tau^{\alpha\beta}{}_{,\beta} = 0$, we find that

$$\frac{dE}{dt} = \int \tau^{00}{}_{,0} d^3x = -\int \tau^{0j}{}_{,j} d^3x = -\oint \tau^{0j} N_j R^2 d\Omega\,, \tag{11.93}$$

where $d\Omega$ is the element of solid angle. By examining carefully the structure of the Landau-Lifshitz and harmonic pseudotensors of Eqs. (5.16) and (5.20) in the far wave zone, and taking Eqs. (11.68) and (11.89) into account, we obtain the energy loss rate

$$\frac{dE}{dt} = -\frac{R^2}{32\pi}\oint \dot{h}^{jk}_{\rm TT}\dot{h}^{jk}_{\rm TT} d\Omega\,, \tag{11.94}$$

where the subscript TT denotes the "transverse tracefree" part of h^{jk}, given by

$$h^{jk}_{\rm TT} \equiv \left(P^j_m P^k_n - \frac{1}{2}P^{jk}P_{mn}\right) h^{mn}\,, \tag{11.95}$$

where P^{jk} is the "projection tensor" given by

$$P^{jk} \equiv \delta^{jk} - N^j N^k\,. \tag{11.96}$$

We substitute Eq. (11.91) into (11.94) and integrate over solid angles, making use of the identities (11.76). The result is the quadrupole formula for the energy loss rate, Eq. (11.53). Substituting the quadrupole moment tensor for a binary system, $I^{jk} = \mu x^j x^k + mX^j X^k$, taking the time derivatives, and using the Newtonian equation of motion $d\boldsymbol{v}/dt = -m\boldsymbol{x}/r^3$, yields Eq. (11.54) for the waveform and Eq. (11.55) for the energy loss rate. Thus, for general relativity, $\kappa_1 = 12$, $\kappa_2 = 11$, and $\kappa_D = 0$.

Analogous formulae for the angular momentum loss rate and the rate of change of linear momentum of the source can be obtained:

$$\frac{dJ^k}{dt} = -\frac{R^2}{16\pi}\epsilon^{kmn}\oint\left[h^{mp}_{\rm TT}\dot{h}^{np}_{\rm TT} - \frac{1}{2}\dot{h}^{pq}_{\rm TT}x^m\frac{\partial}{\partial x^n}h^{pq}_{\rm TT}\right]d\Omega$$
$$= \frac{2}{5}\epsilon^{kmn}\ddot{I}^{\langle mp\rangle}\dddot{I}^{\langle np\rangle}\,, \tag{11.97a}$$
$$\frac{dP^k}{dt} = -\frac{R^2}{32\pi}\oint \dot{h}^{mn}_{\rm TT}\dot{h}^{mn}_{\rm TT}N^k d\Omega$$
$$= \frac{2}{63}\overset{(4)}{I}{}^{\langle kmn\rangle}\dddot{I}^{mn} - \frac{16}{45}\epsilon^{kmn}\dddot{J}^{mp}\dddot{I}^{np}\,, \tag{11.97b}$$

where I^{kmn} and J^{mp} are the mass octupole and current quadrupole moments, given to lowest PN order by

$$I^{kmn} \equiv \int \rho^* x^k x^m x^n d^3x \,,$$
$$J^{mp} \equiv \epsilon^{mab} \int \rho^* x^a v^b x^p d^3x \,. \tag{11.98}$$

Note that, at purely quadrupole order, the linear momentum change vanishes, because the number of unit vectors N^j in the angular integral will always be odd. To get the final expression in Eq. (11.97b) it is necessary to compute the 0.5PN corrections to the waveform h^{jk}, resulting in the appearance of the higher-order multipole moments of Eq. (11.98). For a binary system at lowest PN order, these loss rates become

$$\frac{d\boldsymbol{J}}{dt} = -\frac{8}{5}\eta^2 \left(\frac{m}{r}\right)^3 \boldsymbol{h} \left(2v^2 - 3\dot{r}^2 + 2\frac{m}{r}\right),$$
$$\frac{d\boldsymbol{P}}{dt} = \frac{8}{105}\eta^2 \Delta \left(\frac{m}{r}\right)^4 \left[\boldsymbol{v}\left(50v^2 - 38\dot{r}^2 + 8\frac{m}{r}\right) - \dot{r}\boldsymbol{n}\left(55v^2 - 45\dot{r}^2 + 12\frac{m}{r}\right)\right], \tag{11.99}$$

where $\boldsymbol{h} = \boldsymbol{x} \times \boldsymbol{v}$ and $\Delta = (m_2 - m_1)/m$.

Substituting Keplerian formulae for the orbital variables and averaging over an orbit, we can obtain the orbit-averaged dE/dt and $d\boldsymbol{J}/dt$. Recalling that for the orbit, $E = -\eta m^2/2a$ and $J = \eta m\sqrt{mp}$, where $p = a(1 - e^2)$, we obtain the rate of change of a and e. The results are

$$\frac{dE}{dt} = -\frac{32}{5}\eta^2 \left(\frac{m}{a}\right)^5 F(e) \,,$$
$$\frac{dJ}{dt} = -\frac{32}{15}\eta^2 m \left(\frac{m}{a}\right)^{7/2} G(e) \,.$$
$$\frac{da}{dt} = -\frac{64}{5}\eta \left(\frac{m}{a}\right)^3 F(e) \,,$$
$$\frac{de}{dt} = -\frac{304}{15}\eta \frac{e}{a} \left(\frac{m}{a}\right)^3 H(e) \,, \tag{11.100}$$

where

$$F(e) = (1 - e^2)^{-7/2} \left(1 + \frac{73}{24}e^2 + \frac{37}{96}e^4\right),$$
$$G(e) = (1 - e^2)^{-2} \left(1 + \frac{7}{8}e^2\right),$$
$$H(e) = (1 - e^2)^{-5/2} \left(1 + \frac{121}{304}e^2\right). \tag{11.101}$$

These equations for da/dt and de/dt can be combined and integrated to yield

$$\frac{a}{a_0} = \left(\frac{e}{e_0}\right)^{12/19} \left(\frac{1 - e_0^2}{1 - e^2}\right) \left(\frac{1 + 121e^2/304}{1 + 121e_0^2/304}\right)^{870/2299}, \tag{11.102}$$

where a_0 and e_0 are the initial values. Note that gravitational radiation causes the orbit to both shrink and circularize. For $e_0 \ll 1$, the evolution is given approximately by $e \simeq e_0(a/a_0)^{19/12}$. From da/dt we can obtain an expression for the rate of change of P_b, given by

$$\dot{P}_b = -\frac{192\pi}{5}\left(\frac{2\pi\mathcal{M}}{P_b}\right)^{5/3} F(e)\,, \tag{11.103}$$

where $\mathcal{M}$ is the chirp mass given by Eq. (11.86). Notice that by making precise measurements of the phase $\Phi(t) = 2\pi\int^t f(t')dt'$ of either the orbit or the gravitational waves (for which $f = 2f_b = 2/P_b$ for the dominant component) as a function of the frequency, one in effect measures the chirp mass of the system. The frequency of the orbit and of any emitted gravitational waves increases with time, varying as $f_b \sim (t - t_0)^{-3/8}$ for circular orbits, where t_0 is the merger time when the binary separation vanishes. The semimajor axis shrinks according to $a \sim (t - t_0)^{1/4}$, and thus the waveform amplitude increases as $h \sim a^{-1} \sim (t - t_0)^{-1/4}$. In acoustics, a signal of increasing amplitude and frequency is known as a "chirp", hence the name "chirp mass" for the quantity $\mathcal{M}$ that controls the process.

From Eq. (11.91), the gravitational waveform for a binary system is given, to lowest PN order by (we drop the subscript $\mathcal{N}$)

$$\begin{aligned} h^{jk} &= \frac{4\eta m}{R}\left(v^j v^k - \frac{m}{r} n^j n^k\right)\,, \\ &\to \frac{4\eta m^2}{Rr}\left(\lambda^j \lambda^k - n^j n^k\right)\,, \end{aligned} \tag{11.104}$$

where the second line is the circular orbit limit, and where r $\boldsymbol{v}$, $\boldsymbol{n}$ and $\boldsymbol{\lambda}$ are given by Eqs. (6.73). We now choose the basis in Figure 6.1 so that the X axis is along the line of nodes and also choose $\omega = 0$ for the pericenter angle. Using the fact that, for $\boldsymbol{N}$ in the Z direction, the projection tensor can be written $P^{jk} = e^j_X e^k_X + e^j_Y e^k_Y$, we can show that the transverse tracefree components $h^{jk}_{\rm TT}$, Eq. (11.95) become

$$h^{jk}_{\rm TT} = \left(e^j_X e^k_X - e^j_Y e^k_Y\right) h_+ + \left(e^j_X e^k_Y + e^j_Y e^k_X\right) h_\times\,, \tag{11.105}$$

where

$$\begin{aligned} h_+ &= \frac{2\eta m}{R}\frac{m}{r}\left[(\boldsymbol{\lambda}\cdot\boldsymbol{e}_X)^2 - (\boldsymbol{\lambda}\cdot\boldsymbol{e}_Y)^2 - (\boldsymbol{n}\cdot\boldsymbol{e}_X)^2 + (\boldsymbol{n}\cdot\boldsymbol{e}_Y)^2\right] \\ &= -\frac{2\eta m}{R}\frac{m}{r}(1 + \cos^2\iota)\cos 2\phi\,, \\ h_\times &= \frac{4\eta m}{R}\frac{m}{r}\left[(\boldsymbol{\lambda}\cdot\boldsymbol{e}_X)(\boldsymbol{\lambda}\cdot\boldsymbol{e}_Y) - (\boldsymbol{n}\cdot\boldsymbol{e}_X)(\boldsymbol{n}\cdot\boldsymbol{e}_Y)\right] \\ &= -\frac{4\eta m}{R}\frac{m}{r}\cos\iota\sin 2\phi\,. \end{aligned} \tag{11.106}$$

It can also be shown that

$$R_{0j0k} = -\frac{1}{2}\frac{d^2}{dt^2}h^{jk}_{\rm TT}\,, \tag{11.107}$$

and thus the amplitudes in Eqs. (11.42) and (11.48) are given by

$$\begin{aligned} A_+ &= -\frac{2\eta m^2}{r}(1 + \cos^2\iota)\cos 2\phi\,, \\ A_\times &= -\frac{4\eta m^2}{r}\cos\iota\sin 2\phi\,. \end{aligned} \tag{11.108}$$

with $A_{\rm S} = A_{\rm V1} = A_{\rm V2} = A_{\rm L} = 0$.

The foregoing discussion has been for the Newtonian waveforms and fluxes, but results are now known to many post-Newtonian orders beyond these lowest order formula. As an illustration, we show the rate of change of energy for a circular orbit through 2PN order:

$$\frac{dE}{dt} = -\frac{32}{5}\eta^2\left(\frac{m}{a}\right)^5\left[1 - \frac{m}{a}\left(\frac{2927}{336} + \frac{5}{4}\eta\right) + 4\pi\left(\frac{m}{a}\right)^{3/2} + \left(\frac{m}{a}\right)^2\left(\frac{293383}{9072} + \frac{380}{9}\eta\right)\right]. \tag{11.109}$$

The 1PN corrections to the waveform and energy flux were calculated by Wagoner and Will (1976). The 1.5PN term $4\pi(m/a)^{3/2}$ is the contribution of tails, the result of scattering of the outgoing gravitational wave off the background curved spacetime produced by the source. This and the 2PN terms were calculated by Blanchet et al. (1995a, 1995b) and Will and Wiseman (1996). Through pioneering work by Blanchet and collaborators, these results are now known through 3.5PN order (see Blanchet (2014) for an up-to-date review). The leading effects of spins of the bodies on the waveform and fluxes were first calculated by Kidder et al. (1993) and Kidder (1995) (see Section 6.7 for the leading contributions to the equations of motion). Numerous higher-order spin contributions have also been calculated (see Blanchet (2014) for a review and references).

11.5.3 Scalar-tensor theories

Because we are interested in gravitational waves from systems containing compact bodies, we will combine the "relaxed" or Landau-Lifshitz formulation of the scalar-tensor field equations developed in Section 5.3.2 with the modified EIH formalism described in Section 10.3.2.

The wave equation for the field $\tilde{h}^{\alpha\beta}$ in harmonic gauge has a form parallel to that of Eq. (11.87),

$$\Box\tilde{h}^{\alpha\beta} = -16\pi\tilde{G}\tilde{\tau}^{\alpha\beta}, \tag{11.110}$$

where $\tilde{\tau}^{\alpha\beta} = (-\tilde{g})(\tilde{T}^{\alpha\beta} + \tilde{t}^{\alpha\beta}_{\phi} + \tilde{t}^{\alpha\beta}_{\rm LL} + \tilde{t}^{\alpha\beta}_{\rm H})$ and $\tilde{G} = G/\phi_0$ (see Section 5.3.2 for definitions). Recall that quantities with tildes are defined using the conformally transformed metric $\tilde{g}_{\mu\nu} = (\phi/\phi_0)g_{\mu\nu}$. Because $\tilde{\tau}^{\alpha\beta}{}_{,\beta} = 0$ by virtue of the gauge condition on $\tilde{h}^{\alpha\beta}$, we can follow the same steps as for general relativity, Eqs. (11.88)–(11.90) to find, in the far zone, that

$$\tilde{h}^{jk} = \frac{2\tilde{G}}{R}\frac{d^2}{dt^2}\int_{\mathcal{M}} \tilde{\tau}^{00}(t-R, \mathbf{x}')x'^j x'^k d^3x'. \tag{11.111}$$

To lowest PN order, $\tilde{\tau}^{00} = \rho^* = \sum_a m_a\delta^3(\mathbf{x} - \mathbf{x}_a)$, where m_a is the mass of each body evaluated for the asymptotic value ϕ_0 of the scalar field, and we obtain

$$\tilde{h}^{jk} = \frac{2(1-\zeta)}{R}\ddot{I}^{jk}(t-R), \tag{11.112}$$

where I^{jk} is the quadrupole moment tensor of the source (see Eq. (11.91)), and we have used units in which $G_{\rm today} = \tilde{G}/(1-\zeta) = 1$. We now turn to the scalar wave equation in its relaxed, compact body form, Eq. (10.55b). To lowest PN order, it is straightforward to show that the equation takes the form $\Box\Psi = -8\pi\zeta\rho^*(1-2s)$, where the sensitivity s arises from the derivative $\partial T/\partial\phi$. Using the multipole expansion (11.88) in the far wave zone, and keeping the monopole and dipole terms, we obtain

$$\Psi = \frac{2\zeta}{R}\sum_a m_a(1-2s_a)\left(1+\boldsymbol{N}\cdot\boldsymbol{v}_a+\dots\right). \tag{11.113}$$

Dropping the constant term and specializing to a two-body system at rest at the origin of coordinates, we obtain

$$\Psi = \frac{4\zeta}{R}\eta m(s_2-s_1)\boldsymbol{N}\cdot\boldsymbol{v}. \tag{11.114}$$

Thus we see the presence of dipole gravitational radiation, generated by the difference in sensitivity or gravitational binding energy, between the two bodies. We also see that $\Psi \sim (m/R)v$, while the leading quadrupole contribution for "tensor" waves is $\tilde{h}^{jk} \sim (m/R)v^2$. According to our convention for labelling the orders of gravitational waves, for compact bodies, the dipole radiation contribution to Ψ is of -0.5PN order. Thus, in order to derive the full gravitational waveform to an order that is consistent with the Newtonian-order radiation of general relativity, we must also include the quadrupole moment contributions to Ψ, as well as the 1PN corrections to the monopole moment, which could have time-varying contributions at order $(m/R)v^2$. Furthermore, as we will see later, the energy flux in scalar-tensor theory is proportional to a sum of terms involving $(\tilde{h}^{jk}_{,0})^2$ and $(\Psi_{,0})^2$. This means that the leading dipole radiation term in Ψ will contribute to the flux at -1PN order compared to quadrupole radiation. Therefore, in addition to including the $O[(m/R)v^2]$ terms in Ψ, we must also include $(m/R)v^3$ terms, i.e. contributions of 0.5PN order, because in $(\Psi_{,0})^2$, there will be a cross-term between the -0.5PN dipole terms and the $+0.5$PN term, leading to a 0PN, or Newtonian-order term, at the same order as the quadrupole flux. Thus we must include the octupole moment in the scalar waveform as well as PN corrections to the dipole moment. In addition, in calculating the time derivatives of Ψ for the energy flux, we must use 1PN equations of motion in the -0.5PN dipole term, as these will lead to contributions at $+0.5$PN order.

For a compact binary system, the waveforms $\tilde{h}^{jk}$ and Ψ are given to the required orders (Lang, 2014, 2015) by

$$\tilde{h}^{jk} = 4(1-\zeta)\frac{\eta m}{R}\left(v^j v^k - \frac{\mathcal{G}m}{r}n^j n^k\right), \tag{11.115}$$

where $\mathcal{G} = 1-\zeta+\zeta(1-2s_1)(1-2s_2)$ [see Eq. (10.65)], and

$$\Psi = 2\mathcal{G}^{1/2}\frac{\zeta\eta m}{R}\left[\Psi_{-0.5\rm PN}+\Psi_{0\rm PN}+\Psi_{+0.5\rm PN}\right], \tag{11.116}$$

where

$$\begin{aligned}
\Psi_{-0.5\rm PN} &= 2\mathcal{S}_-\boldsymbol{N}\cdot\boldsymbol{v}, \\
\Psi_{0\rm PN} &= -\left(\mathcal{S}_+ + \Delta\mathcal{S}_-\right)\left[\frac{1}{2}v^2 - (\boldsymbol{N}\cdot\boldsymbol{v})^2 + \frac{\mathcal{G}m}{r}(\boldsymbol{N}\cdot\boldsymbol{n})^2\right] \\
&\quad - 2\frac{\mathcal{G}m}{r}\left[\mathcal{S}_+ - \frac{4}{\bar{\gamma}}\left(\mathcal{S}_+\bar{\beta}_+ + \mathcal{S}_-\bar{\beta}_-\right)\right],
\end{aligned}$$

$$\begin{aligned}
\Psi_{+0.5\text{PN}} = & \frac{1}{2}(\Delta\mathcal{S}_+ + (1-2\eta)\mathcal{S}_-)\left[\frac{\mathcal{G}m}{r}(\boldsymbol{N}\cdot\boldsymbol{n})^2(3\dot{r}\boldsymbol{N}\cdot\boldsymbol{n} - 7\boldsymbol{N}\cdot\boldsymbol{v}) + 2(\boldsymbol{N}\cdot\boldsymbol{v})^3\right] \\
& + \frac{\mathcal{G}m}{2r}\dot{r}\boldsymbol{N}\cdot\boldsymbol{n}\,(5\Delta\mathcal{S}_+ + 3\mathcal{S}_-) - v^2(\boldsymbol{N}\cdot\boldsymbol{v})\,(\Delta\mathcal{S}_+ + \eta\mathcal{S}_-) \\
& - \frac{\mathcal{G}m}{2r}(\boldsymbol{N}\cdot\boldsymbol{v})\,(\Delta\mathcal{S}_+ + (3-4\eta)\mathcal{S}_-) \\
& + 4\frac{\mathcal{G}m}{r}(\boldsymbol{N}\cdot\boldsymbol{v} - \dot{r}\boldsymbol{N}\cdot\boldsymbol{n})\frac{1}{\bar{\gamma}}\left[\Delta\left(\mathcal{S}_+\bar{\beta}_+ + \mathcal{S}_-\bar{\beta}_-\right) + \left(\mathcal{S}_-\bar{\beta}_+ + \mathcal{S}_-\bar{\beta}_-\right)\right],
\end{aligned} \tag{11.117}$$

where $\bar{\gamma}$, $\bar{\beta}_+$, and $\bar{\beta}_-$ are defined in Eqs. (10.71), and

$$\begin{aligned}
\mathcal{S}_- &\equiv \mathcal{G}^{-1/2}(s_2 - s_1), \\
\mathcal{S}_+ &\equiv \mathcal{G}^{-1/2}(1 - s_1 - s_2).
\end{aligned} \tag{11.118}$$

Note that $\bar{\gamma} = -2\zeta(\mathcal{S}_+^2 - \mathcal{S}_-^2)$.

Because of the presence of the scalar field, one can formulate a number of different conservation laws in scalar-tensor theory (Nutku, 1969; Lee, 1974). In the formulation used here, the conservation law $\tilde{\tau}^{\alpha\beta}_{\ ,\beta} = 0$ is related to the "tensor mass-energy" of the system, $P^0_{\rm T}$, given by

$$P^0_{\rm T} \equiv \int \tilde{\tau}^{00} d^3x. \tag{11.119}$$

One can then show that

$$\begin{aligned}
\frac{dP^0_{\rm T}}{dt} &= -\lim_{R\to\infty}\int \tilde{\tau}^{0j} d^2S_j \\
&= -\lim_{R\to\infty}\frac{R^2}{32\pi\tilde{G}}\oint\left[\tilde{h}^{jk}_{{\rm TT},0}\tilde{h}^{jk}_{{\rm TT},0} + (6+4\omega_0)\Psi_{,0}\Psi_{,0}\right]d\Omega.
\end{aligned} \tag{11.120}$$

Evaluating Eq. (11.119) to 1PN order, we find that $P^0_{\rm T}$ is given precisely by Eq. (11.81) for an N-body system and by Eq. (11.82) for a binary system. This is the total mass energy or the *inertial mass* of the system. As we discovered in Section 6.3, Eq. (6.59), in these theories, the inertial mass is *not* the same as the active gravitational, or Kepler-measured mass. Lee (1974) formulated a conservation law for the scalar mass $P^0_{\rm S}$ and from that defined a conservation law for the active or Kepler mass $P^0_{\rm A} = P^0_{\rm T} + P^0_{\rm S}$. Even though the Kepler mass is considered the defining mass for an isolated system, the tensor mass can be directly related via Eq. (11.82) to measurable quantities such as the semimajor axis, which determines the orbital period and the frequencies of gravitational waves. Thus we will use the expression (11.120). Substituting Eqs. (11.115), (11.116), and (11.117) into (11.120) and integrating over solid angle, we obtain Eq. (11.59), with $\mathcal{S} = \mathcal{S}_-$ and

$$\begin{aligned}
\kappa_1 = & 12 + 5\bar{\gamma} - 5\zeta\mathcal{S}_-^2(3 + \bar{\gamma} + 2\bar{\beta}_+ + 2\Delta\bar{\beta}_-) \\
& + 10\zeta\mathcal{S}_-\left(\frac{\mathcal{S}_-\bar{\beta}_+ + \mathcal{S}_+\bar{\beta}_-}{\bar{\gamma}}\right) + 10\zeta\Delta\mathcal{S}_-\left(\frac{\mathcal{S}_+\bar{\beta}_+ + \mathcal{S}_-\bar{\beta}_-}{\bar{\gamma}}\right),
\end{aligned}$$

$$
\begin{aligned}
\kappa_2 &= 11 + \frac{45}{4}\bar{\gamma} - 40\bar{\beta}_+ - 5\zeta\mathcal{S}_-^2\left[17 + 6\bar{\gamma} + \eta + 8\bar{\beta}_+ + 8\Delta\bar{\beta}_-\right] \\
&\quad + 90\zeta\mathcal{S}_-\left(\frac{\mathcal{S}_-\bar{\beta}_+ + \mathcal{S}_+\bar{\beta}_-}{\bar{\gamma}}\right) - 30\zeta\Delta\mathcal{S}_-\left(\frac{\mathcal{S}_+\bar{\beta}_+ + \mathcal{S}_-\bar{\beta}_-}{\bar{\gamma}}\right) \\
&\quad - 120\zeta\left(\frac{\mathcal{S}_+\bar{\beta}_+ + \mathcal{S}_-\bar{\beta}_-}{\bar{\gamma}}\right)^2, \\
\kappa_D &= 4\zeta\,. \qquad (11.121)
\end{aligned}
$$

These results are in complete agreement with the total energy flux to -1PN and 0PN orders, as calculated by Damour and Esposito-Farèse (1992a) in the DEF formulation of scalar-tensor theories, and with the energy loss derived from radiation reaction terms in the equations of motion (Mirshekari and Will, 2013). They disagree with the flux formula of Will and Zaglauer (1989) (see also Alsing et al. (2012)), who failed to take into account PN corrections to the dipole term induced by PN corrections in the equations of motion, and the dipole-octupole cross term in the scalar energy flux. To date, the analogous expressions for angular momentum change have not been calculated.

From Eqs. (5.36), (5.40) and (5.41), we obtain the leading contributions to the physical metric in the far zone,

$$
\begin{aligned}
g_{00} &= -1 + \frac{1}{2}\left(\tilde{h}^{00} + \tilde{h}^{kk} + 2\Psi\right), \\
g_{0j} &= -\tilde{h}^{0j}, \\
g_{jk} &= \delta_{jk} + \tilde{h}^{jk} + \frac{1}{2}\delta_{jk}\left(\tilde{h}^{00} - \tilde{h}^{mm} - 2\Psi\right), \qquad (11.122)
\end{aligned}
$$

from which we obtain

$$
R_{0j0k} = -\frac{1}{2}\frac{d^2}{dt^2}\tilde{h}^{jk}_{\rm TT} + \frac{1}{2}\left(\delta_{jk} - N_jN_k\right)\frac{d^2\Psi}{dt^2}\,. \qquad (11.123)
$$

Substituting Eqs. (11.115), (11.116), and (11.117) and dropping constant terms, we obtain the amplitudes for a circular orbit

$$
\begin{aligned}
A_+ &= -\frac{2(1-\zeta)\eta\mathcal{G}m^2}{r}(1+\cos^2\iota)\cos 2\phi\,, \\
A_\times &= -\frac{4(1-\zeta)\eta\mathcal{G}m^2}{r}\cos\iota\sin 2\phi\,, \\
A_{\rm S} &= 4\zeta\eta m[\mathcal{G}^{1/2}\mathcal{S}_-]\left(\frac{\mathcal{G}m}{r}\right)^{1/2}\sin\iota\cos\phi \\
&\quad + \frac{2\zeta\eta\mathcal{G}m^2}{r}[\mathcal{G}^{1/2}(\mathcal{S}_+ + \Delta\mathcal{S}_-)]\sin^2\iota\cos 2\phi\,. \qquad (11.124)
\end{aligned}
$$

It is worth noting that, for binary black holes, $s_1 = s_2 = 1/2$, and thus $\mathcal{S}_- = \mathcal{S}_+ = 0$, $\mathcal{G} = 1-\zeta$, $\bar{\gamma} = \bar{\beta}_i = 0$ (also $\bar{\beta}_\pm/\bar{\gamma} = 0$). Thus $\kappa_1 = 12$, $\kappa_2 = 11$, $\kappa_D = 0$, and if we replace the mass m with $m/(1-\zeta)$, the waveform and energy flux become identical to those of general relativity, just as we found for the PN equations of motion in Section 10.3.2. For a black-hole neutron-star binary system, the energy flux parameters are $\mathcal{G} = 1-\zeta$,

$\kappa_1 = 12 - 15\mathcal{Q}/4$, $\kappa_2 = 11 - 5\mathcal{Q}(17+\eta)/4$ and in the dipole flux term, $\kappa_D \mathcal{S}_-^2 = \mathcal{Q}$, where $\mathcal{Q} = \zeta(1-\zeta)^{-1}(1-2s_1)^2$, and s_1 is the sensitivity of the neutron star.

These waveforms and fluxes were extended to 2PN order for the tensor waves (Lang, 2014) and to 1PN order for the scalar waves (Lang, 2015). Completing the calculation of the scalar flux to 2PN order requires calculating the monopole and dipole moments of the scalar field to 3PN order, and knowing the equations of motion to 3PN order!

11.5.4 Other theories

Very few other theories of gravity have had their gravitational-wave predictions for compact binaries analyzed to comparable levels. Yagi et al. (2014a) found the Newtonian-order energy flux for compact binaries in Einstein-Æther and Khronometric theories. They did not derive all the sensitivity-dependent contributions in the vector-field sector required for a complete and consistent solution, but they argued that the neglected terms would be negligible for the applications to binary pulsar data that they carried out.

Many of the theories that were studied during the 1970s as either alternatives to general relativity or as "straw-man" examples, were found to predict dipole gravitational radiation. Even worse, in most cases the dipole radiation, and sometimes also the quadrupole radiation carried negative energy (Will, 1977). For example, Rosen's bimetric theory predicted $\kappa_1 = -21/2$, $\kappa_2 = -23/2$ and $\kappa_D = -20/3$, and the theory suffered a fatal blow when data showed that the orbit of the binary pulsar was shrinking, not expanding (Will and Eardley, 1977).

Because of the real possibility of carrying out tests of alternative theories of gravity using recent detections of gravitational waves, the effort to calculate the gravitational radiation predictions in a range of alternative theories will surely intensify.

12 Strong-Field and Dynamical Tests of Relativistic Gravity

We now take the theoretical frameworks for discussing compact body dynamics and gravitational radiation that we developed in the preceding two chapters, and apply them to study current and future tests of general relativity in the strong-field and dynamical regimes. This effort began with the discovery of binary pulsars beginning in 1974, and in Section 12.1, we will discuss these remarkable systems and their implications for testing relativistic gravity. The detection of gravitational waves from an inspiraling binary black hole system in 2015 brought to fruition the ability to test gravitational theories in the strong-field dynamical regime using the gravitational waves themselves. In Section 12.2 we will describe how information about the sources of the waves and about the underlying theory can be extracted from the detected signal, and will discuss the tests that have been carried out since 2015. Another kind of test of strong-field gravity involves exploring the spacetime near otherwise stationary compact objects using probes such as stars, small black holes and gas. Many of these tests are still works in progress; we will give some examples in Section 12.3. Finally, Section 12.4 will close this chapter with a brief description of cosmological tests.

12.1 Binary Pulsars

The summer of 1974 was an eventful one for Joseph Taylor and Russell Hulse. Using the Arecibo radio telescope in Puerto Rico, they had spent the time engaged in a systematic survey for new pulsars. During that survey, they detected 50 pulsars, of which 40 were not previously known, and made a variety of observations, including measurements of their pulse periods to an accuracy of one microsecond. But one of these pulsars, denoted B1913+16,[1] was peculiar: besides having a pulsation period of 59 ms, shorter than that of any known pulsar except the one in the Crab Nebula, it also defied any attempts to measure its period to $\pm 1\,\mu$s, by making "apparent period changes of up to 80 μs, from day to day, and sometimes by as much as 8 μs, over 5 minutes" (Hulse and Taylor, 1975). Such behavior is uncharacteristic of pulsars, and Hulse and Taylor rapidly concluded that the observed period changes were the result of Doppler shifts due to orbital motion of the pulsar about a companion (for a popular account of the day-by-day detective work

[1] At the time of the discovery, pulsars were denoted PSR, followed by the right ascension and declination. PSR was ultimately replaced by "B", for pulsars whose coordinates were referred to the 1950.0 epoch. Most pulsars discovered since 1993 are labeled with "J", have coordinates referred to the 2000.0 epoch, and include minutes in the declination. We will use the current conventions.

involved see Will (1986)). By the end of September, 1974, Hulse and Taylor had obtained an accurate "velocity curve" of this "single line spectroscopic binary." The velocity curve was a plot of apparent period of the pulsar as a function of time. By a detailed fit of this curve to a Keplerian two-body orbit, they obtained the following elements of the orbit of the system: K_1, the semiamplitude of the variation of the radial velocity of the pulsar; P_b, the period of the binary orbit; e, the eccentricity of the orbit; ω_0, the longitude of periastron at a chosen epoch (September 1974); $a_1 \sin \iota$, the projected semimajor axis of the pulsar orbit, where ι is the inclination of the orbit relative to the plane of the sky; and $f_1 = (m_2 \sin \iota)^3/(m_1+m_2)^2$, the mass function, where m_1 and m_2 are the mass of the pulsar and companion, respectively. In addition, they obtained the "rest" period P_p of the pulsar, corrected for orbital Doppler shifts at a chosen epoch.

However, at the end of September 1974, the observers switched to an observation technique that yielded vastly improved accuracy (Taylor et al., 1976). That technique measures the absolute arrival times of pulses (as opposed to the period, or the difference between adjacent pulses) and compares those times to arrival times predicted using the best available pulsar and orbit parameters. The parameters are then improved by means of a least-squares analysis of the arrival-time residuals. With this method, it proved possible to keep track of the precise phase of the pulsar over intervals as long as six months between observations. This was partially responsible for the improvement in accuracy. Other improvements over the years included the sophisticated use of multi-wavelength data to suppress the effects of interstellar dispersion on the radio signals, the use of GPS time transfer to improve the timing precision, and a major upgrade of instrumentation at the Arecibo radio telescope between 1993 and 1997. The results of these analyses using data reported in 2016 are shown in column 3 of Table 12.1 (Weisberg and Huang, 2016); in Section 12.1.1 we will define the Keplerian and post-Keplerian parameters.

The discovery of B1913+16 caused considerable excitement in the relativity community, because it was realized that the system could provide a new laboratory for studying relativistic gravity. Post-Newtonian orbital effects would have magnitudes of order $v_1^2 \sim K_1^2 \sim 5 \times 10^{-7}$, or $m/r \sim f_1/a_1 \sin \iota \sim 3 \times 10^{-7}$, a factor of ten larger than the corresponding quantities for Mercury's orbit, and the shortness of the orbital period ($\sim$ 8 hours) would amplify any secular effect such as the periastron shift. This expectation was confirmed by the announcement in December 1974 (Taylor, 1975) that the periastron shift had been measured to be $4.0 \pm 1.5\,^\circ\,\mathrm{yr}^{-1}$ (compare with Mercury's 43 arcseconds per century!), implying (see below) that the total mass of the system was about $2.8\,M_\odot$. Moreover, the system appeared to be a "clean" laboratory, unaffected by complex astrophysical processes such as mass transfer.

The pulsar radio signal was never eclipsed by the companion, placing limits on the geometrical size of the companion, and the dispersion of the pulsed radio signal showed little change over an orbit, indicating an absence of dense plasma in the system, as would occur if there were mass transfer from the companion onto the pulsar. These data effectively ruled out a main-sequence star as a companion: although such a star could conceivably fit the geometrical constraints placed by the eclipse and dispersion measurements, it would produce an enormous periastron shift ($>5000\,^\circ\,\mathrm{yr}^{-1}$) generated by tidal deformations due to the pulsar's gravitational field. Other candidates for the

Table 12.1 Arrival-time parameters for binary pulsars, and their values in B1913+16. Numbers in parentheses denote errors in the last digit. Data taken from Weisberg and Huang (2016).

Parameter	Symbol (units)	Value in B1913+16
(i) Astrometric and pulsar parameters		
Right ascension	α	$19^{h}15^{m}27.^{s}99942(3)$
Declination	δ	$16^{\circ}06'27.''3868(5)$
Pulsar period	P_p (ms)	59.030003217813(11)
Derivative of period	$\dot{P}_p$	$8.6183(3) \times 10^{-18}$
(ii) Keplerian parameters		
Projected semimajor axis	$a_1 \sin \iota$ (s)	2.341776(2)
Eccentricity	e	0.6171340(4)
Orbital period	P_b (day)	0.322997448918(3)
Longitude of periastron	ω_0 (°)	292.54450(8)
Julian date of periastron passage	T_0 (MJD)	52144.90097849(3)
(iii) Post-Keplerian parameters		
Mean rate of periastron advance	$\langle\dot{\omega}\rangle$ (° yr^{-1})	4.226585(4)
Redshift/time dilation	γ' (ms)	4.307(4)
Derivative of orbital period	$\dot{P}_b$ (10^{-12})	−2.423(1)
Range of Shapiro delay	r (μs)	$9.6^{+2.7}_{-3.5}$
Shape of Shapiro delay	$s = \sin \iota$	$0.68^{+0.10}_{-0.06}$

companion that were considered early on were a helium main-sequence star, a white dwarf, a neutron star and a black hole. Any of these would be consistent with the evolutionary models for binary systems of two massive stars that were popular at the time. In these models, one massive star evolves more rapidly, undergoing a supernova explosion and leaving a neutron star remnant. Mass transfer from the companion star serves to spin up the neutron star to its present 59 ms rotation period (this is the so-called pulsar recycling model, believed to be responsible for the class of millisecond pulsars). Subsequently, the massive companion star evolved rapidly, possibly undergoing its own explosion, leaving one of the four remnants listed earlier.

The first two companion candidates, the helium star and the white dwarf, fell out of favor because of the complete absence of evidence for orbital perturbations that would be induced by tidal or rotational deformation effects. A black hole companion was disfavored when the observed pericenter advance indicated that the total mass of the system was about $2.8\,M_\odot$; most evolutionary scenarios leading to black holes, and data on X-ray binaries containing black holes suggest that black holes in such systems are significantly more massive than $1.4\,M_\odot$. A neutron star remnant from a supernova explosion of the companion star was viewed as the most compatible with the near equality of the two masses, with the large orbital eccentricity (mass loss from the second supernova explosion can easily lead to disruption of the system), and with the utter orbital "cleanliness" of the system. In this scenario, the companion neutron star is the younger of the two. No evidence was ever found for pulsed radiation from the companion, so it is either a pulsar whose signal does not intersect the Earth, or more likely is a dead pulsar, having spun down to a stage where it is no longer able to generate the magnetic fields required to emit radio waves. The main pulsar is an old, weakly magnetized, recycled pulsar, with very weak radio emission (indeed it was barely above the detection threshold set by Hulse in 1974), and a very low spin-down rate.

Within weeks of the discovery, it was recognized that the system might be a testing ground for gravitational radiation damping (Wagoner, 1975; Damour and Ruffini, 1974). The observable effect of this damping is a secular decrease in the period of the orbit (see Section 11.5). However, the timescale for this change, according to general relativity, is so long ($\sim 10^9$ yr) that it was thought that 10–15 years of data would be needed to detect it. However, with improved data acquisition equipment and continued ability to "keep in phase" with the pulsar with the arrival-time method, Taylor and his collaborators surpassed all expectations, and in December 1978 they announced a measurement of the rate of change of the orbital period in an amount consistent with the prediction of gravitational radiation damping in general relativity (Taylor et al., 1979; Taylor and McCulloch, 1980). In 1993, Hulse and Taylor were awarded the Nobel Prize in Physics for this discovery.

In 1990, two more binary pulsars were detected, one eerily similar to B1913+16 in its orbital parameters. Today, more than 220 binary pulsars have been detected, of which 70 have orbital periods shorter than one day (the shorter the period, the more relativistic the system). These include the famous "double pulsar," where, for a time, pulses from *both* neutron stars were also detected; several pulsars with a white dwarf companion; and most recently, a pulsar in a triple system, orbited by two white dwarfs. There are also a number of pulsars with planets revolving around them; while these are fascinating in their own right, they are less interesting for relativity. We will describe the more important denizens of this zoo of binary pulsars and their implications for testing relativistic gravity in Section 12.1.2, but first we turn to the arrival-time analysis for studying these systems.

12.1.1 Arrival-time analysis for binary pulsars

The analysis and interpretation of data from binary pulsars is based on an "arrival-time" framework, originally carried out by Blandford and Teukolsky (1976) and extended by Epstein (1977), Haugan (1985), Damour and Deruelle (1986) and Damour and Taylor (1992). Here we present a simplified version that illustrates the basic parameters that can be measured, and how they can be interpreted in order to test theories of gravity.

We begin by setting up a suitable coordinate system (see Figure 12.1). We choose quasi-Cartesian coordinates $(t, \boldsymbol{x})$ in which the physical metric is of post-Newtonian order everywhere, except in the neighborhood and interior of the pulsar and its companion (if the latter is also compact), and is asymptotically flat. The origin of the coordinate system coincides with a suitably chosen center of mass of the binary system, assumed to be at rest with respect to the solar system. The reference $X-Y$ plane is the plane of the sky, with the X-axis conventionally pointing toward the North celestial pole. The observer is chosen to be at great distance on the negative Z-axis. The orbit of the pulsar is characterized by the standard osculating orbit elements described in Section 6.4.2, and illustrated in

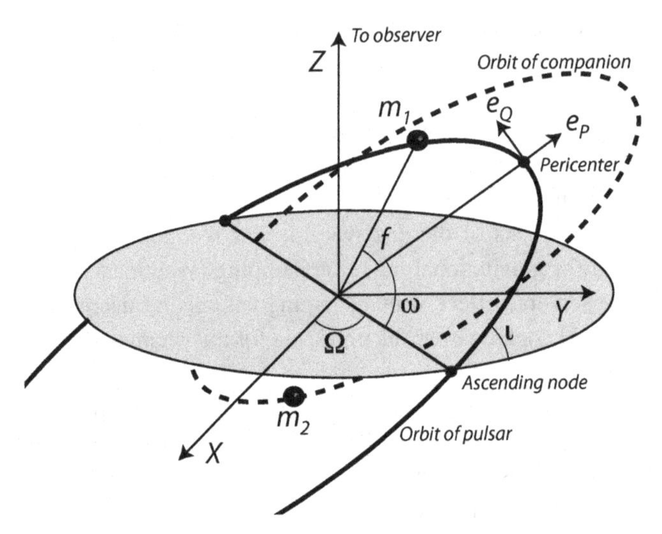

Fig. 12.1 Orbits of a binary pulsar system.

Figure 12.1. However, instead of using the definitions of Eq. (6.73) based on the true anomaly f, we use definitions based on the so-called "eccentric anomaly" u (see PW, section 3.2.4 for detailed discussion), which is more useful in an arrival-time analysis. It is directly related to time by the equation

$$u - e \sin u = n(t - T_0) , \tag{12.1}$$

where n is the "mean anomaly," defined by $n \equiv (\mathcal{G}m/a^3)^{1/2} = 2\pi/P_b$, where a is the semimajor axis, P_b is the orbital period of the binary, and T_0 is the time of pericenter passage. We include the modified EIH parameter $\mathcal{G}$, because that parameter (including possible contributions from the sensitivities of the compact bodies) will govern the quasi-Newtonian limit used to describe the osculating elements in a given theory. The true and eccentric anomalies are related by the equations

$$\cos f = \frac{\cos u - e}{1 - e \cos u} , \quad \sin f = \frac{\sqrt{1 - e^2} \sin u}{1 - e \cos u} . \tag{12.2}$$

With the center of mass $\boldsymbol{X} = (m_1 \boldsymbol{x}_1 + m_2 \boldsymbol{x}_2)/m$ chosen to be at rest at the origin, the orbits of the pulsar (1) and companion (2) are given by

$$\boldsymbol{x}_1 = \frac{m_2}{m} \boldsymbol{x} , \quad \boldsymbol{x}_2 = -\frac{m_1}{m} \boldsymbol{x} , \tag{12.3}$$

where $\boldsymbol{x} = \boldsymbol{x}_1 - \boldsymbol{x}_2$ is given by

$$\boldsymbol{x} = a \left[(\cos u - e) \boldsymbol{e}_P + \sqrt{1 - e^2} \sin u \, \boldsymbol{e}_Q \right] , \tag{12.4}$$

where $\boldsymbol{e}_P$ and $\boldsymbol{e}_Q$ are unit vectors defined by Eq. (8.29); they point in the direction of the pericenter of body 1 and perpendicular to that vector within the orbital plane (Figure 12.1). Note that the unit vector in the direction of the orbital angular momentum is given by $\hat{\boldsymbol{h}} = \boldsymbol{e}_P \times \boldsymbol{e}_Q$. The distance between the two bodies is given by $r = a(1 - e \cos u)$.

Any perturbation of the two-body system will induce periodic and secular changes in the orbit elements that can be calculated using the Lagrange planetary equations (6.74).

We next consider the emission of the radio signals by the pulsar. Let τ be proper time as measured by a hypothetical clock in an inertial frame that is momentarily comoving with respect to the pulsar. The time of the Nth rotation of the pulsar is given in terms of the rotation frequency ν of the neutron star by

$$N = N_0 + \nu\tau + \frac{1}{2}\dot{\nu}\tau^2 + \frac{1}{6}\ddot{\nu}\tau^3 + \dots , \tag{12.5}$$

where N_0 is an arbitrary integer, and $\dot{\nu} = d\nu/d\tau|_{\tau=0}$, $\ddot{\nu} = d^2\nu/d\tau^2|_{\tau=0}$. We will ignore the possibility of discontinuous jumps, or "glitches," in the rotation frequency of the pulsar, or of drifts of the phase of the radio beam with respect to the neutron star's rotation phase. These are complicated issues that are still not fully understood. We ultimately wish to determine the arrival time of the Nth pulse on Earth.

Outside the pulsar and its companion, the metric in our chosen coordinate system is given by the modified EIH metric, Eq. (10.28). Because we are interested in the propagation of the radio signal away from the system, we will ignore the possibility of large strong-field corrections to the metric in the close vicinity of the pulsar or its companion. The main result of such effects will be either to add a constant term in the emission time formula (12.5) that can be absorbed into the arbitrary value of N_0, or to multiply terms by a constant factor, such as the redshift at the surface of the neutron star, that can be absorbed into the unknown intrinsic value of ν. Modulo such factors, proper time τ in the local comoving inertial frame, evaluated at the pulsar's center of mass, can be related to coordinate time t by

$$d\tau = dt\left[1 - \alpha_2^*\frac{m_2}{r} - \frac{1}{2}v_1^2 + O(\epsilon^2)\right] , \tag{12.6}$$

where we have dropped the constant contribution of the pulsar's gravitational potential at the point in the inertial frame chosen to be the point of "emission" of the signal, $\alpha_1^* m_1/|\boldsymbol{x}_{\rm em} - \boldsymbol{x}_1|$, and have ignored the difference between the velocity of the emission point and the pulsar's center of mass $\boldsymbol{x}_1$. The two correction terms in Eq. (12.6) are the gravitational redshift and the second-order Doppler shift, or time dilation.

Using Eqs. (12.1) and (12.4), it is simple to show that $v_1^2 = \mathcal{G}(m_2^2/m)(2/r - 1/a)$, and thus that

$$\frac{d\tau}{dt} = 1 - \frac{m_2}{a}\frac{\alpha_2^* + \mathcal{G}m_2/m}{1 - e\cos u} + \frac{\mathcal{G}m_2^2}{2ma} . \tag{12.7}$$

Integrating with respect to t, we obtain

$$\tau = \left[1 - \frac{m_2}{a}\left(\alpha_2^* + \mathcal{G}\frac{m_2}{2m}\right)\right]t_e - \frac{m_2}{a}\left(\alpha_2^* + \mathcal{G}\frac{m_2}{m}\right)\frac{P_b}{2\pi}e\,(\sin u - \sin u_0) , \tag{12.8}$$

where u and u_0 are the values of the eccentric anomaly at $t = t_e$ and $t = 0$, respectively. The factor multiplying t_e can be absorbed into the definition of the frequency ν, and the constant term involving u_0 can be dropped. Although these constants may actually undergo secular or periodic variations in time due to orbital perturbations or other effects, these will be small perturbations of post-Newtonian corrections, and will have negligible effect. Thus we obtain

$$\tau = t_e - \frac{m_2}{a}\left(\alpha_2^* + \mathcal{G}\frac{m_2}{m}\right)\frac{P_b}{2\pi} e \sin u \,. \tag{12.9}$$

After emission, the pulsar signal travels along a null geodesic. We can therefore use the method in Sections 6.1 and 7.2 to calculate the coordinate time taken for the signal to travel from the pulsar to the solar system barycenter $\boldsymbol{x}_0$, with the result

$$t_{arr} - t_e = |\boldsymbol{x}_0(t_{arr}) - \boldsymbol{x}_1(t_e)| + (\alpha_2^* + \gamma_2^*) m_2 \ln\left[\frac{2r_0(t_{arr})}{r(t_e) + \boldsymbol{n}\cdot\boldsymbol{x}(t_e)}\right], \tag{12.10}$$

where $r_0 = |\boldsymbol{x}_0|$, $\boldsymbol{n} = \boldsymbol{x}_0/r_0$ and we have used the fact that $r_0 \gg r_1$. The second term in Eq. (12.10) is the Shapiro time delay of the pulsar signal in the gravitational field of the companion. The time delay due to the pulsar's own field is constant to the required accuracy and has been dropped. The effect of the companion's motion during the propagation of the signal across the orbit is a higher post-Newtonian effect, and thus has been ignored.

In practice, one must take into account the fact that the measured arrival time is that at the Earth and not at the barycenter of the solar system, and will therefore be affected by the Earth's position in its orbit and by its own gravitational redshift and Doppler-shift corrections. In fact, it is the effect of the Earth's orbital position on the arrival times that permits accurate determinations of the pulsar's position on the sky. It is also necessary to take into account the effects of interstellar dispersion on the radio signal. These effects can be handled in a standard manner and will not be treated here.

Now, because $r_0 \gg r$, we may write

$$|\boldsymbol{x}_0(t_{arr}) - \boldsymbol{x}_1(t_e)| = r_0(t_{arr}) - \boldsymbol{x}_1(t_e)\cdot\boldsymbol{n} + O(r/r_0)\,. \tag{12.11}$$

Combining Eqs. (12.10) and (12.11), resetting the arrival time by the constant offset r_0, evaluating r and $\boldsymbol{x}_1$ at t_{arr} instead of at t_e, and dropping higher-order terms, we obtain

$$t_e = t_{arr} + (\boldsymbol{x}_1(t_{arr})\cdot\boldsymbol{n})(1 + \boldsymbol{v}_1(t_{arr})\cdot\boldsymbol{n}) - \Delta t_S(t_{arr})\,, \tag{12.12}$$

where Δt_S is the Shapiro term in Eq. (12.10).

We can now combine Eqs. (12.9) and (12.12) and substitute the result into Eq. (12.5) to relate t_{arr} to the pulse number N. Here we must include another effect that can occur in theories of gravity that violate SEP (Eardley, 1975), whereby the local gravitational constant at the location of the pulsar may depend on the gravitational potential of the companion, that is,

$$G_L = G_0\left(1 - \eta_2^*\frac{m_2}{r}\right), \tag{12.13}$$

where η^* is a parameter that depends on the theory, and possibly on the sensitivities of the two bodies. As G_L varies during the orbital motion, the structure of the pulsar, its moment of inertia, and thus its intrinsic rotation rate will vary, according to

$$\frac{\Delta\nu}{\nu} = -\frac{\Delta I}{I} = \kappa_{(1)}\frac{\Delta G_L}{G_L} = -\kappa_{(1)}\eta_2^*\frac{m_2}{r}, \tag{12.14}$$

where $\kappa_{(1)}$ is the sensitivity of the moment of inertia of body 1 to variations in G_L. Thus Eq. (12.5) should be rewritten

$$N = N_0 + \nu T + \int \Delta\nu dT + \frac{1}{2}\dot{\nu}T^2 + \frac{1}{6}\ddot{\nu}T^3 + \dots, \tag{12.15}$$

where to the necessary order, and modulo constants,

$$\int \Delta\nu dT = -\nu\kappa_{(1)}\eta_2^* \frac{m_2}{a}\frac{P_b}{2\pi} e \sin u, \tag{12.16}$$

and where we have not applied a similar correction to the smaller terms involving $\dot{\nu}$ and $\ddot{\nu}$. Substituting Eqs. (12.9), (12.12), and (12.16) into (12.15), and dropping the constant N_0 we obtain the timing formula

$$t_{\rm arr} = \nu^{-1}N - \Delta_{\rm R}(u) - \Delta_{\rm E}(u) + \Delta_{\rm S}(u) - \frac{1}{2}\dot{\nu}\nu^{-3}N^2 - \frac{1}{6}\ddot{\nu}\nu^{-4}N^3 + \dots, \tag{12.17}$$

where the three "delay" terms are given by

$$\begin{aligned} \Delta_{\rm R}(u) &= xF(e,\omega,u)\left[1 + x\dot{F}(e,\omega,u)\right], \\ \Delta_{\rm E}(u) &= \gamma' \sin u, \\ \Delta_{\rm S}(u) &= 2r \ln\left[1 - e\cos u - sF(e,\omega,u)\right], \end{aligned} \tag{12.18}$$

where

$$\begin{aligned} x &\equiv a_1 \sin\iota = \frac{m_2}{m} a \sin\iota, \\ \gamma' &\equiv e\, m_2 \left(\frac{P_b}{2\pi\mathcal{G}m}\right)^{1/3} \left(\alpha_2^* + \mathcal{G}\frac{m_2}{m} + \kappa_{(1)}\eta_2^*\right), \\ r &\equiv \frac{1}{2}(\alpha_2^* + \gamma_2^*)m_2, \\ s &\equiv \sin\iota. \end{aligned} \tag{12.19}$$

The function $F(e,\omega,u)$ is given by

$$F(e,\omega,u) \equiv \sin\omega(\cos u - e) + \cos\omega(1-e^2)^{1/2}\sin u, \tag{12.20}$$

where $\dot{F} = dF/dt$, and u is related to $t_{\rm arr}$ by

$$u - e\sin u = n(t_{\rm arr} - T_0). \tag{12.21}$$

Equation (12.17) gives $t_{\rm arr}$ in terms of the pulse number N, a set of orbital and relativistic parameters, and the intrinsic spin parameters ν, $\dot{\nu}$ and $\ddot{\nu}$. From an initial guess for the values of these parameters, a prediction for the arrival time of the Nth pulse can be made. The difference between the predicted arrival time and the observed arrival time is then used to correct the parameters using a suitable parameter estimation technique, such as least-squares.

The term $\Delta_{\rm R}$ in Eq. (12.17) is called the "Roemer" delay. It is simply the shift in arrival time caused by the change in the pulsar's location relative to the center of mass. The amplitude is controlled by the parameter x, the projected semimajor axis of the pulsar on the plane of the sky. The evolution of this term with respect to $t_{\rm arr}$ is governed by the orbital period P_b, the eccentricity e, the pericenter angle ω and the pericenter time T_0. Because this is a purely geometric effect coupled with Newtonian gravity, these five parameters are known as the "Keplerian" parameters of the system, as listed in Table 12.1. This is the time-domain version of the standard technique for determining orbits in Newtonian single-line spectroscopic binary systems whereby one measures the Doppler shift of the pulse period or spectral line. Notice that the combination

$$f_1 \equiv x^3 \left(\frac{2\pi}{P_b}\right)^2 = \mathcal{G}\frac{(m_2 \sin\iota)^3}{m^2}, \tag{12.22}$$

is the standard "mass function" of spectroscopic binaries.

The Roemer delay is the dominant correction, of order $r/\tau \sim v \sim \epsilon^{1/2}$. The Einstein and Shapiro terms are of order ϵ. Thus, if the orbit elements a, e, ω, ι experience any secular changes, they will be observable in the Roemer term, if at all (a change in the angle of nodes Ω will be unobservable, as it merely rotates the system about the line of sight). It is conventional to define

$$\begin{aligned} \omega &= \omega_0 + \langle\dot\omega\rangle(t - t_0) + \ldots, \\ P_b &= P_{b0} + \frac{1}{2}\dot P_b(t - t_0) + \ldots, \end{aligned} \tag{12.23}$$

where ω_0 and P_{b0} and their time derivatives are defined at a chosen epoch t_0. Note that the factor $1/2$ in the expression for P_b comes from the formal definition of P_b in terms of the semimajor axis a. One then substitutes Eqs. (12.23) into Δ_R and includes $\dot\omega$ and $\dot P_b$ as parameters to be estimated. The parameters $\dot\omega$ and $\dot P_b$ are two of the post-Keplerian parameters listed in Table 12.1.

In principle one could include secular variations in e and ι; such variations have recently been detected in B1913+16 (Weisberg and Huang, 2016), resulting from a spin-orbit induced precession of the orbital plane. Furthermore, as we saw in Sections 8.2 and 8.4, and will reiterate below, it has been possible to search for periodic variations in these parameters in some systems, and to place important limits on violations of SEP.

The term Δ_E is the Einstein term, also called the Redshift/time dilation term, with amplitude γ'. The parameter γ' (not to be confused with the PPN parameter γ) is another post-Keplerian parameter. Note that, once e and P_b are determined from the Roemer term, it depends only on the masses of the bodies and possibly on their sensitivities, through α_2^*, $\mathcal{G}$, κ_1, and η_2^*.

The final relativistic term is the Shapiro delay Δ_S, dependent upon the "range" parameter r, and the "shape" parameter s. The former depends on the mass of the companion and possibly on sensitivities, while the latter is simply $\sin\iota$.

The final two post-Keplerian parameters in our discussion are $\langle\dot\omega\rangle$ and $\dot P_b$, whose predicted forms are given by Eqs. (10.49) and (11.85). Box 12.1 summarizes these predictions for the post-Keplerian parameters.

Box 12.1 Predictions for post-Keplerian parameters

Here we summarize the predictions for the post-Keplerian parameters in alternative theories of gravity.

Redshift/time dilation:

$$\gamma' = e\, m_2 \left(\frac{P_b}{2\pi \mathcal{G} m} \right)^{1/3} \left(\alpha_2^* + \mathcal{G} \frac{m_2}{m} + \kappa_{(1)} \eta_2^* \right),$$

Shapiro delay range:

$$r = \frac{1}{2} (\alpha_2^* + \gamma_2^*) m_2 .$$

Shapiro delay shape:

$$s = \sin \iota .$$

Rate of advance of periastron

$$\langle \dot{\omega} \rangle = \frac{6\pi}{P_b (1 - e^2)} \left(\frac{2\pi \mathcal{G} m}{P_b} \right)^{2/3} \mathcal{P} \mathcal{G}^{-2} .$$

Derivative of orbital period

$$\dot{P}_b = -\frac{192\pi}{5} \left(\frac{2\pi \mathcal{G} \mathcal{M}}{P_b} \right)^{5/3} F(e) - 2\pi \kappa_D \eta \mathcal{S}^2 \left(\frac{2\pi \mathcal{G} m}{P_b} \right) G(e) . \tag{12.24}$$

The coefficients α_2^* and γ_2^* appear in the post-Newtonian metric of body 2 [Eqs. (10.28)], while the parameter η_2^* and the sensitivity $\kappa_{(1)}$ refer to the effect of the field of body 2 on the moment of inertia of body 1 [Eqs. (12.13)–(12.16)]. The quantity $\mathcal{G}$ appears in the quasi Newtonian limit of the modified EIH equations of motion, $\boldsymbol{a} = -\mathcal{G} m \boldsymbol{x}/r^3$, while $\mathcal{P}$ is given by Eq. (10.48). The functions $F(e)$ and $G(e)$ are given by Eq. (11.84); κ_D is the dipole radiation parameter of the theory, and $\mathcal{S}$ is related to the difference in sensitivities between the two bodies.

In general relativity, the post-Keplerian parameters simplify to

$$\gamma' = e m_2 \left(\frac{P_b}{2\pi m} \right)^{1/3} \left(1 + \frac{m_2}{m} \right),$$

$$r = m_2 ,$$

$$s = \sin \iota ,$$

$$\langle \dot{\omega} \rangle = \frac{6\pi}{P_b (1 - e^2)} \left(\frac{2\pi m}{P_b} \right)^{2/3},$$

$$\dot{P}_b = -\frac{192\pi}{5} \left(\frac{2\pi \mathcal{M}}{P_b} \right)^{5/3} F(e) , \tag{12.25}$$

where

$$F(e) = (1 - e^2)^{-7/2} \left(1 + \frac{73}{24} e^2 + \frac{37}{96} e^4 \right). \tag{12.26}$$

The formulae for $\dot{P}_b$ include gravitational radiation contributions only through quadrupole order. They ignore other sources of energy loss, such as tidal dissipation, mass loss from the system, energy loss from the pulsar emission or via magnetic interactions between the two bodies. To date, there has not been an example of a binary pulsar system where these mechanisms come close to the energy loss via gravitational radiation. If such an system were found, it would very likely be deemed too "dirty" to provide quantitative tests of relativistic gravity, though it might well yield interesting physics of other types.

However there is a contribution to $\dot{P}_b$ that cannot be ignored. If the center of mass of the binary system is accelerating relative to that of the solar system, then both the orbital and pulsar periods will change at a rate given by

$$\frac{\dot{P}_b}{P_b} = \frac{\dot{P}_p}{P_p} = \ddot{r}_0 = \boldsymbol{a} \cdot \boldsymbol{n} + \frac{1}{r_0}\left[v^2 - (\boldsymbol{v} \cdot \boldsymbol{n})^2\right], \tag{12.27}$$

where $\boldsymbol{v}$ and $\boldsymbol{a}$ are the relative velocity and acceleration, respectively, between the binary system and the solar system. The first term is the projection of the acceleration along the line of sight, while the second, called the Shklovskii effect, represents the effect of variation of the line of sight. If we assume that the binary system (b) and the solar system ($\odot$) are on circular orbits around the galaxy with angular velocities Ω_b and $\Omega_\odot$, distances from the galactic center r_b and $r_\odot$, and longitudes relative to the galactic center ϕ_b and $\phi_\odot$, then Eq. (12.27) takes the form

$$\frac{\dot{P}_b}{P_b} = \frac{\dot{P}_p}{P_p} = (\Omega_b - \Omega_\odot)^2 \frac{r_b r_\odot}{r_0}\left(\cos\phi - \frac{r_b r_\odot}{r_0^2}\sin^2\phi\right), \tag{12.28}$$

where $\phi \equiv \phi_b - \phi_\odot$. We will see that this effect is significant for some binary pulsars, such as B1913+16, but not important for others, such as the double pulsar, depending on their distance from the solar system and precise location in the galaxy.

12.1.2 A zoo of binary pulsars

Here we describe some of the most interesting binary pulsar systems from the point of view of testing relativistic gravity. For the most part, we will discuss the implications of the system for general relativity. In Section 12.1.3, we will discuss bounds placed by binary pulsar observations on various alternative theories of gravity.

The Hulse-Taylor binary pulsar B1913+16

The measured values of $\langle\dot{\omega}\rangle$, γ' and $\dot{P}_b$ shown in Table 12.1 provide three constraints on the two masses m_1 and m_2, given by Eqs. (12.25) in Box 12.1. It is conventional to plot these three constraints, including their uncertainties, on an $m_1 - m_2$ plane. If general relativity is correct, they must overlap at a single point, within the uncertainties. The plot is shown in the left panel of Figure 12.2. The $\langle\dot{\omega}\rangle$ constraint fixes the total mass of the system to be $2.8284\, M_\odot$, accurate to about a part in 10^6; this is the black line in Figure 12.2, with the width of the line much larger than the actual uncertainty. The γ' constraint fixes the quantity $(m_2/m)(1 + m_2/m)$; this is the dark grey line in Figure 12.2. From the intersection of these

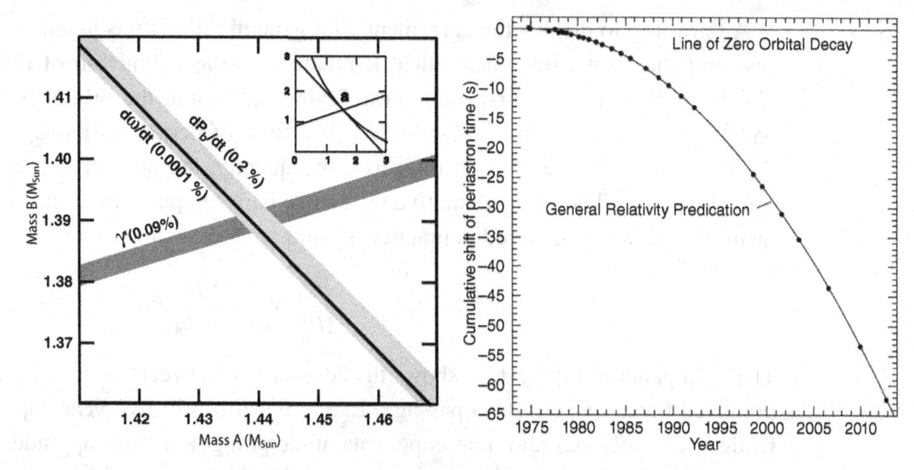

Fig. 12.2 Left: $m_1 - m_2$ plot for B1913+16. Right: Time of periastron passage as a function of time from 1975 to 2013. The curve is the prediction of general relativity for $\dot{P}_b$ given by Eq. (12.25); the points are the measured values with error bars multiplied by a factor of 400. The large gap in data during the middle 1990s occurred during the upgrade of the Arecibo Radio Telescope. Image reproduced with permission from Weisberg and Huang (2016), copyright by AAS.

two constraints, we obtain the values for the individual masses, $m_1 = 1.438 \pm 0.001\, M_\odot$ and $m_2 = 1.390 \pm 0.001\, M_\odot$ (the error is dominated by the uncertainty in γ').

The $\dot{P}_b$ constraint fixes the chirp mass of the system $\mathcal{M} = \eta^{3/5} m$; because m is known, this constrains η. However, in this case, the galactic acceleration effect is important. Using data on the location and proper motion of the pulsar, combined with the best information available on galactic rotation; the current value of this effect is $\dot{P}_b^{gal} = -(0.025 \pm 0.004) \times 10^{-12}$. Subtracting this from the measured post-Keplerian parameter $\dot{P}_b$ shown in Table 12.1 gives the corrected value $\dot{P}_b^{corr} = -\, (2.398 \pm 0.004) \times 10^{-12}$. This value and its uncertainty are used to plot the dP_b/dt curve in Figure 12.2. It is the hyperbolic curve shown in the inset in Figure 12.2 which ranges from zero to three solar masses, and the light grey band parallel to the $\langle \dot{\omega} \rangle$ constraint in the blow-up of the intersection point "a," which ranges over about $0.05\, M_\odot$. The three curves overlap at a common point, within the uncertainties.

Another way to check the agreement with general relativity is to use the masses inferred from the intersection of the $\langle \dot{\omega} \rangle$ and γ' constraints along with the expression for $\dot{P}_b$ in Eq. (12.25) to predict the value $\dot{P}_b^{GR} = -(2.40263 \pm 0.00005) \times 10^{-12}$. This agrees with the measured value after the galactic correction, in other words,

$$\frac{\dot{P}_b^{coor}}{\dot{P}_b^{GR}} = 0.9983 \pm 0.0016 \,. \tag{12.29}$$

Although the uncertainties in the measured post-Keplerian parameter $\dot{P}_b$ continue to decrease, in part because of the decrease of statistical errors with observation time T as $T^{-3/2}$, the uncertainties in the parameters that go into the galactic correction are now the limiting factor in the accuracy of the test of gravitational wave damping in general relativity. In fact, if one assumes that general relativity is correct, then the binary pulsar is providing improved data on the galactic rotation curve.

A third way to display the agreement with general relativity is to compare the observed phase of the orbit with a theoretical template phase as a function of time. If P_b varies slowly in time, then to first order in a Taylor expansion, the orbital phase is given by $\Phi_b(t) = 2\pi t/P_{b0} - \pi \dot{P}_{b0} t^2/P_{b0}^2 + \dots$. The time of periastron passage T_0 is given by $\Phi(T_0) = 2\pi N$, where N is an integer. Consequently, the periastron time will not grow linearly with N. Thus the cumulative difference between periastron time T_0 and NP_{b0}, the quantities actually measured in practice, should vary according to

$$T_0 - NP_{b0} \approx \frac{\dot{P}_{b0}}{2P_{b0}^3} N^2 \approx \frac{1}{2} \frac{\dot{P}_{b0}}{P_{b0}} t^2 . \tag{12.30}$$

The right panel of Figure 12.2 shows the comparison between the measured times and the predicted times of periastron passage. Even after a nearly five-year gap in observations while the Arecibo radio telescope was undergoing a major upgrade, the measured periastron times landed right on top of the predicted curve.

The consistency among the constraints provides a test of the assumption that the two bodies behave as "point" masses, without complicated tidal effects, obeying the general relativistic equations of motion including gravitational radiation. This supports the evolutionary model whereby the pulsar is an old recycled pulsar and the companion is a young but dead pulsar. It is also a test of the adherency of general relativity to the SEP in the presence of strong gravity, in that the highly relativistic internal structure of the neutron stars does *not* influence their orbital motion.

Observations indicate that the pulse profile is varying with time (Kramer, 1998; Weisberg and Taylor, 2002), which suggests that the pulsar is undergoing geodetic precession on a 300-year timescale as it moves through the curved spacetime generated by its companion (see Section 9.1.1). The amount is consistent with GR, assuming that the pulsar's spin is suitably misaligned with the orbital angular momentum. Unfortunately, the evidence suggests that the pulsar beam may precess out of our line of sight by 2025.

The precession of the pulsar's spin is accompanied by a precession of the orbital plane, since the total angular momentum of the system is constant up to the changes induced by gravitational radiation reaction. One consequence of this is that the orbital inclination has increased enough that, with the aid of improved measurement accuracy, the Shapiro delay has become measurable (see Table 12.1). Another consequence is that variations in the projected semimajor axis x and eccentricity e must now be included in the analysis, leading to somewhat larger errors in the redshift/time dilation parameters γ' than those presented in earlier analyses (see, e.g., Weisberg et al. (2010)).

The "double" pulsar J0737-3039A,B

This binary pulsar system, discovered by Burgay et al. (2003), was already remarkable for its extraordinarily short orbital period (0.1 days) and large periastron advance ($16.8995^\circ \ \mathrm{yr}^{-1}$), but then the companion was also detected as a pulsar (Lyne et al., 2004). Because two projected semimajor axes could be measured, the mass ratio was obtained directly and to high precision from the ratio of the values of the Keplerian parameters $x_1 = a_1 \sin\iota$ and $x_2 = a_2 \sin\iota$, since $x_1/x_2 = m_2/m_1$ modulo post-Newtonian corrections

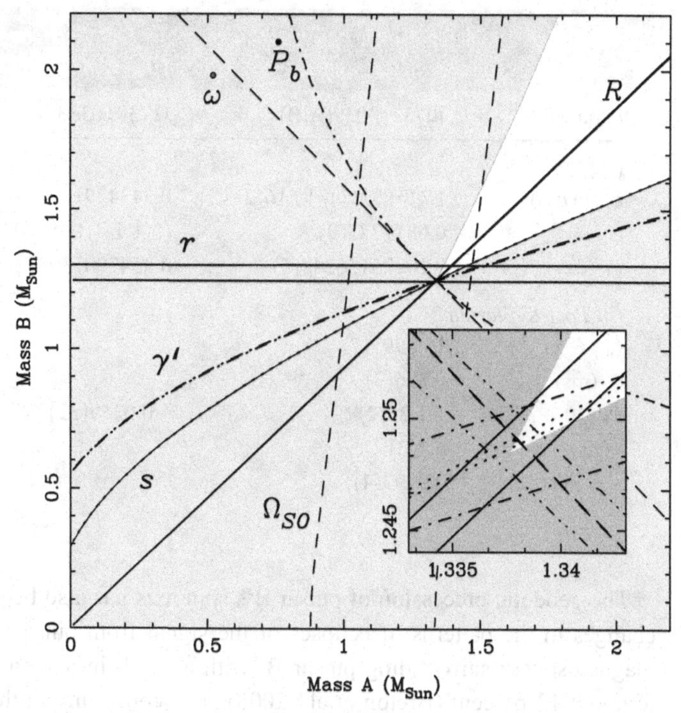

Fig. 12.3 Mass-mass plots for the double pulsar. Image courtesy of Michael Kramer.

(see Table 12.2). In Figure 12.3, the line labeled "R" is that mass ratio, and the white area is the allowed region for the masses, set by the measurement of the two Keplerian mass functions $f_1 = x_1^3(2\pi/P_b)^2$ and $f_2 = x_2^3(2\pi/P_b)^2$ combined with the condition that $|\sin\iota| \leq 1$. Then, the individual masses were obtained by combining the mass ratio with the periastron advance, assuming general relativity to be valid. The results are $m_A = 1.3381 \pm 0.0007\,M_\odot$ and $m_B = 1.2489 \pm 0.0007\,M_\odot$, where A denotes the primary (first) pulsar. From these values, one finds that the orbit is nearly edge-on, with $\sin\iota = 0.9997$, a value which is completely consistent with that inferred from the Shapiro delay shape parameter $s = \sin\iota$. In fact, the five measured post-Keplerian parameters plus the ratio of the projected semimajor axes give six constraints on the masses. As seen in Figure 12.3, all six overlap within their measurement errors (Kramer et al., 2006). The fact that the overlap region, shown in detail in the inset, is so close to the vertex of the allowed region is another indication that the orbit is nearly edge-on. Note that Figure 12.3 is based on more recent data than that quoted in Kramer et al. (2006), described in this discussion and listed in Table 12.2.

Because of the location of the system, galactic proper-motion effects play a significantly smaller role in the interpretation of $\dot{P}_b$ measurements than they did in B1913+16; this and the reduced effect of interstellar dispersion means that the accuracy of measuring the gravitational-wave damping will eventually beat that from the Hulse-Taylor system. It may ultimately be necessary for the data analysis to include 2PN corrections, for example in the periastron advance.

Table 12.2 Parameters of other binary pulsars. Values for $\dot{P}_b$ include corrections for galactic kinematic effects. Numbers in parentheses denote errors in the last digit. See the text for references.

Parameter	J0737–3039(A,B)	J1738+0333	J1141–6545
(i) Keplerian			
$a_1 \sin \iota$ (s)	1.415032(1)/1.516(2)	0.34342913(2)	1.858922(6)
e	0.0877775(9)	$(3.4 \pm 1.1) \times 10^{-7}$	0.171884(2)
P_b (day)	0.10225156248(5)	0.354790739872(1)	0.1976509593(1)
(ii) Post-Keplerian			
$\langle\dot{\omega}\rangle$ ($^\circ$ yr^{-1})	16.8995(7)		5.3096(4)
γ' (ms)	0.386(3)		0.77(1)
$\dot{P}_b$ (10^{-12})	−1.25(2)	−0.0259(32)	−0.401(25)
r (μs)	6.2(3)		
$s = \sin \iota$	0.9997(4)		

The geodetic precession of pulsar B's spin axis has also been measured by monitoring changes in the patterns of eclipses of the signal from pulsar A as it passes through the magnetosphere surrounding pulsar B, with a result in agreement with general relativity to about 13 percent (Breton et al., 2008). The constraint on the masses from that effect, denoted Ω_{SO} is also shown in Figure 12.3. In fact, pulsar B has precessed so much that its signal no longer sweeps by the Earth, so it has gone "silent." For a recent overview of the double pulsar, see Burgay (2012).

J1738+0333: A white-dwarf companion

This is an ultra-low-eccentricity, 8.5-hour period system in which the white-dwarf companion is bright enough to permit detailed spectroscopy, allowing the companion mass to be determined directly to be $0.181\,M_\odot$. The mass ratio is determined from Doppler shifts of the pulsar signal and of the spectral lines of the companion, giving the pulsar mass $1.46\,M_\odot$. Ten years of observation of the system yielded both a measurement of the apparent orbital period decay, and enough information about its parallax and proper motion to account for the substantial galactic effect to give a value of the intrinsic period decay of $\dot{P}_b = (-25.9 \pm 3.2) \times 10^{-15}$ in agreement with the predicted effect of general relativity (Freire et al., 2012). But because of the asymmetry in the sensitivities of the bodies in the system, the result also places a significant bound on the existence of dipole radiation, predicted by many alternative theories of gravity. Data from this system were also used to place a tight bound on the PPN parameter α_1 (see Section 12.1.3 for discussion of these tests of alternative theories).

J1141–6545: A white-dwarf companion

This system is similar in some ways to the Hulse-Taylor binary: short orbital period (0.2 days), significant orbital eccentricity (0.172), rapid periastron advance (5.3° yr^{-1}) and massive components ($m_p = 1.27 \pm 0.01\,M_\odot$, $m_c = 1.02 \pm 0.01\,M_\odot$). The key difference is that the companion is again a white dwarf. The intrinsic $\dot{P}_b$ has been measured in agreement with general relativity to about 6 percent, again placing limits on dipole gravitational radiation (Bhat et al., 2008).

J0348+0432: The most massive neutron star

Discovered in 2011 (Lynch et al., 2013; Antoniadis et al., 2013), this is another neutron-star white-dwarf system, in a very short-period (0.1 day), low-eccentricity (2×10^{-6}) orbit. Timing of the neutron star and spectroscopy of the white dwarf have led to mass values of $0.172\,M_\odot$ for the white dwarf and $2.01 \pm 0.04\,M_\odot$ for the pulsar, making it the most massive accurately measured neutron star yet. This supported an earlier discovery of a $2\,M_\odot$ pulsar (Demorest et al., 2010); such large masses rule out a number of heretofore viable soft equations of state for nuclear matter, assuming general relativity to be correct. The orbit period decay agrees with the general relativistic prediction within 20 percent and is expected to improve steadily with time.

J0337+1715: Two white-dwarf companions

This remarkable system was reported by Ransom et al. (2014). It consists of a 2.73 millisecond pulsar of $1.4378(13)\,M_\odot$, with extremely good timing precision, accompanied by two white dwarfs in coplanar circular orbits. The inner white dwarf ($m = 0.19751(15)\,M_\odot$) has an orbital period of 1.629 days, with $e = 6.9177(2) \times 10^{-4}$, and the outer white dwarf ($m = 0.4101(3)\,M_\odot$) has a period of 327.26 days, with $e = 3.5356196(4) \times 10^{-2}$. This is an ideal system for testing the Nordtvedt effect in the strong-field regime. Here the inner system is the analogue of the Earth–Moon system, and the outer white dwarf plays the role of the Sun. Because the outer semimajor axis is about 1/3 of an astronomical unit, the basic driving perturbation is comparable to that provided by the Sun on the Earth–Moon system. However, the self-gravitational binding energy per unit mass (sensitivity) of the neutron star is almost a billion times larger than that of the Earth, greatly amplifying the potential size of the Nordtvedt effect. In 2018, Archibald et al (Nature, in press) reported a bound of 2.6 parts per million on any difference in acceleration between the neutron star and the white dwarf, representing an improvement over lunar laser ranging of about a factor of 10.

Other binary pulsars

Two of the earliest binary pulsars, B1534+12 and B2127+11C, discovered in 1990, failed to live up to their early promise despite being similar to the Hulse-Taylor system in most respects (both were believed to be double neutron-star systems). The main reason was the significant uncertainty in the kinematic effect on $\dot{P}_b$ of local accelerations, that of the galaxy in the case of B1534+12, and that of the host globular cluster in the case of B2127+11C.

A class of wide-orbit binary millisecond pulsar (WBMSP) systems, containing a pulsar and a white dwarf in low-eccentricity orbits has been used to place interesting bounds on the Nordtvedt effect. See Section 8.1 for a discussion.

12.1.3 Tests of alternative theories

Soon after the discovery of the binary pulsar B1913+16 it was widely hailed as a new testing ground for relativistic gravitational effects. As we have seen in the case of general relativity, in most respects, the system has lived up to, indeed exceeded, the early expectations. In many ways, the double pulsar topped the Hulse-Taylor binary in this regard.

In another respect, however, B1913+16 only partially lived up to its promise, namely as a direct testing ground for alternative theories of gravity. The origin of this promise was the discovery (Eardley, 1975; Will, 1977) that alternative theories of gravity generically predict the emission of dipole gravitational radiation from binary star systems. As one fulfillment of this promise, Will and Eardley (1977) worked out in detail the effects of dipole gravitational radiation in the bimetric theory of Rosen (1974), and when the first observation of the decrease of the orbital period was announced in 1979, the Rosen theory suffered a fatal blow. A wide class of alternative theories of that period also failed the binary pulsar test because of dipole gravitational radiation (Will, 1977).

On the other hand, the early observations of B1913+16 already indicated that, in general relativity, the masses of the two bodies were nearly equal, so that, in theories of gravity that are in some sense "close" to general relativity, dipole gravitational radiation would not be a strong effect, because of the apparent symmetry of the system. The Rosen theory, and others like it, are not "close" to general relativity, except in their predictions for the weak-field, slow-motion regime of the solar system. When relativistic neutron stars are present, theories like these can predict strong effects on the motion of the bodies resulting from their internal highly relativistic gravitational structure (violations of SEP). As a consequence, the masses inferred from observations of the periastron shift and γ' may be significantly different from those inferred using general relativity, and may be different from each other, leading to strong dipole gravitational radiation damping.

By contrast, the Brans-Dicke theory is close to general relativity, roughly speaking within $1/\omega_{\rm BD}$ of the predictions of the latter, for large values of the coupling constant $\omega_{\rm BD}$. Thus, despite the presence of dipole gravitational radiation, the Hulse-Taylor binary pulsar provides only a weak test of pure Brans-Dicke theory, not competitive with solar-system tests.

However, the discovery of binary pulsar systems with a white dwarf companion, such as J1738+0333, J1141–6545, and J0348+0432 has made it possible to perform strong tests of the existence of dipole radiation. This is because such systems are necessarily asymmetrical, since the gravitational binding energy per unit mass, or sensitivity of white dwarfs is of order 10^{-4}, much less than that of the neutron star. Already, significant bounds have been placed on dipole radiation using J1738+0333 and J1141–6545.

Because the gravitational-radiation and strong-field properties of alternative theories of gravity can be quite different from those of general relativity and each other, it is difficult to parametrize these aspects of the theories in the manner of the PPN framework. In addition, because of the generic violation of the Strong Equivalence Principle in these theories, the results can be very sensitive to the equation of state and mass of the neutron star(s) in the system. In the end, there is no way around having to analyze every theory in turn. On the other hand, because of their relative simplicity, scalar-tensor theories provide an illustration of the essential effects, and so we will begin by discussing binary pulsars within this class of theories.

Scalar-tensor theories

We combine the results of Section 10.3.2, notably Eqs. (10.48), (10.49), (10.59), and (10.65) to obtain the metric coefficients α^* and γ^* and the periastron advance rate $\langle\dot{\omega}\rangle$. We then combine Eqs. (11.84), (11.85), and (11.121) to obtain $\dot{P}_b$. Recalling that $P_b = 2\pi(a^3/\mathcal{G}m)^{1/2}$ where

$$\mathcal{G} = 1 - \zeta + \zeta(1-2s_1)(1-2s_2)\,, \tag{12.31}$$

and $\zeta = (4+2\omega_0)^{-1}$, we obtain expressions for the post-Keplerian parameters in scalar-tensor theories:

$$\begin{aligned}
\gamma' &= e\,m_2\left(\frac{P_b}{2\pi\mathcal{G}m}\right)^{1/3}\left(1 - 2\zeta s_2 + \mathcal{G}\frac{m_2}{m} + 2\zeta(1-2s_2)(1+2\lambda)\kappa_{(1)}\right).\\
r &= m_2(1-\zeta)\,,\\
s &= \sin\iota\,,\\
\langle\dot{\omega}\rangle &= \frac{6\pi}{P_b(1-e^2)}\left(\frac{2\pi\mathcal{G}m}{P_b}\right)^{2/3}\left[1 + \frac{1}{3}\left(2\bar{\gamma} - \bar{\beta}_+ - \Delta\bar{\beta}_-\right)\right],\\
\dot{P}_b &= -\frac{192\pi}{5}\left(\frac{2\pi\mathcal{G}\mathcal{M}}{P_b}\right)^{5/3}F(e) - 8\pi\zeta\eta\mathcal{S}^2\left(\frac{2\pi\mathcal{G}m}{P_b}\right)G(e)\,.
\end{aligned} \tag{12.32}$$

In the limit $\zeta \to 0$, we recover general relativity, and all structure dependence disappears. The first term in $\dot{P}_b$ is the combined effect of quadrupole and monopole gravitational radiation, post-Newtonian corrections to dipole radiation, and a dipole-octupole coupling term, all contributing at 0PN order, while the second term is the effect of dipole radiation, contributing at the dominant −1PN order.

Unfortunately, because of the near equality of neutron star masses in typical double neutron star binary pulsars, dipole radiation is somewhat suppressed, and the bounds obtained are typically not competitive with the Cassini bound on scalar-tensor theories. Figure 12.4 uses the α_0-β_0 parametrization of scalar-tensor theories of Damour and Esposito-Farèse, to display bounds from a variety of systems and experiments. The bounds from the three binary neutron star systems B1913+16, J0737–3039, and B1534+12 are not close to being competitive with the Cassini bound on α_0, or with lunar laser ranging (LLR) bounds, except for those generalized scalar-tensor theories with $\beta_0 < 0$, where the strong gravity of the neutron stars induces spontaneous scalarization effects (Damour and Esposito-Farèse, 1998). Recall that $\alpha_0 = (3+2\omega_0)^{-1/2} = [\zeta/(1-\zeta)]^{1/2}$ and $\beta_0 = 2\omega_0'\phi_0/(3+2\omega_0)^2 = 2\lambda/(1-\zeta)$.

On the other hand, a binary pulsar system with dissimilar objects, such as one with a white dwarf or black hole companion, is a more promising testing ground for dipole radiation. As a result, the neutron-star-white-dwarf systems J1141–6545 and J1738+0333 yield much more stringent bounds. Indeed, the latter system surpasses the Cassini bound for $\beta_0 > 1$ and $\beta_0 < -2$, and is close to that bound for the pure Brans-Dicke case $\beta_0 = 0$ (Freire et al., 2012).

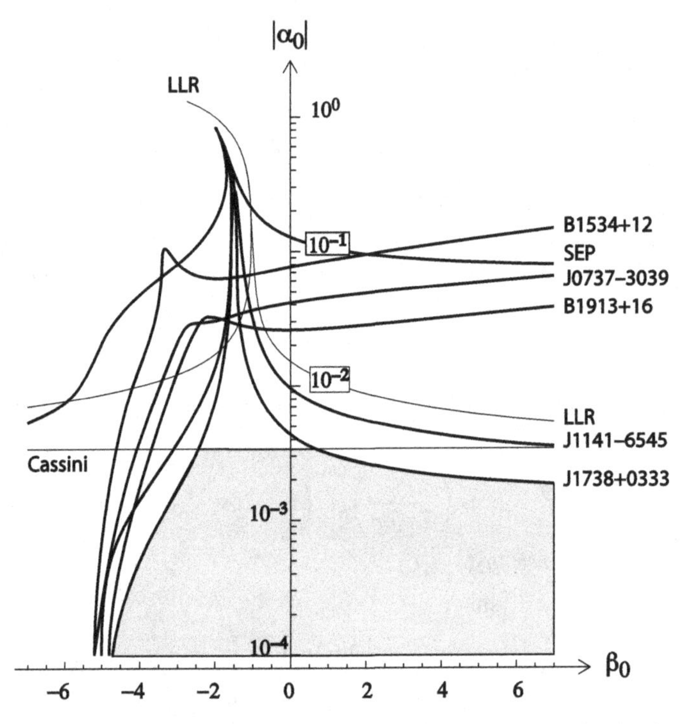

Fig. 12.4 Bounds on scalar-tensor theories from solar-system and binary-pulsar tests. Image reproduced with permission from Freire et al. (2012), copyright by Oxford University Press.

It is worth pointing out that the bounds displayed in Figure 12.4 have been calculated using a specific choice of scalar-tensor theory, in which the function $A(\varphi)$ is given by

$$A(\varphi) = \exp\left[\alpha_0(\varphi - \varphi_0) + \frac{1}{2}\beta_0(\varphi - \varphi_0)^2\right], \tag{12.33}$$

where α_0 and β_0 are arbitrary parameters, and φ_0 is the asymptotic value of the scalar field. In the notation for scalar tensor theories used here, this theory corresponds to the choice

$$\omega(\phi) = -\frac{3}{2} + \frac{1}{2(\alpha_0^2 - \beta_0 \ln \phi)}, \tag{12.34}$$

where $\phi_0 = A(\varphi_0)^{-2} = 1$. The parameters ζ and λ are given by $\zeta = \alpha_0^2/(1 + \alpha_0^2)$, and $\lambda = \beta_0/2(1 + \alpha_0^2)$. It is useful to note that the PPN parameter combination $4\beta - \gamma - 3$, which governs the Nordtvedt effect and other violations of SEP in the post-Newtonian limit is given by

$$4\beta - \gamma - 3 = 2\zeta(1 + 2\lambda) = \frac{2\alpha_0^2}{1 + \alpha_0^2}\left(1 + \frac{\beta_0}{1 + \alpha_0^2}\right), \tag{12.35}$$

which vanishes when $\beta_0 \simeq -1$. This partially explains the distorted "spike" in the curve for lunar laser ranging, where the violation of SEP is suppressed, and no useful bound can be obtained. Something similar is occurring in the other curves, albeit complicated by the strong gravity effects taking place in the neutron stars.

The bounds shown in Figure 12.4 were calculated using a polytropic equation of state, which tends to give lower maximum masses for neutron stars than do more realistic equations of state. It is not known how much the use of stiffer equations of state would alter the bounds on this class of theories.

Other theories

Bounds on various versions of TeVeS theories have also been established, with the tightest constraints again coming from neutron-star-white-dwarf binaries (Freire et al., 2012); in the case of TeVeS, the theory naturally predicts $\gamma = 1$ in the post-Newtonian limit, so the bounds from Cassini are irrelevant here.

Strong constraints on the Einstein-Æther and Khronometric theories were also set using binary pulsar measurements, exploiting both gravitational-wave damping data, and data related to preferred-frame effects (Yagi et al., 2014a, 2014b). For Einstein-Æther theory, for example, the approximate bounds were $c_+ < 0.03$ and $c_- < 0.003$.

12.2 Inspiralling Compact Binaries and Gravitational Waves

A new era for testing general relativity began on September 14, 2015, with the first detection of gravitational waves by LIGO (Abbott et al., 2016c). The signal, denoted GW150914, was the final burst of gravitational waves from the inspiral and merger of two black holes. Even though the signal was detectable above the noise for only 0.2 seconds, three phases could be clearly delineated: a "chirp" phase of increasing amplitude and frequency, corresponding to the late inspiral phase, a "merger" phase, where the two black holes were becoming one, and a "ringdown" phase, where the final, perturbed black emitted exponentially damped radiation from its quasinormal mode oscillations before settling down to a stationary black hole (see Figure 12.5 for an illustration). Analysis of the chirp radiation showed that the black holes had masses 26 and $39\,M_\odot$, while analysis of the ringdown radiation revealed that the final black hole had a mass of $62\,M_\odot$ and a spin parameter $\chi \simeq 0.67$. The three solar masses converted to energy during the process corresponded to an approximate luminosity of 3.6×10^{56} erg s^{-1}, larger than the luminosity of all the stars in the observable universe combined.

Already with the first detection, important tests of general relativity were carried out. One test checked the consistency between the observed signal and theoretical template waveforms based on combining post-Newtonian theory with numerical relativity within general relativity. Another test placed a bound on the mass of the graviton that improved upon the bound derived from solar system dynamics. The fourth detection, GW170814, which included the Virgo detector (Abbott et al., 2017c), resulted in an interesting test of the polarization content of the gravitational waves. The most recent detection GW170817 was a binary neutron-star inspiral (Abbott et al., 2017d). This observation was notable because it was also observed across the entire electromagnetic spectrum, with major implications for gamma-ray bursts, the neutron-star equation of state, and the synthesis of

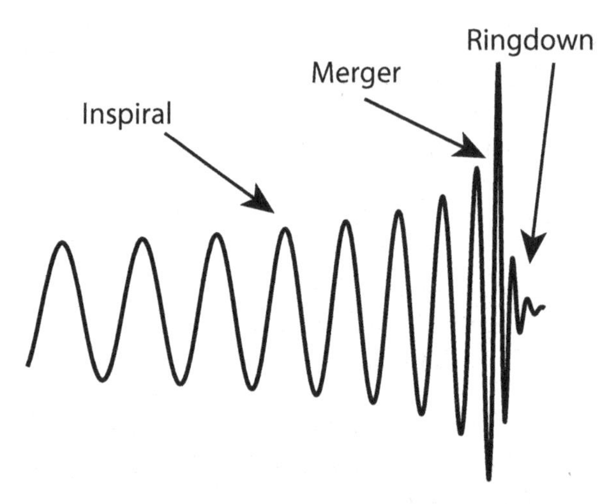

Fig. 12.5 Schematic gravitational wave from binary inspiral, showing the inspiral, merger and ringdown parts of the waveform.

the heaviest elements (Abbott et al., 2017e). It also produced a very precise direct bound on the difference between the speeds of gravitational waves and light. When the analysis is complete, it may also contribute additional tests of relativistic gravity.

In this section we will lay the groundwork for discussing gravitational waves from binary inspiral within general relativity and scalar-tensor theory, and then will discuss a few parametrized frameworks for analysing alternative theories. Finally we will describe the results from specific tests carried out so far. This field is truly just beginning, and so it is likely that much of what is presented here will soon be out of date!

12.2.1 Compact binary inspiral in general relativity

To keep the discussion simple, we will assume that general relativity is correct and that the two compact bodies are in quasi-circular orbits, which means orbits that are circular apart from the adiabatic inspiral induced by gravitational radiation. For compact binaries that have evolved in isolation for considerable time, this turns out to be an excellent approximation. As we learned in Section 11.5.2, the orbital eccentricity decreases with semimajor axis approximately as $e = e_0(a/a_0)^{19/12} = e_0(P_b/P_{b0})^{19/18}$. Thus, for example, by the time the Hulse-Taylor binary pulsar B1913+16 has spiralled inward until its orbital period is around 0.1 seconds, so that its gravitational waves will be entering the LIGO-Virgo sensitive band, its eccentricity will be 10^{-6}. On the other hand, for an "extreme mass-ratio inspiral," in which a stellar-mass compact object is injected from a surrounding star cluster into a highly eccentric orbit around a supermassive black hole, there may not be enough time for the orbit to circularize, and thus it may be important to include eccentricity.

We will also assume that the bodies are spinning, but we will assume that the spins are either aligned or anti-aligned with the orbital angular momentum. This means that we will ignore the precessions of the spins and of the orbital plane that result from spin-orbit and spin-spin coupling (see Section 9.1 for discussion). These precessions have the effect of modulating the amplitude and phase of the gravitational waveform in complex ways. There has not been sufficient information in the short binary black hole waveforms observed so far to detect precession effects, but they will surely be important for binary black hole inspirals in the future. Similarly, it is not yet known whether spin precession played a role in the binary neutron-star inspiral event. For simplicity, we will ignore such complications, and treat only the aligned spin cases.

We will ignore tidal interactions. For binary black holes, these effects are unimportant until the onset of the actual merger of two event horizons. For systems containing neutron stars they will also be unimportant until the final few orbits, but then the signature of tidal distortion and even disruption of the neutron star as imprinted on the gravitational wave signal could carry important information about the equation of state of the matter in the neutron star. In order to test alternative theories of gravity, it is desirable to avoid regimes where such complex phenomena could occur.

Finally, we will treat only the part of the inspiral that is governed by the post-Newtonian approximation.

To obtain the relevant conditions for a circular orbit, we take the PPN two-body equation of motion (6.70) without the J_2 term, choose the PPN parameters for general relativity and impose the constraint $\dot{r} = \ddot{r} = 0$. This yields a condition on the orbital angular velocity ω given by $\omega^2 = (m/r^3)[1-(3-\eta)m/r]$. Including the spin-orbit and spin-spin contributions to the equations of motion, Eqs. (9.4a) and (9.4b), along with 2PN terms in the equations of motion (see, for example, Blanchet et al. (1995a)), we obtain

$$\omega^2 = \frac{m}{r^3}\left[1 - \frac{m}{r}(3-\eta) - \left(\frac{m}{r}\right)^{3/2}\sum_a\left(2\frac{m_a^2}{m^2} + 3\eta\right)\hat{\boldsymbol{h}}\cdot\boldsymbol{\chi}_a + \left(\frac{m}{r}\right)^2\left\{6 + \frac{41}{4}\eta + \eta^2 - \frac{3}{2}\eta\left(\boldsymbol{\chi}_1\cdot\boldsymbol{\chi}_2 - 3\hat{\boldsymbol{h}}\cdot\boldsymbol{\chi}_1\hat{\boldsymbol{h}}\cdot\boldsymbol{\chi}_2\right)\right\}\right], \tag{12.36}$$

where $\boldsymbol{\chi}_a = \boldsymbol{S}_a/m_a^2$, and $\hat{\boldsymbol{h}}$ is the unit vector in the direction of the orbital angular momentum $\boldsymbol{x}\times\boldsymbol{v}$. Similarly, the energy of the circular orbit is given by

$$E = -\eta\frac{m^2}{2r}\left[1 - \frac{1}{4}\frac{m}{r}(7-\eta) + \left(\frac{m}{r}\right)^{3/2}\sum_a\left(2\frac{m_a^2}{m^2} + \eta\right)\hat{\boldsymbol{h}}\cdot\boldsymbol{\chi}_a - \frac{1}{8}\left(\frac{m}{r}\right)^2\left\{7 - 49\eta - \eta^2 - 4\eta\left(\boldsymbol{\chi}_1\cdot\boldsymbol{\chi}_2 - 3\hat{\boldsymbol{h}}\cdot\boldsymbol{\chi}_1\hat{\boldsymbol{h}}\cdot\boldsymbol{\chi}_2\right)\right\}\right]. \tag{12.37}$$

The rate of loss of energy is given by Eq. (11.109), with spin-orbit and spin-spin effects added (Kidder et al., 1993; Kidder, 1995),

$$\frac{dE}{dt} = -\frac{32}{5}\eta^2\left(\frac{m}{r}\right)^5\left[1 - \frac{m}{r}\left(\frac{2927}{336} + \frac{5}{4}\eta\right) + \left(\frac{m}{r}\right)^{3/2}\left\{4\pi - \frac{1}{12}\sum_a\left(73\frac{m_a^2}{m^2} + 75\eta\right)\hat{\boldsymbol{h}}\cdot\boldsymbol{\chi}_a\right\} + \left(\frac{m}{a}\right)^2\left\{\frac{293383}{9072} + \frac{380}{9}\eta - \frac{\eta}{48}\left(223\boldsymbol{\chi}_1\cdot\boldsymbol{\chi}_2 - 649\hat{\boldsymbol{h}}\cdot\boldsymbol{\chi}_1\hat{\boldsymbol{h}}\cdot\boldsymbol{\chi}_2\right)\right\}\right]. \tag{12.38}$$

The orbital angular frequency $\omega = 2\pi f_b$, where f_b is the binary orbit frequency. But for quadrupole radiation from a circular orbit, the gravitational-wave frequency F is given by $F = 2f_b$. Combining Eqs. (12.36), (12.37), and (12.38) we can obtain an equation for the evolution of the gravitational-wave frequency with time as a function of F, given by $dF/dt = (dE/dt)/(dE/dF)$, or

$$\frac{dF}{dt} = \frac{96\pi}{5} F^2 (\pi \mathcal{M} F)^{5/3} \left[1 - \left(\frac{743}{336} + \frac{11}{4}\eta \right) (\pi m F)^{2/3} + (4\pi - \beta)(\pi m F) \right. \\ \left. + \left(\frac{34103}{18144} + \frac{13661}{2016}\eta + \frac{59}{18}\eta^2 + \sigma \right) (\pi m F)^{4/3} \right] , \tag{12.39}$$

where β and σ are the spin-orbit and spin-spin contributions, given by

$$\beta \equiv \frac{1}{12} \sum_a \left(113 \frac{m_a^2}{m^2} + 75\eta \right) \hat{\boldsymbol{h}} \cdot \boldsymbol{\chi}_a , \\ \sigma \equiv \frac{\eta}{48} \left(-247 \boldsymbol{\chi}_1 \cdot \boldsymbol{\chi}_2 + 721 \hat{\boldsymbol{h}} \cdot \boldsymbol{\chi}_1 \hat{\boldsymbol{h}} \cdot \boldsymbol{\chi}_2 \right) . \tag{12.40}$$

Note that $(\pi m F)^{2/3} \sim v^2 \sim m/r$. If the source is at a sufficiently great distance that cosmological redshifts become significant, then the observed frequency $F_{\rm obs} = F/(1+Z)$ and the observed time interval is $dt_{\rm obs} = (1+Z)dt$. Then Eq. (12.39) applies equally well to the observed quantities provided that we define the "observed" or redshifted masses $\mathcal{M}_{\rm obs} = \mathcal{M}(1+Z)$ and $m_{\rm obs} = m(1+Z)$.

Combining Eqs. (11.47) and (11.108) for a circular orbit, and substituting $r = m^{1/3}(2\pi F)^{-2/3}[1 + O(\epsilon)]$ into the wave amplitude, we can write the response function $S(t)$ of a laser interferometer in the general form

$$S(t) = \frac{\mathcal{M}}{R} Q(\text{angles}) (\pi \mathcal{M} F)^{2/3} \cos \Phi(t) , \tag{12.41}$$

where $\Phi(t) = 2\phi(t) = 2\pi \int^t F(t')dt'$ is the gravitational-wave phase. Integrating Eq. (12.39) to obtain F as a function of t, and then integrating to obtain the phase, we obtain

$$\Phi(F) = \Phi_c - \frac{1}{16} (\pi \mathcal{M} F)^{-5/3} \left[1 + \frac{5}{3} \left(\frac{743}{336} + \frac{11}{4}\eta \right) (\pi m F)^{2/3} - \frac{5}{2}(4\pi - \beta)(\pi m F) \right. \\ \left. + 5 \left(\frac{3058673}{1016064} + \frac{5429}{1008}\eta + \frac{617}{144}\eta^2 - \sigma \right) (\pi m F)^{4/3} \right] , \tag{12.42}$$

where Φ_c is a constant. Thus we have an accurate prediction (under the chosen assumptions) for the gravitational-wave signal at the detector. This is essential for confirming a detection and for the measurement of the source parameters (Cutler et al., 1993), which include distance, position in the sky, orientation of the orbital plane, and the masses and spins of the companions. Roughly speaking, the measured signal (which includes detector noise) is passed through a linear filter constructed from the theoretical signal $S(t; \boldsymbol{\theta})$ and the spectral density of the detector noise (Wainstein and Zubakov, 1962). The theoretical signal is expressed as a function of an abstract vector $\boldsymbol{\theta}$, which collectively represents the source parameters, such as R, t_c, $\mathcal{M}$, η, and so on. The signal-to-noise ratio S/N is then computed (see below). The actual values of these parameters $\tilde{\boldsymbol{\theta}}$ are unknown prior to the measurement. When $\boldsymbol{\theta} = \tilde{\boldsymbol{\theta}}$, the linear filter becomes the Wiener optimum filter which is well known to yield the largest possible signal-to-noise ratio (Wainstein and Zubakov, 1962). A detection can be confirmed and the source parameters determined by maximizing S/N over a broad collection of expected signals $S(t;\boldsymbol{\theta})$, loosely referred to as "templates." This method is called "matched filtering"; see Finn and Chernoff (1993), Cutler and Flanagan (1994), and Poisson and Will (1995) for the foundations of this approach for binary inspiral.

It has been established that it is the *phasing* of the signal that plays the largest role in parameter estimation. This is because a slight variation in the parameters can quickly cause $h(t;\boldsymbol{\theta})$ to get out of phase with respect to the true signal $h(t;\tilde{\boldsymbol{\theta}})$, thus seriously reducing S/N from its maximum possible value. Therefore a good match between the phases of the template and the measured signal throughout the observed cycles singles out, to a large extent, the value of the source parameters.

The leading term in $\Phi(F)$ gives the chirp mass. If there is sufficient sensitivity to measure the 1PN correction term, then because of the dependence on η and m, one can measure the individual masses m_1 and m_2. Measuring higher-order terms can give information about the spins, and can also yield tests of general relativity.

The key ingredient in matched filtering is the Fourier transform of $S(t)$, given by $\tilde{S}(f) = (2\pi)^{-1}\int_{-\infty}^{\infty} S(t)e^{2\pi i f t}dt$. Using the stationary phase approximation, and confining attention to positive frequencies, it is straightforward to show that

$$\tilde{S}(f) = \mathcal{A}f^{-7/6}e^{i\psi(f)}\,, \tag{12.43}$$

where $\mathcal{A} \propto \mathcal{M}^{5/6}Q(\text{angles})/R$, and

$$\psi(f) = 2\pi f t_c - \frac{\pi}{4} + f\int^f f'^{-2}\Phi(f')df'\,, \tag{12.44}$$

where t_c is the time corresponding to the phase Φ_c, and the gravitational-wave frequency F is to be replaced by the Fourier frequency f in Eq. (12.42). In general relativity, the phase $\psi(f)$ is given by

$$\begin{aligned}\psi(f) = 2\pi f t_c - \Phi_c - \frac{\pi}{4} + \frac{3}{128}(\pi\mathcal{M}f)^{-5/3}\Bigg[1 + \frac{20}{9}\left(\frac{743}{336} + \frac{11}{4}\eta\right)(\pi m f)^{2/3} \\ -4(4\pi - \beta)(\pi m f) + 10\left(\frac{3058673}{1016064} + \frac{5429}{1008}\eta + \frac{617}{144}\eta^2 - \sigma\right)(\pi m f)^{4/3}\Bigg]\,.\end{aligned} \tag{12.45}$$

Then, for a detector characterized by a Gaussian noise spectral density $S_n(f)$, the signal-to-noise ratio associated with a measurement of a signal $S(t)$ is given by

$$S/N \equiv 4\int_0^\infty \frac{|\tilde{S}(f)|^2}{S_n(f)}df. \tag{12.46}$$

Defining the Fisher information matrix by

$$\Gamma_{ab} \equiv 2\int_0^\infty \frac{\tilde{S}^*_{,a}\tilde{S}_{,b} + \tilde{S}_{,a}\tilde{S}^*_{,b}}{S_n(f)}df, \tag{12.47}$$

where subscripts $,a$ and $,b$ denote partial derivatives with respect to one of the parameters θ^a characterizing the signal, it can be shown that the error σ_a in measuring the parameter

θ^a, and the correlation coefficients c^{ab} between parameters θ^a and θ^b are obtained from the inverse of the Fisher matrix, according to the prescription

$$\sigma_a^2 = \langle(\Delta\theta^a)^2\rangle = (\Gamma^{-1})^{aa} \,,$$
$$c^{ab} = \frac{\langle\Delta\theta^a \Delta\theta^b\rangle}{\sigma_a \sigma_b} = \frac{(\Gamma^{-1})^{ab}}{\sigma_a \sigma_b} \,. \tag{12.48}$$

This sensitivity to phase is one of the keys to testing general relativity with gravitational waves. In an alternative theory of gravity, the coefficients that appear in Eq. (12.45) may differ from those in general relativity, and if those differences depend on a parameter, such as the scalar-tensor coupling parameter ω_0, one can use matched filtering to place a bound on such parameters.

12.2.2 Compact binary inspiral in scalar-tensor theories

At present, the evolution of the gravitational-wave frequency in scalar-tensor theory cannot be written down to the same PN order as that displayed in Eq. (12.39) for general relativity. This is because dipole gravitational radiation contributes to the waveform at -0.5PN order and to the energy flux at -1PN order, and therefore to obtain terms in dF/dt at nPN order beyond the quadrupole term, one needs to evaluate the scalar wave field and the equations of motion to $(n+1)$PN order. In addition, the clean expansion in powers of $(\pi m F)^{2/3} \sim v^2$ shown in Eq. (12.39) is no longer so clean when dipole radiation is present. Even though the dipole term is formally of leading order, its contribution to the waveform and energy flux is proportional to ζ. Since solar-system experiments already place the constraint $\zeta < 10^{-5}$, and $\mathcal{S}$ is typically less than 0.5, the dipole flux is likely to be small compared to the quadrupole flux in the late inspiral phase, where $v > 10^{-2}$, except for situations where spontaneous scalarization yields extremely large values of $\mathcal{S}$. Thus a double expansion of quantities such as dF/dt and $\Phi(F)$ must be carried out, with one parameter being $(\pi m F)^{2/3}$, the other being the ratio of the two dominant types of flux (see Sennett et al. (2016) for discussion). The expansions will look very different depending on whether the evolution is dipole-radiation dominated or quadrupole-radiation dominated.

Here we will keep things simple, and work only to the equivalent of quadrupole order, mainly to demonstrate how dipole radiation alters the frequency evolution of the inspiral (Will, 1994). From the binary equation of motion (10.69) without spins, we follow the method described in Section 12.2.1 to obtain the quasicircular orbit expressions (compare Eqs. (12.36) and (12.37))

$$\omega^2 = \frac{\mathcal{G}m}{r^3}\left[1 - \frac{\mathcal{G}m}{r}\left(3 - \eta + \bar{\gamma} + 2\bar{\beta}_+ + 2\Delta\bar{\beta}_-\right)\right] ,$$
$$E = -\eta\frac{\mathcal{G}m^2}{2r}\left[1 - \frac{1}{4}\frac{\mathcal{G}m}{r}\left(7 - \eta + 4\bar{\gamma}\right)\right] . \tag{12.49}$$

From Eqs. (11.59), the energy loss rate from a quasicircular orbit is given by

$$\frac{dE}{dt} = -\frac{8}{15}\eta^2\kappa_1\frac{m}{r}\left(\frac{\mathcal{G}m}{r}\right)^4 - \frac{4}{3}\zeta\eta^2\frac{m}{r}\left(\frac{\mathcal{G}m}{r}\right)^3 \mathcal{S}_-^2 \,, \tag{12.50}$$

where κ_1 is given by Eq. (11.121). The final result for the evolution of the gravitational-wave frequency is

$$\frac{dF}{dt} = \frac{96\pi}{5} F^2 (\pi \mathcal{G} \mathcal{M} F)^{5/3} \left[\xi + \frac{5}{24} (\pi \mathcal{G} m F)^{-2/3} \zeta \mathcal{S}_-^2 \right], \tag{12.51}$$

where

$$\xi \equiv 1 + \frac{5}{12}\bar{\gamma} - \frac{5}{144}\zeta \mathcal{S}_-^2 \left(3 + 7\eta - 4\bar{\gamma} + 16\bar{\beta}_+ + 16\Delta\bar{\beta}_-\right) + \frac{5}{6}\zeta \mathcal{S}_- \left(\frac{\mathcal{S}_-\bar{\beta}_+ + \mathcal{S}_+\bar{\beta}_-}{\bar{\gamma}} \right) + \frac{5}{6}\zeta \Delta \mathcal{S}_- \left(\frac{\mathcal{S}_+\bar{\beta}_+ + \mathcal{S}_-\bar{\beta}_-}{\bar{\gamma}} \right). \tag{12.52}$$

Whereas in general relativity, the post-Newtonian corrections to the leading quadrupole frequency evolution are given by a power series in increasing powers of $(\pi \mathcal{G} m F)^{2/3}$, the dipole correction term depends on a negative power that parameter, reflecting its -1PN nature. But by measuring the evolution of F in an inspiral, one can place a bound on the scalar-tensor parameter ζ. Unfortunately, for the sensitivities of the advanced LIGO-Virgo instruments, the bounds are not likely to be competitive with the solar-system bound (Will, 1994), except possibly for scalar-tensor theories that admit spontaneous scalarization, leading to anomalously large values of $\mathcal{S}_\pm$. On the other hand, such bounds would be derived from tests of the strong-field dynamical regime, not the weak field, slow motion regime of the solar system. Note, however, that for binary black hole inspirals, with $s_1 = s_2 = 1/2$, the gravitational-wave signal is identical to that of general relativity, after rescaling the masses by the factor $1 - \zeta$, so no test of basic scalar-tensor theory is possible for such sources. One needs a neutron star inspiral, such as the event GW170817, to test these theories.

Sennett et al. (2016) have analyzed the next-order contributions to the frequency and phase evolution.

12.2.3 Compact binary inspiral in other theories

Compact binary inspiral in other theories of gravity has not been analyzed systematically in the same manner as have general relativity and scalar-tensor theories, so only limited conclusions about alternative theories can be drawn from the gravitational-wave detections made to date (Yunes et al., 2016). On the other hand, a number of phenomenological parametrizations have been developed that make it possible to place some preliminary bounds on general classes of theories.

Bounds on the graviton mass and the speed of gravitational waves

If gravitation is propagated by a massive field, as in massive gravity theories (Section 5.7), then the velocity of gravitational waves will depend upon their wavelength as $v_g^2 = 1 - (\lambda/\lambda_g)^2$ [Eq. (11.10)], where $\lambda_g = h/m_g$ is the graviton Compton wavelength. In the case of inspiralling compact binaries, gravitational waves emitted at low frequency early in the inspiral will thus travel slightly more slowly than those emitted at high frequency later, resulting in an offset in the relative arrival times at a detector. This modifies the observed frequency evolution of the observed inspiral gravitational waveform, similar to that caused by post-Newtonian corrections to the quadrupole evolution (Will, 1998).

To make this quantitative, we consider the propagation of a massive "graviton" in a background Friedmann-Robertson-Walker (FRW) homogeneous and isotropic spacetime, with the line element

$$ds^2 = -dt^2 + a^2(t)[d\chi^2 + \Sigma^2(\chi)(d\theta^2 + \sin^2\theta d\phi^2)], \tag{12.53}$$

where $a(t)$ is the scale factor of the universe and $\Sigma(\chi)$ is equal to χ, $\sin\chi$ or $\sinh\chi$ if the universe is spatially flat, closed or open, respectively. For a graviton moving radially from an emitter at $\chi = \chi_e$ to a receiver at $\chi = 0$, it is straightforward to show that the component of the graviton's four-momentum p_χ = constant. Using the fact that $m_g^2 = -p^\alpha p^\beta g_{\alpha\beta} = E^2 - a^{-2}p_\chi^2$, where $E = p^0$, together with $p^\chi/E = d\chi/dt$, we obtain

$$\frac{d\chi}{dt} = -\frac{1}{a}\left(1 + \frac{m_g^2 a^2}{p_\chi^2}\right)^{-1/2}, \tag{12.54}$$

where $p_\chi^2 = a^2(t_e)(E_e^2 - m_g^2)$. Assuming that $E_e \gg m_g$, expanding Eq. (12.54) to first order in $(m_g/E_e)^2$ and integrating, we obtain

$$\chi_e = \int_{t_e}^{t_r} \frac{dt}{a(t)} - \frac{1}{2}\frac{m_g^2}{a^2(t_e)E_e^2}\int_{t_e}^{t_r} a(t)dt\,. \tag{12.55}$$

Consider gravitons emitted at two different times t_e and t_e', with energies E_e and E_e', and received at corresponding arrival times (χ_e is the same for both). Assuming that $\Delta t_e \equiv t_e - t_e' \ll a/\dot{a}$, and noting that $m_g/E_e = (\lambda_g F_e)^{-1}$, where F_e is the emitted frequency, we obtain, after eliminating χ_e,

$$t_r = t_r' + (1+Z)\left[t_e - t_e' + \frac{D}{2\lambda_g^2}\left(\frac{1}{F_e^2} - \frac{1}{F_e'^2}\right)\right], \tag{12.56}$$

where $Z \equiv a_0/a(t_e) - 1$ is the cosmological redshift, and

$$\begin{aligned} D &\equiv \frac{(1+Z)}{a_0}\int_{t_e}^{t_a} a(t)dt \\ &= \frac{1+Z}{H_0}\int_0^Z \frac{dZ'}{(1+Z')^2[\Omega_M(1+Z')^3 + \Omega_\Lambda]^{1/2}}, \end{aligned} \tag{12.57}$$

where $a_0 = a(t_r)$ is the present value of the scale factor. The expression in the second line corresponds to a spatially flat universe with matter density parameter Ω_M and dark-energy parameter Ω_Λ, with $\Omega_M + \Omega_\Lambda = 1$. Note that D is not a conventional cosmological distance measure, like the luminosity distance given by

$$\begin{aligned} D_L &\equiv (1+Z)a_0\int_{t_e}^{t_a} \frac{dt}{a(t)} \\ &= \frac{1+Z}{H_0}\int_0^Z \frac{dZ'}{[\Omega_M(1+Z')^3 + \Omega_\Lambda]^{1/2}}. \end{aligned} \tag{12.58}$$

For a standard ΛCDM cosmology, D/D_L ranges from unity at $Z = 0$ to 0.3 at $Z = 3$.

Because the arrival time of the signal now depends on frequency, the *observed* frequency evolution is given by

$$\frac{dF}{dt} = \frac{dF_r}{dt_r} = (1+Z)^{-1}\frac{dF_e}{dt_e}\frac{dt_e}{dt_r} \, . \tag{12.59}$$

where we have inserted the redshift factor linking F_r with F_e. We now calculate dt_r/dt_e using Eq. (12.56), and insert the general relativistic quadrupole evolution for dF_e/dt_e, Eq. (12.39), including the 1PN correction. We are assuming here that, in whatever massive gravity theory is being employed, the intrinsic frequency evolution of the system is given by the formulae from general relativity, ignoring corrections of fractional order $(r/\lambda_g)^2$, where r is the size of the binary system. Since solar-system bounds already imply that $\lambda_g > 10^{12}$ km, these are likely to be small corrections. We then obtain, for the observed frequency evolution,

$$\frac{dF}{dt} = \frac{96\pi}{5}F^2(\pi \mathcal{M} F)^{5/3}\left[1 - \left(\frac{743}{336} + \frac{11}{4}\eta + \beta_g\right)(\pi m F)^{2/3} + \dots\right] , \tag{12.60}$$

where F, m, and $\mathcal{M}$ are the observed frequency and masses, and where

$$\beta_g = \frac{96}{5}\frac{\pi^2 D\eta m}{\lambda_g^2(1+Z)} \, . \tag{12.61}$$

The effect of the graviton mass is to modify the coefficient of the 1PN term in the frequency evolution. Thus with sufficiently accurate measurements of the frequency evolution, it is possible to measure the individual masses and to place a bound on β_g. Early studies estimated that advanced LIGO-Virgo detectors could place a bound $\lambda_g > 10^{12}$ km, while the LISA space detector could place a bound $\lambda > 10^{16}$ km (Will, 1998; Berti et al., 2005; Stavridis and Will, 2009; Arun and Will, 2009).

The dispersion relation $E^2 = p^2 + m_g^2$ for a massive graviton is Lorentz invariant. Mirshekari et al. (2012) considered extensions to theories of gravity that effectively violate Lorentz invariance, in which the dispersion relation could take the form

$$E^2 = p^2 + m_g^2 + Ap^\alpha \, , \tag{12.62}$$

where A and α are two parameters characterizing the Lorentz violation. For example, in some extra-dimension theories, $\alpha = 4$ and A is a parameter of the order of the square of the Planck length. Mirshekari et al. (2012) studied the bounds that could be placed on m_g and A for various values of α.

Tests of the massive graviton hypothesis have now been carried out using data from gravitational-wave detections. Data from the discovery event GW150914, led to the bound $\lambda_g > 10^{13}$ km (Abbott et al., 2016c,d). This was improved to 1.6×10^{13} km by combining data from the three events GW150914, GW151226, and GW170104. (Abbott et al., 2017b).

This limit is comparable to the solar-system limit, which comes from assuming that massive gravity also implies a modification of the Newtonian potential by the inclusion of a Yukawa factor, that is, $U = (m/r)e^{-r/\lambda_g}$. Data on Mercury's perihelion advance and on orbits of the outer planets imply that $\lambda > 10^{12}$ km (Talmadge et al., 1988), although recent improvements in knowledge of Mercury's orbit described in Section 7.3 probably push that bound above 10^{13} km. Some have argued for a larger bound on λ_g from galactic and cluster dynamics (Hiida and Yamaguchi, 1965; Hare, 1973; Goldhaber and Nieto, 1974), noting that the evidence of bound clusters and of clear tidal interactions between galaxies argues for a range λ_g at least as large as a few megaparsecs (6×10^{19} km). However, in view of the uncertainties related to the amount of dark matter in the universe, and the fact that massive gravity theories frequently include other modifications of gravity on large scales, these bounds should be viewed with caution. Even the dynamical arguments used to establish these large-scale bounds have been questioned recently (Christodoulou and Kazanas, 2017; Mukherjee and Sounda, 2017). Bounds on the Lorentz-violating dispersion parameter A were also placed using combined data from GW150914, GW151226, and GW170104 (Abbott et al., 2017b).

A very strong bound was placed on the difference in speed between light and gravitational waves using the binary neutron-star event GW170817 (Abbott et al., 2017a). Because the gravitational-wave event could be associated with the gamma-ray burst event GRB170817A in the galaxy NGC 4993, the observed time difference between the two signals, 1.74 ± 0.05 s, combined with the estimate of the distance of 42.9 ± 3.2 Mpc, produced the bound

$$-3 \times 10^{-15} < c_g - 1 < 7 \times 10^{-16} . \tag{12.63}$$

Bounds on polarizations of gravitational waves

A restricted test of gravitational-wave polarizations was carried out in 2017, when the Virgo detector joined LIGO for several weeks toward the end of LIGO's second observing run. Using data from the binary black hole inspiral GW170814, seen in all three detectors, a Bayesian analysis was carried out in which a fit to the data was carried out for pure tensor response (i.e., including only F_+ and $F_\times$ in the response), pure scalar response (only F_S and F_L) and pure vector response (only F_{V1} and F_{V2}). The quadrupole response was strongly favored over the two alternatives by Bayes factors of 200 and 1000, respectively (Abbott et al., 2017c).

Parametrizations of binary inspiral signals

Several frameworks have been developed, modeled on the PPN formalism, in which arbitrary parameters are introduced into the predicted gravitational-wave signal from binary inspiral in order to encompass alternative theories of gravity. Arun et al. (2006a) wrote the Fourier phase $\psi(f)$ of Eq. (12.45) in the form

$$\psi(f) = 2\pi f t_c - \phi_c - \frac{\pi}{4} + \sum_k \frac{3}{128\eta} (\pi \mathcal{M} f)^{(k-5)/3} \alpha_k , \tag{12.64}$$

where α_k are arbitrary parameters. The general relativistic values of the parameters can be read off from the coefficients inside the square brackets of Eq. (12.45) (see also Arun et al. (2006b) and Mishra et al. (2010)).

Yunes and Pretorius (2009) extended this post-Newtonian parametrization to include nongeneral relativistic frequency dependences, in the form

$$\psi(f) = 2\pi f t_c - \phi_c - \frac{\pi}{4} + \sum_{k=0}^{N-1} \phi_k (\pi \mathcal{M} f)^{b_k}, \tag{12.65}$$

where ϕ_k and b_k are each a set of N parameters. This parametrization allows for the presence of dipole radiation, which begins at order $(\pi \mathcal{M} f)^{-7/3}$. They also included parametrized waveform templates for both the merger and ringdown phase. This is known as the parametrized post-Einsteinian (PPE) formalism.

These parameterizations were extended and adapted for specific use within the LIGO-Virgo data analysis pipeline, based on Bayesian model selection (Li et al., 2012; Agathos et al., 2014). This "test infrastructure for general relativity" (TIGER) was used in a joint analysis of the data from GW150914 and GW151226 to place limits on the various parameters of the model (Abbott et al., 2016a). For example, the 0PN, 1PN, and 1.5PN coefficients in the model were consistent with general relativity to between 10 and 20 percent; higher PN-order coefficients were more poorly constrained. With more detections, these constraints are likely to improve.

12.3 Exploring Spacetime near Compact Objects

One of the difficulties of testing GR in the strong-field regime is the possibility of contamination by uncertain or complex physics. In the solar system, weak-field gravitational effects can in most cases be measured cleanly and separately from nongravitational effects. The remarkable cleanliness of many binary pulsars permits precise measurements of gravitational phenomena in a strong-field context. The gravitational waves from inspiraling compact binaries detected by LIGO and Virgo appear to be likewise amazingly clean, not entirely surprising, since four of the first five detections involved binary black holes.

Unfortunately, nature is rarely so kind. Still, under suitable conditions, qualitative and even quantitative strong-field tests of GR are possible. The combination of improved astronomical observations in all wavelength bands, from radio to gamma rays, and better theoretical and computational modelling of complex physical processes have offered the promise of performing striking tests of strong-field predictions of GR. This is a rapidly evolving field, and therefore we will be unable to give a thorough account here. Instead we will give a few examples of interesting arenas where possible astrophysical tests of general relativity might occur, and will otherwise refer readers to a number of recent reviews (Psaltis, 2008; Berti et al., 2015; Johannsen, 2016).

12.3.1 SgrA*: A black hole in the galactic center

Lynden-Bell and Rees (1971) suggested that there might be a supermassive black hole in the center of the Milky Way, and Balick and Brown (1974) observed a bright, unresolved synchroton radio source at the precise galactic center, naming it Sagittarius A*, or SgrA* for short. While evidence mounted for the existence of massive black holes in the centers of quasars and active galactic nuclei, a black hole in the Milky Way remained only an intriguing possibility until advances in infrared interferometry and adaptive optics made it possible to detect stars very near the location of SgrA* and to observe their orbital motions (Eckart and Genzel, 1996; Ghez et al., 1998). A major breakthrough came with the unambiguous determination of the orbit of the star denoted S2 with an orbital period of about 16 years, and projected semimajor axis of 0.1 arcseconds (Schödel et al., 2002; Ghez et al., 2003). This made it possible, using Kepler's third law, to determine the mass of the black hole directly. Combined with more recent data on S2 and on other stars in the central cluster, this has led to an estimate for the black-hole mass of $4.4 \pm 0.4 \times 10^6 M_\odot$ and an improved value for the distance to the galactic center of 8.3 ± 0.4 kiloparsecs (see Genzel et al. (2010) for a thorough review of SgrA*). In addition to opening a window on the innermost region of the galactic center, the discovery of these stars has made it possible to contemplate using orbital dynamics to probe the curved spacetime of a rotating black hole, with the potential to test general relativity in the strong-field regime.

Soon after the detection of stars orbiting SgrA*, numerous authors pointed out the relativistic effects that are potentially observable, including the gravitational redshift and time-dilation effects (Zucker et al., 2006) and the pericenter advance (Jaroszynski, 1998; Fragile and Mathews, 2000; Rubilar and Eckart, 2001; Weinberg et al., 2005; Kraniotis, 2007). Although these effects were not detectable with the instrumentation at the time, notably the infrared interferometer at the Keck Observatory in Hawaii and the Very Large Telescope infrared Interferometer (VLTI) at the European Southern Observatory, the next generation of the instrumentation may bring relativistic tests within reach. The upgrade of the VLTI, known as GRAVITY, made its first observations of the galactic center in June 2016 (GRAVITY Collaboration et al., 2017). The Thirty Meter Telescope (TMT), currently under development, will also have the capability to make such measurements.

For example, the gravitational redshift of spectra of the stars S2 and S102 will be detectable during their next pericenter passages in 2018 and 2021, respectively. The pericenter precession of S2 is another target for detection in 2018. In general relativity, the leading contribution to the pericenter advance rate is given by $\dot{\omega} = 6\pi m/P_b a(1-e^2)$. However, if the precession of the orbit is being measured astrometrically from Earth, then we must project the orbit onto the plane of the sky and determine the rate of change of the relevant point on the orbit as seen from Earth. Thus the projected pericenter shift is given by $\dot{\omega}_{\rm proj} = [a(1 \pm e)/R]\dot{\omega}\cos\iota$ where the plus (minus) sign corresponds to measuring the shift at apocenter (pericenter), ι is the inclination of the orbit, and R is the distance to the galactic center. Using the mass and distance of SgrA* quoted earlier, the result is

$$\dot{\omega}_{\rm proj} = 98.3\,\mu{\rm as\,yr}^{-1}\left(\frac{1\,{\rm yr}}{P_b}\right)\frac{\cos\iota}{1 \mp e}, \tag{12.66}$$

where the minus sign now corresponds to a measurement at apocenter. For S2 and S102, the rates are 37 and 23 microarcseconds per year. Current observations of S2 and S102 are within a factor of 10 of measuring the precession (Hees et al., 2017).

Among the goals of second-generation projects such as GRAVITY and the TMT are astrometric precisions of tens of microarcseconds per year, and sensitivities enabling the detection of stars orbiting much closer to the black hole than the S-stars (if such stars exist). This makes it possible to consider doing more than merely detect relativistic effects, but

Table 12.3 Orbital parameters of selected stars orbiting the galactic center black hole SgrA*. Data taken from Gillessen et al. (2009) and Meyer et al. (2012).

Star	a (a.u)	e	$\iota(°)$	P_b (yr)
S2	1020 ± 8	0.880 ± 0.003	135.25 ± 0.47	15.8 ± 0.1
S4	2470 ± 160	0.406 ± 0.022	77.83 ± 0.32	59.5 ± 2.6
S5	2080 ± 350	0.842 ± 0.017	143.7 ± 4.7	45.7 ± 6.9
S9	2430 ± 430	0.825 ± 0.020	81.0 ± 0.7	58.0 ± 9.5
S14	2125 ± 83	0.963 ± 0.006	99.4 ± 1.0	47.3 ± 2.9
S38	1160 ± 340	0.802 ± 0.041	166 ± 22	18.9 ± 5.8
S102	812 ± 32	0.68 ± 0.02	151 ± 3	11.5 ± 0.3

rather to provide the first test of the black hole "no-hair" or uniqueness theorems of general relativity (Will, 2008). According to those theorems, an electrically neutral black hole is completely characterized by its mass m and angular momentum J. As a consequence, all the multipole moments of its external spacetime are functions of m and J. Specifically, the quadrupole moment $Q_2 = -J^2/m$.

To see how such a test might be carried out, we work in the post-Newtonian limit, and write down the equation of motion for a test body in the field of a body with mass m, angular momentum $\boldsymbol{J}$, and quadrupole moment Q_2. This can be obtained from Eq. (6.70), using general relativistic values of the PPN parameters, setting $\eta = 0$, and including the fact that the dimensionless quadrupole moment J_2 and Q_2 are related by $mR^2J_2 = -Q_2$. We also include the spin-orbit terms from Eq. (9.4a), with $\gamma = 1$, $\alpha_1 = 0$, and $\boldsymbol{S}_1 = m_1 = 0$, hence $\boldsymbol{\sigma} = 0$. The result is

$$\begin{aligned} \frac{d\boldsymbol{v}}{dt} &= -\frac{m\boldsymbol{n}}{r^2}\left[1 + v^2 - \frac{4m}{r}\right] + \frac{4m\boldsymbol{v}\dot{r}}{r^2} \\ &\quad - \frac{2J}{r^3}\left[2\boldsymbol{v} \times \boldsymbol{e} - 3\dot{r}\boldsymbol{n} \times \boldsymbol{e} - 3r^{-1}\boldsymbol{n}(\boldsymbol{h} \cdot \boldsymbol{e})\right] \\ &\quad - \frac{3}{2}\frac{Q_2}{r^4}\left[5\boldsymbol{n}(\boldsymbol{n} \cdot \boldsymbol{e})^2 - 2\boldsymbol{e}(\boldsymbol{n} \cdot \boldsymbol{e}) - \boldsymbol{n}\right], \end{aligned} \tag{12.67}$$

where $\boldsymbol{v}$ is the velocity of the body, $\boldsymbol{n} = \boldsymbol{x}/r$, $\boldsymbol{h} = \boldsymbol{x} \times \boldsymbol{v}$, and $\boldsymbol{e} = \boldsymbol{J}/J$ is a unit vector along the symmetry axis of the black hole. If we define a dimensionless spin parameter χ by $\chi \equiv J/m^2$, then, for a Kerr black hole, $0 \leq \chi \leq 1$ and $Q_2 = -M^3\chi^2$.

We consider a body with orbital elements a, e, ι, Ω and ω defined using the plane of the sky as the X–Y plane of the reference system (see Figure 6.1). The axis of the black hole has an unknown orientation relative to the reference system. Substituting the perturbing acceleration of Eq. (12.67) into the Lagrange planetary equations (6.74) and integrating over an orbit, we obtain the changes $\Delta e = \Delta a = 0$, and

$$\begin{aligned} \Delta\varpi &= A_S - 2A_J\cos\alpha - \frac{1}{2}A_Q(1 - 3\cos^2\alpha), \\ \Delta\Omega &= \frac{\sin\alpha\sin\beta}{\sin\iota}(A_J - A_Q\cos\alpha), \\ \Delta i &= \sin\alpha\cos\beta(A_J - A_Q\cos\alpha), \end{aligned} \tag{12.68}$$

where

$$\begin{aligned} A_S &= 6\pi m/p\,, \\ A_J &= 4\pi J/(mp^3)^{1/2}\,, \\ A_Q &= 3\pi Q_2/mp^2\,, \end{aligned} \tag{12.69}$$

where $\Delta\varpi = \Delta\omega + \cos\iota\,\Delta\Omega$, and $p = a(1-e^2)$. The angles α and β are the polar angles of the black hole's symmetry axis $\boldsymbol{e}$ with respect to the star's orbital plane defined by the line of nodes $\boldsymbol{e}_\Omega$, and the vector in the orbital plane $\boldsymbol{e}_\perp$ orthogonal to $\boldsymbol{e}_\Omega$ and $\mathbf{h}$, according to $\cos\alpha \equiv \boldsymbol{e}\cdot\hat{\boldsymbol{h}}$, $\sin\alpha\cos\beta \equiv \boldsymbol{e}\cdot\boldsymbol{e}_\Omega$ and $\sin\alpha\sin\beta \equiv \boldsymbol{e}\cdot\boldsymbol{e}_\perp$.

The structure of the expressions for $\Delta\Omega$ and $\Delta\iota$ can be understood as follows: Eq. (12.67) implies that the orbital angular momentum $\boldsymbol{h}$ precesses according to $d\boldsymbol{h}/dt = \boldsymbol{\omega}_{\rm P}\times\mathbf{h}$, where the orbit-averaged precession vector is given by $\boldsymbol{\omega}_{\rm P} = \boldsymbol{e}(A_J - A_{Q_2}\cos\alpha)$. The orbit element variations are given by $d\iota/dt = \boldsymbol{\omega}_{\rm P}\cdot\boldsymbol{e}_\Omega$ and $\sin\iota\, d\Omega/dt = \boldsymbol{\omega}_{\rm P}\cdot\boldsymbol{e}_\perp$. As a consequence, we have the purely geometric relationship,

$$\frac{\sin\iota\, d\Omega/dt}{di/dt} = \tan\beta\,. \tag{12.70}$$

To get rough idea of the astrometric size of these precessions, we define an angular precession rate amplitude $\dot\Theta_n = (a/R)A_n/P_{\rm b}$, where R is the distance to the galactic center and $P_{\rm b}$ is the orbital period. Using $m = 4.4\times10^6\,M_\odot$, $D = 8.3$ kpc, we obtain the rates, in microarcseconds per year,

$$\begin{aligned} \dot\Theta_S &\approx 98.3\,P_{\rm b}^{-1}(1-e^2)^{-1}\,, \\ \dot\Theta_J &\approx 1.07\,\chi P_{\rm b}^{-4/3}(1-e^2)^{-3/2}\,, \\ \dot\Theta_Q &\approx 1.3\times10^{-2}\,\chi^2 P_{\rm b}^{-5/3}(1-e^2)^{-2}\,, \end{aligned} \tag{12.71}$$

where $P_{\rm b}$ is measured in years. The observable precessions will be reduced somewhat from these raw rates because the orbit itself must be projected onto the plane of the sky. As we noted earlier, the contribution to $\Delta\varpi$ is reduced by $\cos\iota$, and the contribution to $\Delta\iota$ is reduced by $\sin\iota$. For an orbit in the plane of the sky, changes in the inclination are unmeasurable, and changes in the nodal angle become degenerate with changes in the pericenter.

For the quadrupole precessions to be observable, it is clear that the black hole must have a decent angular momentum ($\chi > 0.5$) and that the star must be in a short-period high-eccentricity orbit. For example, for $\chi = 0.7$, $P_{\rm b} = 0.1$ yr, and $e = 0.9$, the three amplitudes listed in Eq. (12.71) have the values 5200, 195 and 8 μ as per year, respectively.

Although the pericenter advance is the largest relativistic orbital effect, it is *not* the most suitable effect for testing the no-hair theorems. The frame-dragging and quadrupole effects are small corrections of the leading Schwarzschild pericenter precession, and thus one would need to know m, a and e to sufficient accuracy to be able to subtract that dominant term to reveal the smaller effects of interest. Furthermore, the pericenter advance is affected by a number of complicating phenomena, including 2PN effects, the disturbing effects of surrounding mass in the form of gas, stars or dark matter, and the effects of tidal distortion of the star near its pericenter.

On the other hand, the precessions of the node and inclination are relatively immune from such effects. Any spherically symmetric distribution of mass has no effect on these orbit elements. As long as any tidal distortion of the star is quasi-equilibrium with negligible tidal lag, the resulting perturbing forces are purely radial, and thus have no effect on the node or inclination. On the other hand, even if a surrounding cluster of stars is spherical on average, the "graininess" of the perturbing forces from a finite number of such stars will have an effect on the orbital plane of a chosen target star. These perturbing effects were studied for a range of hypothetical distributions of cluster stars by Merritt et al. (2010) and Sadeghian and Will (2011)

In order to test the no-hair theorem using the precessions $\Delta\Omega$ and $\Delta\iota$, we need to measure four quantities, the magnitudes J and Q_2, and the angles α and β. Thus it will be necessary to measure $\Delta\Omega$ and $\Delta\iota$ for at least two stars whose orbital planes are appropriately nondegenerate. Even if the precision is not sufficient to be sensitive to the small quadrupole effect, it would still be possible to measure the spin magnitude and direction of the black hole, which would shed light on how it evolved and grew during the lifetime of the galaxy.

Observing stars is not the only way to explore the spacetime near SgrA*. If a pulsar were observed orbiting sufficiently close to the black hole, observations of its orbital precession using pulsar timing rather than astrometry could also contribute to a test of the no-hair theorem (Wex and Kopeikin, 1999; Liu et al., 2012). In addition to stars, there are a number of gaseous disks orbiting the central black hole (Genzel et al., 2010). Working at submillimeter wavelengths and linking a worldwide set of radio telescopes to form a giant VLBI array, a collaboration known as the Event Horizon Telescope (EHT) is approaching the capability of imaging SgrA* with event-horizon-scale angular resolution (Doeleman et al., 2009). Observation of accretion phenomena in the innermost disk at these angular resolutions could provide tests of the spacetime geometry very close to the black hole. (Johannsen et al., 2016a,b). And combining data from stellar astrometry, pulsar timing and EHT could have significant advantages for testing the no-hair theorem (Psaltis et al., 2016).

The observations needed to explore the strong-field region around SgrA* are very challenging, but steady progress is being made, notably with GRAVITY and the EHT. In addition, some luck will be called for: it is not known whether a population of sufficiently bright young stars exists close enough to the black hole to make no-hair tests feasible. Nor is it known if suitable pulsars exist sufficiently close to SgrA*, although the discovery of a magnetar (a pulsar with an extremely large magnetic field) in the neigborhood of the black hole gives reason for hope.

12.3.2 Neutron stars and black holes

Neutron stars and stellar-mass black holes may also be important arenas for testing strong-field gravity.

Studies of certain kinds of accretion known as advection-dominated accretion flow (ADAF) in low-luminosity binary X-ray sources have yielded hints of the signature of the black hole event horizon (Narayan and McClintock, 2008). The spectrum of frequencies of quasi-periodic oscillations (QPO) from accretion onto black holes and neutron stars in binaries may permit measurement of the spins of the compact objects. Aspects of strong-field gravity and frame-dragging may be revealed in spectral shapes of iron fluorescence lines and continuum emission from the inner regions of accretion disks (Reynolds, 2013; Miller and Miller, 2015). See Abramowicz and Fragile (2013) for a review of accretion onto black holes and neutron stars.

The structure of neutron stars depends strongly on both the equation of state of nuclear matter and the theory of gravity, and there is considerable degeneracy between these two ingredients in such parameters as the mass and radius of the neutron star. However, the discovery of the "I-Love-Q" phenomenon, a relation between the moment of inertia, the Love number and the quadrupole moment of rotating neutron star models that is remarkably insensitive to the equation of state (Yagi and Yunes, 2013), may break the degeneracy and open up ways to test alternative theories of gravity.

For detailed reviews of strong-field tests of GR involving neutron stars and black holes using electromagnetic observations, see Psaltis (2008) and Johannsen (2016).

12.4 Cosmological Tests

From a few seconds after the Big Bang until the present, the underlying physics of the universe is well understood, in terms of a standard model of a nearly spatially flat universe, 13.6 billion years old, dominated by cold dark matter and dark energy, called the ΛCDM model. Notwithstanding some observational tensions, such as slightly differing values of the Hubble parameter coming from different observational techniques, difficulties accounting for the distribution of structures over all galactic scales, and the failure to date to detect the fundamental particle that is presumed to constitute dark matter, the general relativistic ΛCDM model agrees remarkably well with a wide range of observations.

Other theories, such as Brans-Dicke theory, are sufficiently close to general relativity (for large enough ω_0) that they conform to all cosmological observations, within the uncertainties. Certain generalized scalar-tensor theories, however, could have small values of ω at early times making aspects of early universe evolution highly non general relativistic, while evolving to large ω today, thereby agreeing with all solar system and astrophysical experiments, and hewing closely to late-time ΛCDM cosmology (Damour and Nordtvedt, 1993a,b).

One way to test such theories is through Big-Bang nucleosynthesis (BBN), since the abundances of the light elements produced when the temperature of the universe was about 1 MeV are sensitive to the rate of expansion at that epoch, which in turn depends on the field equations of the theory. Because the universe is radiation-dominated at that epoch, uncertainties in the amount of cold dark matter or dark energy (Λ) are unimportant. The nuclear reaction rates are reasonably well understood from laboratory experiments and theory, and the number of light neutrino families (3) conforms with evidence from particle accelerators. Thus, within modest uncertainties, one can assess the quantitative difference between the BBN predictions of general relativity and other theories of gravity under strong-field conditions and compare with observations. For recent analyses, primarily within scalar-tensor theories see Santiago et al. (1997), Damour and Pichon (1999), Clifton et al. (2005), and Coc et al. (2006).

A different class of theories has been developed in part to provide an alternative to the dark energy of the standard ΛCDM model, by modifying gravity on large, cosmological scales, while preserving the conventional solar and stellar-system phenomenology of general relativity. The $f(R)$ theories are examples. Since we are now in a period of what may be called "precision cosmology," one can begin to test alternative theories in this class using the accumulation of data on many fronts, including the growth of large-scale structure, cosmic background radiation fluctuations, galactic rotation curves, BBN, weak lensing, baryon acoustic oscillations, and so on.

Apart from direct measurements of the expansion rate of the universe through the observations of standard candles such as Type II supernovae, most information about the universe comes from studying deviations from a homogeneous, isotropic Friedmann-Robertson-Walker background spacetime, expressed roughly in the form

$$ds^2 = -dt^2 + a(t)^2 \left[\frac{dr^2}{1 - kr^2} + r^2 (d\theta^2 + \sin^2\theta d\phi^2) \right] + h_{\mu\nu} dx^\mu dx^\nu \,, \tag{12.72}$$

where t is proper time as measured by a clock at rest, $a(t)$ is the scale factor, here with units of distance, r, θ, ϕ, and x^μ are dimensionless coordinates, and $k = 0$ or ± 1 determines the curvature of the spatial sections. Unlike the PPN formalism, where the background spacetime was flat and it was relatively simple to characterize the deviations represented by $h_{\mu\nu}$, in the cosmological case, there are many different approaches, with a range of possible gauge choices. In addition, the evolution of the background will itself depend on the theory of gravity being studied. Finally the split between "background" and "perturbation" may not be as clean as it was in the post-Newtonian case.

This has led to a rich variety of approaches to treating cosmological tests of gravitational theories, including Amin et al. (2008), Daniel et al. (2010), Dossett et al. (2011), Dossett and Ishak (2012), Zuntz et al. (2012), Hojjati et al. (2012), Baker et al. (2013), and Sanghai and Clifton (2017). For a comprehensive review, see Ishak (2018). This is another rapidly evolving field, and further details are beyond the scope of this book.

References

Abbott, B. P., Abbott, R., Abbott, T. D., Abernathy, M. R., et al. 2016a. Binary black hole mergers in the first Advanced LIGO observing run. *Phys. Rev. X*, **6**, 041015, ArXiv e-prints 1606.04856.

Abbott, B. P., Abbott, R., Abbott, T. D., Abernathy, M. R., et al. 2016b. GW151226: Observation of gravitational waves from a 22-solar-mass binary black hole coalescence. *Phys. Rev. Lett.*, **116**, 241103, ArXiv e-prints 1606.04855.

Abbott, B. P., Abbott, R., Abbott, T. D., Abernathy, M. R., et al. 2016c. Observation of gravitational waves from a binary black hole merger. *Phys. Rev. Lett.*, **116**, 061102, ArXiv e-prints 1602.03837.

Abbott, B. P., Abbott, R., Abbott, T. D., Abernathy, M. R., et al. 2016d. Tests of general relativity with GW150914. *Phys. Rev. Lett.*, **116**, 221101, ArXiv e-prints 1602.03841.

Abbott, B. P., Abbott, R., Abbott, T. D., Acernese, F., et al. 2017a. Gravitational waves and gamma-rays from a binary neutron star merger: GW170817 and GRB 170817A. *Astrophys. J. Lett.*, **848**, L13, ArXiv e-prints 1710.05834.

Abbott, B. P., Abbott, R., Abbott, T. D., Acernese, F., et al. 2017b. GW170104: Observation of a 50-solar-mass binary black hole coalescence at redshift 0.2. *Phys. Rev. Lett.*, **118**, 221101, ArXiv e-prints 1706.01812.

Abbott, B. P., Abbott, R., Abbott, T. D., Acernese, F., et al. 2017c. GW170814: A three-detector observation of gravitational waves from a binary black hole coalescence. *Phys. Rev. Lett.*, **119**, 141101, ArXiv e-prints 1709.09660.

Abbott, B. P., Abbott, R., Abbott, T. D., Acernese, F., et al. 2017d. GW170817: Observation of gravitational waves from a binary neutron star inspiral. *Phys. Rev. Lett.*, **119**, 161101, ArXiv e-prints 1710.05832.

Abbott, B. P., Abbott, R., Abbott, T. D., Acernese, F., et al. 2017e. Multi-messenger observations of a binary neutron star merger. *Astrophys. J. Lett.*, **848**, L12, ArXiv e-prints 1710.05833.

Abramowicz, M. A., and Fragile, P. C. 2013. Foundations of black hole accretion disk theory. *Living Rev. Relativ.*, **16**, 1, ArXiv e-prints 1104.5499.

Adelberger, E. G. 2001. New tests of Einstein's equivalence principle and Newton's inverse-square law. *Class. Quantum Grav.*, **18**, 2397–2405.

Adelberger, E. G., Heckel, B. R., Stubbs, C. W., and Rogers, W. F. 1991. Searches for new macroscopic forces. *Ann. Rev. Nucl. Particle Sci.*, **41**, 269–320.

Adelberger, E. G., Heckel, B. R., and Nelson, A. E. 2003. Tests of the gravitational inverse-square law. *Ann. Rev. Nucl. Particle Sci.*, **53**, 77–121, ArXiv e-prints hep-ph/0307284.

Adelberger, E. G., Heckel, B. R., Hoedl, S., Hoyle, C. D., et al. 2007. Particle-physics implications of a recent test of the gravitational inverse-square law. *Phys. Rev. Lett.*, **98**, 131104, ArXiv e-prints hep-ph/0611223.

Agathos, M., Del Pozzo, W., Li, T. G. F., Van Den Broeck, C., et al. 2014. TIGER: A data analysis pipeline for testing the strong-field dynamics of general relativity with gravitational wave signals from coalescing compact binaries. *Phys. Rev. D*, **89**, 082001, ArXiv e-prints 1311.0420.

Alexander, S., and Yunes, N. 2009. Chern-Simons modified general relativity. *Phys. Rep.*, **480**, 1–55, ArXiv e-prints 0907.2562.

Ali-Haïmoud, Y., and Chen, Y. 2011. Slowly rotating stars and black holes in dynamical Chern-Simons gravity. *Phys. Rev. D*, **84**, 124033, ArXiv e-prints 1110.5329.

Alsing, J., Berti, E., Will, C. M., and Zaglauer, H. 2012. Gravitational radiation from compact binary systems in the massive Brans-Dicke theory of gravity. *Phys. Rev. D*, **85**, 064041, ArXiv e-prints 1112.4903.

Altschul, B. 2009. Bounding isotropic Lorentz violation using synchrotron losses at LEP. *Phys. Rev. D*, **80**, 091901, ArXiv e-prints 0905.4346.

Altschul, B., Bailey, Q. G., Blanchet, L., Bongs, K., et al. 2015. Quantum tests of the Einstein Equivalence Principle with the STE-QUEST space mission. *Adv. Space Res.*, **55**, 501–524, ArXiv e-prints 1404.4307.

Alväger, T., Farley, F. J. M., Kjellman, J., and Wallin, L. 1964. Test of the second postulate of special relativity in the GeV region. *Phys. Lett.*, **12**, 260–262.

Alves, M. E. D. S., and Tinto, M. 2011. Pulsar timing sensitivities to gravitational waves from relativistic metric theories of gravity. *Phys. Rev. D*, **83**, 123529, ArXiv e-prints 1102.4824.

Amaro-Seoane, P., Aoudia, S., Babak, S., Binétruy, P., et al. 2012. Low-frequency gravitational-wave science with eLISA/NGO. *Class. Quantum Grav.*, **29**, 124016, ArXiv e-prints 1202.0839.

Amin, M. A., Wagoner, R. V., and Blandford, R. D. 2008. A subhorizon framework for probing the relationship between the cosmological matter distribution and metric perturbations. *Mon. Not. R. Astron. Soc.*, **390**, 131–142, ArXiv e-prints 0708.1793.

Anderson, D., and Yunes, N. 2017. Solar system constraints on massless scalar-tensor gravity with positive coupling constant upon cosmological evolution of the scalar field. *Phys. Rev. D*, **96**, 064037, ArXiv e-prints 1705.06351.

Anderson, D., Yunes, N., and Barausse, E. 2016. Effect of cosmological evolution on solar system constraints and on the scalarization of neutron stars in massless scalar-tensor theories. *Phys. Rev. D*, **94**, 104064, ArXiv e-prints 1607.08888.

Anderson, J. D., Laing, P. A., Lau, E. L., Liu, A. S., et al. 1998. Indication, from Pioneer 10/11, Galileo, and Ulysses data, of an apparent anomalous, weak, long-range acceleration. *Phys. Rev. Lett.*, **81**, 2858–2861, ArXiv e-prints gr-qc/9808081.

Anderson, J. L. 1987. Gravitational radiation damping in systems with compact components. *Phys. Rev. D*, **36**, 2301–2313.

Andersson, N., and Comer, G. L. 2007. Relativistic fluid dynamics: Physics for many different scales. *Living Rev. Relativ.*, **10**, 1, ArXiv e-prints gr-qc/0605010.

Anninos, P., Hobill, D., Seidel, E., Smarr, L., and Suen, W.-M. 1993. Collision of two black holes. *Phys. Rev. Lett.*, **71**, 2851–2854, ArXiv e-prints gr-qc/9309016.

Antia, H. M., Chitre, S. M., and Gough, D. O. 2008. Temporal variations in the Sun's rotational kinetic energy. *Astron. Astrophys.*, **477**, 657–663, ArXiv e-prints 0711.0799.

Antoniadis, I., Arkani-Hamed, N., Dimopoulos, S., and Dvali, G. 1998. New dimensions at a millimeter to a fermi and superstrings at a TeV. *Phys. Lett. B*, **436**, 257–263, ArXiv e-prints hep-ph/9804398.

Antoniadis, J., Freire, P. C. C., Wex, N., Tauris, T. M., et al. 2013. A massive pulsar in a compact relativistic binary. *Science*, **340**, 448, ArXiv e-prints 1304.6875.

Antonini, P., Okhapkin, M., Göklü, E., and Schiller, S. 2005. Test of constancy of speed of light with rotating cryogenic optical resonators. *Phys. Rev. A*, **71**, 050101, ArXiv e-prints gr-qc/0504109.

Arkani-Hamed, N., Dimopoulos, S., and Dvali, G. 1998. The hierarchy problem and new dimensions at a millimeter. *Phys. Lett. B*, **429**, 263–272, ArXiv e-prints hep-ph/9803315.

Armano, M., Audley, H., Auger, G., Baird, J. T., et al. 2016. Sub-femto-g free fall for space-based gravitational wave observatories: LISA Pathfinder results. *Phys. Rev. Lett.*, **116**, 231101.

Armstrong, J. W., Estabrook, F. B., and Tinto, M. 1999. Time-delay interferometry for space-based gravitational wave searches. *Astrophys. J.*, **527**, 814–826.

Arnett, W. D., and Bowers, R. L. 1977. A microscopic interpretation of neutron star structure. *Astrophys. J. Suppl.*, **33**, 415.

Arun, K. G., and Will, C. M. 2009. Bounding the mass of the graviton with gravitational waves: Effect of higher harmonics in gravitational waveform templates. *Class. Quantum Grav.*, **26**, 155002, ArXiv e-prints 0904.1190.

Arun, K. G., Iyer, B. R., Qusailah, M. S. S., and Sathyaprakash, B. S. 2006a. Letter to the editor: Testing post-Newtonian theory with gravitational wave observations. *Class. Quantum Grav.*, **23**, L37–L43, ArXiv e-prints gr-qc/0604018.

Arun, K. G., Iyer, B. R., Qusailah, M. S. S., and Sathyaprakash, B. S. 2006b. Probing the nonlinear structure of general relativity with black hole binaries. *Phys. Rev. D*, **74**, 024006, ArXiv e-prints gr-qc/0604067.

Ashby, N. 2002. Relativity and the Global Positioning System. *Physics Today*, **55**, 41–47.

Ashby, N. 2003. Relativity in the Global Positioning System. *Living Rev. Relativ.*, **6**, 1.

Ashby, N., Bender, P. L., and Wahr, J. M. 2007. Future gravitational physics tests from ranging to the BepiColombo Mercury planetary orbiter. *Phys. Rev. D*, **75**, 022001.

Baade, W., and Zwicky, F. 1934. On super-novae. *Proc. Nat. Acad. Sci. (US)*, **20**, 254–259.

Babichev, E., and Langlois, D. 2009. Relativistic stars in $f(R)$ gravity. *Phys. Rev. D*, **80**, 121501, ArXiv e-prints 0904.1382.

Babichev, E., and Langlois, D. 2010. Relativistic stars in $f(R)$ and scalar-tensor theories. *Phys. Rev. D*, **81**, 124051, ArXiv e-prints 0911.1297.

Baeßler, S., Heckel, B. R., Adelberger, E. G., Gundlach, J. H., et al. 1999. Improved test of the equivalence principle for gravitational self-energy. *Phys. Rev. Lett.*, **83**, 3585–3588.

Baierlein, R. 1967. Testing general relativity with laser ranging to the Moon. *Phys. Rev.*, **162**, 1275–1287.

Bainbridge, M. B., and Webb, J. K. 2017. Artificial intelligence applied to the automatic analysis of absorption spectra: Objective measurement of the fine structure constant. *Mon. Not. R. Astron. Soc.*, **468**, 1639–1670, ArXiv e-prints 1606.07393.

Baker, J. G., Centrella, J., Choi, D.-I., Koppitz, M., et al. 2006. Gravitational-wave extraction from an inspiraling configuration of merging black holes. *Phys. Rev. Lett.*, **96**, 111102, ArXiv e-prints gr-qc/0511103.

Baker, T., Ferreira, P. G., and Skordis, C. 2013. The parameterized post-Friedmann framework for theories of modified gravity: Concepts, formalism, and examples. *Phys. Rev. D*, **87**, 024015, ArXiv e-prints 1209.2117.

Balick, B., and Brown, R. L. 1974. Intense sub-arcsecond structure in the galactic center. *Astrophys. J.*, **194**, 265–270.

Bambi, C., Giannotti, M., and Villante, F. L. 2005. Response of primordial abundances to a general modification of G_N and/or of the early universe expansion rate. *Phys. Rev. D*, **71**, 123524, ArXiv e-prints astro-ph/0503502.

Barausse, E., and Sotiriou, T. P. 2013a. Black holes in Lorentz-violating gravity theories. *Class. Quantum Grav.*, **30**, 244010, ArXiv e-prints 1307.3359.

Barausse, E., and Sotiriou, T. P. 2013b. Slowly rotating black holes in Hořava-Lifshitz gravity. *Phys. Rev. D*, **87**, 087504, ArXiv e-prints 1212.1334.

Barausse, E., Jacobson, T., and Sotiriou, T. P. 2011. Black holes in Einstein-Æther and Hořava-Lifshitz gravity. *Phys. Rev. D*, **83**, 124043, ArXiv e-prints 1104.2889.

Barausse, E., Palenzuela, C., Ponce, M., and Lehner, L. 2013. Neutron-star mergers in scalar-tensor theories of gravity. *Phys. Rev. D*, **87**, 081506, ArXiv e-prints 1212.5053.

Barausse, E., Sotiriou, T. P., and Vega, I. 2016. Slowly rotating black holes in EinsteinÆther theory. *Phys. Rev. D*, **93**, 044044, ArXiv e-prints 1512.05894.

Barker, B. M., and O'Connell, R. F. 1974. Nongeodesic motion in general relativity. *Gen. Relativ. Gravit.*, **5**, 539–554.

Bartel, N., Bietenholz, M. F., Lebach, D. E., Ransom, R. R., et al. 2015. VLBI for Gravity Probe B: The guide star, IM Pegasi. *Class. Quantum Grav.*, **32**, 224021, ArXiv e-prints 1509.07529.

Bartlett, D. F., and van Buren, D. 1986. Equivalence of active and passive gravitational mass using the moon. *Phys. Rev. Lett.*, **57**, 21–24.

Bauch, A., and Weyers, S. 2002. New experimental limit on the validity of local position invariance. *Phys. Rev. D*, **65**, 081101.

Baumgarte, T. W., and Shapiro, S. L. 2010. *Numerical Relativity: Solving Einstein's Equations on the Computer*. Cambridge: Cambridge University Press.

Baym, G., and Pethick, C. 1979. Physics of neutron stars. *Ann. Rev. Astron. Astrophys.*, **17**, 415–443.

Bekenstein, J. D. 2004. Relativistic gravitation theory for the modified Newtonian dynamics paradigm. *Phys. Rev. D*, **70**, 083509, ArXiv e-prints astro-ph/0403694.

Belinfante, F. J., and Swihart, J. C. 1957a. Phenomenological linear theory of gravitation: Part I. Classical mechanics. *Ann. Phys. (N.Y.)*, **1**, 168–195.

Belinfante, F. J., and Swihart, J. C. 1957b. Phenomenological linear theory of gravitation: Part II. Interaction with the maxwell field. *Ann. Phys. (N.Y.)*, **1**, 196–212.

Belinfante, F. J., and Swihart, J. C. 1957c. Phenomenological linear theory of gravitation: Part III: Interaction with the spinning electron. *Ann. Phys. (N.Y.)*, **2**, 81–99.

Bell, J. F., and Damour, T. 1996. A new test of conservation laws and Lorentz invariance in relativistic gravity. *Class. Quantum Grav.*, **13**, 3121–3127, ArXiv e-prints gr-qc/9606062.

Benacquista, M., and Nordtvedt Jr., K. 1988. A many-body Lagrangian for celestial body dynamics to second post-Newtonian linear field order. *Astrophys. J.*, **328**, 588–593.

Benacquista, M. J. 1992. Second-order parametrized-post-Newtonian Lagrangian. *Phys. Rev. D*, **45**, 1163–1173.

Benkhoff, J., van Casteren, J., Hayakawa, H., Fujimoto, M., et al. 2010. BepiColombo—Comprehensive exploration of Mercury: Mission overview and science goals. *Planet. Space Sci.*, **58**, 2–20.

Bennett, C. L., Larson, D., Weiland, J. L., Jarosik, N., et al. 2013. Nine-year Wilkinson Microwave Anisotropy Probe (WMAP) observations: Final maps and results. *Astrophys. J. Suppl.*, **208**, 20, ArXiv e-prints 1212.5225.

Berry, C. P. L., and Gair, J. R. 2011. Linearized $f(R)$ gravity: Gravitational radiation and solar system tests. *Phys. Rev. D*, **83**, 104022, ArXiv e-prints 1104.0819.

Berti, E., Buonanno, A., and Will, C. M. 2005. Estimating spinning binary parameters and testing alternative theories of gravity with LISA. *Phys. Rev. D*, **71**, 084025, ArXiv e-prints gr-qc/0411129.

Berti, E., Barausse, E., Cardoso, V., Gualtieri, L., et al. 2015. Testing general relativity with present and future astrophysical observations. *Class. Quantum Grav.*, **32**, 243001, ArXiv e-prints 1501.07274.

Bertotti, B., Brill, D. R., and Krotkov, R. 1962. Experiments on gravitation. Pages 1–48 in Witten, L. (ed), *Gravitation: An Introduction to Current Research*. New York: Wiley.

Bertotti, B., Iess, L., and Tortora, P. 2003. A test of general relativity using radio links with the Cassini spacecraft. *Nature*, **425**, 374–376.

Bessel, F. 1832. Versuche über die Kraft, mit welcher die Erde Körper von verschiedener Beschaffenheit anzieht. *Ann. Phys. (Leipzig)*, **101**, 401–417.

Bezerra, V. B., Klimchitskaya, G. L., Mostepanenko, V. M., and Romero, C. 2011. Constraints on non-Newtonian gravity from measuring the Casimir force in a configuration with nanoscale rectangular corrugations. *Phys. Rev. D*, **83**, 075004, ArXiv e-prints 1103.0993.

Bhat, N. D. R., Bailes, M., and Verbiest, J. P. W. 2008. Gravitational-radiation losses from the pulsar white-dwarf binary PSR J1141-6545. *Phys. Rev. D*, **77**, 124017, ArXiv e-prints 0804.0956.

Bi, X.-J., Cao, Z., Li, Y., and Yuan, Q. 2009. Testing Lorentz invariance with the ultrahigh energy cosmic ray spectrum. *Phys. Rev. D*, **79**, 083015, ArXiv e-prints 0812.0121.

Biller, S. D., Breslin, A. C., Buckley, J., Catanese, M., et al. 1999. Limits to quantum gravity effects on energy dependence of the speed of light from observations of TeV flares in active galaxies. *Phys. Rev. Lett.*, **83**, 2108–2111, ArXiv e-prints gr-qc/9810044.

Bize, S., Diddams, S. A., Tanaka, U., Tanner, C. E., et al. 2003. Testing the stability of fundamental constants with the $^{199}Hg^+$ single-ion optical clock. *Phys. Rev. Lett.*, **90**, 150802, ArXiv e-prints physics/0212109.

Blanchet, L. 2014. Gravitational radiation from post-Newtonian sources and inspiralling compact binaries. *Living Rev. Relativ.*, **17**, 2, ArXiv e-prints 1310.1528.

Blanchet, L., and Novak, J. 2011a. External field effect of modified Newtonian dynamics in the solar system. *Mon. Not. R. Astron. Soc.*, **412**, 2530–2542, ArXiv e-prints 1010.1349.

Blanchet, L., and Novak, J. 2011b. Testing MOND in the solar system, ArXiv e-prints 1105.5815.

Blanchet, L., Damour, T., Iyer, B. R., Will, C. M., et al. 1995a. Gravitational-radiation damping of compact binary systems to second post-Newtonian order. *Phys. Rev. Lett.*, **74**, 3515–3518, ArXiv e-prints gr-qc/9501027.

Blanchet, L., Damour, T., and Iyer, B. R. 1995b. Gravitational waves from inspiralling compact binaries: Energy loss and waveform to second-post-Newtonian order. *Phys. Rev. D*, **51**, 5360–5386, ArXiv e-prints gr-qc/9501029.

Blandford, R., and Teukolsky, S. A. 1976. Arrival-time analysis for a pulsar in a binary system. *Astrophys. J.*, **205**, 580–591.

Blas, D., Pujolàs, O., and Sibiryakov, S. 2010. Consistent extension of Hořava gravity. *Phys. Rev. Lett.*, **104**, 181302, ArXiv e-prints 0909.3525.

Blas, D., Pujolàs, O., and Sibiryakov, S. 2011. Models of non-relativistic quantum gravity: The good, the bad and the healthy. *J. High Energy Phys.*, **4**, 18, ArXiv e-prints 1007.3503.

Blatt, S., Ludlow, A. D., Campbell, G. K., Thomsen, J. W., et al. 2008. New limits on coupling of fundamental constants to gravity using ^{87}Sr optical lattice clocks. *Phys. Rev. Lett.*, **100**, 140801, ArXiv e-prints 0801.1874.

Blázquez-Salcedo, J. L., Cardoso, V., Ferrari, V., Gualtieri, L., et al. 2017. Black holes in Einstein-Gauss-Bonnet-dilaton theory. Pages 265–272 in *New Frontiers in Black Hole Astrophysics, Proceedings of the International Astronomical Union Symposium*, vol. 324.

Bollini, C. G., Giambiagi, J. J., and Tiomno, J. 1970. A linear theory of gravitation. *Lett. Nuovo Cimento*, **3**, 65–70.

Bolton, A. S., Rappaport, S., and Burles, S. 2006. Constraint on the post-Newtonian parameter γ on galactic size scales. *Phys. Rev. D*, **74**, 061501, ArXiv e-prints astro-ph/0607657.

Bolton, C. T. 1972. Identification of Cygnus X-1 with HDE 226868. *Nature*, **235**, 271–273.

Bondi, H. 1957. Negative mass in general relativity. *Rev. Mod. Phys.*, **29**, 423–428.

Braginsky, V. B., and Panov, V. I. 1972. Verification of the equivalence of inertial and gravitational mass. *J. Exp. Theor. Phys.*, **34**, 463.

Brans, C., and Dicke, R. H. 1961. Mach's Principle and a relativistic theory of gravitation. *Phys. Rev.*, **124**, 925–935.

Brault, J. W. 1962. *The gravitational red shift in the Solar spectrum*. Ph.D. thesis, Princeton University.

Brecher, K. 1977. Is the speed of light independent of the velocity of the source. *Phys. Rev. Lett.*, **39**, 1051–1054.

Breton, R. P., Kaspi, V. M., Kramer, M., McLaughlin, M. A., et al. 2008. Relativistic spin precession in the double pulsar. *Science*, **321**, 104, ArXiv e-prints 0807.2644.

Brillet, A., and Hall, J. L. 1979. Improved laser test of the isotropy of space. *Phys. Rev. Lett.*, **42**, 549–552.

Brune, Jr., R. A., Cobb, C. L., Dewitt, B. S., Dewitt-Morette, C., et al. 1976. Gravitational deflection of light: Solar eclipse of 30 June 1973. I. Description of procedures and final result. *Astron. J.*, **81**, 452–454.

Brunetti, M., Coccia, E., Fafone, V., and Fucito, F. 1999. Gravitational wave radiation from compact binary systems in the Jordan-Brans-Dicke theory. *Phys. Rev. D*, **59**, 044027, ArXiv e-prints gr-qc/9805056.

Bruns, D. 2018. Gravitational starlight deflection measurements during the 21 August 2017 total solar eclipse. *Class. Quantum Grav.*, **35**, 075009, ArXiv e-prints 1802.00343.

Buonanno, A., and Damour, T. 1999. Effective one-body approach to general relativistic two-body dynamics. *Phys. Rev. D*, **59**, 084006, ArXiv e-prints gr-qc/9811091.

Burgay, M. 2012. The double pulsar system in its 8th anniversary, ArXiv e-prints 1210.0985.

Burgay, M., D'Amico, N., Possenti, A., Manchester, R. N., et al. 2003. An increased estimate of the merger rate of double neutron stars from observations of a highly relativistic system. *Nature*, **426**, 531–533, ArXiv e-prints astro-ph/0312071.

Burrage, C., and Sakstein, J. 2017. Tests of Chameleon Gravity. *Living Rev. Relativ.*, **21**, 1, ArXiv e-prints 1709.09071.

Campanelli, M., Lousto, C. O., Marronetti, P., and Zlochower, Y. 2006. Accurate evolutions of orbiting black-hole binaries without excision. *Phys. Rev. Lett.*, **96**, 111101, ArXiv e-prints gr-qc/0511048.

Caves, C. M. 1980. Gravitational radiation and the ultimate speed in Rosen's bimetric theory of gravity. *Ann. Phys. (N.Y.)*, **125**, 35–52.

Celotti, A., Miller, J. C., and Sciama, D. W. 1999. Astrophysical evidence for the existence of black holes. *Class. Quantum Grav.*, **16**, A3–A21, ArXiv e-prints astro-ph/9912186.

Chamberlin, S. J., and Siemens, X. 2012. Stochastic backgrounds in alternative theories of gravity: Overlap reduction functions for pulsar timing arrays. *Phys. Rev. D*, **85**, 082001, ArXiv e-prints 1111.5661.

Champeney, D. C., Isaak, G. R., and Khan, A. M. 1963. An "aether drift" experiment based on the Mössbauer effect. *Phys. Lett.*, **7**, 241–243.

Chand, H., Petitjean, P., Srianand, R., and Aracil, B. 2005. Probing the time-variation of the fine-structure constant: Results based on Si IV doublets from a UVES sample. *Astron. Astrophys.*, **430**, 47–58, ArXiv e-prints astro-ph/0408200.

Chandrasekhar, S. 1965. The post-Newtonian equations of hydrodynamics in general relativity. *Astrophys. J.*, **142**, 1488.

Chandrasekhar, S., and Contopoulos, G. 1967. On a post-Galilean transformation appropriate to the post-Newtonian theory of Einstein, Infeld and Hoffmann. *Proc. R. Soc. A*, **298**, 123–141.

Chatziioannou, K., Yunes, N., and Cornish, N. 2012. Model-independent test of general relativity: An extended post-Einsteinian framework with complete polarization content. *Phys. Rev. D*, **86**, 022004, ArXiv e-prints 1204.2585.

Chiaverini, J., Smullin, S. J., Geraci, A. A., Weld, D. M., et al. 2003. New experimental constraints on non-Newtonian forces below 100 μm. *Phys. Rev. Lett.*, **90**, 151101, ArXiv e-prints hep-ph/0209325.

Chou, C. W., Hume, D. B., Rosenband, T., and Wineland, D. J. 2010. Optical clocks and relativity. *Science*, **329**, 1630–1633.

Christodoulou, D. M., and Kazanas, D. 2017. New bound closed orbits in spherical potentials. *ArXiv e-prints*, ArXiv e-prints 1707.04937.

Chupp, T. E., Hoare, R. J., Loveman, R. A., Oteiza, E. R., et al. 1989. Results of a new test of local Lorentz invariance: A search for mass anisotropy in ^{21}Ne. *Phys. Rev. Lett.*, **63**, 1541–1545.

Ciufolini, I. 1986. Measurement of the Lense-Thirring drag on high-altitude, laser-ranged artificial satellites. *Phys. Rev. Lett.*, **56**, 278–281.

Ciufolini, I. 2000. The 1995-99 measurements of the Lense-Thirring effect using laser-ranged satellites. *Class. Quantum Grav.*, **17**, 2369–2380.

Ciufolini, I., Chieppa, F., Lucchesi, D., and Vespe, F. 1997. Test of Lense-Thirring orbital shift due to spin. *Class. Quantum Grav.*, **14**, 2701–2726.

Ciufolini, I., Pavlis, E., Chieppa, F., Fernandes-Vieira, E., et al. 1998. Test of general relativity and measurement of the Lense-Thirring effect with two Earth satellites. *Science*, **279**, 2100.

Ciufolini, I., Paolozzi, A., Pavlis, E. C., Ries, J., et al. 2011. Testing gravitational physics with satellite laser ranging. *Eur. Phys. J. Plus*, **126**, 72.

Ciufolini, I., Paolozzi, A., Pavlis, E. C., Koenig, R., et al. 2016. A test of general relativity using the LARES and LAGEOS satellites and a GRACE Earth gravity model. Measurement of Earth's dragging of inertial frames. *Eur. Phys. J. C*, **76**, 120, ArXiv e-prints 1603.09674.

Clifton, T., Barrow, J. D., and Scherrer, R. J. 2005. Constraints on the variation of G from primordial nucleosynthesis. *Phys. Rev. D*, **71**, 123526, ArXiv e-prints astro-ph/0504418.

Coc, A., Olive, K. A., Uzan, J.-P., and Vangioni, E. 2006. Big bang nucleosynthesis constraints on scalar-tensor theories of gravity. *Phys. Rev. D*, **73**, 083525, ArXiv e-prints astro-ph/0601299.

Cocconi, G., and Salpeter, E. 1958. A search for anisotropy of inertia. *Nuovo Cimento*, **10**, 646–651.

Colladay, D., and Kostelecký, V. A. 1997. CPT violation and the standard model. *Phys. Rev. D*, **55**, 6760–6774, ArXiv e-prints hep-ph/9703464.

Colladay, D., and Kostelecký, V. A. 1998. Lorentz-violating extension of the standard model. *Phys. Rev. D*, **58**, 116002, ArXiv e-prints hep-ph/9809521.

Cook, G. B., Shapiro, S. L., and Teukolsky, S. A. 1994. Rapidly rotating neutron stars in general relativity: Realistic equations of state. *Astrophys. J.*, **424**, 823–845.

Cooney, A., Dedeo, S., and Psaltis, D. 2010. Neutron stars in $f(R)$ gravity with perturbative constraints. *Phys. Rev. D*, **82**, 064033, ArXiv e-prints 0910.5480.

Copi, C. J., Davis, A. N., and Krauss, L. M. 2004. New nucleosynthesis constraint on the variation of *G*. *Phys. Rev. Lett.*, **92**, 171301, ArXiv e-prints astro-ph/0311334.

Corinaldesi, E., and Papapetrou, A. 1951. Spinning test-particles in general relativity. II. *Proc. R. Soc. A*, **209**, 259–268.

Creighton, J., and Anderson, W. 2011. *Gravitational-Wave Physics and Astronomy: An Introduction to Theory, Experiment and Data Analysis*. Weinheim, Germany: Wiley-VCH.

Crelinsten, J. 2006. *Einstein's Jury: The Race to Test Relativity*. Princeton: Princeton University Press.

Creminelli, P., Nicolis, A., Papucci, M., and Trincherini, E. 2005. Ghosts in massive gravity. *J. High Energy Phys.*, **9**, 003, ArXiv e-prints hep-th/0505147.

Cutler, C. 1998. Angular resolution of the LISA gravitational wave detector. *Phys. Rev. D*, **57**, 7089–7102, ArXiv e-prints gr-qc/9703068.

Cutler, C., and Flanagan, É. É. 1994. Gravitational waves from merging compact binaries: How accurately can one extract the binary's parameters from the inspiral waveform? *Phys. Rev. D*, **49**, 2658–2697, ArXiv e-prints gr-qc/9402014.

Cutler, C., Apostolatos, T. A., Bildsten, L., Finn, L. S., et al. 1993. The last three minutes - Issues in gravitational-wave measurements of coalescing compact binaries. *Phys. Rev. Lett.*, **70**, 2984–2987, ArXiv e-prints astro-ph/9208005.

Damour, T. 1987. The problem of motion in Newtonian and Einsteinian gravity. Pages 128–198 in Hawking, S. W., and Israel, W. (eds), *Three Hundred Years of Gravitation*. New York: Cambridge University Press.

Damour, T., and Deruelle, N. 1986. General relativistic celestial mechanics of binary systems. II. The post-Newtonian timing formula. *Ann. Inst. Henri Poincaré A*, **44**, 263–292.

Damour, T., and Dyson, F. 1996. The Oklo bound on the time variation of the fine-structure constant revisited. *Nucl. Phys. B*, **480**, 37–54, ArXiv e-prints hep-ph/9606486.

Damour, T., and Esposito-Farèse, G. 1992a. Tensor-multi-scalar theories of gravitation. *Class. Quantum Grav.*, **9**, 2093–2176.

Damour, T., and Esposito-Farèse, G. 1992b. Testing local Lorentz invariance of gravity with binary-pulsar data. *Phys. Rev. D*, **46**, 4128–4132.

Damour, T., and Esposito-Farèse, G. 1993. Nonperturbative strong-field effects in tensor-scalar theories of gravitation. *Phys. Rev. Lett.*, **70**, 2220–2223.

Damour, T., and Esposito-Farèse, G. 1996. Tensor-scalar gravity and binary-pulsar experiments. *Phys. Rev. D*, **54**, 1474–1491, ArXiv e-prints gr-qc/9602056.

Damour, T., and Esposito-Farèse, G. 1998. Gravitational-wave versus binary-pulsar tests of strong-field gravity. *Phys. Rev. D*, **58**, 042001, ArXiv e-prints gr-qc/9803031.

Damour, T., and Nordtvedt, Jr., K. 1993a. General relativity as a cosmological attractor of tensor-scalar theories. *Phys. Rev. Lett.*, **70**, 2217–2219.

Damour, T., and Nordtvedt, Jr., K. 1993b. Tensor-scalar cosmological models and their relaxation toward general relativity. *Phys. Rev. D*, **48**, 3436–3450.

Damour, T., and Pichon, B. 1999. Big bang nucleosynthesis and tensor-scalar gravity. *Phys. Rev. D*, **59**, 123502, ArXiv e-prints astro-ph/9807176.

Damour, T., and Polyakov, A. M. 1994. The string dilation and a least coupling principle. *Nucl. Phys. B*, **423**, 532–558, ArXiv e-prints hep-th/9401069.

Damour, T., and Ruffini, R. 1974. Certain new verifications of general relativity made possible by the discovery of a pulsar belonging to a binary system. *C. R. Acad. Sci. Ser. A*, **279**, 971–973.

Damour, T., and Schaefer, G. 1991. New tests of the Strong Equivalence Principle using binary-pulsar data. *Phys. Rev. Lett.*, **66**, 2549–2552.

Damour, T., and Taylor, J. H. 1992. Strong-field tests of relativistic gravity and binary pulsars. *Phys. Rev. D*, **45**, 1840–1868.

Damour, T., Piazza, F., and Veneziano, G. 2002a. Runaway dilaton and equivalence principle violations. *Phys. Rev. Lett.*, **89**, 081601, ArXiv e-prints gr-qc/0204094.

Damour, T., Piazza, F., and Veneziano, G. 2002b. Violations of the equivalence principle in a dilaton-runaway scenario. *Phys. Rev. D*, **66**, 046007, ArXiv e-prints hep-th/0205111.

Daniel, S. F., Linder, E. V., Smith, T. L., Caldwell, R. R., et al. 2010. Testing general relativity with current cosmological data. *Phys. Rev. D*, **81**, 123508, ArXiv e-prints 1002.1962.

De Felice, A., and Tsujikawa, S. 2010. *f*(*R*) theories. *Living Rev. Relativ.*, **13**, 3, ArXiv e-prints 1002.4928.

de Rham, C. 2014. Massive gravity. *Living Rev. Relativ.*, **17**, 7, ArXiv e-prints 1401.4173.

de Sitter, W. 1916. On Einstein's theory of gravitation and its astronomical consequences. Second paper. *Mon. Not. R. Astron. Soc.*, **77**, 155–184.

D'Eath, P. D. 1975. Interaction of two black holes in the slow-motion limit. *Phys. Rev. D*, **12**, 2183–2199.

Deffayet, C., Dvali, G., Gabadadze, G., and Vainshtein, A. 2002. Nonperturbative continuity in graviton mass versus perturbative discontinuity. *Phys. Rev. D*, **65**, 044026, ArXiv e-prints hep-th/0106001.

Deller, A. T., Verbiest, J. P. W., Tingay, S. J., and Bailes, M. 2008. Extremely high precision VLBI astrometry of PSR J0437-4715 and implications for theories of gravity. *Astrophys. J. Lett.*, **685**, L67–L70, ArXiv e-prints 0808.1594.

Delva, P., Hees, A., Bertone, S., Richard, E., et al. 2015. Test of the gravitational redshift with stable clocks in eccentric orbits: Application to Galileo satellites 5 and 6. *Class. Quantum Grav.*, **32**, 232003, ArXiv e-prints 1508.06159.

Demorest, P. B., Pennucci, T., Ransom, S. M., Roberts, M. S. E., et al. 2010. A two-solar-mass neutron star measured using Shapiro delay. *Nature*, **467**, 1081–1083, ArXiv e-prints 1010.5788.

Deser, S. 1970. Self-interaction and gauge invariance. *Gen. Relativ. Gravit.*, **1**, 9–18, ArXiv e-prints gr-qc/0411023.

Deser, S., and Laurent, B. E. 1968. Gravitation without self-interaction. *Ann. Phys. (N.Y.)*, **50**, 76–101.

Dicke, R. H. 1964. Experimental relativity. In: DeWitt, C., and DeWitt, B. (eds), *Relativity, Groups and Topology*. New York: Gordon and Breach.

Dicke, R. H. 1964. Remarks on the observational basis of general relativity. Pages 1–16 in Chiu, H.-Y., and Hoffman, W. F. (eds), *Gravitation and Relativity*. New York: Benjamin.

Dicke, R. H. 1969. *Gravitation and the Universe.* Philadelphia: American Philosophical Society.

Dicke, R. H., and Goldenberg, H. M. 1974. The oblateness of the Sun. *Astrophys. J. Suppl.*, **27**, 131.

Dickey, J. O., Bender, P. L., Faller, J. E., Newhall, X X, et al. 1994. Lunar laser ranging: A continuing legacy of the Apollo Program. *Science*, **265**, 482–490.

Dixon, W. G. 1979. Extended bodies in general relativity: Their description and motion. Pages 156–219 in Ehlers, J. (ed), *Isolated Gravitating Systems in General Relativity*. Amsterdam: North-Holland.

Doeleman, S., Agol, E., Backer, D., Baganoff, F., et al. 2009. Imaging an event horizon: Submm-VLBI of a supermassive black hole. Page 68 in *Astro2010: The Astronomy and Astrophysics Decadal Survey*. Washington, DC: National Academy Press.

Doser, M., Amsler, C., Belov, A., Bonomi, G., et al. 2012. Exploring the WEP with a pulsed cold beam of antihydrogen. *Class. Quantum Grav.*, **29**, 184009.

Dossett, J. N., and Ishak, M. 2012. Spatial curvature and cosmological tests of general relativity. *Phys. Rev. D*, **86**, 103008, ArXiv e-prints 1205.2422.

Dossett, J. N., Ishak, M., and Moldenhauer, J. 2011. Testing general relativity at cosmological scales: Implementation and parameter correlations. *Phys. Rev. D*, **84**, 123001, ArXiv e-prints 1109.4583.

Drever, R. W. P. 1961. A search for anisotropy of inertial mass using a free precession technique. *Philos. Mag.*, **6**, 683–687.

Droste, J. 1917. The field of *N* moving centres in Einstein's theory of gravitation. *Koninklijke Nederlandse Akademie van Wetenschappen Proceedings Series B Physical Sciences*, **19**, 447–455.

Dyda, S., Flanagan, É. É., and Kamionkowski, M. 2012. Vacuum instability in Chern-Simons gravity. *Phys. Rev. D*, **86**, 124031, ArXiv e-prints 1208.4871.

Dyson, F. J. 1972. The fundamental constants and their time variation. Pages 213–236 in Salam, A., and Wigner, E. P. (eds), *Aspects of Quantum Theory*. Cambridge: Cambridge University Press.

Dyson, F. W., Eddington, A. S., and Davidson, C. 1920. A determination of the deflection of light by the Sun's gravitational field, from observations made at the total eclipse of May 29, 1919. *Phil. Trans. R. Soc. A*, **220**, 291–333.

Eardley, D. M. 1975. Observable effects of a scalar gravitational field in a binary pulsar. *Astrophys. J. Lett.*, **196**, L59–L62.

Eardley, D. M., Lee, D. L., Lightman, A. P., Wagoner, R. V., et al. 1973a. Gravitational-wave observations as a tool for testing relativistic gravity. *Phys. Rev. Lett.*, **30**, 884–886.

Eardley, D. M., Lee, D. L., and Lightman, A. P. 1973b. Gravitational-wave observations as a tool for testing relativistic gravity. *Phys. Rev. D*, **8**, 3308–3321.

Earman, J., and Glymour, C. 1980. Relativity and eclipses: The British eclipse expeditions of 1919 and their predecessors. *Historical Studies in the Physical Sciences*, **11**, 49–85.

Eckart, A., and Genzel, R. 1996. Observations of stellar proper motions near the galactic centre. *Nature*, **383**, 415–417.

Eddington, A. S. 1922. *The Mathematical Theory of Relativity*. Cambridge: Cambridge University Press.

Eddington, A. S. 1922. The propagation of gravitational waves. *Proc. R. Soc. A*, **102**, 268–282.

Eddington, A. S., and Clark, G. L. 1938. The problem of *N* bodies in general relativity theory. *Proc. R. Soc. A*, **166**, 465–475.

Ehlers, J. 1971. General relativity and kinetic theory. Pages 1–70 in Sachs, R. K. (ed), *General Relativity and Cosmology: Proceedings of Course 47 of the International School of Physics "Enrico Fermi"*. New York: Academic Press.

Ehlers, J., Rosenblum, A., Goldberg, J. N., and Havas, P. 1976. Comments on gravitational radiation damping and energy loss in binary systems. *Astrophys. J. Lett.*, **208**, L77–L81.

Einstein, A. 1908. Über das Relativitätsprinzip und die aus demselben gezogenen Folgerungen. *Jahrbuch der Radioaktivität und Elektronik*, **4**, 411 – 62.

Einstein, A. 1916. Näherungsweise Integration der Feldgleichungen der Gravitation. *Sitzungsberichte der Königlich Preußischen Akademie der Wissenschaften (Berlin)*, 688–696.

Einstein, A. 1918. Über Gravitationswellen. *Sitzungsberichte der Königlich Preußischen Akademie der Wissenschaften (Berlin), Seite 154-167.*, 154–167.

Einstein, A., and Rosen, N. 1937. On gravitational waves. *J. Franklin Inst.*, **223**, 43–54.

Einstein, A., Infeld, L., and Hoffmann, B. 1938. The gravitational equations and the problem of motion. *Ann. Math.*, **39**, 65–100.

Eisenstaedt, J. 2006. *The Curious History of Relativity: How Einstein's Theory Was Lost and Found Again*. Princeton: Princeton University Press.

Eling, C., and Jacobson, T. 2004. Static post-Newtonian equivalence of general relativity and gravity with a dynamical preferred frame. *Phys. Rev. D*, **69**, 064005, ArXiv e-prints gr-qc/0310044.

Eling, C., and Jacobson, T. 2006. Spherical solutions in Einstein-Æther theory: Static Æther and stars. *Class. Quantum Grav.*, **23**, 5625–5642, ArXiv e-prints gr-qc/0603058.

Eling, C., Jacobson, T., and Miller, M. C. 2007. Neutron stars in Einstein-Æther theory. *Phys. Rev. D*, **76**, 042003, ArXiv e-prints 0705.1565.

Elliott, J. W., Moore, G. D., and Stoica, H. 2005. Constraining the new aether: Gravitational Čerenkov radiation. *Journal of High Energy Physics*, **8**, 066, ArXiv e-prints hep-ph/0505211.

Eötvös, R. V., Pekár, D., and Fekete, E. 1922. Beiträge zum Gesetze der Proportionalität von Trägheit und Gravität. *Ann. Phys. (Leipzig)*, **373**, 11–66.

Epstein, R. 1977. The binary pulsar—Post-Newtonian timing effects. *Astrophys. J.*, **216**, 92–100.

Epstein, R., and Shapiro, I. I. 1980. Post-post-Newtonian deflection of light by the Sun. *Phys. Rev. D*, **22**, 2947–2949.

Epstein, R., and Wagoner, R. V. 1975. Post-Newtonian generation of gravitational waves. *Astrophys. J.*, **197**, 717–723.

Everitt, C. W. F., Debra, D. B., Parkinson, B. W., Turneaure, J. P., and et al. 2011. Gravity Probe B: Final results of a space experiment to test general relativity. *Phys. Rev. Lett.*, **106**, 221101, ArXiv e-prints 1105.3456.

Everitt, C. W. F., Muhlfelder, B., DeBra, D. B., Parkinson, B. W., et al. 2015. The Gravity Probe B test of general relativity. *Class. Quantum Grav.*, **32**, 224001.

Faber, J. A., and Rasio, F. A. 2012. Binary neutron star mergers. *Living Rev. Relativ.*, **15**, 8, ArXiv e-prints 1204.3858.

Famaey, B., and McGaugh, S. S. 2012. Modified Newtonian Dynamics (MOND): Observational phenomenology and relativistic extensions. *Living Rev. Relativ.*, **15**, ArXiv e-prints 1112.3960.

Farley, F. J. M., Bailey, J., Brown, R. C. A., Giesch, M., et al. 1966. The anomalous magnetic moment of the negative muon. *Nuovo Cimento A*, **45**, 281–286.

Fienga, A., Laskar, J., Kuchynka, P., Manche, H., et al. 2011. The INPOP10a planetary ephemeris and its applications in fundamental physics. *Cel. Mech. Dyn. Astron.*, **111**, 363–385, ArXiv e-prints 1108.5546.

Fienga, A., Laskar, J., Exertier, P., Manche, H., et al. 2015. Numerical estimation of the sensitivity of INPOP planetary ephemerides to general relativity parameters. *Cel. Mech. Dyn. Astron.*, **123**, 325–349.

Finkelstein, D. 1958. Past-future asymmetry of the gravitational field of a point particle. *Phys. Rev.*, **110**, 965–967.

Finn, L. S., and Chernoff, D. F. 1993. Observing binary inspiral in gravitational radiation: One interferometer. *Phys. Rev. D*, **47**, 2198–2219, ArXiv e-prints gr-qc/9301003.

Fischbach, E., and Freeman, B. S. 1980. Second-order contribution to the gravitational deflection of light. *Phys. Rev. D*, **22**, 2950–2952.

Fischbach, E., and Talmadge, C. 1992. Six years of the fifth force. *Nature*, **356**, 207–215.

Fischbach, E., and Talmadge, C. L. 1999. *The Search for Non-Newtonian Gravity*. New York: Springer-Verlag.

Fischbach, E., Sudarsky, D., Szafer, A., Talmadge, C., et al. 1986. Reanalysis of the Eötvös experiment. *Phys. Rev. Lett.*, **56**, 3–6.

Fischbach, E., Gillies, G. T., Krause, D. E., Schwan, J. G., et al. 1992. Non-Newtonian gravity and new weak forces: An index of measurements and theory. *Metrologia*, **29**, 213–260.

Fischer, M., Kolachevsky, N., Zimmermann, M., Holzwarth, R., et al. 2004. New limits on the drift of fundamental constants from laboratory measurements. *Phys. Rev. Lett.*, **92**, 230802, ArXiv e-prints physics/0312086.

Fock, V. A. 1964. *The Theory of Space, Time and Gravitation*. New York: Macmillan.

Foffa, S., and Sturani, R. 2014. Effective field theory methods to model compact binaries. *Class. Quantum Grav.*, **31**, 043001, ArXiv e-prints 1309.3474.

Fomalont, E. B., Kopeikin, S. M., Lanyi, G., and Benson, J. 2009. Progress in measurements of the gravitational bending of radio waves using the VLBA. *Astrophys. J.*, **699**, 1395–1402, ArXiv e-prints 0904.3992.

Foster, B. Z. 2007. Strong field effects on binary systems in Einstein-Æther theory. *Phys. Rev. D*, **76**, 084033, ArXiv e-prints 0706.0704.

Foster, B. Z., and Jacobson, T. 2006. Post-Newtonian parameters and constraints on Einstein-Æther theory. *Phys. Rev. D*, **73**, 064015, ArXiv e-prints gr-qc/0509083.

Fragile, P. C., and Mathews, G. J. 2000. Reconstruction of stellar orbits close to Sagittarius A*: Possibilities for testing general relativity. *Astrophys. J.*, **542**, 328–333, ArXiv e-prints astro-ph/9904177.

Freire, P. C. C., Kramer, M., and Wex, N. 2012. Tests of the universality of free fall for strongly self-gravitating bodies with radio pulsars. *Class. Quantum Grav.*, **29**, 184007, ArXiv e-prints 1205.3751.

Freire, P. C. C., Wex, N., Esposito-Farèse, G., Verbiest, J. P. W., et al. 2012. The relativistic pulsar-white dwarf binary PSR J1738+0333 - II. The most stringent test of scalar-tensor gravity. *Mon. Not. R. Astron. Soc.*, **423**, 3328–3343, ArXiv e-prints 1205.1450.

Froeschlé, M., Mignard, F., and Arenou, F. 1997. Determination of the PPN parameter γ with the Hipparcos data. Pages 49–52 in Bonnet, R. M., Høg, E., Bernacca, P. L., Emiliani, L., et al. (ed), *Proceedings of the Hipparcos Venice Symposium*, vol. 402. Noordwijk, Netherlands: ESA.

Fujii, Y. 2004. Oklo constraint on the time-variability of the fine-structure constant. Pages 167–185 in Karshenboim, S. G., and Peik, E. (eds), *Astrophysics, Clocks and Fundamental Constants*. Lecture Notes in Physics, Berlin Springer Verlag, vol. 648.

Fujii, Y., and Maeda, K.-I. 2007. *The Scalar-Tensor Theory of Gravitation*. Cambridge: Cambridge University Press.

Gabriel, M. D., and Haugan, M. P. 1990. Testing the Einstein Equivalence Principle: Atomic clocks and Local Lorentz Invariance. *Phys. Rev. D*, **41**, 2943–2955.

Gaia Collaboration, Prusti, T., de Bruijne, J. H. J., Brown, A. G. A., et al. 2016. The Gaia mission. *Astron. Astrophys.*, **595**, A1, ArXiv e-prints 1609.04153.

Gasperini, M. 1999. On the response of gravitational antennas to dilatonic waves. *Phys. Lett. B*, **470**, 67–72, ArXiv e-prints gr-qc/9910019.

Genzel, R., Eisenhauer, F., and Gillessen, S. 2010. The galactic center massive black hole and nuclear star cluster. *Rev. Mod. Phys.*, **82**, 3121–3195, ArXiv e-prints 1006.0064.

Geraci, A. A., Smullin, S. J., Weld, D. M., Chiaverini, J., et al. 2008. Improved constraints on non-Newtonian forces at 10 microns. *Phys. Rev. D*, **78**, 022002, ArXiv e-prints 0802.2350.

Ghez, A. M., Klein, B. L., Morris, M., and Becklin, E. E. 1998. High proper-motion stars in the vicinity of Sagittarius A*: Evidence for a supermassive black hole at the center of our galaxy. *Astrophys. J.*, **509**, 678–686, ArXiv e-prints astro-ph/9807210.

Ghez, A. M., Duchêne, G., Matthews, K., Hornstein, S. D., et al. 2003. The first measurement of spectral lines in a short-period star bound to the galaxy's central black hole: A paradox of youth. *Astrophys. J. Lett.*, **586**, L127–L131, ArXiv e-prints astro-ph/0302299.

Giannios, D. 2005. Spherically symmetric, static spacetimes in a tensor-vector-scalar theory. *Phys. Rev. D*, **71**, 103511, ArXiv e-prints gr-qc/0502122.

Gibbons, G., and Will, C. M. 2008. On the multiple deaths of Whitehead's theory of gravity. *Studies Hist. Philos. Mod. Phys.*, **39**, 41–61, ArXiv e-prints gr-qc/0611006.

Gillessen, S., Eisenhauer, F., Trippe, S., Alexander, T., et al. 2009. Monitoring stellar orbits around the massive black hole in the galactic center. *Astrophys. J.*, **692**, 1075–1109, ArXiv e-prints 0810.4674.

Gleiser, R. J., and Kozameh, C. N. 2001. Astrophysical limits on quantum gravity motivated birefringence. *Phys. Rev. D*, **64**, 083007, ArXiv e-prints gr-qc/0102093.

Godone, A., Novero, C., and Tavella, P. 1995. Null gravitational redshift experiment with nonidentical atomic clocks. *Phys. Rev. D*, **51**, 319–323.

Goldberger, W. D., and Rothstein, I. Z. 2006. Effective field theory of gravity for extended objects. *Phys. Rev. D*, **73**, 104029, ArXiv e-prints hep-th/0409156.

Goldhaber, A. S., and Nieto, M. M. 1974. Mass of the graviton. *Phys. Rev. D*, **9**, 1119–1121.

Gonzalez, M. E., Stairs, I. H., Ferdman, R. D., Freire, P. C. C., et al. 2011. High-precision timing of five millisecond pulsars: Space velocities, binary evolution, and equivalence principles. *Astrophys. J.*, **743**, 102, ArXiv e-prints 1109.5638.

Gralla, S. E. 2010. Motion of small bodies in classical field theory. *Phys. Rev. D*, **81**, 084060, ArXiv e-prints 1002.5045.

Gralla, S. E. 2013. Mass, charge, and motion in covariant gravity theories. *Phys. Rev. D*, **87**, 104020, ArXiv e-prints 1303.0269.

GRAVITY Collaboration, Abuter, R., Accardo, M., Amorim, A., et al. 2017. First light for GRAVITY: Phase referencing optical interferometry for the Very Large Telescope Interferometer. *Astron. Astrophys.*, **602**, A94, ArXiv e-prints 1705.02345.

Greenstein, J. L., and Matthews, T. A. 1963. Redshift of the radio source 3C 48. *Astron. J.*, **68**, 279.

Grieb, J. N., Sánchez, A. G., Salazar-Albornoz, S., Scoccimarro, R., et al. 2017. The clustering of galaxies in the completed SDSS-III Baryon Oscillation Spectroscopic Survey: Cosmological implications of the Fourier space wedges of the final sample. *Mon. Not. R. Astron. Soc.*, **467**, 2085–2112, ArXiv e-prints 1607.03143.

Guéna, J., Abgrall, M., Rovera, D., Rosenbusch, P., et al. 2012. Improved tests of Local Position Invariance using ^{87}Rb and ^{133}Cs fountains. *Phys. Rev. Lett.*, **109**, 080801, ArXiv e-prints 1205.4235.

Guenther, D. B., Krauss, L. M., and Demarque, P. 1998. Testing the constancy of the gravitational constant using helioseismology. *Astrophys. J.*, **498**, 871–876.

Hafele, J. C., and Keating, R. E. 1972a. Around-the-world atomic clocks: Observed relativistic time gains. *Science*, **177**, 168–170.

Hafele, J. C., and Keating, R. E. 1972b. Around-the-world atomic clocks: Predicted relativistic time gains. *Science*, **177**, 166–168.

Hahn, S. G., and Lindquist, R. W. 1964. The two-body problem in geometrodynamics. *Ann. Phys. (N.Y.)*, **29**, 304–331.

Hare, M. G. 1973. Mass of the graviton. *Can. J. Phys.*, **51**, 431.

Harvey, G. M. 1979. Gravitational deflection of light. *The Observatory*, **99**, 195–198.

Haugan, M. P. 1978. *Foundations of gravitation theory: The principle of equivalence*. Ph.D. thesis, Stanford Univ., CA.

Haugan, M. P. 1979. Energy conservation and the principle of equivalence. *Ann. Phys. (N.Y.)*, **118**, 156–186.

Haugan, M. P. 1985. Post-Newtonian arrival-time analysis for a pulsar in a binary system. *Astrophys. J.*, **296**, 1–12.

Haugan, M. P., and Will, C. M. 1976. Weak interactions and Eötvös experiments. *Phys. Rev. Lett.*, **37**, 1–4.

Haugan, M. P., and Will, C. M. 1977. Principles of equivalence, Eötvös experiments, and gravitational redshift experiments—The free fall of electromagnetic systems to post-post-Coulombian order. *Phys. Rev. D*, **15**, 2711–2720.

Haugan, M. P., and Will, C. M. 1987. Modern tests of special relativity. *Phys. Today*, **40**, 69–86.

Havas, P. 1989. The early history of the "problem of motion" in general relativity. Pages 234–276 in Howard, D., and Stachel, J. (eds), *Einstein and the History of General Relativity*. Birkhäuser.

Hawking, S. W. 1972. Black holes in the Brans-Dicke theory of gravitation. *Commun. Math. Phys.*, **25**, 167–171.

Hees, A., Folkner, W. M., Jacobson, R. A., and Park, R. S. 2014. Constraints on modified Newtonian dynamics theories from radio tracking data of the Cassini spacecraft. *Phys. Rev. D*, **89**, 102002, ArXiv e-prints 1402.6950.

Hees, A., Do, T., Ghez, A. M., Martinez, G. D., et al. 2017. Testing general relativity with stellar orbits around the supermassive black hole in our galactic center. *Phys. Rev. Lett.*, **118**, 211101, ArXiv e-prints 1705.07902.

Helbig, T. 1991. Gravitational effects of light scalar particles. *Astrophys. J.*, **382**, 223–232.

Hellings, R. W., and Nordtvedt, Jr., K. 1973. Vector-metric theory of gravity. *Phys. Rev. D*, **7**, 3593–3602.

Herrmann, S., Dittus, H., Lämmerzahl, C., and the QUANTUS and PRIMUS Teams. 2012. Testing the equivalence principle with atomic interferometry. *Class. Quantum Grav.*, **29**, 184003.

Hewish, A., Bell, S. J., Pilkington, J. D. H., Scott, P. F., et al. 1968. Observation of a rapidly pulsating radio source. *Nature*, **217**, 709–713.

Hiida, K., and Yamaguchi, Y. 1965. Gravitation physics. *Prog. Theor. Phys. Suppl.*, **65**, 261–297.

Hillebrandt, W., and Heintzmann, H. 1974. Neutron stars and incompressible fluid spheres in the Jordan-Brans-Dicke theory of gravitation. *Gen. Relativ. Gravit.*, **5**, 663–672.

Hinterbichler, K. 2012. Theoretical aspects of massive gravity. *Rev. Mod. Phys.*, **84**, 671–710, ArXiv e-prints 1105.3735.

Hofmann, F., Müller, J., and Biskupek, L. 2010. Lunar laser ranging test of the Nordtvedt parameter and a possible variation in the gravitational constant. *Astron. Astrophys.*, **522**, L5.

Hojjati, A., Zhao, G.-B., Pogosian, L., Silvestri, A., et al. 2012. Cosmological tests of general relativity: A principal component analysis. *Phys. Rev. D*, **85**, 043508, ArXiv e-prints 1111.3960.

Horbatsch, M. W., and Burgess, C. P. 2011. Semi-analytic stellar structure in scalar-tensor gravity. *J. Cosmol. Astropart. Phys.*, **8**, 027, ArXiv e-prints 1006.4411.

Hořava, P. 2009. Quantum gravity at a Lifshitz point. *Phys. Rev. D*, **79**, 084008, ArXiv e-prints 0901.3775.

Hoyle, C. D., Schmidt, U., Heckel, B. R., Adelberger, E. G., et al. 2001. Submillimeter test of the gravitational inverse-square law: A search for "large" extra dimensions. *Phys. Rev. Lett.*, **86**, 1418–1421, ArXiv e-prints hep-ph/0011014.

Hoyle, C. D., Kapner, D. J., Heckel, B. R., Adelberger, E. G., et al. 2004. Submillimeter tests of the gravitational inverse-square law. *Phys. Rev. D*, **70**, 042004, ArXiv e-prints hep-ph/0405262.

Hughes, V. W., Robinson, H. G., and Beltran-Lopez, V. 1960. Upper limit for the anisotropy of inertial mass from nuclear resonance experiments. *Phys. Rev. Lett.*, **4**, 342–344.

Hulse, R. A., and Taylor, J. H. 1975. Discovery of a pulsar in a binary system. *Astrophys. J. Lett.*, **195**, L51–L53.

Imperi, L., and Iess, L. 2017. The determination of the post-Newtonian parameter γ during the cruise phase of BepiColombo. *Class. Quantum Grav.*, **34**, 075002.

Ishak, M. 2018. Testing general relativity on cosmological scales. *Living Rev. Relativ.*, **to be published**.

Israel, W. 1987. Dark stars: The evolution of an idea. Pages 199–276 in Hawking, S. W., and Israel, W. (eds), *Three Hundred Years of Gravitation*. Cambridge: Cambridge University Press.

Itoh, Y., Futamase, T., and Asada, H. 2000. Equation of motion for relativistic compact binaries with the strong field point particle limit: Formulation, the first post-Newtonian order, and multipole terms. *Phys. Rev. D*, **62**, 064002, ArXiv e-prints gr-qc/9910052.

Ives, H. E., and Stilwell, G. R. 1938. An experimental study of the rate of a moving atomic clock. *J. Opt. Soc. Am.*, **28**, 215.

Jackiw, R., and Pi, S.-Y. 2003. Chern-Simons modification of general relativity. *Phys. Rev. D*, **68**, 104012, ArXiv e-prints gr-qc/0308071.

Jacobson, T. 2014. Undoing the twist: The Hořava limit of Einstein-Æther theory. *Phys. Rev. D*, **89**, 081501, ArXiv e-prints 1310.5115.

Jacobson, T., and Mattingly, D. 2001. Gravity with a dynamical preferred frame. *Phys. Rev. D*, **64**, 024028, ArXiv e-prints gr-qc/0007031.

Jacobson, T., and Mattingly, D. 2004. Einstein-Æther waves. *Phys. Rev. D*, **70**, 024003, ArXiv e-prints gr-qc/0402005.

Jaime, L. G., Patiño, L., and Salgado, M. 2011. Robust approach to $f(R)$ gravity. *Phys. Rev. D*, **83**, 024039, ArXiv e-prints 1006.5747.

Jaroszynski, M. 1998. Relativistic effects in proper motions of stars surrounding the galactic center. *Acta Astron.*, **48**, 653–665, ArXiv e-prints astro-ph/9812314.

Jaseja, T. S., Javan, A., Murray, J., and Townes, C. H. 1964. Test of special relativity or of the isotropy of space by use of infrared masers. *Phys. Rev.*, **133**, 1221–1225.

Johannsen, T. 2016. Testing the no-hair theorem with observations of black holes in the electromagnetic spectrum. *Class. Quantum Grav.*, **33**, 124001, ArXiv e-prints 1602.07694.

Johannsen, T., Wang, C., Broderick, A. E., Doeleman, S. S., et al. 2016a. Testing general relativity with accretion-flow imaging of SgrA*. *Phys. Rev. Lett.*, **117**, 091101, ArXiv e-prints 1608.03593.

Johannsen, T., Broderick, A. E., Plewa, P. M., Chatzopoulos, S., et al. 2016b. Testing general relativity with the shadow size of SgrA*. *Phys. Rev. Lett.*, **116**, 031101, ArXiv e-prints 1512.02640.

Jones, B. F. 1976. Gravitational deflection of light: Solar eclipse of 30 June 1973. II. Plate reductions. *Astron. J.*, **81**, 455–463.

Kanekar, N., Langston, G. I., Stocke, J. T., Carilli, C. L., et al. 2012. Constraining fundamental constant evolution with H I and OH lines. *Astrophys. J. Lett.*, **746**, L16–L20, ArXiv e-prints 1201.3372.

Kapner, D. J., Cook, T. S., Adelberger, E. G., Gundlach, J. H., et al. 2007. Tests of the gravitational inverse-square law below the dark-energy length scale. *Phys. Rev. Lett.*, **98**, 021101, ArXiv e-prints hep-ph/0611184.

Kates, R. E. 1980. Motion of a small body through an external field in general relativity calculated by matched asymptotic expansions. *Phys. Rev. D*, **22**, 1853–1870.

Katz, J. I. 1999. Comment on "Indication, from Pioneer 10/11, Galileo, and Ulysses data, of an apparent anomalous, weak, long-range acceleration." *Phys. Rev. Lett.*, **83**, 1892, ArXiv e-prints gr-qc/9809070.

Kennefick, D. 2005. Einstein versus the Physical Review. *Phys. Today*, **58**, 43.

Kennefick, D. 2007. *Traveling at the Speed of Thought: Einstein and the Quest for Gravitational Waves*. Princeton; Woodstock, UK: Princeton University Press.

Kennefick, D. 2009. Testing relativity from the 1919 eclipse – A question of bias. *Phys. Today*, **62**, 37.

Kerr, R. P. 1963. Gravitational field of a spinning mass as an example of algebraically special metrics. *Phys. Rev. Lett.*, **11**, 237–238.

Khoury, J., and Weltman, A. 2004. Chameleon fields: Awaiting surprises for tests of gravity in space. *Phys. Rev. Lett.*, **93**, 171104, ArXiv e-prints astro-ph/0309300.

Kidder, L. E. 1995. Coalescing binary systems of compact objects to (post)$^{5/2}$-Newtonian order. V. Spin effects. *Phys. Rev. D*, **52**, 821–847, ArXiv e-prints gr-qc/9506022.

Kidder, L. E., Will, C. M., and Wiseman, A. G. 1993. Spin effects in the inspiral of coalescing compact binaries. *Phys. Rev. D*, **47**, R4183–R4187, ArXiv e-prints gr-qc/9211025.

King, J. A., Webb, J. K., Murphy, M. T., Flambaum, V. V., et al. 2012. Spatial variation in the fine-structure constant: New results from VLT/UVES. *Mon. Not. R. Astron. Soc.*, **422**, 3370–3414, ArXiv e-prints 1202.4758.

Kleihaus, B., Kunz, J., and Radu, E. 2011. Rotating black holes in dilatonic Einstein-Gauss-Bonnet theory. *Phys. Rev. Lett.*, **106**, 151104, ArXiv e-prints 1101.2868.

Klimchitskaya, G. L., Mohideen, U., and Mostepanenko, V. M. 2013. Constraints on corrections to Newtonian gravity from two recent measurements of the Casimir interaction between metallic surfaces. *Phys. Rev. D*, **87**, 125031, ArXiv e-prints 1306.4979.

Klinkhamer, F. R., and Risse, M. 2008. Addendum: Ultrahigh-energy cosmic-ray bounds on nonbirefringent modified Maxwell theory. *Phys. Rev. D*, **77**, 117901, ArXiv e-prints 0806.4351.

Klinkhamer, F. R., and Schreck, M. 2008. New two-sided bound on the isotropic Lorentz-violating parameter of modified Maxwell theory. *Phys. Rev. D*, **78**, 085026, ArXiv e-prints 0809.3217.

Kobayashi, T., and Maeda, K.-I. 2008. Relativistic stars in *f*(*R*) gravity, and absence thereof. *Phys. Rev. D*, **78**, 064019, ArXiv e-prints 0807.2503.

Konno, K., Matsuyama, T., and Tanda, S. 2009. Rotating black hole in extended Chern-Simons modified gravity. *Prog. Theor. Phys.*, **122**, 561–568, ArXiv e-prints 0902.4767.

Konopliv, A. S., Asmar, S. W., Folkner, W. M., Karatekin, Ö., et al. 2011. Mars high resolution gravity fields from MRO, Mars seasonal gravity, and other dynamical parameters. *Icarus*, **211**, 401–428.

Kormendy, J., and Richstone, D. 1995. Inward bound—The search for supermassive black holes in galactic nuclei. *Ann. Rev. Astron. Astrophys.*, **33**, 581.

Kostelecký, V. A., and Mewes, M. 2002. Signals for Lorentz violation in electrodynamics. *Phys. Rev. D*, **66**, 056005, ArXiv e-prints hep-ph/0205211.

Kostelecký, V. A., and Russell, N. 2011. Data tables for Lorentz and CPT violation. *Rev. Mod. Phys.*, **83**, 11–32, ArXiv e-prints 0801.0287.

Kostelecký, V. A., and Samuel, S. 1989. Gravitational phenomenology in higher-dimensional theories and strings. *Phys. Rev. D*, **40**, 1886–1903.

Kramer, M. 1998. Determination of the geometry of the PSR B1913+16 system by geodetic precession. *Astrophys. J.*, **509**, 856–860, ArXiv e-prints astro-ph/9808127.

Kramer, M., Stairs, I. H., Manchester, R. N., McLaughlin, M. A., et al. 2006. Tests of general relativity from timing the double pulsar. *Science*, **314**, 97–102, ArXiv e-prints arXiv:astro-ph/0609417.

Kraniotis, G. V. 2007. Periapsis and gravitomagnetic precessions of stellar orbits in Kerr and Kerr de Sitter black hole spacetimes. *Class. Quantum Grav.*, **24**, 1775–1808, ArXiv e-prints gr-qc/0602056.

Kreuzer, L. B. 1968. Experimental measurement of the equivalence of active and passive gravitational mass. *Phys. Rev.*, **169**, 1007–1012.

Krisher, T. P., Anderson, J. D., and Campbell, J. K. 1990a. Test of the gravitational redshift effect at Saturn. *Phys. Rev. Lett.*, **64**, 1322–1325.

Krisher, T. P., Maleki, L., Lutes, G. F., Primas, L. E., et al. 1990b. Test of the isotropy of the one-way speed of light using hydrogen-maser frequency standards. *Phys. Rev. D*, **42**, 731–734.

Krisher, T. P., Morabito, D. D., and Anderson, J. D. 1993. The Galileo solar redshift experiment. *Phys. Rev. Lett.*, **70**, 2213–2216.

Kruskal, M. D. 1960. Maximal extension of Schwarzschild metric. *Phys. Rev.*, **119**, 1743–1745.

Lambert, S. B., and Le Poncin-Lafitte, C. 2009. Determining the relativistic parameter γ using very long baseline interferometry. *Astron. Astrophys.*, **499**, 331–335, ArXiv e-prints 0903.1615.

Lambert, S. B., and Le Poncin-Lafitte, C. 2011. Improved determination of γ by VLBI. *Astron. Astrophys.*, **529**, A70.

Lamoreaux, S. K., Jacobs, J. P., Heckel, B. R., Raab, F. J., et al. 1986. New limits on spatial anisotropy from optically-pumped ^{201}Hg and ^{199}Hg. *Phys. Rev. Lett.*, **57**, 3125–3128.

Landau, L. D., and Lifshitz, E. M. 1962. *The Classical Theory of Fields*. Reading, Massachusetts: Addison-Wesley.

Lang, R. N. 2014. Compact binary systems in scalar-tensor gravity. II. Tensor gravitational waves to second post-Newtonian order. *Phys. Rev. D*, **89**, 084014, ArXiv e-prints 1310.3320.

Lang, R. N. 2015. Compact binary systems in scalar-tensor gravity. III. Scalar waves and energy flux. *Phys. Rev. D*, **91**, 084027, ArXiv e-prints 1411.3073.

Laplace, P. S. 1799. Beweis des Satzes, dass die anziehende Kraft bey einem Weltkörper so groß seyn könne, dass das Licht davon nicht ausströmen kann. *Allgemeine Geographische Ephemeriden, Vol. 4, Issue 1, p. 1-6*, **4**, 1–6.

Laplace, P. S. 1808. *Exposition du Système du Monde. Part II.* 3rd edn. Courcier.

Lasky, P. D. 2009. Black holes and neutron stars in the generalized tensor-vector-scalar theory. *Phys. Rev. D*, **80**, 081501, ArXiv e-prints 0910.0240.

Lasky, P. D., Sotani, H., and Giannios, D. 2008. Structure of neutron stars in tensor-vector-scalar theory. *Phys. Rev. D*, **78**, 104019, ArXiv e-prints 0811.2006.

Lattimer, J. M., and Prakash, M. 2007. Neutron star observations: Prognosis for equation of state constraints. *Phys. Rep.*, **442**, 109–165, ArXiv e-prints astro-ph/0612440.

Lazaridis, K., Wex, N., Jessner, A., Kramer, M., et al. 2009. Generic tests of the existence of the gravitational dipole radiation and the variation of the gravitational constant. *Mon. Not. R. Astron. Soc.*, **400**, 805–814, ArXiv e-prints 0908.0285.

Lebach, D. E., Corey, B. E., Shapiro, I. I., Ratner, M. I., et al. 1995. Measurement of the solar gravitational deflection of radio waves using Very-Long-Baseline Interferometry. *Phys. Rev. Lett.*, **75**, 1439–1442.

Lee, D. L. 1974. Conservation laws, gravitational waves, and mass losses in the Dicke-Brans-Jordan theory of gravity. *Phys. Rev. D*, **10**, 2374–2383.

Lee, D. L., Lightman, A. P., and Ni, W.-T. 1974. Conservation laws and variational principles in metric theories of gravity. *Phys. Rev. D*, **10**, 1685–1700.

Lee, D. L., Ni, W.-T., Caves, C. M., and Will, C. M. 1976. Theoretical frameworks for testing relativistic gravity. V - Post-Newtonian limit of Rosen's theory. *Astrophys. J.*, **206**, 555–558.

Lee, K. J., Jenet, F. A., and Price, R. H. 2008. Pulsar timing as a probe of non-Einsteinian polarizations of gravitational waves. *Astrophys. J.*, **685**, 1304–1319.

Leefer, N., Weber, C. T. M., Cingöz, A., Torgerson, J. R., and Budker, D. 2013. New limits on variation of the fine-structure constant using atomic Dysprosium. *Phys. Rev. Lett.*, **111**, 060801, ArXiv e-prints 1304.6940.

Lentati, L., Carilli, C., Alexander, P., Maiolino, R., et al. 2013. Variations in the fundamental constants in the QSO host J1148+5251 at $z = 6.4$ and the BR1202-0725 system at $z = 4.7$. *Mon. Not. R. Astron. Soc.*, **430**, 2454–2463, ArXiv e-prints 1211.3316.

Levi-Civita, T. 1937. Astronomical consequences of the relativistic two-body problem. *Am. J. Math.*, **59**, 225–334.

Levi-Civita, T. 1965. *The n-Body Problem in General Relativity.* Dordrecht, Holland: D. Reidel.

Li, T. G. F., Del Pozzo, W., Vitale, S., Van Den Broeck, C., et al. 2012. Towards a generic test of the strong field dynamics of general relativity using compact binary coalescence. *Phys. Rev. D*, **85**, 082003, ArXiv e-prints 1110.0530.

Liberati, S. 2013. Tests of Lorentz invariance: A 2013 update. *Class. Quantum Grav.*, **30**, 133001, ArXiv e-prints 1304.5795.

Lightman, A. P., and Lee, D. L. 1973. Restricted proof that the Weak Equivalence Principle implies the Einstein Equivalence Principle. *Phys. Rev. D*, **8**, 364–376.

Lipa, J. A., Nissen, J. A., Wang, S., Stricker, D. A., et al. 2003. New limit on signals of Lorentz violation in electrodynamics. *Phys. Rev. Lett.*, **90**, 060403, ArXiv e-prints physics/0302093.

Liu, K., Wex, N., Kramer, M., Cordes, J. M., et al. 2012. Prospects for probing the spacetime of SgrA* with pulsars. *Astrophys. J.*, **747**, 1, ArXiv e-prints 1112.2151.

Long, J. C., Chan, H. W., and Price, J. C. 1999. Experimental status of gravitational-strength forces in the sub-centimeter regime. *Nucl. Phys. B*, **539**, 23–34, ArXiv e-prints hep-ph/9805217.

Long, J. C., Chan, H. W., Churnside, A. B., Gulbis, E. A., et al. 2003. Upper limits to submillimetre-range forces from extra space-time dimensions. *Nature*, **421**, 922–925, ArXiv e-prints hep-ph/0210004.

LoPresto, J. C., Schrader, C., and Pierce, A. K. 1991. Solar gravitational redshift from the infrared oxygen triplet. *Astrophys. J.*, **376**, 757–760.

Lorentz, H. A., and Droste, J. 1917. The motion of a system of bodies under the influence of their mutual attraction, according to Einstein's theory. *Versl. K. Akad. Wetensch. Amsterdam*, **26**, 392. English translation in Lorentz, H. A. 1937. Collected papers, Vol. 5, edited by Zeeman, P. and Fokker, A. D. Martinus Nijhoff.

Lorimer, D. R., and Kramer, M. 2012. *Handbook of Pulsar Astronomy*. Cambridge: Cambridge University Press.

Lucchesi, D. M., and Peron, R. 2010. Accurate measurement in the field of the Earth of the general-relativistic precession of the LAGEOS II pericenter and new constraints on non-Newtonian gravity. *Phys. Rev. Lett.*, **105**, 231103, ArXiv e-prints 1106.2905.

Lucchesi, D. M., and Peron, R. 2014. LAGEOS II pericenter general relativistic precession (1993-2005): Error budget and constraints in gravitational physics. *Phys. Rev. D*, **89**, 082002.

Lynch, R. S., Boyles, J., Ransom, S. M., Stairs, I. H., et al. 2013. The Green Bank Telescope 350 MHz drift-scan survey II: Data analysis and the timing of 10 new pulsars, including a relativistic binary. *Astrophys. J.*, **763**, 81, ArXiv e-prints 1209.4296.

Lynden-Bell, D., and Rees, M. J. 1971. On quasars, dust and the galactic centre. *Mon. Not. R. Astron. Soc.*, **152**, 461.

Lyne, A. G., Burgay, M., Kramer, M., Possenti, A., et al. 2004. A double-pulsar system: A rare laboratory for relativistic gravity and plasma physics. *Science*, **303**, 1153–1157, ArXiv e-prints astro-ph/0401086.

MacArthur, D. W. 1986. Special relativity: Understanding experimental tests and formulations. *Phys. Rev. A*, **33**, 1–5.

Maeda, K.-I. 1988. On time variation of fundamental constants in superstring theories. *Mod. Phys. Lett. A*, **3**, 243–249.

Maggiore, M. 2007. *Gravitational Waves. Volume 1: Theory and Experiments*. Oxford: Oxford University Press.

Magueijo, J. 2003. New varying speed of light theories. *Rep. Prog. Phys.*, **66**, 2025–2068, ArXiv e-prints astro-ph/0305457.

Malaney, R. A., and Mathews, G. J. 1993. Probing the early universe: A review of primordial nucleosynthesis beyond the standard big bang. *Phys. Rep.*, **229**, 145–219.

Manchester, R. N. 2015. Pulsars and gravity. *Int. J. Mod. Phys. D*, **24**, 1530018, ArXiv e-prints 1502.05474.

Mansouri, R., and Sexl, R. U. 1977a. A test theory of special relativity. I - Simultaneity and clock synchronization. *Gen. Relativ. Gravit.*, **8**, 497–513.

Mansouri, R., and Sexl, R. U. 1977b. A test theory of special relativity: II. First order tests. *Gen. Relativ. Gravit.*, **8**, 515–524.

Mansouri, R., and Sexl, R. U. 1977c. A test theory of special relativity: III. Second-order tests. *Gen. Relativ. Gravit.*, **8**, 809–814.

Marion, H., Pereira Dos Santos, F., Abgrall, M., Zhang, S., et al. 2003. Search for variations of fundamental constants using atomic fountain clocks. *Phys. Rev. Lett.*, **90**, 150801, ArXiv e-prints physics/0212112.

Martins, C. J. A. P. 2017. The status of varying constants: A review of the physics, searches and implications, ArXiv e-prints 1709.02923.

Mathisson, M. 1937. Neue Mechanik materieller Systeme. *Acta Phys. Polon.*, **6**, 163–200.

Matthews, T. A., and Sandage, A. R. 1963. Optical identification of 3C 48, 3C 196, and 3C 286 with stellar sbjects. *Astrophys. J.*, **138**, 30.

Mattingly, D. 2005. Modern tests of Lorentz invariance. *Living Rev. Relativ.*, **8**, 5, ArXiv e-prints gr-qc/0502097.

Mattingly, D., and Jacobson, T. 2002. Relativistic gravity with a dynamical preferred frame. Pages 331–335 in Kostelecký, V. A. (ed), *CPT and Lorentz Symmetry*. Singapore: World Scientific.

Matzner, R. A., Seidel, H. E., Shapiro, S. L., Smarr, L., et al. 1995. Geometry of a black hole collision. *Science*, **270**, 941–947.

Mecheri, R., Abdelatif, T., Irbah, A., Provost, J., et al. 2004. New values of gravitational moments J_2 and J_4 deduced from helioseismology. *Solar Phys.*, **222**, 191–197, ArXiv e-prints 0911.5055.

Mercuri, S., and Taveras, V. 2009. Interaction of the Barbero-Immirzi field with matter and pseudoscalar perturbations. *Phys. Rev. D*, **80**, 104007, ArXiv e-prints 0903.4407.

Merlet, S., Bodart, Q., Malossi, N., Landragin, A., et al. 2010. Comparison between two mobile absolute gravimeters: Optical versus atomic interferometers. *Metrologia*, **47**, L9–L11, ArXiv e-prints 1005.0357.

Merritt, D., Alexander, T., Mikkola, S., and Will, C. M. 2010. Testing properties of the galactic center black hole using stellar orbits. *Phys. Rev. D*, **81**, 062002, ArXiv e-prints 0911.4718.

Meyer, L., Ghez, A. M., Schödel, R., Yelda, S., et al. 2012. The shortest-known-period star orbiting our galaxy's supermassive black hole. *Science*, **338**, 84, ArXiv e-prints 1210.1294.

Meylan, G., Jetzer, P., North, P., Schneider, P., et al. (eds). 2006. *Gravitational Lensing: Strong, Weak and Micro*. Berlin: Springer-Verlag.

Michell, J. 1784. On the means of discovering the distance, magnitude, etc. of the fixed stars, in consequence of the diminution of the velocity of their light, in case such a diminution should be found to take place in any of them, and such other data should be procured from observations, as would be farther necessary for that purpose. *Philos. Trans. R. Soc. London*, **74**, 35 – 57.

Michelson, A. A., and Morley, E. W. 1887. On the relative motion of the Earth and the luminiferous ether. *Am. J. Science*, **34**, 333.

Mignard, F., and Klioner, S. A. 2010. GAIA: Relativistic modelling and testing. Pages 306–314 in Klioner, S. A., Seidelmann, P. K., and Soffel, M. H. (eds), *Relativity in Fundamental Astronomy: Dynamics, Reference Frames, and Data Analysis*. IAU Symposium, vol. 261. Cambridge: Cambridge University Press.

Milani, A., Vokrouhlický, D., Villani, D., Bonanno, C., et al. 2002. Testing general relativity with the BepiColombo radio science experiment. *Phys. Rev. D*, **66**, 082001.

Milgrom, M. 1983. A modification of the Newtonian dynamics as a possible alternative to the hidden mass hypothesis. *Astrophys. J.*, **270**, 365–370.

Milgrom, M. 2009. MOND effects in the inner solar system. *Mon. Not. R. Astron. Soc.*, **399**, 474–486, ArXiv e-prints 0906.4817.

Miller, M. C., and Miller, J. M. 2015. The masses and spins of neutron stars and stellar-mass black holes. *Phys. Rep.*, **548**, 1–34, ArXiv e-prints 1408.4145.

Mirshekari, S., and Will, C. M. 2013. Compact binary systems in scalar-tensor gravity: Equations of motion to 2.5 post-Newtonian order. *Phys. Rev. D*, **87**, 084070, ArXiv e-prints 1301.4680.

Mirshekari, S., Yunes, N., and Will, C. M. 2012. Constraining Lorentz-violating, modified dispersion relations with gravitational waves. *Phys. Rev. D*, **85**, 024041, ArXiv e-prints 1110.2720.

Mishra, C. K., Arun, K. G., Iyer, B. R., and Sathyaprakash, B. S. 2010. Parametrized tests of post-Newtonian theory using Advanced LIGO and Einstein Telescope. *Phys. Rev. D*, **82**, 064010, ArXiv e-prints 1005.0304.

Misner, C. W., Thorne, K. S., and Wheeler, J. A. 1973. *Gravitation*. San Francisco. W.H. Freeman and Co.

Mitchell, T., and Will, C. M. 2007. Post-Newtonian gravitational radiation and equations of motion via direct integration of the relaxed Einstein equations. V. Evidence for the Strong Equivalence Principle to second post-Newtonian order. *Phys. Rev. D*, **75**, 124025, ArXiv e-prints 0704.2243.

Modenini, D., and Tortora, P. 2014. Pioneer 10 and 11 orbit determination analysis shows no discrepancy with Newton-Einstein laws of gravity. *Phys. Rev. D*, **90**, 022004, ArXiv e-prints 1311.4978.

Moffat, J. W. 2006. Scalar tensor vector gravity theory. *J. Cosmol. Astropart. Phys.*, **3**, 004, ArXiv e-prints gr-qc/0506021.

Mohr, P. J., and Taylor, B. N. 2005. CODATA recommended values of the fundamental physical constants: 2002. *Rev. Mod. Phys.*, **77**, 1–107.

Møller, C. 1952. *The Theory of Relativity*. Oxford: Oxford University Press.

Montgomery, C., Orchiston, W., and Whittingham, I. 2009. Michell, Laplace and the origin of the black hole concept. *J. Astron. Hist. Heritage*, **12**, 90–96.

Moura, F., and Schiappa, R. 2007. Higher-derivative-corrected black holes: Perturbative stability and absorption cross section in heterotic string theory. *Class. Quantum Grav.*, **24**, 361–386, ArXiv e-prints hep-th/0605001.

Mukherjee, R., and Sounda, S. 2017. Single particle closed orbits in Yukawa potential. *Indian J. Phys., Online First*, Sept., ArXiv e-prints 1705.02444.

Müller, H., Herrmann, S., Braxmaier, C., Schiller, S., et al. 2003. Modern Michelson-Morley experiment using cryogenic optical resonators. *Phys. Rev. Lett.*, **91**, 020401, ArXiv e-prints physics/0305117.

Müller, H., Peters, A., and Chu, S. 2010. A precision measurement of the gravitational redshift by the interference of matter waves. *Nature*, **463**, 926–929.

Müller, J., Nordtvedt, Jr., K., and Vokrouhlický, D. 1996. Improved constraint on the α_1 PPN parameter from lunar motion. *Phys. Rev. D*, **54**, R5927–R5930.

Müller, J., Schneider, M., Nordtvedt, Jr., K., and Vokrouhlický, D. 1999. What can LLR provide to relativity? Page 1151 in Piran, T., and Ruffini, R. (eds), *Recent Developments in Theoretical and Experimental General Relativity, Gravitation, and Relativistic Field Theories*. Singapore: World Scientific.

Müller, J., Williams, J. G., and Turyshev, S. G. 2008. Lunar laser ranging contributions to relativity and geodesy. Page 457 in Dittus, H., Lammerzahl, C., and Turyshev, S. G. (eds), *Lasers, Clocks and Drag-Free Control: Exploration of Relativistic Gravity in Space*. Astrophysics and Space Science Library, vol. 349.

Murphy, M. T., Webb, J. K., Flambaum, V. V., Dzuba, V. A., et al. 2001. Possible evidence for a variable fine-structure constant from QSO absorption lines: Motivations, analysis and results. *Mon. Not. R. Astron. Soc.*, **327**, 1208–1222, ArXiv e-prints astro-ph/0012419.

Murphy, T. W., Adelberger, E. G., Battat, J. B. R., Hoyle, C. D., et al. 2011. Laser ranging to the lost Lunokhod 1 reflector. *Icarus*, **211**, 1103–1108, ArXiv e-prints 1009.5720.

Murphy, Jr., T. W., Adelberger, E. G., Battat, J. B. R., Hoyle, C. D., et al. 2012. APOLLO: Millimeter lunar laser ranging. *Class. Quantum Grav.*, **29**, 184005.

Narayan, R., and McClintock, J. E. 2008. Advection-dominated accretion and the black hole event horizon. *New Astron. Rev.*, **51**, 733–751, ArXiv e-prints 0803.0322.

Newman, E., and Penrose, R. 1962. An approach to gravitational radiation by a method of spin coefficients. *J. Math. Phys.*, **3**, 566–578.

Newton, I. 1686. *Philosophiae Naturalis Principia Mathematica*. London: Benjamin Motte.

Ni, W.-T. 1973. A new theory of gravity. *Phys. Rev. D*, **7**, 2880–2883.

Ni, W.-T. 1977. Equivalence principles and electromagnetism. *Phys. Rev. Lett.*, **38**, 301–304.

Niebauer, T. M., McHugh, M. P., and Faller, J. E. 1987. Galilean test for the fifth force. *Phys. Rev. Lett.*, **59**, 609–612.

Nishizawa, A., Taruya, A., Hayama, K., Kawamura, S., et al. 2009. Probing nontensorial polarizations of stochastic gravitational-wave backgrounds with ground-based laser interferometers. *Phys. Rev. D*, **79**, 082002, ArXiv e-prints 0903.0528.

Nishizawa, A., Taruya, A., and Kawamura, S. 2010. Cosmological test of gravity with polarizations of stochastic gravitational waves around 0.1-1 Hz. *Phys. Rev. D*, **81**, 104043, ArXiv e-prints 0911.0525.

Nobili, A. M., and Will, C. M. 1986. The real value of Mercury's perihelion advance. *Nature*, **320**, 39–41.

Nobili, A. M., Shao, M., Pegna, R., Zavattini, G., et al. 2012. "Galileo Galilei" (GG): Space test of the Weak Equivalence Principle to 10^{-17} and laboratory demonstrations. *Class. Quantum Grav.*, **29**, 184011.

Nordström, G. 1913. Zur Theorie der Gravitation vom Standpunkt des Relativitätsprinzips. *Ann. Phys. (Leipzig)*, **347**, 533–554.

Nordtvedt, Jr., K. 1968a. Equivalence principle for massive bodies. I. Phenomenology. *Phys. Rev.*, **169**, 1014–1016.

Nordtvedt, Jr., K. 1968b. Equivalence principle for massive bodies. II. Theory. *Phys. Rev.*, **169**, 1017–1025.

Nordtvedt, Jr., K. 1968c. Testing relativity with laser ranging to the Moon. *Phys. Rev.*, **170**, 1186–1187.

Nordtvedt, Jr., K. 1970. Post-Newtonian metric for a general class of scalar-tensor gravitational theories and observational consequences. *Astrophys. J.*, **161**, 1059.

Nordtvedt, Jr., K. 1975. Quantitative relationship between clock gravitational 'red-shift' violations and nonuniversality of free-fall rates in nonmetric theories of gravity. *Phys. Rev. D*, **11**, 245–247.

Nordtvedt, Jr., K. 1985. A post-Newtonian gravitational Lagrangian formalism for celestial body dynamics in metric gravity. *Astrophys. J.*, **297**, 390–404.

Nordtvedt, Jr., K. 1987. Probing gravity to the second post-Newtonian order and to one part in 10^7 using the spin axis of the sun. *Astrophys. J.*, **320**, 871–874.

Nordtvedt, Jr., K. 1990. $\dot{G}/G$ and a cosmological acceleration of gravitationally compact bodies. *Phys. Rev. Lett.*, **65**, 953–956.

Nordtvedt, Jr., K. 2001. Testing Newton's third law using lunar laser ranging. *Class. Quantum Grav. Lett.*, **18**, L133–L137.

Nordtvedt, Jr., K., and Will, C. M. 1972. Conservation laws and preferred frames in relativistic gravity. II. Experimental evidence to rule out preferred-frame theories of gravity. *Astrophys. J.*, **177**, 775.

Nutku, Y. 1969. The energy-momentum complex in the Brans-Dicke theory. *Astrophys. J.*, **158**, 991.

Olive, K. A., Pospelov, M., Qian, Y.-Z., Manhès, G., et al. 2004. Reexamination of the ^{187}Re bound on the variation of fundamental couplings. *Phys. Rev. D*, **69**, 027701, ArXiv e-prints astro-ph/0309252.

Oppenheimer, J. R., and Snyder, H. 1939. On continued gravitational contraction. *Phys. Rev.*, **56**, 455–459.

Overduin, J., Everitt, F., Worden, P., and Mester, J. 2012. STEP and fundamental physics. *Class. Quantum Grav.*, **29**, 184012, ArXiv e-prints 1401.4784.

Palenzuela, C., Barausse, E., Ponce, M., and Lehner, L. 2014. Dynamical scalarization of neutron stars in scalar-tensor gravity theories. *Phys. Rev. D*, **89**, 044024, ArXiv e-prints 1310.4481.

Pani, P., and Cardoso, V. 2009. Are black holes in alternative theories serious astrophysical candidates? The case for Einstein-Dilaton-Gauss-Bonnet black holes. *Phys. Rev. D*, **79**, 084031, ArXiv e-prints 0902.1569.

Pani, P., Berti, E., Cardoso, V., and Read, J. 2011. Compact stars in alternative theories of gravity: Einstein-Dilaton-Gauss-Bonnet gravity. *Phys. Rev. D*, **84**, 104035, ArXiv e-prints 1109.0928.

Papapetrou, A. 1951. Spinning Test-Particles in General Relativity. I. *Proc. R. Soc. A*, **209**, 248–258.

Pati, M. E., and Will, C. M. 2000. Post-Newtonian gravitational radiation and equations of motion via direct integration of the relaxed Einstein equations: Foundations. *Phys. Rev. D*, **62**, 124015, ArXiv e-prints gr-qc/0007087.

Pati, M. E., and Will, C. M. 2002. Post-Newtonian gravitational radiation and equations of motion via direct integration of the relaxed Einstein equations. II. Two-body equations of motion to second post-Newtonian order, and radiation reaction to 3.5 post-Newtonian order. *Phys. Rev. D*, **65**, 104008, ArXiv e-prints gr-qc/0201001.

Peik, E., Lipphardt, B., Schnatz, H., Schneider, T., et al., 2004. Limit on the present temporal variation of the fine structure constant. *Phys. Rev. Lett.*, **93**, 170801, ArXiv e-prints physics/0402132.

Peil, S., Crane, S., Hanssen, J. L., Swanson, T. B., and Ekstrom, C. R. 2013. Tests of local position invariance using continuously running atomic clocks. *Phys. Rev. A*, **87**, 010102, ArXiv e-prints 1301.6145.

Penrose, R. 1960. A spinor approach to general relativity. *Ann. Phys. (N.Y.)*, **10**, 171–201.

Perez, P., and Sacquin, Y. 2012. The GBAR experiment: Gravitational behaviour of antihydrogen at rest. *Class. Quantum Grav.*, **29**, 184008.

Peters, P. C., and Mathews, J. 1963. Gravitational radiation from point masses in a Keplerian orbit. *Phys. Rev.*, **131**, 435–440.

Petrov, Y. V., Nazarov, A. I., Onegin, M. S., Petrov, V. Y., et al. 2006. Natural nuclear reactor at Oklo and variation of fundamental constants: Computation of neutronics of a fresh core. *Phys. Rev. C*, **74**, 064610, ArXiv e-prints hep-ph/0506186.

Pitjeva, E. V. 2005. Relativistic effects and solar oblateness from radar observations of planets and spacecraft. *Astron. Lett.*, **31**, 340–349.

Pitjeva, E. V., and Pitjev, N. P. 2013. Relativistic effects and dark matter in the solar system from observations of planets and spacecraft. *Mon. Not. R. Astron. Soc.*, **432**, 3431–3437, ArXiv e-prints 1306.3043.

Planck Collaboration, Ade, P. A. R., Aghanim, N., Alves, M. I. R., et al. 2014. Planck 2013 results. I. Overview of products and scientific results. *Astron. Astrophys.*, **571**, A1, ArXiv e-prints 1303.5062.

Poisson, E., and Will, C. M. 1995. Gravitational waves from inspiraling compact binaries: Parameter estimation using second-post-Newtonian waveforms. *Phys. Rev. D*, **52**, 848–855, ArXiv e-prints gr-qc/9502040.

Poisson, E., and Will, C. M. 2014. *Gravity: Newtonian, Post-Newtonian, Relativistic*. Cambridge: Cambridge University Press.

Potter, Harold H. 1923. Some experiments on the proportionality of mass and weight. *Proc. R. Soc. A*, **104**, 588–610.

Pound, R. V., and Rebka, G. A. 1960. Apparent weight of photons. *Phys. Rev. Lett.*, **4**, 337–341.

Pound, R. V., and Snider, J. L. 1965. Effect of gravity on gamma radiation. *Phys. Rev.*, **140**, 788–803.

Prestage, J. D., Bollinger, J. J., Itano, W. M., and Wineland, D. J. 1985. Limits for spatial anisotropy by use of nuclear-spin-polarized $^9Be^+$ ions. *Phys. Rev. Lett.*, **54**, 2387–2390.

Prestage, J. D., Tjoelker, R. L., and Maleki, L. 1995. Atomic clocks and variations of the fine structure constant. *Phys. Rev. Lett.*, **74**, 3511–3514.

Pretorius, F. 2005. Evolution of binary black-hole spacetimes. *Phys. Rev. Lett.*, **95**, 121101, ArXiv e-prints gr-qc/0507014.

Psaltis, D. 2008. Probes and tests of strong-field gravity with observations in the electromagnetic spectrum. *Living Rev. Relativ.*, **11**, ArXiv e-prints 0806.1531.

Psaltis, D., Wex, N., and Kramer, M. 2016. A quantitative test of the no-hair theorem with SgrA* using stars, pulsars, and the Event Horizon Telescope. *Astrophys. J.*, **818**, 121, ArXiv e-prints 1510.00394.

Quast, R., Reimers, D., and Levshakov, S. A. 2004. Probing the variability of the fine-structure constant with the VLT/UVES. *Astron. Astrophys.*, **415**, L7–L11, ArXiv e-prints astro-ph/0311280.

Randall, L., and Sundrum, R. 1999a. An alternative to compactification. *Phys. Rev. Lett.*, **83**, 4690–4693, ArXiv e-prints hep-th/9906064.

Randall, L., and Sundrum, R. 1999b. Large mass hierarchy from a small extra dimension. *Phys. Rev. Lett.*, **83**, 3370–3373, ArXiv e-prints hep-ph/9905221.

Ransom, S. M., Stairs, I. H., Archibald, A. M., Hessels, J. W. T., et al. 2014. A millisecond pulsar in a stellar triple system. *Nature*, **505**, 520–524, ArXiv e-prints 1401.0535.

Reasenberg, R. D., Shapiro, I. I., MacNeil, P. E., Goldstein, R. B., et al. 1979. Viking relativity experiment—Verification of signal retardation by solar gravity. *Astrophys. J. Lett.*, **234**, L219–L221.

Reasenberg, R. D., Patla, B. R., Phillips, J. D., and Thapa, R. 2012. Design and characteristics of a WEP test in a sounding-rocket payload. *Class. Quantum Grav.*, **29**, 184013, ArXiv e-prints 1206.0028.

Reeves, H. 1994. On the origin of the light elements ($Z < 6$). *Rev. Mod. Phys.*, **66**, 193–216.

Reynaud, S., Salomon, C., and Wolf, P. 2009. Testing general relativity with atomic clocks. *Space Sci. Rev.*, **148**, 233–247, ArXiv e-prints 0903.1166.

Reynolds, C. S. 2013. The spin of supermassive black holes. *Class. Quantum Grav.*, **30**, 244004, ArXiv e-prints 1307.3246.

Richter, G. W., and Matzner, R. A. 1982. Second-order contributions to gravitational deflection of light in the parametrized post-Newtonian formalism. *Phys. Rev. D*, **26**, 1219–1224.

Rievers, B., and Lämmerzahl, C. 2011. High precision thermal modeling of complex systems with application to the flyby and Pioneer anomaly. *Ann. Phys. (Berlin)*, **523**, 439–449, ArXiv e-prints 1104.3985.

Riis, E., Andersen, L.-U. A., Bjerre, N., Poulsen, O., et al. 1988. Test of the isotropy of the speed of light using fast-beam laser spectroscopy. *Phys. Rev. Lett.*, **60**, 81–84.

Robertson, H. P. 1938. The two-body problem in general relativity. *Ann. Math.*, **39**, 101–104.

Robertson, H. P. 1962. Relativity and cosmology. Page 228 in Deutsch, A. J., and Klemperer, W. B. (eds), *Space Age Astronomy*. New York: Academic Press.

Roca Cortés, T., and Pallé, P. L. 2014. The Mark-I helioseismic experiment—I. Measurements of the solar gravitational redshift (1976-2013). *Mon. Not. R. Astron. Soc.*, **443**, 1837–1848, ArXiv e-prints 1406.5944.

Rohlf, J. W. 1994. *Modern Physics from α to Z^0*. New York: Wiley.

Roll, P. G., Krotkov, R., and Dicke, R. H. 1964. The equivalence of inertial and passive gravitational mass. *Ann. Phys. (N.Y.)*, **26**, 442–517.

Rosen, N. 1973. A bi-metric theory of gravitation. *Gen. Relativ. Gravit.*, **4**, 435–447.

Rosen, N. 1974. A theory of gravitation. *Ann. Phys. (N.Y.)*, **84**, 455–473.

Rosen, N. 1977. Bimetric gravitation and cosmology. *Astrophys. J.*, **211**, 357–360.

Rossi, B., and Hall, D. B. 1941. Variation of the rate of decay of mesotrons with momentum. *Phys. Rev.*, **59**, 223–228.

Rozelot, J.-P., and Damiani, C. 2011. History of solar oblateness measurements and interpretation. *Eur. Phys. J. H*, **36**, 407–436.

Rubilar, G. F., and Eckart, A. 2001. Periastron shifts of stellar orbits near the galactic center. *Astron. Astrophys.*, **374**, 95–104.

Rudolph, E., and Börner, G. 1978a. The importance of gravitational self-field effects in binary systems with compact objects. I. The "static two-body problem" and the attraction law in a post-Newtonian approximation of general relativity. *Gen. Relativ. Gravit.*, **9**, 809–820.

Rudolph, E., and Börner, G. 1978b. The importance of gravitational self-field effects in binary systems with compact objects. II. The "static two-body problem" and the attraction law in a post-post-Newtonian approximation of general relativity. *Gen. Relativ. Gravit.*, **9**, 821–833.

Sadeghian, L., and Will, C. M. 2011. Testing the black hole no-hair theorem at the galactic center: Perturbing effects of stars in the surrounding cluster. *Class. Quantum Grav.*, **28**, 225029, ArXiv e-prints 1106.5056.

Sagi, E. 2009. Preferred frame parameters in the tensor-vector-scalar theory of gravity and its generalization. *Phys. Rev. D*, **80**, 044032, ArXiv e-prints 0905.4001.

Sagi, E. 2010. Propagation of gravitational waves in the generalized tensor-vector-scalar theory. *Phys. Rev. D*, **81**, 064031, ArXiv e-prints 1001.1555.

Salgado, M., Bonazzola, S., Gourgoulhon, E., and Haensel, P. 1994. High precision rotating neutron star models. I. Analysis of neutron star properties. *Astron. Astrophys.*, **291**, 155–170.

Salmona, A. 1967. Effect of gravitational scalar field on high-density star structure. *Phys. Rev.*, **154**, 1218–1223.

Sanghai, V. A. A., and Clifton, T. 2017. Parametrized post-Newtonian cosmology. *Class. Quantum Grav.*, **34**, 065003, ArXiv e-prints 1610.08039.

Santiago, D. I., Kalligas, D., and Wagoner, R. V. 1997. Nucleosynthesis constraints on scalar-tensor theories of gravity. *Phys. Rev. D*, **56**, 7627–7637, ArXiv e-prints gr-qc/9706017.

Sarmiento, A. F. 1982. Parametrized post-post-Newtonian (PP^2N) formalism for the solar system. *Gen. Relativ. Gravit.*, **14**, 793–805.

Sathyaprakash, B. S., and Schutz, B. F. 2009. Physics, astrophysics and cosmology with gravitational waves. *Living Rev. Relativ.*, **12**, ArXiv e-prints 0903.0338.

Saulson, P. R. 1994. *Fundamentals of Interferometric Gravitational Wave Detectors*. Singapore: World Scientific.

Schiff, L. I. 1960. On experimental tests of the general theory of relativity. *Am. J. Phys.*, **28**, 340–343.

Schiff, L. I. 1967. Comparison of theory and observation in general relativity. Page 105 in Ehlers, J. (ed), *Relativity Theory and Astrophysics. Vol. 1: Relativity and Cosmology*. Providence, RI: American Mathematical Society.

Schilpp, P.A. (ed). 1949. *Albert Einstein: Philosopher—Scientist*. Evanston: Library of Living Philosophers.

Schlamminger, S., Choi, K.-Y., Wagner, T. A., Gundlach, J. H., et al. 2008. Test of the equivalence principle using a rotating torsion balance. *Phys. Rev. Lett.*, **100**, 041101, ArXiv e-prints 0712.0607.

Schmidt, M. 1963. 3C 273: A star-like object with large redshift. *Nature*, **197**, 1040.

Schödel, R., Ott, T., Genzel, R., Hofmann, R., et al. 2002. A star in a 15.2-year orbit around the supermassive black hole at the centre of the Milky Way. *Nature*, **419**, 694–696, ArXiv e-prints astro-ph/0210426.

Schutz, B. F. 2009. *A First Course in General Relativity*. Cambridge: Cambridge University Press.

Schwarzschild, K. 1916. Über das Gravitationsfeld eines Massenpunktes nach der Einsteinschen Theorie. *Sitzungsberichte der Königlich Preußischen Akademie der Wissenschaften (Berlin), 1916, Seite 189-196.*

Sennett, N., Marsat, S., and Buonanno, A. 2016. Gravitational waveforms in scalar-tensor gravity at 2PN relative order. *Phys. Rev. D*, **94**, 084003, ArXiv e-prints 1607.01420.

Shankland, R. S., McCuskey, S. W., Leone, F. C., and Kuerti, G. 1955. New analysis of the interferometer observations of Dayton C. Miller. *Rev. Mod. Phys.*, **27**, 167–178.

Shao, L. 2016. Testing the Strong Equivalence Principle with the triple pulsar PSR J 0337 +1715. *Phys. Rev. D*, **93**, 084023, ArXiv e-prints 1602.05725.

Shao, L., and Wex, N. 2012. New tests of local Lorentz invariance of gravity with small-eccentricity binary pulsars. *Class. Quantum Grav.*, **29**, 215018, ArXiv e-prints 1209.4503.

Shao, L., and Wex, N. 2013. New limits on the violation of local position invariance of gravity. *Class. Quantum Grav.*, **30**, 165020, ArXiv e-prints 1307.2637.

Shao, L., Caballero, R. N., Kramer, M., Wex, N., Champion, D. J., and Jessner, A. 2013. A new limit on local Lorentz invariance violation of gravity from solitary pulsars. *Class. Quantum Grav.*, **30**, 165019, ArXiv e-prints 1307.2552.

Shapiro, I. I. 1964. Fourth test of general relativity. *Phys. Rev. Lett.*, **13**, 789–791.

Shapiro, I. I., Pettengill, G. H., Ash, M. E., Stone, M. L., et al. 1968. Fourth test of general relativity: Preliminary results. *Phys. Rev. Lett.*, **20**, 1265–1269.

Shapiro, I. I., Ash, M. E., Ingalls, R. P., Smith, W. B., et al. 1971. Fourth test of general relativity: New radar result. *Phys. Rev. Lett.*, **26**, 1132–1135.

Shapiro, S. S., Davis, J. L., Lebach, D. E., and Gregory, J. S. 2004. Measurement of the solar gravitational deflection of radio waves using geodetic very-long-baseline interferometry data, 1979-1999. *Phys. Rev. Lett.*, **92**, 121101.

Shibata, M., and Taniguchi, K. 2011. Coalescence of black hole-neutron star binaries. *Living Rev. Relativ.*, **14**, 6.

Shibata, M., Taniguchi, K., Okawa, H., and Buonanno, A. 2014. Coalescence of binary neutron stars in a scalar-tensor theory of gravity. *Phys. Rev. D*, **89**, 084005, ArXiv e-prints 1310.0627.

Shlyakhter, A. I. 1976. Direct test of the constancy of fundamental nuclear constants. *Nature*, **264**, 340.

Skordis, C. 2008. Generalizing tensor-vector-scalar cosmology. *Phys. Rev. D*, **77**, 123502, ArXiv e-prints 0801.1985.

Skordis, C. 2009. The tensor-vector-scalar theory and its cosmology. *Class. Quantum Grav.*, **26**, 143001, ArXiv e-prints 0903.3602.

Smarr, L. 1977 (Dec.). Space-time generated by computers: Black holes with gravitational radiation. Page 569 in Papagiannis, M. D. (ed), *Eighth Texas Symposium on Relativistic Astrophysics*. Ann. N.Y. Acad. Sci., vol. 302.

Smarr, L., Čadež, A., Dewitt, B., and Eppley, K. 1976. Collision of two black holes: Theoretical framework. *Phys. Rev. D*, **14**, 2443–2452.

Smiciklas, M., Brown, J. M., Cheuk, L. W., Smullin, S. J., and Romalis, M. V. 2011. New test of Local Lorentz Invariance using a ^{21}Ne-Rb-K comagnetometer. *Phys. Rev. Lett.*, **107**, 171604, ArXiv e-prints 1106.0738.

Smith, W. B. 1963. Radar observations of Venus, 1961 and 1959. *Astron. J.*, **68**, 15.

Snider, J. L. 1972. New measurement of the solar gravitational redshift. *Phys. Rev. Lett.*, **28**, 853–856.

Sotiriou, T. P., and Faraoni, V. 2010. *f*(*R*) theories of gravity. *Rev. Mod. Phys.*, **82**, 451–497, ArXiv e-prints 0805.1726.

Sotiriou, T. P., and Faraoni, V. 2012. Black holes in scalar-tensor gravity. *Phys. Rev. Lett.*, **108**, 081103, ArXiv e-prints 1109.6324.

Srianand, R., Chand, H., Petitjean, P., and Aracil, B. 2004. Limits on the time variation of the electromagnetic fine-structure constant in the low energy limit from absorption lines in the spectra of distant quasars. *Phys. Rev. Lett.*, **92**, 121302, ArXiv e-prints astro-ph/0402177.

St. John, C. E. 1917. The principle of generalized relativity and the displacement of Fraunhofer lines toward the red. *Astrophys. J.*, **46**, 249.

Stairs, I. H., Faulkner, A. J., Lyne, A. G., Kramer, M., et al. 2005. Discovery of three wide-orbit binary pulsars: Implications for binary evolution and equivalence principles. *Astrophys. J.*, **632**, 1060–1068, ArXiv e-prints astro-ph/0506188.

Stanwix, P. L., Tobar, M. E., Wolf, P., Susli, M., et al. 2005. Test of Lorentz invariance in electrodynamics using rotating cryogenic sapphire microwave oscillators. *Phys. Rev. Lett.*, **95**, 040404, ArXiv e-prints hep-ph/0506074.

Stavridis, A., and Will, C. M. 2009. Bounding the mass of the graviton with gravitational waves: Effect of spin precessions in massive black hole binaries. *Phys. Rev. D*, **80**, 044002, ArXiv e-prints 0906.3602.

Stecker, F. W., and Scully, S. T. 2009. Searching for new physics with ultrahigh energy cosmic rays. *New J. Phys.*, **11**, 085003, ArXiv e-prints 0906.1735.

Stergioulas, N. 2003. Rotating stars in relativity. *Living Rev. Relativ.*, **6**, 3, ArXiv e-prints gr-qc/0302034.

Su, Y., Heckel, B. R., Adelberger, E. G., Gundlach, J. H., et al. 1994. New tests of the universality of free fall. *Phys. Rev. D*, **50**, 3614–3636.

Sushkov, A. O., Kim, W. J., Dalvit, D. A. R., and Lamoreaux, S. K. 2011. New experimental limits on non-Newtonian forces in the micrometer range. *Phys. Rev. Lett.*, **107**, 171101, ArXiv e-prints 1108.2547.

Synge, J. L. 1960. *Relativity: The General Theory*. Amsterdam: North-Holland.

Talmadge, C., Berthias, J.-P., Hellings, R. W., and Standish, E. M. 1988. Model-independent constraints on possible modifications of Newtonian gravity. *Phys. Rev. Lett.*, **61**, 1159–1162.

Tamaki, T., and Miyamoto, U. 2008. Generic features of Einstein-Æther black holes. *Phys. Rev. D*, **77**, 024026, ArXiv e-prints 0709.1011.

Taveras, V., and Yunes, N. 2008. Barbero-Immirzi parameter as a scalar field: K-inflation from loop quantum gravity? *Phys. Rev. D*, **78**, 064070, ArXiv e-prints 0807.2652.

Taylor, J. H. 1975. Discovery of a pulsar in a binary system. Pages 490–492 in Bergman, P. G., Fenyves, E. J., and Motz, L. (eds), *Seventh Texas Symposium on Relativistic Astrophysics*. Annals of the New York Academy of Sciences, vol. 262.

Taylor, J. H. 1987. Astronomical and space experiments to test relativity. Page 209 in MacCallum, M. A. H. (ed), *General Relativity and Gravitation*. New York: Cambridge University Press.

Taylor, J. H., and McCulloch, P. M. 1980 (Feb.). Evidence for the existence of gravitational radiation from measurements of the binary pulsar PSR 1913+16. Pages 442–446 in Ehlers, J., Perry, J. J., and Walker, M. (eds), *Ninth Texas Symposium on Relativistic Astrophysics*. Ann. N.Y. Acad. Sci., vol. 336.

Taylor, J. H., Hulse, R. A., Fowler, L. A., Gullahorn, G. E., et al. 1976. Further observations of the binary pulsar PSR 1913+16. *Astrophys. J. Lett.*, **206**, L53–L58.

Taylor, J. H., Fowler, L. A., and McCulloch, P. M. 1979. Measurements of general relativistic effects in the binary pulsar PSR 1913+16. *Nature*, **277**, 437–440.

Taylor, S., and Poisson, E. 2008. Nonrotating black hole in a post-Newtonian tidal environment. *Phys. Rev. D*, **78**, 084016, ArXiv e-prints 0806.3052.

Taylor, T. R., and Veneziano, G. 1988. Dilaton couplings at large distances. *Phys. Lett. B*, **213**, 450–454.

Thorne, K. S. 1980. Multipole expansions of gravitational radiation. *Rev. Mod. Phys.*, **52**, 299–340.

Thorne, K. S., and Dykla, J. J. 1971. Black holes in the Dicke-Brans theory of gravity. *Astrophys. J. Lett.*, **166**, L35–L38.

Thorne, K. S., and Hartle, J. B. 1985. Laws of motion and precession for black holes and other bodies. *Phys. Rev. D*, **31**, 1815–1837.

Thorne, K. S., Lee, D. L., and Lightman, A. P. 1973. Foundations for a theory of gravitation theories. *Phys. Rev. D*, **7**, 3563–3578.

Tinto, M., and Alves, M. E. D. S. 2010. LISA sensitivities to gravitational waves from relativistic metric theories of gravity. *Phys. Rev. D*, **82**, 122003, ArXiv e-prints 1010.1302.

Touboul, P., Métris, G., Rodrigues, M., André, Y., et al. 2017. The MICROSCOPE mission: First results of a space test of the equivalence principle. *Phys. Rev. Lett.*, **119**, 231101, ArXiv e-prints 1712.01176.

Trautman, A. 1962. Conservation laws in general relativity. Pages 169–198 in Witten, L. (ed), *Gravitation: An Introduction to Current Research*. New York: Wiley.

Tu, L.-C., Guan, S.-G., Luo, J., Shao, C.-G., et al. 2007. Null test of Newtonian inverse-square law at submillimeter range with a dual-modulation torsion pendulum. *Phys. Rev. Lett.*, **98**, 201101.

Turneaure, J. P., Will, C. M., Farrell, B. F., Mattison, E. M., et al. 1983. Test of the principle of equivalence by a null gravitational redshift experiment. *Phys. Rev. D*, **27**, 1705–1714.

Turyshev, S. G., and Toth, V. T. 2010. The Pioneer anomaly. *Living Rev. Relativ.*, **13**, ArXiv e-prints 1001.3686.

Turyshev, S. G., Toth, V. T., Kinsella, G., Lee, S.-C., et al. 2012. Support for the thermal origin of the Pioneer anomaly. *Phys. Rev. Lett.*, **108**, 241101, ArXiv e-prints 1204.2507.

Upadhye, A., and Hu, W. 2009. Existence of relativistic stars in *f*(*R*) gravity. *Phys. Rev. D*, **80**, 064002, ArXiv e-prints 0905.4055.

Uzan, J.-P. 2011. Varying constants, gravitation and cosmology. *Living Rev. Relativ.*, **14**, 2, ArXiv e-prints 1009.5514.

Vainshtein, A. I. 1972. To the problem of nonvanishing gravitation mass. *Phys. Lett. B*, **39**, 393–394.

van Dam, H., and Veltman, M. 1970. Massive and mass-less Yang-Mills and gravitational fields. *Nucl. Phys. B*, **22**, 397–411.

van Flandern, T. C. 1975. A determination of the rate of change of *G*. *Mon. Not. R. Astron. Soc.*, **170**, 333–342.

van Patten, R. A., and Everitt, C. W. F. 1976. Possible experiment with two counter-orbiting drag-free satellites to obtain a new test of Einstein's general theory of relativity and improved measurements in geodesy. *Phys. Rev. Lett.*, **36**, 629–632.

Verma, A. K., Fienga, A., Laskar, J., Manche, H., et al. 2014. Use of MESSENGER radioscience data to improve planetary ephemeris and to test general relativity. *Astron. Astrophys.*, **561**, A115, ArXiv e-prints 1306.5569.

Vessot, R. F. C., Levine, M. W., Mattison, E. M., Blomberg, E. L., et al. 1980. Test of relativistic gravitation with a space-borne hydrogen maser. *Phys. Rev. Lett.*, **45**, 2081–2084.

Visser, M. 1998. Mass for the graviton. *Gen. Relativ. Gravit.*, **30**, 1717–1728, ArXiv e-prints gr-qc/9705051.

von Klüber, H. 1960. The determination of Einstein's light-deflection in the gravitational field of the Sun. *Vistas in Astronomy*, **3**, 47–77.

Wagner, T. A., Schlamminger, S., Gundlach, J. H., and Adelberger, E. G. 2012. Torsion-balance tests of the Weak Equivalence Principle. *Class. Quantum Grav.*, **29**, 184002, ArXiv e-prints 1207.2442.

Wagoner, R. V. 1970. Scalar-tensor theory and gravitational waves. *Phys. Rev. D*, **1**, 3209–3216.

Wagoner, R. V. 1975. Test for the existence of gravitational radiation. *Astrophys. J. Lett.*, **196**, L63–L65.

Wagoner, R. V., and Kalligas, D. 1997. Scalar-tensor theories and gravitational radiation. Pages 433–446 in Marck, J.-A., and Lasota, J.-P. (eds), *Relativistic Gravitation and Gravitational Radiation*. Cambridge: Cambridge University Press.

Wagoner, R. V., and Will, C. M. 1976. Post-Newtonian gravitational radiation from orbiting point masses. *Astrophys. J.*, **210**, 764–775.

Wainstein, L.A., and Zubakov, V.D. 1962. *Extraction of Signals from Noise*. Englewood Cliffs, NJ: Prentice–Hall.

Walsh, D., Carswell, R. F., and Weymann, R. J. 1979. 0957 + 561 A, B - Twin quasistellar objects or gravitational lens. *Nature*, **279**, 381–384.

Warburton, R. J., and Goodkind, J. M. 1976. Search for evidence of a preferred reference frame. *Astrophys. J.*, **208**, 881–886.

Webb, J. K., Flambaum, V. V., Churchill, C. W., Drinkwater, M. J., et al. 1999. Search for time variation of the fine structure constant. *Phys. Rev. Lett.*, **82**, 884–887, ArXiv e-prints astro-ph/9803165.

Webster, B. L., and Murdin, P. 1972. Cygnus X-1: A spectroscopic binary with a heavy companion? *Nature*, **235**, 37–38.

Weinberg, N. N., Milosavljević, M., and Ghez, A. M. 2005. Stellar dynamics at the galactic center with an Extremely Large Telescope. *Astrophys. J.*, **622**, 878–891, ArXiv e-prints astro-ph/0404407.

Weinberg, S. 1965. Photons and gravitons in perturbation theory: Derivation of Maxwell's and Einstein's equations. *Phys. Rev.*, **138**, 988–1002.

Weinberg, S. 1972. *Gravitation and Cosmology: Principles and Applications of the General Theory of Relativity*. New York: Wiley.

Weinberg, S. 2008. Effective field theory for inflation. *Phys. Rev. D*, **77**, 123541, ArXiv e-prints 0804.4291.

Weisberg, J. M., and Huang, Y. 2016. Relativistic Measurements from Timing the Binary Pulsar PSR B1913+16. *Astrophys. J.*, **829**, 55, ArXiv e-prints 1606.02744.

Weisberg, J. M., and Taylor, J. H. 2002. General relativistic geodetic spin precession in binary pulsar B1913+16: Mapping the emission beam in two dimensions. *Astrophys. J.*, **576**, 942–949, ArXiv e-prints astro-ph/0205280.

Weisberg, J. M., Nice, D. J., and Taylor, J. H. 2010. Timing measurements of the relativistic binary pulsar PSR B1913+16. *Astrophys. J.*, **722**, 1030–1034, ArXiv e-prints 1011.0718.

Wen, L., and Schutz, B. F. 2005. Coherent network detection of gravitational waves: The redundancy veto. *Class. Quantum Grav.*, **22**, S1321–S1335, ArXiv e-prints gr-qc/0508042.

Wex, N. 2000. Small-eccentricity binary pulsars and relativistic gravity. Page 113 in Kramer, M., Wex, N., and Wielebinski, R. (eds), *IAU Colloq. 177: Pulsar Astronomy - 2000 and Beyond*. Astronomical Society of the Pacific Conference Series, vol. 202. San Francisco: Astronomical Society of the Pacific.

Wex, N., and Kopeikin, S. M. 1999. Frame dragging and other precessional effects in black hole pulsar binaries. *Astrophys. J.*, **514**, 388–401, ArXiv e-prints astro-ph/9811052.

Whitrow, G. J., and Morduch, G. E. 1965. Relativistic theories of gravitation: A comparative analysis with particular reference to astronomical tests. *Vistas Astron.*, **6**, 1–67.

Will, C. M. 1971a. Relativistic gravity in the solar system. I. Effect of an anisotropic gravitational mass on the Earth-Moon distance. *Astrophys. J.*, **165**, 409.

Will, C. M. 1971b. Relativistic gravity in the solar system. II. Anisotropy in the Newtonian gravitational constant. *Astrophys. J.*, **169**, 141.

Will, C. M. 1971c. Theoretical frameworks for testing relativistic gravity. II. Parametrized post-Newtonian hydrodynamics, and the Nordtvedt effect. *Astrophys. J.*, **163**, 611.

Will, C. M. 1971d. Theoretical frameworks for testing relativistic gravity. III. Conservation laws, Lorentz invariance, and values of the PPN parameters. *Astrophys. J.*, **169**, 125.

Will, C. M. 1973. Relativistic gravity in the solar system. III. Experimental disproof of a class of linear theories of gravitation. *Astrophys. J.*, **185**, 31–42.

Will, C. M. 1974. Gravitational redshift measurements as tests of nonmetric theories of gravity. *Phys. Rev. D*, **10**, 2330–2337.

Will, C. M. 1976a. A test of post-Newtonian conservation laws in the binary system PSR 1913+16. *Astrophys. J.*, **205**, 861–867.

Will, C. M. 1976b. Active mass in relativistic gravity: Theoretical interpretation of the Kreuzer experiment. *Astrophys. J.*, **204**, 224–234.

Will, C. M. 1977. Gravitational radiation from binary systems in alternative metric theories of gravity: Dipole radiation and the binary pulsar. *Astrophys. J.*, **214**, 826–839.

Will, C. M. 1986. *Was Einstein Right? Putting General Relativity to the Test*. New York: Basic Books.

Will, C. M. 1990. Twilight time for the fifth force? *Sky and Telescope*, **80**, 472.

Will, C. M. 1992a. Clock synchronization and isotropy of the one-way speed of light. *Phys. Rev. D*, **45**, 403–411.

Will, C. M. 1992b. Is momentum conserved? A test in the binary system PSR 1913+16. *Astrophys. J. Lett.*, **393**, L59–L61.

Will, C. M. 1994. Testing scalar-tensor gravity with gravitational-wave observations of inspiralling compact binaries. *Phys. Rev. D*, **50**, 6058–6067, ArXiv e-prints gr-qc/9406022.

Will, C. M. 1998. Bounding the mass of the graviton using gravitational-wave observations of inspiralling compact binaries. *Phys. Rev. D*, **57**, 2061–2068, ArXiv e-prints gr-qc/9709011.

Will, C. M. 2000. Einstein's relativity and everyday life. Physics Central: American Physical Society, http://physicscentral.com/explore/writers/will.cfm.

Will, C. M. 2005. Post-Newtonian gravitational radiation and equations of motion via direct integration of the relaxed Einstein equations. III. Radiation reaction for binary systems with spinning bodies. *Phys. Rev. D*, **71**, 084027, ArXiv e-prints gr-qc/0502039.

Will, C. M. 2006. Special relativity: A centenary perspective. Page 33 in Damour, T., Darrigol, O., Duplantier, B., and Rivasseau, V. (eds), *Einstein, 1905-2005: Poincaré Seminar 2005*. Basel: Birkhäuser Verlag.

Will, C. M. 2008. Testing the general relativistic "no-hair" theorems using the galactic center black hole Sagittarius A*. *Astrophys. J. Lett.*, **674**, L25, ArXiv e-prints 0711.1677.

Will, C. M. 2011. On the unreasonable effectiveness of the post-Newtonian approximation in gravitational physics. *Proc. Nat. Acad. Sci. (US)*, **108**, 5938–5945, ArXiv e-prints 1102.5192.

Will, C. M., and Eardley, D. M. 1977. Dipole gravitational radiation in Rosen's theory of gravity: Observable effects in the binary system PSR 1913+16. *Astrophys. J. Lett.*, **212**, L91–L94.

Will, C. M., and Nordtvedt, Jr., K. 1972. Conservation laws and preferred frames in relativistic gravity. I. Preferred-frame theories and an extended PPN formalism. *Astrophys. J.*, **177**, 757.

Will, C. M., and Wiseman, A. G. 1996. Gravitational radiation from compact binary systems: Gravitational waveforms and energy loss to second post-Newtonian order. *Phys. Rev. D*, **54**, 4813–4848, ArXiv e-prints gr-qc/9608012.

Will, C. M., and Zaglauer, H. W. 1989. Gravitational radiation, close binary systems, and the Brans-Dicke theory of gravity. *Astrophys. J.*, **346**, 366–377.

Williams, J. G., Newhall, X X, and Dickey, J. O. 1996. Relativity parameters determined from lunar laser ranging. *Phys. Rev. D*, **53**, 6730–6739.

Williams, J. G., Turyshev, S. G., and Murphy, T. W. 2004a. Improving LLR Tests of gravitational theory. *Int. J. Mod. Phys. D*, **13**, 567–582, ArXiv e-prints gr-qc/0311021.

Williams, J. G., Turyshev, S. G., and Boggs, D. H. 2004b. Progress in lunar laser ranging tests of relativistic gravity. *Phys. Rev. Lett.*, **93**, 261101, ArXiv e-prints gr-qc/0411113.

Williams, J. G., Turyshev, S. G., and Boggs, D. H. 2009. Lunar laser ranging tests of the equivalence principle with the Earth and Moon. *Int. J. Mod. Phys. D*, **18**, 1129–1175, ArXiv e-prints gr-qc/0507083.

Wiseman, A. G., and Will, C. M. 1991. Christodoulou's nonlinear gravitational-wave memory: Evaluation in the quadrupole approximation. *Phys. Rev. D*, **44**, R2945–R2949.

Wolf, P., and Blanchet, L. 2016. Analysis of Sun/Moon gravitational redshift tests with the STE-QUEST space mission. *Class. Quantum Grav.*, **33**, 035012, ArXiv e-prints 1509.02854.

Wolf, P., Bize, S., Clairon, A., Luiten, A. N., et al. 2003. Tests of Lorentz invariance using a microwave resonator. *Phys. Rev. Lett.*, **90**, 060402, ArXiv e-prints gr-qc/0210049.

Wolf, P., Chapelet, F., Bize, S., and Clairon, A. 2006. Cold atom clock test of Lorentz invariance in the matter sector. *Phys. Rev. Lett.*, **96**, 060801, ArXiv e-prints hep-ph/0601024.

Wolf, P., Blanchet, L., Bordé, C. J., Reynaud, S., et al. 2011. Does an atom interferometer test the gravitational redshift at the Compton frequency? *Class. Quantum Grav.*, **28**, 145017, ArXiv e-prints 1012.1194.

Wolfe, A. M., Brown, R. L., and Roberts, M. S. 1976. Limits on the variation of fundamental atomic quantities over cosmic time scales. *Phys. Rev. Lett.*, **37**, 179–181.

Yagi, K., and Yunes, N. 2013. I-Love-Q: Unexpected universal relations for neutron stars and quark stars. *Science*, **341**, 365–368, ArXiv e-prints 1302.4499.

Yagi, K., Yunes, N., and Tanaka, T. 2012. Slowly rotating black holes in dynamical Chern-Simons gravity: Deformation quadratic in the spin. *Phys. Rev. D*, **86**, 044037, ArXiv e-prints 1206.6130.

Yagi, K., Stein, L. C., Yunes, N., and Tanaka, T. 2013. Isolated and binary neutron stars in dynamical Chern-Simons gravity. *Phys. Rev. D*, **87**, 084058, ArXiv e-prints 1302.1918.

Yagi, K., Blas, D., Barausse, E., and Yunes, N. 2014a. Constraints on Einstein-Æther theory and Hořava gravity from binary pulsar observations. *Phys. Rev. D*, **89**, 084067, ArXiv e-prints 1311.7144.

Yagi, K., Blas, D., Yunes, N., and Barausse, E. 2014b. Strong binary pulsar constraints on Lorentz violation in gravity. *Phys. Rev. Lett.*, **112**, 161101, ArXiv e-prints 1307.6219.

Yang, S.-Q., Zhan, B.-F., Wang, Q.-L., Shao, C.-G., et al. 2012. Test of the gravitational inverse square law at millimeter ranges. *Phys. Rev. Lett.*, **108**, 081101.

Yunes, N., and Pretorius, F. 2009. Dynamical Chern-Simons modified gravity: Spinning black holes in the slow-rotation approximation. *Phys. Rev. D*, **79**, 084043, ArXiv e-prints 0902.4669.

Yunes, N., and Pretorius, F. 2009. Fundamental theoretical bias in gravitational wave astrophysics and the parametrized post-Einsteinian framework. *Phys. Rev. D*, **80**, 122003, ArXiv e-prints 0909.3328.

Yunes, N., and Stein, L. C. 2011. Nonspinning black holes in alternative theories of gravity. *Phys. Rev. D*, **83**, 104002, ArXiv e-prints 1101.2921.

Yunes, N., Psaltis, D., Özel, F., and Loeb, A. 2010. Constraining parity violation in gravity with measurements of neutron-star moments of inertia. *Phys. Rev. D*, **81**, 064020, ArXiv e-prints 0912.2736.

Yunes, N., Yagi, K., and Pretorius, F. 2016. Theoretical physics implications of the binary black-hole mergers GW150914 and GW151226. *Phys. Rev. D*, **94**, 084002, ArXiv e-prints 1603.08955.

Zaglauer, H. W. 1990. *Phenomenological aspects of scalar fields in astrophysics, cosmology and particle physics*. Ph.D. thesis, Washington Univ., St. Louis.

Zaglauer, H. W. 1992. Neutron stars and gravitational scalars. *Astrophys. J.*, **393**, 685–696.

Zakharov, V. I. 1970. Linearized gravitation theory and the graviton mass. *Sov. Phys. JETP Lett.*, **12**, 312.

Zucker, S., Alexander, T., Gillessen, S., Eisenhauer, F., et al. 2006. Probing post-Newtonian physics near the galactic black hole with stellar redshift measurements. *Astrophys. J. Lett.*, **639**, L21–L24, ArXiv e-prints astro-ph/0509105.

Zuntz, J., Baker, T., Ferreira, P. G., and Skordis, C. 2012. Ambiguous tests of general relativity on cosmological scales. *J. Cosm. Astropart. Phys.*, **6**, 32, ArXiv e-prints 1110.3830.

Index

Printed in the United States
by Baker & Taylor Publisher Services

Printed in the United States
by Baker & Taylor Publisher Services